AF530710

ADVANCES IN MICROBIOLOGY

Editor

Pravin Chandra Trivedi

Ph.D., Post-Doct. (U.S.A.), F.L.S., F.B.S., F.P.S.I., F.N.S.I., F.B.R.S., F.M.A., F.E.S.

Professor of Botany,
University of Rajasthan,
Jaipur - 302 004, India

Published by:

SCIENTIFIC PUBLISHERS (INDIA)
5A, New Pali Road, P.O. Box 91
Jodhpur 342 001, India
E-mail: info@scientificpub.com
Website: www.scientificpubonline.com

Print : 2026

ISBN : 978-81-72333-32-4
eISBN: 978-93-87741-69-0

Printed in India

Dedicated to
My Teacher
Dr. SUDHAKAR MISHRA

PREFACE

Microbiology has become increasingly useful to our society now, and it has emerged as one of the most important branch of the Life sciences. Microbiologist have made significant contributions to basic biological science as well as in the applied areas of public health, medical sciences, agriculture, industry and environmental sciences. The most dramatic current development in applied microbiology is due to development of genetic engineering and recombinant DNA technology. Using these techniques, microorganism can be engineered through modification of its DNA to produce new substances such as human proteins. Bacteria have been modified to produce human insulin and interferon. Genetically engineered microorganisms hold great potentials for the production of drugs and vaccines, for improvement of agricultural crops and for other products and processes. There is growing recognition of the potential of microorganisms in many applied areas. The ability of micro-oganisms to decompose materials such as — herbicides, pesticides and oils in oil spills, the potential of microorganism as food supplement, the exploitation of microbial activity to produce methane gas as energy sources for rural consumption and the potential of new therapeutic substances produced by microorganisms — these are and the other uses of microorganisms are also becoming attractive. Bio-remediation processes use microorganisms to clean-up toxic wastes. Microorganisms are being used as biological control agent for pests. In gene therapy, viruses are used to carry replacements for defective or missing genes into human cells. Genetically engineered bacteria are used in agriculture to protect plants from frost and insects and to improve the shelf-life of produce.

This book "Advances in Microbiology" provides a comprehensive and critical review of the work done on different areas of microbiology including agriculture, industry, medical science, bioremediation etc. The book contains 24 chapters, which include informations on the status of microbial diversity, applications of biosensors, *Azolla* as biofertilizer, *Frankia* - nitrogen fixing actinomycetes, extraction of metals from ores using bacteria, alkaliphiles, citric acid fermentation, biodiversity of cyanobacteria, microbial degradation of xenobiotics etc. Aspects, covering biotechnological applications of microbes for improved plant productivity and new approaches for development of vaccines have been specially included to project out their role and use in the twenty-first century. Comprehensive account of microbes in the management of soil borne diseases and plant parasitic nematodes throw light on the importance of microbes in the management of plant pests.

This festschrift — Advances in Microbiology is a humble dedication to Dr. S. Mishra and to commemorate his outstanding contribution in the field of General Botany and Phytoplasma in particular. Dr. Mishra is a dedicated teacher and an accomplished researcher, which is testified by students who got their doctoral degree under his able supervision. He has published many research papers, completed research projects and participated in many national and international conferences. Dr. Mishra has been the sole worker in the field of Mycoplasma at the University of Rajasthan. He is a voracious reader and has a rich personal library. Dr. Mishra is not an armchair academician and is actively associated with science education centre and has extensive field knowledge. He taught me microbiology and molecular biology during my M.Sc., I wish and pray for a long, sustained creative life for him for many many more years to come.

The publication of this book could not have been possible without the sincere cooperation and hard work of the contributors. They are all specialists in their respective fields. Therefore, I have tried to honour their ideas in the original shape. While dealing with such a voluminous work, errors are likely to occur despite my best efforts. However, the onus of technical contents rests with the contributors.

I am highly thankful to all the learned contributors for their cooperation in compiling useful information on various facets of microbiology. I am sure, this detailed account on a wide variety of subjects will be of great help to researchers and teachers of microbiology and for planning future strategies for the development world over.

I wish to thank my wife Kusum, daughter Priyanka and son Rohit, who extended their cooperation in many invisible ways to me. I highly appreciate the all-round cooperation and support of Mr. Pawan Kumar, Scientific Publishers (India), Jodhpur for printing and publishing this book with patience, care and interest.

JAIPUR **P.C. Trivedi**

LIST OF CONTRIBUTORS

A.K.Mathur
Atomic Minerals Division for Exploration and Research, BARC, Hyderabad

A.N. Balakrishna
Dept. of Agricultural Microbiology, University of Agricultural Sciences, G.K.V.K. Campus, Bangalore - 560 065

Arun Sataraddi
College of Agriculture, Raichur - 584 101

B.V. Singh
National Centre for Conservation and Utilisation of Blue Green Algae, IARI, New Delhi - 110 012

C.K.M. Tripathi
Fermentation Technology Division, Central Drug Research Institute, Chattar Manzil Palace, Lucknow - 226 001

Chandra B. Sharma
Dept. of Microbiology, SBS (PG) Institute of Biomedical Sciences & Research, Balawala, Dehradun - 248 161

D.A. Sumana
Dept. of Agricultural Microbiology, University of Agricultural Sciences, G.K.V.K. Campus, Bangalore - 560 065

D.J. Bagyaraj
Dept. of Agricultural Microbiology, University of Agricultural Sciences, G.K.V.K. Campus, Bangalore - 560 065

Deepak Chand Sharma
Department of Microbiology, University of Delhi, South Campus, Benito Juarez Road, New Delhi - 110 021

Devyani, S. Nath
Division of Biochemical Sciences, National Chemical Laboratory, Pune - 411 008

Dolly Wattal Dhar
National Centre for Conservation and Utilisation of Blue Green Algae, IARI, New Delhi - 110 012

G. Prasad
Dept. of Veterinary Microbiology, CCS Haryana Agricultural University, Hisar - 125 004

Geeta Singh
Division of Microbiology, Indian Agricultural Research Institute, New Delhi - 110 012

H. Polasa
Department of Microbiology, Osmania University, Hyderabad - 500 007

I.K. Sharma
Central Ground Water Board, State Unit Office, Kangra, H.P.

Jyoti Saxena
Dept. of Bioscience and Biotechnology, Banasthali Vidyapith, P.O. Banasthali Vidyapith - 304 022 (Raj.)

K. Kumar
Azolla Laboratory, Tamilnadu Agricultural University, Coimbatore - 641 003 (TN)

M.K. Naik
College of Agriculture, Raichur - 584 101

M.N. Sreenivasa
College of Agriculture, Raichur - 584 101 (Karnataka)

Mahadeva Swamy
College of Agriculture, Raichur - 584 101 (Karnataka)

Mala B. Rao
Division of Biochemical Sciences, National Chemical Laboratory, Pune - 411 008

Minakshi
Dept. of Veterinary Microbiology, CCS Haryana Agricultural University, Hisar - 125 004

N. Anuradha
Department of Microbiology, Osmania University, Hyderabad - 500 007

N.S. Sashidhar
Department of Microbiology, Osmania University, Hyderabad - 500 007

Nandita Ashtaputre
Central Institute for Research on Cotton Technology, ICAR, Adenwala Road, Matunga, Mumbai - 400 019

O.N. Tiwari
National Centre for Conservation and Utilisation of Blue Green Algae, IARI, New Delhi - 110 012

P. Kaushik
Department of Botany and Microbiology, Gurukul Kangri University, Haridwar - 249 404

P.C. Trivedi
Department of Botany, University of Rajasthan, Jaipur - 302 004

P.K. Singh
National Centre for Conservation and Utilisation of Blue Green Algae, IARI, New Delhi - 110 012

P.S. Jain
Department of Botany, University of Rajasthan, Jaipur - 302 004

Padmanabh Dwivedi
Dept. of Botany, Arunachal University, Rono Hills, Itanagar - 791 111

R.H. Balasubramanya
Central Institute for Research on Cotton Technology, ICAR, Adenwala Road, Matunga, Mumbai - 400 019

R.K.S. Kushwaha
Department of Botany, Christ Church College, Kanpur - 208 001

S. Bhosle
Dept. of Microbiology, Goa University, Taleigao Plateau, Goa

S. Borkar
Dept. of Microbiology, P.E.S. College, Farmagudi, Ponda, Goa

S. Kannaiyan
Azolla Laboratory, Tamilnadu Agricultural University, Coimbatore - 641 003 (TN)

S.K. Dwivedi
Department of Environmental Science, B.B. Ambedkar University,
Rae Bareli Road, Lucknow - 226 025

S.S. Kanwar
Department of Microbiology, College of Basic Sciences,
Himachal Pradesh Agricultual University, Palampur - 176 062

Sanjay Gupta
Dept. of Microbiology, SBS (PG) Institute of Biomedical Sciences & Research, Balawala, Dehradun - 248 161

Saroj Bhosle
Department of Microbiology, Goa University, Taleigao, Plateau, Goa - 403 206

Sudeep P. George
Division of Biochemical Sciences, National Chemical Laboratory, Pune - 411 008

Sunita Devi
Department of Microbiology, College of Basic Sciences,
Himachal Pradesh Agricultual University, Palampur - 176 062

T. Satyanarayana
Department of Microbiology, University of Delhi, South Campus,
Benito Juarez Road, New Delhi - 110 021

Upal Roy
Department of Microbiology, Goa University, Taleigao, Plateau, Goa - 403 206

Vinay Sharma
Dept. of Bioscience and Biotechnology, Banasthanli Vidyapith,
P.O. Banasthali Vidyapith - 304 022 (Raj.)

Vinod Bihari
Fermentation Technology Division, Central Drug Research Institute,
Chattar Manzil Palace, Lucknow - 226 001

Yashpal Malik
Dept. of Veterinary Microbiology, CCS Haryana Agricultural University, Hisar - 125 004

CONTENTS

CHAPTER – 1

SIGNIFICANCE AND STATUS OF MICROBIAL DIVERSITY

Deepak Chand Sharma and T. Satyanarayana

Department of Microbiology, University of Delhi, South Campus, Benito Juarez Road, New Delhi- 110 021

1. INTRODUCTION

In the last decade the term 'Biodiversity' came in to the focus of all branches of biological sciences. Today it is one of the most commonly used terms in the biological sciences, and has become a household word. It was born during the National Forum on Biodiversity, held in Washington, on September 21-24, 1986. The credit of coining the term biodiversity goes to E.O. Wilson, who largely orchestrated the event and proceedings. It was a key topic in the UNICED conference (Convention on Biodiversity) held in Reo de Janeiro in 1992, and now it has become a critical issue of scientific and political concerns worldwide.

Biodiversity means the total genetic variation, species variation and ecosystem variation in the world or a particular area (Wand *et al.*, 1997). Another definition of biodiversity appeared in the Article 2 of the 1992 convention : "The variability among living organisms from all sources including, inter alia, terrestrial, marine and other aquatic ecosystems and the ecological complexes of which they are part. This includes diversity within species, between species and of ecosystem." The International Union for Conservation of Nature and Natural Resources defines biodiversity as that encompasses all life forms, ecosystems and ecological processes and acknowledges the hierarchy at genetic, taxon and ecosystem levels (Ludwig and Schlleifer, 1999). Or in short we can say the variety and abundance of life forms that live on Earth is called biodiversity.

All the above definitions of biodiversity are also applicable to the microbial diversity. It is a well-documented fact that increasing number of plant and animal species are getting extinct day by day. Therefore, this is a major issue of the convention. Microbes, on the other hand, are only mentioned as potential sources of valuable items, processes, aids and genetic material, but no strong step has been taken in the direction of their conservation. The reason for this may be that there is no clear evidence of extinction of microbial species (except obligate symbionts of threatened plants or animals) and the rate of evolutionary adaptation by microbes is high and the concept of species is more diffuse for microbes. Even though the environment is rich in microbes, the driving forces of microbial (bacteria, viruses, viroids, filamentous fungi, yeasts, micro-algae, and protozoa) diversity include the genetic constitution of these organisms, the environment in which they are found and ecological interaction with other components of the biosphere. The result is extraordinary richness of microbial diversity, most of which remains to be explored. An attempt has been made to describe various methods employed in determining microbial diversity, its significance and present status in this chapter.

2. SIGNIFICANCE OF MICROBIAL DIVERSITY

Microorganisms are the most ancient and numerous biotic entities on earth. Through their metabolic diversity and genetic adaptability, microbes have successfully colonized even those ecological niches that are thought to be inhospitable for any forms of life (*Kane* et al., 1993). They play an important and often critical role in the functioning of ecosystem by interacting with it and maintain a sustainable biosphere. The role of microorganisms in maintaining the dynamic equilibrium and integrity of biosphere is so critical that the continued existence of life is dependent on the sustained microbially mediated transformation of terrestrial, aquatic and atmospheric environments. Almost all biological processes in the environment, either directly or indirectly, involve microorganisms, even though the potential benefits of regulating, optimizing and exploiting microbial activity is largely unexplored (Woese *et al.*, 1990). Some of the important examples that show the importance of microbial diversity are described below.

In marine ecosystem, many unique compounds of potential medical value are retrieved from the tissues of a range of marine invertebrate and vertebrate organisms that are produced by marine bacteria reside in their tissues. An example of such type of compound produced by marine microorganisms is tetrodotoxin. This toxin is found in the fatty tissues and reproductive organs of the puffer fish, and was thought to be produced by the fish. Tetrodotoxin is a very powerful analgesic (about 300,000 times more powerful than cocaine) and resembles procaine in its ability to inhibit transmission of nerve signals through nerve cells. It is used in Japan to treat pain in neurogenic leprosy and cancer. Thus tetrodotoxin is a compound produced by bacteria inhabiting the puffer fish, and more recently from the bacteria associated with benthic invertebrates. This is a good example of the compound with a wide pharmaceutical value in the future.

Sclerosponges of Caribbean coral reefs harbor the bacterial species that produce unusual antimicrobial compounds. Microscopic observation of the cross section of sclerosponge revealed that 50% of the sponge mass comprised prokaryotic structures confirmed to be bacteria. Most of these bacterial species have not yet been cultured (Sambrook *et al.*, 1989). Out of them, which are culturable found to be producing compounds of potential pharmacological value. In general, marine sponges are widely known for the production of chemicals that harm predators or even cause ill effects in diverse forms exposed to them. The occurrence of *Cenarchaeum symbiosum*, a phychrophilc archeabacterium was reported recently in association with a marine sponge *Axinella mexicana* (Pizarro *et al.*, 1996; Sayler *et al.*, 1985). The diverse bacteria living within the tissues of the sponges may produce many such compounds.

Agrobacterium tumefaciens is a well-known bacterium that has been used extensively in plant genetic engineering. *Agrobacterium tumefaciens* carries a plasmid that normally causes plant disease. However, by inserting specific genes into the bacterial plasmid, desirable characteristics can be transferred to the plant genome where plasmid DNA gets integrated. Thus herbicide and pesticide resistant transgenic plants have been produced.

Another application of microbial diversity is the production of enzymes active at high temperatures from the bacterium *Thermus aquaticus* isolated from the hot springs of Yellow Stone National Park many years ago. Using this enzyme, the polymerase chain reaction (PCR) was developed. This is an enormously powerful molecular genetic tool that amplifies DNA fragment and increase its concentration that allows detailed examination of gene sequences. Later several extreme and hyperthermophiles isolated from thermal vents and terrestrial hot springs have been shown to be sources of thermostable DNA polymerases with proof-reading exonuclease activity. Application of PCR in criminology is well known.

More importantly microbes can be used as 'Janitors' of the planet, i.e., in bioremediation. Microorganisms are useful in restoring habitats to a functional ecological state because they are capable of degrading pollutants, purifying water and soil in this process. Genetically engineered microorganisms degrade toxic compounds more rapidly, and offer a great promise for bioremediation and biorestoration in the future.

3. METHODS FOR STUDYING MICROBIAL DIVERSITY

The methods employed for studying microbial diversity are described below:

(a) Classical Method for Culturable Microbes

In these methods, microbes have to be cultured on certain media to check their colony characters, morphological characters, staining reactions, growth and nutritional characteristics, and biochemical and physiological characteristics. But it is a fact that only less than 5% of the estimated microbes are culturable on the media routinely used in microbiological laboratory (Table 1). This inability to recover the most numerous organisms from natural habitats by using cultural approaches has been called the 'enumeration anomaly' or the 'Great Plate Count Anomaly' (Staley *et al.*, 1997). Despite their extraordinary genetic as well as phenetic diversity and ecological importance, the study of unculturable microbes was just impossible before a few years. With the help of modern methods, now it is possible to detect them.

Table 1. Estimated Proportion of unculturable microorganisms in various terrestrial and aquatic biotopes (Amann *et al.* 1995)

Biotopes	Proportion of culturability (%)
Seawater	0.001-0.100
Freshwater	0.25
Mesotrophic lake	0.1-1.0
Unpolluted estuarine water	0.1-3.0
Activated sludge	1-15
Sediments	0.25
Soil	0.3

(b) Modern Methods for Non-culturable Microbes

The characterization of an organism in terms of its phylotype requires only a gene sequence and not a functioning cell. Genes can be obtained by cloning nucleic acids, which can be isolated directly from the environment using standard technique (Rifai, 1989). Several specialized methods are available for the extraction and purification of nucleic acids from a wide range of environmental samples, including tissues, water, soil and even rock. Methods are usually based on chemical and/or physical disruption of cells, combined with treatments to remove contaminating materials such as humic acids and metals that poison enzymatic steps. Number of strategies can be used to obtain rRNA gene clones from 'total community' nucleic acids isolated from the environmental sample.

(i) Nucleic acid hybridization techniques

The most straight forward way of detecting specific nucleic acid sequences or genes of interest is through direct hybridization of DNA/RNA probes to microbial nucleic acid extracts. Whole cell DNA/RNA is extracted from the microbes of an experimental sample such as soil or water and applied to a positively charged nitrocellulose or nylon membrane.

Colonies of cloned bacteria can also be replica-plated from agar plates to any of these membranes. The cells are lysed and their nucleic acids exposed in situ for subsequent hybridization. This membrane bound nucleic acid is then probed with DNA/RNA containing a sequence complementary to the target sequence. These probes are labeled with P^{32} (Radioactive Phosphorus) or a chemoluminescent compound or digoxigenin (for antibody peroxidase detection) before hybridization. DNA probes could also be used to detect genes in the bacterial genome or on plasmids (Southern blots) or to detect mRNA or rRNA (Northern blots). The nucleic acid hybridization is a quantitative test; positive bands or spots on the membrane can be directly compared through densitometry.

The structural genes for naphthalene degradation (the *nah* operon) have been detected and quantified by using different types of nucleic acid hybridization techniques. In the earliest study direct hybridization technique was used to monitor the microbes with TOL and NAH gene, these genes are responsible for the degradation of toluene and naphthalene, respectively (Santany *et al.*, 1990). Colony hybridization with entire plasmids as probes allowed the counting of cells containing these catabolic plasmids. High plasmid concentration was found to have high rate of mineralization, and exposure to these aromatic substrates caused an increase in plasmid levels. In one study the same method was used for monitoring the *xyl*E and *ndo*B gene, which are involved in creosote degradation in soil (Horikoshi and Grant, 1998). Standard Southern blot hybridizations were used to follow the bacterial population responsible for naphthalene degradation in seeded microcosms induced with salicylate (Myers *et al.*, 1987). In this study, probes specific for the *nah* operon provided a measure of the naphthalene degradation potential of the microbial population. *Nah*A transcripts have also been quantified with a different technique i.e. ribonuclease protection assays (Erhart *et al.*, 1997). The extracted RNA was hybridized in solution to radiolabelled anti-sense RNA. Treatment with RNase destroys unhybridized transcripts, permitting quantification of the *nah*A mRNA. It was shown that *nah*A transcription was positively correlated with naphthalene mineralization in polycyclic aromatic hydrocarbon contaminated soils.

These variations of direct hybridization techniques have been used to determine contamination caused by toxic chemical in natural systems and industrial bioreactors. For example the presence of *mer*A gene, which codes for mercuric reductase, was detected with the help of radiolabelled RNA probe to trace transcript RNA. The level of *mer*A gene expression was found to have direct correlation with mercury-contaminated aquatic microbial communities (Ausubel *et al.*, 1992). Similarly, dot-blot hybridizations were used to detect several genes involved in the degradation of aromatic compounds in soil (Goodfellow, 2000). The most important indicator of contamination with petroleum is the *xyl*E gene. Total DNA from a variety of anaerobic sludge reactors was hybridized with genus-specific probes for methanogens and sulphate-reducting bacteria to describe these complex microbial communities.

In some studies the results of molecular probing have been correlated with traditional most probable number (MPN) techniques. In the following case, focusing on microbes that are capable of dechlorinating aromatic compounds, a combination of MPN and colony hybridization was used. To follow the microbial communities of a flow-through lake, microcosms are seeded with a chlorobenzoate-degrading *Alcaligenes* strain (Fleming *et al.*, 1993). It is found that there is a direct relation between the size and activity of the specific catabolic population during exposure to various concentrations of 3-chlorobenzoate.

(ii) PCR and associated techniques

PCR provides a powerful tool for the analysis of nucleic acid sequences by amplifying (making many copies of) them, that might be in low abundance in a complex mixture of whole-cell extracts (Amann *et al.*, 1995; Hosein *et al.*, 1997). Due to low concentration of these sequences, they may not be detectable with direct hybridi- zation techniques. In PCR, a thermostable DNA polymerase (Taq/Vent DNA polymerase) is used to amplify the desired DNA sequences. These amplified DNA sequences can be purified and examined by agarose gel electrophoresis. In addition, PCR-amplified products can easily be cloned and used for transformation or sequencing.

There are many variations of the PCR technique. In nested PCR, first a larger piece of the target DNA is amplified by using primers that are not organism-specific and then re-amplify these products with primers specific for an organism of interest. Nested PCR has been used to monitor the population of exogenous 3-chlorobenzoate-dechlorinating bacteria in nonsterile soil slurry microcosm (Davis, 1998). Nowadays nested PCR is used to monitor *Desulfitobacterium* in a methanogenic consortium degrading pentachlorophenol (Khoshoo, 1994).

MPN method coupled with PCR (MPN-PCR) is used to estimate the number of *nah*Ac, *alk*B and *xyl*E genes present in soils contaminated with jet fuel (Brink *et al.*, 1994). In MPN-PCR, DNA extracts of the sample are serially diluted before PCR amplification and limits can be set on the number of genes in the sample in relation to known control dilutions. In other application, microbial populations of copper bioleaching plant were monitored by PCR amplifications of the variable-length spacer region between the 16 S and 23 S rDNA species (Persoon *et al.*, 1995). The resulting patterns were compared with a library of spacer regions made from the main isolates obtained from the bioleaching plant for which 16S rDNA species were previously sequenced. Now the traditional culture methods are combined with PCR and colony hybridizations to trace the release of phenol-degrading *Pseudomonas* in an aquatic ecosystem (Vainio *et al.*, 1997).

Reverse transcriptase-mediated PCR (RT-PCR) has changed the face of PCR technique. In this technique, before amplifying, the DNA in a sample is destroyed with DNase. Then reverse transcriptase and random primers (usually hexamers) are added to the reaction mixture, the RNA in the sample (both mRNA and rRNA) is transcribed into DNA. PCR is then used to amplify specific sequences of interest from this cDNA. With the help of RT-PCR, the detection, quantification and the expression of individual structural genes are possible. RT-PCR could also be used to measure the level of transcription of the *dmp*N gene responsible for the degradation of phenol. In a similar study transcription of the *tfd*B genes was measured by RT-PCR in activated-sludge bioreactors augmented with a 3-chlorobenzoate degrading *Pseudomonas* (Schlper *et al.*, 1997).

There are several procedures that can be used in conjunction with PCR for further examination of the PCR-amplified products. The combination of PCR and Southern blot hybridization with 16S rDNA phylogenetic probes was used to detect *Azoarcus tolulyticus*, a toluene-degrading denitrifier, in seeded microcosms and field sites (Zedan, 1993). This organism was then demonstrated to be present in toluene-contaminated groundwater samples, where it degraded toluene anaerobically.

PCR is a widely used technique for examining mixed microbial communities in natural samples. However, a care should be taken to get right results. Very little contamination can lead to false positive results, and false negative amplifications are often seen (Wilson, 1986).

(iii) Polymorphism-based procedures

These techniques are generally used in separating and identifying PCR-amplified products when they have same size but slightly different nucleotide sequence and can not be resolved by simple gel electrophoresis techniques. Sometime the amplified portions of 16S rDNA genes from a mixed microbial population may have similar sizes with a particular set of oligonucleotide primers, but can have small differences within the PCR-amplified product at the nucleotide level. Then these differences are detected by restricting the PCR-amplified product with endonucleases and examining the pattern of the restriction fragments. There are several types of polymorphism-based procedures that are used in microbial ecological systems. The detection of small differences in specific DNA sequences gives important information about community structure and the diversity of microbes containing a critical gene. These techniques are generally coupled to a PCR reaction to amplify low abundant sequences. PCR-amplified products are examined to detect single base substitutions in the nucleotide sequence. The PCR-amplified product can be either end-labeled or uniformly labeled.

In one study, the presence of the *nah*Ac gene sequences were tested in natural sediments by using PCR (Guo *et al.*, 1997). In gene sequence, polymorphisms were detected after restricting the PCR-amplified product. In another case, PCR amplified *bph*C genes from natural soils allowed further investigation of the PCR degradation pathway (Fantrouss *et al.*, 1997). But no restriction fragment length polymorphism (RFLP) was seen in the PCR-amplified products. This shows the presence of limited biodiversity in the PCB-degrading population at the site of sampling. In the same way, RFLP analysis was used with PCR to examine the 16S rDNA and catabolic genes of organisms in a contaminated aquifer (Stoner *et al.*, 1996). The presence of less biodiversity at copper contaminated site in comparison to non-contaminated site was proved by RFLP (Selvaratnam *et al.*, 1997).

In another protocol, a single primer is used to amplify DNA fragments with PCR before their RFLP examination. In randomly amplified polymorphic DNA (RAPD) or arbitrarily primed PCR, non-specific primers are used to amplify fragments. After running the gel, organism-specific fingerprints are obtained. The fingerprints so obtained are analysed with the help of computer software and identify patterns representative of different known organisms. The genomic fingerprints obtained during various treatments are matched with the dominant microorganisms isolated by traditional methods. RAPD analysis has been used to investigate the microbial diversity of a variety of bioreactors harboring mixed microbial populations (Becker *et al.,* 1997). The genomic fingerprints so obtained were used to compare population changes with time in a reactor and in comparing bench-scale reactors with that of pilot-scale, but community biodiversity could not be assessed. Another example is the study of bacterial flora of laboratory scale oily sludge composts with RAPD (Pace, 1999). The composts amended with oily sludges were found to contain a different microbial population than unamended sludges. By the use of denaturing gradient gel electrophoresis (DGGE), the single-base substitutions in PCR-amplified products can be detected. To make a denaturing gradient gel, a vertical polyacrylamide gel is poured with an increasing concentration of denaturant (fomamide and/or urea) (Moller *et al.*, 1996; Muyzer *et al.*, 1993). As double helical DNA fragments are subjected to DGGE, partial melting occurs at the point of equidensity, which leads to partial denaturation due to which the movement of DNA fragments are ceased leading to resolution in a single nucleotide base pair. The DGGE of PCR-amplified 16S rDNA genes has been used to compare the microbial complexity of 3-chlorobenzoate-contaminated soil (Barkay *et al.*, 1989).

Denaturing Gradient Gel Electrophoretic analysis of extracted 5S rRNA is used to characterise mixed microbial populations in acidic mining environments (Staley and Konopka, 1985).

(iv) RNA-Based analysis or 16S rDNA sequencing

The probes based on 5S, 16S and 23S rRNA are used together with the isolation, separation, and sequencing of RNA molecules, and this has made possible the study of non-culturable microbes. Among all the above RNA species, 16S rRNA is most important for the phylogenetic identification of microbes because it has highly conserved regions in nucleotide sequence. While 18S rRNA is employed for establishing similarity among groups or individual. The analysis of 16S rRNA has tremendously altered the classification of microorganisms into three domains, the Bacteria, Archaea, and Eukarya (Wilson, 1997). By choosing the appropriate probe for a particular rRNA sequence, one can analyze members of communities at domain, kingdom, genus, or species-specific levels of resolution (Smit *et al.*, 1997).

Microbial communities are directly examined through the direct sequencing of rDNA genes (Ogunseitan *et al.,* 1991). This is generally done with the help of 16S rDNA, although sometimes 23S rDNA is also used. This is a very laborious task, and it is made easier through the use of PCR. The 16S rDNA species of a mixed population is amplified by PCR, and then amplified product is purified. In this way organisms can be examined and identified in a mixed population even if the organism is non-culturable. In a study, the selected 16S rDNA species were sequenced from a group of 516 isolates, and was used to investigate community structures in an activated sludge (Udvardy, 1984). The advantage of 16S rDNA sequencing technique is that it gives an unambiguous identification of the organisms in the community. The important disadvantage of this technique is that it is very laborious and costly.

(v) Fluorescent in situ hybridization (FISH)

To study the role of microorganism in an ecological system, it is very important to obtain information about the position of microorganisms in biotic components, i.e. how they interact with each other and what role they play in the organization of the system. These systems are studied by preparing biofilms and flocs, but it is very difficult to locate the various organisms (and their activities). The distribution of microbes within an architectural framework can be seen by using FISH, in which tagged DNA oligomers are used as probes. This technique enhances the study of intact microbial communities (Amann, 1995). Biofilms or flocs are prepared on gelatin coated glass slides by dipping them in a pond or any other site of interest. These slides are then treated with paraformaldehyde (the paraformaldehyde makes the cell permeable by dissolving protein of the cell membrane and fixes it to the slide). The fixed and permeabilized cells are then exposed to a DNA probe, generally is an oligonucleotide, tagged with a fluorescent dye such as fluorescein. These DNA probes bind to complementary DNA of the permeable slide-bound cells, which glow when observed under an epifluorescence microscope. If FISH is coupled with confocal scanning laser microscope, the labeled microbes can be seen in three-dimensional space with the intact microbial community. This method is problematic in soil systems because organic matter present in the soil binds with the dye and gives false result.

Population-specific 16S rRNA probes were used for monitoring the enrichment and isolation of *Desulfovibrio vulgaris* from a mixed anaerobic bioreactor (Innis *et al.*, 1990). The 16S rDNA of a species isolated from the bioreactor was sequenced and used to design FISH and PCR probes. The other isolates were identified with the help of probe hybridization, and those showing similarities were further investigated by sequencing their 16S rRNA genes. PCR amplification and FISH of 16S rDNA sequences accurately provide phylogenetic information about mixed microbial populations. *Desulfobacter* spp. were enriched from oilfield samples and detected with chemiluminiscent 16S rRNA-specific probe. (Breen *et al.*, 1995). Then microbial community structure was compared with those obtained by standard plating techniques. FISH was used to monitor *Microthrix parvicella*, an organism implicated in the bulking of activated sludge (Erb, 1993).

The three-dimensional distribution of toluene degrading *Pseudomonas putida* was shown by using scanning confocal laser microscopy and fluorescently labeled 16S rRNA targeted probes in waste-gas treatment biofilter (McNeely *et al.*, 1990). FISH was also used to determine the vertical distribution of sulfate-reducing bacteria in a biofilm, from a trickling-filter bioreactor, with phylogenetic probes. (Preston *et al.*, 1996).

4. OTHER IMPORTANT METHODS

Fatty acid analysis is a modern method used to identify culturable microbes. In this method cellular lipid profile of microbes are analysed for identifying individual strain or to characterize whole community. Phospholipid fatty acids (PLFAs) are the primary components of cell membranes found in all living organisms. Fatty acids of different organisms differ in their chain length, number and location of unsaturation and functional groups. These are specific for an individual organism or a group. The membrane sample is analysed by gas chromatography or gas liquid chromatography.

5. INDICES OF MICROBIAL DIVERSITY

Many efforts have been made to estimate the microbial diversity in a particular area. The counting of species is sufficient for its determination, although some times it is very difficult to identify the microbial species. All species are not equally abundant in an ecosystem, and indices of biodiversity such as richness and equitability are usually considered. Richness means number of species, and equitability is their evenness of abundance.

Many factors like pollution and invasion by man affect biodiversity to a great extent. The ecosystem rich in nutrients support less microbial diversity. The Shannon-Weaver index is widely used to measure the biodiversity. The diversity at strain level is less important than

Table 2. Sources of taxonomic information

Cell Component	Analysis	Taxonomic rank		
		Genus or Above	Species	Strain
Chromosomal DNA	Base composition	←	→	
	DNA: DNA hybridization		←	→
	Restriction pattern		←	→
DNA segments	DNA probe	←		→
	DNA sequencing	←		→
	PCR based DNA fingerprinting		←	→
	(RFLP, BOX-PCR, ERIC-PCR, RAPD, ARDRA)			
Ribosomal RNA	DNA: rRNA hybridization	←	→	
	Nucleotide sequences	←	→	
Proteins	Amino acid sequences	←	→	
	Electrophoretic patterns		←	→
	Multilocus enzyme electrophoresis			↔
	Serological comparisons	←		→
Chemical marker	Peptidoglycan	↔		
	Fatty acid	←	→	
	Isoprenoid Quinones	←	→	
	Mycolic acid	←	→	
	Polar lipids	←	→	
	Poly amines	↔		
	Polysaccharide	←	→	
	Teichoic acid	←	→	
Whole organism	Pyrolysis mass-spectrometry		←	→
	Rapid enzyme test		←	→
Expressed features	Morphology	←	→	
	Physiology	←	→	

Modified From Goodfellow (2000)

Abbreviations: PCR, Polymerase Chain Reaction. RFLP, Restriction Fragment Length Polymorphism. RAPD, Randomly Amplified Polymorphic DNA Fingerprints. ARDRA, Amplified Ribosomal DNA Restriction Analysis diversity at species level. The information obtained from various sources is useful in systematics of microbes (Table 2).

6. PRESENT STATUS OF MICROBIAL DIVERSITY

There is no right and reliable measure of microbial diversity. Therefore, there is no clear-cut idea as to how much biodiversity exists, and how much is becoming extinct day by day. Consequently, the statements about loss of microbial diversity are at best anecdotal (Ram *et al.*, 1993). Biodiversity on earth (eight broad realms with 193 bio-geographical provinces) is composed of ecosystems that are constituted by communities of living species existing in an ecological region. (Theim *et al.*, 1994).

The developing countries, located in sub-tropical/tropical belt, are far richer in biodiversity than the industrial countries in the temperate region. The estimated number of various species of microbes in India and the world are shown in Tables 3 & 4. Among 1,08,276 total biological species described in India, microbial species constitute 24.3% (Karner, 2001).

Table 3. Known and estimated numbers of microorganisms (Bull *et al.* 1992)

Group	Known species	Estimated total species	% of known species
Viruses	5000	130,000	4
Bacteria	4760	40,000	12
Fungi	69000	1,500,000	5
Protozoa	30,800	100,000	31
Total	140360	**2130000**	**52**

Table 4. Number of known microbial species in India

Taxon	Number of species	Percentage
Bacteria	8,50	0.8
Fungi	2,300	21.2
Algae	2,500	2.3
Total	5,650	24.3

Conservative estimates place the total number of bacterial species to be about 50,000 to 3,000,000 (Stahl and Amann, 1991). As of 1999, approximately 5000 bacterial species had been described (Olsen *et al.*, 1986). The microbes inhabiting extreme environments have been called extremophiles. The realisation that extreme environments harbor very different kind of prokaryotic microbes has led to intensive work on various aspects of extremophiles including their diversity [Herric *et al.*, 1993; Chandler and Brockman, 1996]. A recent investigation by Karner *et al.* (2001) using genetic probes suggested archaea to make up a significant fraction of picoplankton, below 150m depth of Pacific Ocean near Hawaii. Extreme environments are expected to provide us with unique physiologies with a range of potentially exploitable biotechnological tools and processes, and have been provided thermostable enzymes for specific applications in industry. From the biotechnological point of view, thermostable enzymes are more robust than their low temperature counterparts, and may show resistance to organic solvents. Other extremophiles have application in the recovery and removal of metals and degradation of toxic pollutants. Yet others are sources of industrial enzymes active at extremes of pH or salinity.

7. CONCLUSIONS

Although a good progress has been made in the field of microbial diversity, much is yet to be explored. The only reliable method for estimating microbial diversity is considerd to be rRNA sequence analysis. This is of course a reliable method, but dependence on a single method is not good, and this method has its own limitation and not always successful in all the cases.

Along with rRNA, other molecules such as elongation factor Tu, 23S rRNA, and f_1f_0 ATPase can also provide phylogenetic information and can serve as alternative markers for inferring relationship (Levesque *et al.*, 1997).

REFERENCES

Amann, R.I. 1995. Molecular Microbial Ecology Manual. Kluwer Academic Publisher, Dordrecht, Netherlands. pp. 1-15.

Ausubel, F. M., Brent, R., Kingston, R. E., Moore, D. D., Seidman, J. G., Smith, J. A. and Struhl, K. (eds) 1992. In *Current Protocols in Molecular Biology*, Vol. 2, John Wiley and Sons, New York.

Barkay, T., Liebert, C. and Gillman, M. 1989. Hybridization of DNA probe with whole-community genome for detection of gene that encode microbial responses to pollutants: *mer* genes and Hg^{2+} resistance. Appl. Enviorn. Microbiol. 55: 1574-1577.

Becker, J. G, Berardesco, G., Rittman, B. E. and Stahl, D. A. 1997. Abstracts of the 97th General meeting of the American Society for Microbiology, American Society for Microbiology, Washington. p. 476.

Breen, A., Rope, A. F., Taylor, D., Loper, J. C. and Sferra, P. R. 1995. Application of DNA amplification fingerprinting (DAF) to mixed culture bioreactors. J. Indust. Microbiol. 14: 10-16.

Brink, D. E., Vance, I and White, D. C. 1994. Detection of *Desulfobacter* in oil field environments by non radioactive DNA probes. Appl. Microbiol. Biotechnol. 42: 469-475.

Chandler, D. P. and Brockman F. J. 1996. Estimating biodegradative gene number at a JP-5 contaminated site using PCR. Appl. Biochem. Biotechnol. 57/58: 971-982.

Davis, M. C., 1998. Making living at the extremes. TIBTECH. 16: 102-104.

EL Fantroussi, S., Mahillon, J., Naveau, H. and Agathos, S. N. 1997. Introduction of anaerobic dechlorinating bacteria into soil slurry microcosms and Nested-PCR monitoring. Appl. Enviorn. Microbiol. 63: 806-811.

Erb, R. W. and Wagner-Dobler, I 1993. Detection of polychlorinated biphenyl degradation genes in polluted sediments by direct DNA extraction and Polymerase Chain Reaction. Appl. Enviorn. Microbiol. 59: 4065-4073.

Erhart, R., Bradford, D., Seviour, R. J., Amann, R. I. and Blackall, L. L. 1997. Development and use of fluorescent in situ hybridization probes for the detection and identification of "Microthrix parvicella" in activated sludge. Syst. Appl. Microbiol. 20: 310-318.

Fleming, J. T., Sanseverino, J. and Sayler, G. S. 1993. Quantitative relationship between naphthalene catabolic gene frequency and expression in predicting PAH degradation in soils. Enviorn. Sci. Technol. 27: 1068-1074.

Fulthorpe, R. R., and Wyndham, R. C. 1989. Survival and activity of 3-Chlorobenzoate-catabolic genotype in a natural system. Appl. Enviorn. Microbiol. 55: 1584-1590.

Goodfellow, M. 2000. Microbial systematics: background and uses. In Applied Microbial Systematics (Fergus G. P. and Goodfellow, M eds.). Kluwer Academic Publisher, Dordrecht, Netherlands, 1: 1-18.

Guo, C., Sun, W., Harsh, J. B., and Ogram, A. 1997. Hybridization analysis of microbial DNA from fuel oil-contaminated and noncontaminated Soil. Microbial Ecol. 34: 178-187.

Herrick, J. B., Madsen, E. L., Batt, C. A. and Ghiorse, W. C. 1993. Polymerase chain reaction amplification of Napthalene catabolic and 16S rRNA gene sequence from indigenous sediment bacteria. Appl. Enviorn. Microbiol. 59: 687-694.

Horikoshi K. and Grant W. D. Eds., 1998. Extremophiles: Microbial life in Extreme environments. Wiley-Liss; New York, pp. 322.

Hosein, S. G., Millette, D., Butler, B. J. and Greer, C. W. 1997. Catabolic gene probe analysis of an aquifer microbial community Degrading. Microbial. Ecol. 34: 81-89.

Innis, M. A., Gelfand, D. H., Sninsky, J. J. and White, T. J. 1990. PCR Protocols, Academic Press, San Diego.

Kane, M. D., Poulsen, L. K. and Stahl, D. A. 1993. Monitoring the enrichment and isolation of sulfate reducing bacteria by using Oligonucleotide hybridization probes designed from environmentally derived 16S rRNA sequences. Appl. Environ. Microbiol. 59: 682-686.

Karl, D. M. 1987. Bacterial production at deep-sea hydrothermal vent and cold seeps: evidence of chemosynthetic primary production. Symp. Soc. Gen. Microbiol. 41: 319-360.

Karner, M. B., Delong, E. F. and Karl, Don. 2001. Archaeal dominance in the mesopelogic zone of the Pacific Ocean. Nature 409: 507-510.

Khoshoo, T. N., 1994. India's biodiversity: tasks ahead. Curr. Sci. 67: 578.

Levesque, M. J., La Boissire, S., Thomas, J. C., Beaudet, R. and Villemur, R. 1997. Rapid method for detecting *Desulfitobacterium frappieri* strain PCP-1 in soil by the polymerase chain reaction. Appl. Microbiol. Biotechnol. 47: 719-725.

Ludwig, W. and Schlleifer, K-H. 1999. Phylogeny of bacteria beyond the 16s rRNA standard. ASM News 65: 752-757.

Mc Neely, J. A., Miller, K. R., Reid, W. V., Mittermeier, R. A., Werner, T. B. 1990. *Conserving the world's Biological diversity*. Gland: Int. Union for Conservation of Nature and Natural Resources, Switzerland..

Moller, S., Pedersen, A. R., Poulsen, L. K., Arvin, E. and Molin, S. 1996. Activity and three-dimensional distribution of Toluene-Degrading *Pseudomonas putida* in a multispecies biofilm assessed by quantitative insitu hybridization and scanning confocal laser microscopy. Appl. Environ. Microbiol. 62: 4632- 4640.

Muyzer, G., De Waal, E. C. and Uitterlinden, A. G. 1993. Profiling of complex microbial population by denaturing gradient gel electrophoresis analysis of polymerase chain reaction amplified genes coding for 16S rRNA. Appl. Environ. Microbiol. 59: 695-700.

Myers, R. M., Maniatis, T. and Lerman, L. S. 1987. Detection and localization of single base change by denaturing gradient gel electrophoresis. Methods Enzymol. 155: 501-527.

Ogunseitan, O. A., Delgado, I. L., Tsai, Y. L. and Olson, B. H. 1991. Effect of 2-Hydroxybenzoate on the maintenance of Naphthalene degrading Pseudomonads in seeded and unseeded soil. Appl. Enviorn. Microbiol. 57: 2873-2879.

Olsen, G. J., Lane, D. J., Giovannoni, S. J., Pace, N. R. and Stahl, D. A. 1986. Microbial ecology and evolution: a ribosomal RNA approach. Ann. Rev. Microbiol. 40: 337-365.

Pace, N. R. 1999. Microbial ecology and diversity. ASM News 65: 328-333.

Persson, A., Quednau, M. and Ahrne, S. 1995. Monitoring and verification of bioremediation (Hinchee, R. E., Douglas, G. S. and Ong. S. K., eds.), Battelle Press, Columbus. pp. 147-155.

Pizarro, J., Jedlicki, E., Orellana, O., Romero, J. and Espejo, R. T. 1996. Bacterial population in samples of bioleached Copper ore as revealed by analysis of DNA obtained before and after cultivation. Appl. Enviorn. Microbiol. 62: 1323-1328.

Preston, C. M. K., Wu, K. Y. Molinski, T. F., and De Long, E. F. 1996. A psychrophlic crenarchaeon inhabits a marine sponge: *Cenarchaeum symbiosum*. Gen. Nov., sp. Nov. Proc. Natl. Acad. Sci. USA 93: 6241- 6246.

Ramsing, N.B., Kuhl, M. Jorgensen, B.B. 1993. Distribution of sulfate-reducing bacteria, O_2 , and H_2S in photosynthetic biofilms determined by oligonucleotide probes and microelectrodes. Appl. Enviorn. Microbiol. 59: 3840-3849.

Rifai, M. A. 1989. Astounding fungal phenomenon as manifestations between tropical plant and microorganisms. In *Interaction between Plant and Microorganisms*, (G. Lim, K. Katsya. eds.). Singapore: Natl. Univ., Singapore, pp. 1-8.

Sambrook, J., Fritsch, E. F. and Maniatis, T., 1989. Molecular Cloning: A Laboratory Manual , Cold spring Harbor Laboratory press.

Santany, D. L., and R. R. Colwell. 1990. Comparison of the symbiont bacterial community associated with the caribbean sclerosponge *Ceratoporella nicholsoni* and ambient sea water. Mar. Ecol. Prog. Ser. 67: 73-82.

Sayler, G. S., Sheilds, M. S., Tedford, E. T., Breen, A., Hooper, S. W., Sirotkin, K. M. and Davis, J. W. 1985. Application of DNA-DNA colony hybridization to the detection of catabolic genotypes in environmental samples. Appl. Enviorn. Microbiol. 49: 1295-1303.

Schlper, C., Swanson, R. V., Mathur, E. J. and De Long, E. F. 1997. Chracterisation of a DNA polymerase from the uncultured psychrophilic archaeon *Cenarchaeum symbiosum* J. Bacteriol. 179: 7803-7811.

Selvaratnam, S., Schoeddel, B. A., Mcfarland, B. L. and Kulpa, C. F. 1997. Application of the polymerase chain reaction (PCR) and reverse transcriptase/ PCR for determining the fate of phenol degrading *Pseudomonas putida* ATTC 11172 in a bioagumented sequencing batch reactor. Appl. Microbiol. Biotechnol. 47: 236-240.

Smit, E., Leeflang, P. and Wernars, K. 1997.Detection of shift in microbial community structure and diversity in soil caused by copper contamination using ribosomal DNA restriction analysis. FEMS Microbiol. Ecol. 23: 249-261.

Stahl, D. A., Amann, R. 1991. Development and application of nucleic acid probe. In nucleic acid Techniques in Bacterial Systematics, (E. Stackebrandt, M. Goodfellow eds.) Chichester: Wiley, pp. 205-248.

Staley *et al.*, 1997. The microbial world: foundation of the biosphere. A report of the American Academy of Microbiology. American society for Microbiology, Washington, D.C.

Staley, J. T., Konopka, A. 1985. Measurement of in situ activities of nonphotosynthetic microorganisms in aquatic and terrestrial habitats. Annu. Rev. Microbiol. 39:321-346.

Stoner, D. L., Browning, C. K., Bulmer, D. K., Ward, T. E. and MacDonell, M. T. 1996. Direct 5s rRNA assay for monitoring mixed-culture bioprocesses. Appl. Environ. Microbiol. 62: 1969-1976.

Thiem, S. M., Krumme, M. L., Smith, R. L. and Tiedje, J. M. 1994. Use of molecular techniques to evaluate the survival of a microorganism injected into an aquifer. Appl. Enviorn. Microbiol. 60: 1059-1067.

Udvardy, M. D. F., 1975, 1984. A classification of the bio-geographical provinces of the world, IUCN, Gland.

Vainio, E. J., Moilanen, A and Koivula, T. T., Bamford D. I., Hantula J. 1997. Comparison of partial 16S rRNA gene sequences obtained from activated sludge bacteria. Appl. Microbiol. Biotechnol. 48: 73-79.

Wand, H., Laht, T., Peters, M., Becker, P. M., Stottmeister, U. and Heinaru, A. 1997. Monitoring of biodegradative *Pseudomonas putida* strains in aquatic environments. Microb. Ecol. 33: 124-133.

Wilson, E.O. 1986. Biodiversity, National Academy, Washington, p. 521.

Wilson, I. G. 1997. Inhibition and facilitation of nucleic acid amplification. Appl. Enviorn. Microbiol. 63: 3741-3751.

Woese, C. R., Kandler, O. and Wheelis, M. L. 1990. Towards a natural system of organisms: Proposal for the domains Archaea, Bacteria, and Eucarya. Proc. Natl. Acad. Sci. USA 87: 4576-79.

Zedan, H. 1993. The economic value of microbial diversity. Biotech. Appl. Microbiol: 43: 178-185.

Zhou, J., Palumbo, A. V. and Tiedje, J. M. 1997. Sensitive detection of a novel class of Tolune- degrading denitrifiers, *Azoarcus tolulyticus,* with small subunit rRNA primers and probes. Appl. Enviorn. Microbiol. 63: 2384-2390.

CHAPTER – 2

BIOSENSORS AND THEIR APPLICATIONS

S.S. Kanwar and Sunita Devi

Department of Microbiology, College of Basic Sciences,
Himachal Pradesh Agricultural University, Palampur-176 062 (India)

1. INTRODUCTION

A biosensor is an analytical device, which employs a biological material to specifically interact with an analyte; this interaction produces some detectable physical change, which is measured and converted into an electrical signal by a transducer. Finally, the electrical signal is amplified, interpreted and displayed as analyte concentration in the solution/ preparation.

An analyte is a compound whose concentration is to be determined, in this case, by the biosensor. The biological materials used are usually enzymes but nucleic acids, antibodies, lectins, whole cells, entire organs or tissue slices are also used.

The nature of interaction between the analyte and the biological material used in the biosensor may be of two types:

1. The analyte may be converted into a new chemical molecule by enzyme; such biosensors are called **catalytic biosensors.**
2. The analyte may simply bind to the biological material present on the surface of biosensor (e.g., to antibodies, nucleic acids, receptors etc.) and such biosensors are called **affinity/non-catalytic biosensors.**

As defined by Gronow (1984), biosensor is an analytical tool or system consisting of an immobilized biological material in intimate contact with a suitable transducer device that converts the biological signal into a quantifiable electrical signal.

The concept of biosensor was pioneered by Clark and Lyons (1962) who proposed that enzyme could be immobilized at electrical detectors to form enzyme electrodes.

2. BIOSENSORS — A DIVERSE TECHNOLOGY

Biosensor technology can be divided into two key components (Griffths and Hall, 1993) :

1. Biological component.
2. Physical component.

1. *Biological component:* It includes biological materials like enzymes, tissues, micro-organisms, antibodies, nucleic acids, receptors etc.The biological component of biosensor technology can be divided into two distinct groups; Catalytic and non-catalytic.

Catalytic group includes enzymes, microorganisms and tissues.

Non-catalytic or affinity class comprises antibodies, receptors and nucleic acids.

2. *Physical component* : Transducer, amplifier etc.

Biological component of biosensor performs two important functions:

(a) It specifically recognizes the analyte.

(b) It interacts with it in such a manner, which produces some physical change, detectable by the transducer. These properties of biological component impart on biosensor its specificity, sensitivity and ability to detect and measure the analyte. The biological component is suitably immobilized on to the transducer.

The biological component interacts specifically to the analyte that produces a physical change close to transducer surface. The physical change may be :

heat released or absorbed by the reaction, production of an electrical potential due to changed distribution of electrons, movement of electrons produced in redox reactions, light produced or absorbed during the reaction and change in the mass of the biological component as a result of the reaction.

The transducer detects and measures this change and converts it into an electrical signal. This signal is necessarily very small and is amplified by an amplifier before it is fed into the microprocessor. The signal is then processed, interpreted and is displayed in suitable units.

Thus biosensor converts a *chemical information flow into electrial information flow* which involves following steps.

(a) The analyte diffuses from the solution to the surface of biosensor.

(b) The analyte reacts specifically and efficiently with the biological component of the biosensor.

(c) This reaction changes the physio-chemical properties of the transducer surface.

(d) This leads to a change in the optical or electronic properties of the transducer surface.

(e) The change in the optical / electronic properties is measured, converted into electrical signal, which is amplified, processed and displayed.

3. FEATURES OF A SUCCESSFUL BIOSENSOR

A successful biosensor must possess atleast some of the following beneficial features:

1. The biocatalyst must be highly specific for the purpose of analysis, stable under normal storage conditions (except in case of colorimetric enzyme strips and dipsticks) and show good stability over a large number of assays.
2. The reaction should be as independent of such physical parameters as stirring, pH and temperature, as is manageable. This would allow the analysis of samples with minimal pretreatment. If

the reaction involves cofactors or coenzymes, these should, preferably be co-immobilized with the enzymes.

3. The response should be accurate, precise, reproducible and linear over the useful analytical range, without dilution or concentration. It should also be free from electrical noise.
4. If the biosensor is to be used for invasive monitoring in clinical situations, the probe must be tiny and biocompatible, having no toxic or antigenic effects. If it is to be used in fermenters, it should be sterilizable. This is preferably performed by autoclaving but no biosensor enzymes can presently withstand such drastic wet heat treatment. In either case, the biosensor should not be prone to fouling or proteolysis.
5. The complete biosensor should be cheap, small, portable and capable of being used by semi-skilled operators.

Principle of biosensor : The principle of biosensor is quite simple. The biological material is immobilized on immobilized support, the permeable membrane in direct vicinity of a sensor. The substances to be measured pass through the membrane, interact with the immobilized material, and yield the product. A product (i.e. the monitored substrate) may be heat, gas (oxygen), electrons, H^+ions or the product of ammonium ions. The product passes through another membrane to the transducer. The transducer converts product into an electrical signal, which is amplified. The signal processing equipment converts the amplified signals into a display that can be read out and recorded.

4. TYPES OF BIOSENSORS

Biosensors are of 5 types :

1. Calorimetric biosensors
2. Potentiometric biosensors
3. Amperometric biosensors
4. Optical biosensors
5. Acoustic wave biosensors

(a) Calorimetric biosensors : Many enzymes–catalyze reactions which produce heat (exothermic). Calorimetric biosensor measures the change in the temperature of the solution containing the analyte following enzyme action and interpret it in terms of analyte concentration in the solution. The temperature changes are usually determined by means of thermistors at the entrance and exist of small packed bed columns containing immobilized enzymes within a constant temperature environment. Under such closely controlled conditions, up to 80% of the heat generated in the reaction may be registered as a temperature change in the sample stream.

This is the most generally applicable type of biosensor. It can be used for turbid and strongly coloured solutions. The greatest disadvantage is to maintain the temperature of the sample stream at a constant say 0.01°C.

The sensitivity (10^{-4} M) and range (10^{-4} to 10^{-2} M) of such biosensors are quite low cost of applications. The sensitivity can be increased by using two or more enzymes of the pathwayin the biosensor to link several reactions to increase the heat output. The sensitivity of the glucose analysis using glucose oxidase can be increased two fold by the co-immobilization of catalase within the column reactor in order to utilizes the H_2O_2 produced.

(b) Potentiometric biosensors : Potentiometric biosensors measures the difference in potential between the active electrode and a second reference electrode. Potentiometric biosensor make use of ion selective electrodes in order to convert the biological reaction into an electrical signal. In the simplest terms it consists of a membrane, containing immobilized enzyme, surrounding the probe from a pH meter, where the catalyzed reaction generates or absorbs H^+ as shown in the following reactions.

(a) H^+ cations

1. Penicillin -------Penicillinase-----------------Penicillinoic acid + H^+

2. $H_2NCONH_2 + H_2O + 2H^+$---------UREASE---------$2NH_4^+ + CO_2$

3. $H_2NCONH_2 + 2H_2O$ ---------UREASE---------- $2NH_3 + HCO_3^- + H^+$

4. Neutral lipids + H_2O--------Lipase-------Glycerol + Fatty acids +H^+

(b) NH_4^+ cations

1. α - Asparagine + H_2O---Asparaginase------ - Aspartate +NH_4^+

2. α - Amino acid + O_2 + H_2O--α-amino acid oxidase---- Keto acid + $NH_4^+ + H_2O_2$

3. $H_2NCONH_2 + 2H_2O + H^+$ ---------UREASE------$2NH_4^+ + HCO_3^-$

(c) I^- anion

$H_2O_2 + 2H^+ + 2I^-$--------Peroxidase-----------$I_2 + 2H_2O$

The reaction occurring next to the thin sensing glass membrane causes a change in pH which may be read directly from the pH meter's display.

There are three types of electrodes which are of use in potentiometric biosensors.

1. Glass electrodes for cations (e.g. normal pH electrodes) in which the sensing element is very thin hydrated glass membrane which generates a transverse electrical potential due to the concentration dependent competition between the cations for specific binding sites. The selectivity of this membrane is determined by the composition of glass. The sensitivity of H+ is greater than that achievable for NH_4+.
2. Glass pH electrodes coated with gas-permeable membrane selective for CO_2, NH_3 or H_2S. The diffusion of the gas through this membrane causes a change in pH of a sensing solution between the membrane and the electrode, which is then determined.
3. Solid –state electrodes where the glass membrane is replaced by a thin membrane of a specific ion conductor made from a mixture of silver sulfide and a silver halide. The iodide electrode is useful for the determination of I^- ion in the peroxidase reaction.

We have seen that in this case enzymes are used as biological material. Recently many electrochemical sensors in which microbial cells are used in conjunction with potentiometric electrodes have been developed. The cells are grown in an appropriate sterile medium, harvested by centrifugation, washed several times and then fixed at the tip of the electrode between the sensing membrane of the probe and a restraining membrane. Dialysis membranes and microporus filter membranes are usually used for this purpose.

The 1st potentiometric cell-based electrode was a sensor for L- arginine described by Rechnitz *et al.*,(1977) which consisted of the bacterium *Streptococcus faecium* coupled to an ammonia gas sensing electrode. *Streptococcus faecium* metabolize L-arginine by the following reaction.

L-arginine + H_2O Arginine deiminase Citrulline + NH_3

The NH_3 produced is detected by the ammonia gas- sensing electrode, giving rise to the response of the sensor to arginine.

(c) Amperometric biosensors : These electrodes function by the production of a current when potential is applied between two electrodes, the magnitude of current is being proportional to the substrate concentration.

The first Amperometric cell based biosensor was reported by Divies (1975) for the determination of ethanol. It consisted of the bacterium *Acetobacter xylinum* immobilized on an amperometric oxygen electrode. The response of the sensor is based on the consumption of oxygen by the

Table 1. Some of the examples of cell based potentiometric biosensors.

Substrate	Micro - organism	Detected substance	Reference
L-arginine	*Streptococcus faecium*	NH_3	Rechnitz et al., (1977)
L-aspartate	*Bacterium cadaveris*	NH_3	Kobos and Rechnitz (1977)
L-glutamine	*Sarcina flava*	NH_3	Rechnitz et al., (1978)
NAD^+	*Escherichia coli*	NH_3	Riechel and Rechnitz (1978)
L-tyrosine	*Aeromonas phenologenes*	NH_3	Dipaolantonia and Rechnitz (1982)
L-Histidine	*Pseudomonas* sp.	NH_3	Walters et al., (1980)
L- serine	*Clostridium acidiurici*	NH_3	Dipaolantonia et al., (1981)
Nitrate	*Azotobacter vinelandii*	NH_3	Kobos et al., (1979)
L-glutamate	*Escherichia coli*	CO_2	Hikuma et al., (1980)
Pyruvate	*Streptococcus faecium*	CO_2	Dipaolantonia et al., (1983)

bacteria to produce acetic acid.

Ethanol + O_2 ----> Acetic acid + H_2O

The decrease in oxygen is measured as a means of assessing electrode response.

Most of the amperometric bacterial electrodes prepared to date have used the oxygen electrode as the electrochemical detectors as shown in Table 2.

A clinically important sensor of this type is for the determination of cholesterol (Wollenberger *et al.*, 1980). This electrode consists of the bacterium *Nocardia erythropolis* immobilized in polyacrilamide or agar gel which is placed on the membrane of an oxygen electrode. The bacteria affect the following reaction.

$$\text{Cholesterol} + O_2 \xrightarrow{\textit{Cholesterol oxidase}} \text{Cholest- 4 en- 3 one} + H_2O_2$$

The amount of cholesterol present is directly proportional to the amount of oxygen consumed in the reaction. Because the classical colorimetric assays for cholesterol involves extraction and separation of lipid and strong acid hydrolysis, they are rather time consuming (Zak, 1977). With cholesterol microbial bio-sensor, total assay time is 15-20 minutes. Using the steady state method and 5-8 minutes using kinetic method offering a significant advantage over more conventional methods of analysis.

Table 2. Some of the examples of amperometric biosensors

Substrate	Microorganism	Detected substance	References
Ethanol	*Acetobacter xylinum*	O_2	Divies (1975)
Glucose	*Pseudomonas flourescens*	O_2	Suzuki and karube (1981)
Acetic acid	*Trichosporon brassicae*	O_2	Suzuki and karube (1981)
Ammonia	Nitrifying bacteria	O_2	Suzuki and karube (1981)
Biochemical oxygen demand	Mixed culture of bacteria from soil	O_2	Suzuki and karube (1981)
Mixtures of Acetate, L-lactate and succinate	*Alcaligenes eutrophus*	O_2	Plegge et al., (2000)
Cholesterol	*Nocardia erythropolis*	O_2	Wollenberger et al., (1980)
Methane	*Methylomonas flagellata*	O_2	Karube et al., (1982)

BOD also known as biological / biochemical oxygen demand is another example of amperometric biosensors. BOD is an indicator of the amount of biodegradable organic compounds and is a widely used parameter for the control of wastewater. The main disadvantage of this parameter is the 5-days needed for measurement (BOD). Therefore, conventional BOD measurements allow only a subsequent evaluation of waste water and are unsuitable for process control. A more rapid estimation of BOD is possible by using microbial sensor containing whole cell immobilized on an oxygen electrode.

The first report of such a microbial BOD sensor, using activated sludge obtained from waste water treatment plant, was published by Karube et al., (1977). However, these biosensors contained an undefined variety of microbial species and did not give reproducible results (Hikuma et al., 1979). Sensors using well-defined single microbial species seem to be more suitable. The pre requisite for the use of microorganisms for BOD sensors is wide substrate spectrum.

BOD sensors have been developed using various microorganisms as depicted in the Table 3.

One problem is that one microbial species has a determined substrate spectrum and can therefore determine only a fraction of, for e.g., organic compounds in waste water. For that reason, it is advantageous to use a combination of two or more microbial strains. Mixed bacterial cultures of closely related strains were used by Tan et al., (1992) i.e *Bacillus subtilis* and *B. licheniformis* and Galindo et al. (1992) i.e. *Enterobacter* sp. + *Citrobacter* sp. Further more it is possible to use microbial species that are not related and with different substrate spectrum, such as in BOD sensor *Issatchenkia orientalis* and *Rhodococuss orientalis* (Riedel et al. 1988)

The most widely used transducer in BOD sensor is the amperometric oxygen electrode. However, the application of an optical oxygen transducer instead of the usual amperometric oxygen electrode is described by Preininger et al., (1994).

Another interesting technique is the use of the luminous bacterium *Photobacterium phosphoreum* (Hyan et al. 1993). This device relies on the relationship between the intensity of luminescence and the cellular assimilation of organic compounds from waste water.

Table 3. Some of the examples of BOD sensors.

Micro- organism	Measuring time (min)	Stability (day)	References
Activated sludge	15-20	10-20	Karube et al., (1977)
Trichosporn cutaneum	18	30	Hikuma et al., (1979) and Harita et al., (1985)
Hansenula anomala	15-20	7	Kulys and kadziauskiene (1980)
Bacillus subtilis	0.1- 0.2	40	Riedel et al., (1988)
Bacillus subtilis + B. licheniformis	4-15	22	Tan et al., (1992)
Pseudomonas putida	15	20	Zhang et al., (1986)
Citrobactor sp. + Enterobactor sp.	8	25	Galindo et al., (1992)
Rhodococuss erythropolis + Issatchenkia orientalis	0.25-0.5	40	Tanaka et al., (1994)

Transducer :- An amperometric oxygen electrode in all the cases.

(i) Stability of BOD biosensor

The stability of the microbial sensor is related to the operating conditions. In general, much measurement results in high stability but few determinations and long pause gives low stability probably due to increased microbial death rates. Normally, microbial sensor should be stored in buffer overnight at room temperature or 4°C. If the activity of the biosensor decreases, the microorganisms can be re-activated by incubation with nutrient medium for some hours until the activity is recovered by growth of new cells. Care should be taken to avoid growing too much biomass, because high loading leads to high diffusional resistance and hence a decreased signal.

(ii) Influence of toxic compounds on BOD sensor

One problem of microbial BOD sensor could be the poisoning of microbial cells by toxic substances in the waste water. The toxic effects of heavy metals on the activity of microorganisms are well known. Therefore, the influence of $Cu,^{++}$ $Zn,^{++}Pb,^{++}$ $Hg,^{++}$and Cd^{++} on the signals of *Trichosporon cutaneum* - containing biosensor is tested by Riedel et al. (1990). $Cu,^{++}$ $Zn,^{++}Pb,^{++}$ have been shown to exhibit little effect whereas Hg++ and Cd++ in the concentration of 50 mg/ml inhibit the activity of the biosensor. A heavy metal-resistance BOD biosensor using *Alcaligenes eutrophus* which makes the estimation of BOD feasible even in the presence of 4mM nickel, copper and zinc is described by Slama et al. (1995).

An arsenic resistant BOD sensor based on *Pseudomonas putida* has been developed by Okhi et al. (1990)

(iii) Prevention of BOD biosensor contamination

Prevention of contamination is important for the stability of a measurement system containing a microbial sensor. Contamination can influence results by degradation of the substrate and by oxygen consumption. It is possible to protect the system with antimicrobial agents, which reduces the contamination by microorganism under anaerobic conditions. The sensor microorganisms are stable under aerobic conditions. The possible antimicrobial reagents are *Kathon CG* and *Euxyl K400* which are used in concentrations of 0.1% (Riedel, 1998). The measuring system is rinsed with the chosen reagent and then the measuring chamber is emptied and filled with buffer. To prevent microbial growth in the tubes of the dosage system, they should be filled with the biocide solution, allow to stand overnight and then thoroughly rinsed with buffer.

(iv) Comparison of conventional method of BOD Vs biosensor BOD_5 values

In general, comparable estimates for BOD are obtained for waste water from fermentation and food plants by biosensor and conventional means. These waste waters contain a high concentration of easily assimilable compounds. However, the biosensor frequently shows low values for domestic wastewater as compared to conventional method preferably due to low concentration of easily assimilable compounds. Therefore the BOD biosensor is applicable to specific wastewater generated by food and fermentation industry with high conc. of easily assimilable compounds, but it is only of limited applicability for domestic waste water as suggested by Tanaka et al. (1994).

Table 4. BOD values estimated by microbial sensor and by the conventional 5-Day test for various wastewaters.

Waste water	Microbial sensor	5-day BOD	sensor BOD	5-day BOD/ biosensor ratio	References
Domestic	*Bacillus subtilis and Bacillus licheniformis*	170	154	1.10	Tan et al., (1992)
	Issatchenkia+ Rhodococcus	285	334	0.85	Tanaka et al., (1994)
	Issatchenkia+ Rhodococcus	366	366	1.00	Tanaka et al., (1994)
	Issatchenkia+ Rhodococcus	112	63	1.78	Tanaka et al., (1994)
	Issatchenkia+ Rhodococcus	153	57	2.68	Tanaka et al., (1994)
Food factory	*Trichosporon cutaneum*	152	155	0.98	Riedel et al., (1990)
	Trichosporon cutaneum	8000	8764	0.91	Riedel et al., (1990)
	Bacillus subtilis and Bacillus licheniformis	151	147	1.03	Tan et al., (1992)
Starch factory	*Trichosporon cutaneum*	4000	4250	0.94	Riedel et al., (1990)
Fermentation factory	*Bacillus subtilis and Bacillus licheniformis*	15040	1564	0.96	Tan et al., (1992)
Industrial	Issatchenkia+ Rhodococuss	253	389	0.65	Tanaka et al., (1994)

(d) Optical Biosensor : There are mainly two main areas of development in optical biosensors. These involve the determining changes in light absorption between the reactant and products of a reaction or measuring the light output by a luminescent process. The former usually involves the widely establis- hed use of colorimetric test strips. They are single-use strips of cellulose coated with the appropriate enzymes and suitable reagents. They are dipped into the sample solution and the presence of the substrate or analyte in the sample is detected by a change in the colour of the strip, which can be compared with a coloured chart to estimate the approximate amount of the analyte. An example of such strip is the one used to detect glucose in blood/urine of diabetic patients (Wilson and Walker, 1994). The strip contains glucose oxide, horse radish peroxidase and a weakly coloured chromogen, which is converted into a highly coloured dye on oxidation. The H_2O_2 produced during aerobic oxidation of glucose by glucose oxidase is used to oxidize the chromogen as follows:

Chromogen (2H) + H_2O_2 --^{Peroxidase}------Dye + H_2O

A wide variety of test strips based on other enzymes are commercially available.

A most promising biosensor involving luminescence uses is firefly luciferase to detect the presence of bacteria in food or clinical samples. Bacteria are specifically lysed and the ATP released (roughly proportional to the number of bacteria present) reacted with D-luciferin and oxygen in a reaction which produces yellow light in high quantum yield.

ATP + D-Leuciferin + O_2-------------- Luciferase --------- Oxyluciferin + AMP + PP_I + CO_2 + Light (562nm)

The light produced may be detected photometrically by the use of high voltage and expensive photomultiplier tubes or low voltage cheap photodiode system. Firefly luciferase is a very expensive enzyme, only obtained from the tails of wild fireflies. Use of immobilized luciferase greatly reduces the cost of these analysis.

(e) Acoustic Wave Biosensors : They are also known as piezoelectric devices. Their surface is usually coated with antibodies, which bind to the complementary antigen present in the sample solution. This leads to increased mass which reduces their vibrational frequency; this change is used to determine the amount of antigen present in the sample solution.

5. APPLICATIONS OF BIOSENSORS

In the beginning, biosensors were applied in the field of medicine and industry. But in recent years, biosensors are becoming popular in many areas due to small size, rapid and easy handling, low cost and greater sensitivity and selectivity.

Applications of biosensors in some of the areas are described as below.

(a) Use in medicine and health

Biosensors have tremendous potential for its application in the field of medical science. In 1979, first glucose analyzer using bio molecules for the detection of blood glucose was commercialized by *Yellow springs instruments* Co. USA. A device, a minipump filled with insulin, has been constructed to deliver insulin to diabetics based on glucose levels of blood. When biosensors provide information, the device delivers accurate amount of insulin required by diabetics. *Mitomycin*, an aflatoxin, causes cancer in inborn infants. Mutagenicity of such chemicals can be detected by using the biosensor. Similarly, any other abnormal toxic substance produced in body due to infectious diseases can also be detected.

(b) Uses in pollution control

Biosensors are very helpful in environmental monitoring and pollution control, since they can be miniaturized and automated. As far as quality control of drinking water is concerned, the monitoring biosensors are useful in monitoring of pesticides in water. In JAPAN, a biosensor coupled with oxygen electrode and immobilized *Trichosporon cutaneum* is used for measuring BOD (Gronow et al., 1988)

(c) Uses in industry

Spectrophotometer and autoanalyser are used to estimate the substrates utilized and the products formed in the fermented broth and generally these are associated with lot of problems. So the biosensors are designed to measure the fermentation products to improve the feedback control. In addition, there are a lot of problems associated with these. So the biosensors can be designed for rapid sampling and rejection of below standard raw materials to improve the efficiency of workers. Karube and coworkers of Tokyo University Research Center of Advance Science and Technology have recently developed an ion sensitive field effect transistor (ISFET). This device is highly sensitive to change the ion concentration. Using this biosensor, it is possible to measure the odour, taste and freshness of foods.

In determining fish freshness, either ATPase, aminooxidase or Putrescine oxidase is used. ATPase detects the presence of ATP in fish muscle. As ATP is not present in staled food, therefore, signals do not occur. Recently, a biosensor has been developed at Cransfield institute of technology. U.K., which measures cholesterol levels in butter. The enzyme cholesterol oxidase, when immobilized on the electrodes, reacts with cholesterol of food.

REFERENCES

Chaplin, M.F. and Bucke, C. 1990. Biosensors. In Enzyme Technology. Cambridge university Press, New York. pp.197-219.

Clark, L.C.Jr. and Lyons, C. 1962. *Annals of Academy of Sciences*, New York 102,29-45. Cited from : Luong, J.H.T., Mulchandani, A. and Guilbault, G. G. 1988. Developments and applications of biosensors. *Tibtech* 6,310 –316.

Corcoran, C.A. and Rechnitz, G.A. 1985. Cell based biosensors. *Trends in Biotechnology* 3(4), 92-96.

Duarte, J.M.C., Archer, L.J., Bull, A.T. and Hott, G. 1987. Perspectives in Biotechnology. Plenum press, New York. pp. 55-65.

Dipaolantonio, C.L., Arnold, M.A. and Rechnitz, G.A. 1981. Serine selected membranes of probe based on immobilized anaerobic bacteria and a potentiometric ammonia gas sensor. *Analytical chimica. Acta* 128, 121-127.

Dipaolantonio, C.L. and Rechnitz, G.A. 1982. Induced bacterial electrode for the potentiometric measurement of tyrosine *Analytical chimica. Acta* 141,1-13.

Dipaolantonio, C.L. and Rechnitz, G.A. 1983. Stabilized bacteria based potentiometric electrode for pyruvate .*Analytical chimica Acta* 148,1-12.

Divies, C. 1975. Remarks on ethanol oxidation by an *Acetobacter xylinum* microbial electrode .*Annals of Microbiology* 126A, 175-186.

Dubey, R.C. 1995. Enzyme technology. In Textbook of Biotechnology. S. Chand and Company, New Delhi. pp. 244-251.

Galindo, E., Garcia, J.L., Torres, L.G. and Quintero, F. 1992. Characterization of microbial membranes used for the estimation of biochemical oxygen demand with a biosensor. *Biotechnological Techniques* 6, 399-404.

Griffths, D. and Hall, G. 1993. Biosensors- What real progress is being made? *Tibtech* 11, 122-130

Gronow, M. 1984. Biosensors.*Trends in Biochemical Science* 9, 336-340.

Gronow, M., Muller, W.H., Russel, I.J. and Anderson, D.J. 1988. In Molecular Biology and Biotechnology (edsWalker, J.M. and Gingold, E.B.), Royal Society of Chemistry, Berlington House, London. pp. 323-347.

Harita, K.,Otani,Y.,Hikuma,M. and Yasuda, T.1985. BOD quick estimating system utilizing a microbial electrode, in Drake, R.A.C. (ed.), Instrumental Control of Water and Waste Water Treatment, Transport systems, Proceedings of IAWPRAC workshop, 529-532.

Hikuma, M.,Obana, H.,Yasuda, T., Karube, I. And Suzuki, S. 1980. Method and sensor for determining ammonia . *Analytical chimica. Acta* 116,61-67.

Hikuma, M., Suzuki, H. and Yasuda, T. 1979. Amperometric estimation of BOD by using living immobilized Yeasts. *European Journal of Applied Microbiology and Biotechnology* 8, 289-297.

Hikuma, M. and Yasuda, T. 1988. *Microbial sensors for estimation of biochemical oxygen demand and determination of glutamate . Methods in Enzymology* 137, 124-131.

Hyan, C.K., Tamiya, E., Takeuchi, T. and Karube, I. 1993. *A novel BOD sensor based on bacterial luminiscence. Biotechnology and Bioenginerring* 41,1107-1111.

Karube, I., Okada, T. and Suzuki, S. 1982. A methane gas sensor based on oxidising bacteria. *Anal. Chim. Acta* 135,61-67.

Karube, I., Matsunaga, T., Mitsuda, S. and Suzuki, S. 1977. *Microbial electrode BOD sensor. Biotechnology and Bioenginerring* 19,1535-1547.

Karube, I. 1988. Hybrid biosensor for clinical analysis and fermentation control. *Methods in Enzymology*137, 133-138.

Kim, Y.C., Lee, K.H., Sasaki, S., Hashimoto, K., Ikebukuro, K. and Karube, I. 2000. Photocatalytic sensor for chemical oxygen demand determination based on oxygen electrode. *Analytical Chemistry* 72,3379-3382.

Kobos, R.K. and Rechnitz, G.A. 1977. Regenerable bacterial membrane electrode for L-aspartate . *Analytical Letters* 10,751-758.

Kobos, R.K., Rice, D.J. and Flourney, D.S. 1979. Bacterial membrane electrode for the determination of nitrate. *Analytical Chemistry* 51,1122-1125.

Konig, A., Reul, T., Harmeling, C., Spener, F., Knoll, M. and Zaborosch, C. 2000. Multimicrobial sensor using microstructured three dimensional electrode based on silicon technology. *Analytical Chemistry* 72,2022-2028.

Kulys, J. and Kadzi, A. 1980. Yeast BOD sensor. *Biotechnology and Bioenginerring* 22, 221-226.

Luong, J.H.T., Mulchandani, A. and Guilbaut, G.G. 1988. Developments and applications of biosensors. *Tibtech* 6, 310-316.

Lowe,C.R. 1985. An introduction to the concepts and technology of biosensors. *Biosensors* 1, 3-16.

Marshall, R.D. and Tipton, K. 1985. Biosensors. *Essays in Biochemistry* 21, 129-134.

Okhi, A., Shinohara, K. and Maedd, S. 1990. Biochemical oxygen demand sensor using an arsenic resistant bacterium. *Analytical Science* 6, 905-906.

Plegge, V., Slama, M., Suselbeck, B., Wienke, D., Spener, F. and Zaborosch, G. 2000. Analysis of ternary mixtures with a single dynamic microbial sensor and chemometrics using a non linear multivariate calibration. *Analytical Chemistry* 72, 2937-2942.

Preininger, C., Klimant, I, and Wolfbeis, O.S. 1994. Optical fiber sensor for biochemical oxygen demand. *Analytical Chemistry* 66, 1841-1846.

Rechnitz, G.A., Kobos, R.K., Riechel, S.J. and Gebaur, C.R. 1977. A bioselective membrane electrode that uses living bacterial cells. *Analytical chimica Acta* 94,357-365.

Rechnitz, G.A., Riechel, S.J., Kobos, R.K. and Mayerhoff, M.E.1978. Glutamine – selective membrane electrode that uses living bacterial cells. *Science* 199, 440-441.

Riechel, T.L. and Rechnitz, G.A. 1978. Hybrid bacterial and enzyme membrane electrode with nicotinamide adenine dinucleotide response. *Journal of Membrane Science* 4, 243-250.

Riedel, K., Renneberg, R., Kuhn, H. and Scheller, M. 1988. A fast estimation of BOD with microbial sensors. *Applied Microbiology and Biotechnology* 28,316-318.

Riedel, K., Lange, K.P., Stein, H.J., Kuhn, m., Ott, P. and Scheller, F. 1990. A microbial sensor for biochemical oxygen demand . *Water Resources*24,883-887.

Riedel, K. 1998. Applications of biosensors to environmental samples. In Commercial Biosensors: Application to Clinical, Bioprocess and Environmental samples (ed. Ramsay, G). John Wiley and sons Inc. pp. 267-294.

Schultz, J.S. 1991. Biosensors .*Scientific American* 265,48-55.

Scouten, W.H., Luong, J.H.T. and Brown, R.S. 1995. Enzyme or Protein immobilization techniques for applications in biosensor design. *Tibtech* 13, 178-185.

Singh, B.D. 2000. Enzyme Technology. In Biotechnology, Kalyani publishers, New Delhi. pp. 574-577.

Slama, M., Zaborosch, C.H. and Spener, F. 1995. Microbial sensor for rapid estimation of biochemical oxygen demand in the presence of heavy metal ions, In Wilken, R.D., Forster, U. and Krodel, A. (eds.), International conference on Heavy Metals in the Environment. CER consultants Ltd. Edingburgh, Scotland 2, 171-174.

Suzuki, S. and Karube, I. 1981. In Applied Biochemistry and Bioengineering (Wingard, L.B., Katchalsi, K.E. and Goldstein, L., eds.), vol. 3, Academic Press.

Tan, T.C., Neoh, K.G. and Lee, Y.K. 1992. Microbial membrane modified dissolved oxygen probe for rapid biochemical demand measurement. *Sens. Actuat.B.* 8,167-172.

Tanaka, H., Nakamura, E., Minamiya, Y. and Toyoda, T. 1994. BOD biosensor for secondary effluent from waste water treatment plants. *Water Science Technology* 30, 215-227.

Walter, R.R., Moriarty, B.E. and Buck, R.P. 1980. *Pseudomonas* bacterial electrode for the determination of L-histidine. *Analytical Chemistry* 52, 1680-1684.

Wilson, K.and Walker, J. 1994. Principles and Techniques of Practical Biochemistry. Cambridge University Press, Cambridge. pp. 565-574.

Zak, B. 1977. Cholesterol methodologies : A review. *Clinical Chemistry* 23, 1201-1214.

Zhang, X., Wang, Z. and Jian, H.1986. Microbial sensor for biochemical oxygen demand estimation (in Chinese), Huanjing Kexue Xuebao 6, 184–192; Chemical *Abstracts* 105, 158-286.

CHAPTER – 3

THERMOPHILIC XYLANASES — MOLECULAR AND BIOTECHNOLOGICAL PERSPECTIVES

Sudeep P. George, Devyani S. Nath and Mala B. Rao

Division of Biochemical Sciences, National Chemical Laboratory, Pune, India-411 008

1. INTRODUCTION

Biomass is an alternative natural source of chemical feedstock with a replacement cycle short enough to meet the demand in the world fuel market. Xylan is a major constituent of hemicellulose and is the second most abundant renewable resource [86]. Eventually, if not in the near future, hemicellulose in combination with cellulose will supply most of the global demand for raw materials. Thus, it is not unrealistic to foresee that coal and crude oil are most likely to be substituted by biomass in another 50 years [Goheen, 1981].

Due to the heterogeneity of xylan, the complete enzymatic degradation to its monomer is a complex process requiring the action of a battery of hydrolytic enzymes. They include Endo-(1→4)-β-xylanase, β-xylosidase, α-L-arabinofuranosidase, α-glucuronidase, acetylxylan esterase and phenolic acid esterases. Endo-(1→4)-β- xylanase, which attacks the polysaccharide backbone is one of the most crucial enzyme involved in xylan hydrolysis. Xylanases stable and active at high temperature and alkaline conditions are suitable for biotechnological applications. Thermally stable enzymes can be used at high temperatures for efficient hydrolysis of substrates and may be effectively recycled. Extremophilic organisms are one of the natural sources for the identification of such enzymes [Nicolson *et al.*, 1988]. Xylanases have potential application in waste treatment, fuel and chemical production, paper manufacture and deinking [Biely, 1985]. Recently interest on cellulase free xylanase has been augmented, due to their action resulting in the increased brightness of pulp and a substantial reduction in the requirement of chlorine as a bleaching agent, which could lead to the development of environmentally friendly technologies [Coughlan & Hazelwood, 1993].

2. SCOPE OF PRESENT REVIEW

Thermozymes are thermostable enzymes that function optimally between 60 and 125°C. They have attracted increasing attention in recent years owing to their biotechnological potential. They are already used in molecular biology (Taq polymerase), detergent industry (proteases) and starch processing (α-amylase) and are excellent catalytic candidates for numerous additional applications that require high temperature stability. The fact that extremozymes have significant financial impact for the companies that exploit them is evident from the example of Taq polymerase, which has sales of $80 million per annum [Persidis, 1998]. Commercial application of xylanases demands identification of highly stable enzymes active under routine handling conditions. Most of the biobleaching of paper is carried out at high temperatures, therefore it is advisable to use enzymes that are active and stable at high temperatures. Xylanases are industrially important enzymes and many reviews have been published covering the various aspects of these enzymes (Biely, 1985; Debeire-Gasselin *et al.*, 1992; Gilbert & Hazlewood, 1993; Kulkarni *et al.*, 1999). However, in recent times the biotechnological application of thermostable xylanases has stimulated research on its biochemical and molecular aspects. The present article is a comprehensive review of the xylanases produced by thermophilic organisms and gives an overall view of the latest developments in this area.

3. MICROBIAL DIVERSITY

Microbial strains capable of growth at high temperatures at which most microorganisms fail to grow are grouped together as thermophiles. The microbial species that produce xylanases and other glycosidases have been found in extremely diverse natural habitats. A host of microorganisms inhabit extreme environmental conditions where they thrive and grow at temperatures above 50°C, at pH values 9 or greater [Uffen, 1997] and/or in high ionic strength aqueous systems containing salt approaching saturating conditions [Wong *et al.*, 1988]. *Bacillus stearothermophilus* [Gilead & Shoham, 1995] is one of the aerobic, thermophilic bacteria shown to actively degrade plant polymers, including xylan. We have isolated an alkalothermophilic *Bacillus* and *Thermomonospora* sp. that grow at 50°C, from hot water spring and self-heating compost respectively [Dey *et al.*, 1992; George *et al.*, 2001]. Hyperthermophilic eubacteria have been isolated that grow at temperatures above 80°C. These microbes include *Thermotoga maritima* [Winlerhalter, *et al.*, 1995], *Thermotoga* sp. [Saul *et al.*, 1995], *Caldocellum saccharolyticum* [Luthi, *et al.*, 1960], *Dictyoglomus* sp. [49], *Pyrolobus fumarii* [Blochl *et al.*, 1997] and *Rhodothermus marinus* [Dahlberg *et al.*, 1993]. Anaerobic spore forming thermophilic cells include *Clostridium thermocellum, Clostridium thermohydrosullfuricum and Clostridium thermosaccharolyticum* [Biely, 1985]. Additional anaerobic cells that grow and thrive at higher temperature with xylan include *Thermoanaerobacterium thanolicus, Thermoanaerobacterium acetigenum* [Nielsen, *et al.*, 1993], *Thermoanaerobium brockii, Thermoanaerobacterium* sp. strain JW/ SL-Y S485 [Shao *et al.*, 1995], and *Thermobacteriodes* species [Biely, 1985]. Xylanolytic enzymes in these hyperthermophilic cells operate around neutral pH. There are many reports of *Bacillus* sp. that degrade xylan at alkaline pH. However, xylanolysis by these cells occurred under modest thermophilic conditions of 50°C.

Table 1. The properties of xylanases purified from thermophiles

Source		Growth		Optimum		pI	MW [kD]	K_m [mg/ml]	References
		Temp [°C]	pH	Temp [°C]	pH				
Thermophilic bacteria									
Bacillus sp. **TAR-1**		50	10.5	Broad	70	>9.3	23	–	Takahashi et al., 2000
Bacillus sp.	W1 I	50	9-10	6	65	8.5	21.5	4.5	Okazaki et al., 1984
	W1 II			7-6	70	3.6	49.5	3.6	
	W2 I			6	65	8.3	22.5	8.3	
	W2 II			7-9.5	70	3.7	50	3.7	
Bacillus acidocaldarius		60	3.5-4	4	80	–	56	1.68	Uchino and Fakuda, 1981
Bacillus stearothermophilus **T6**		60	7-7.3	6.5	75	9	43	1.63	Khasin et al., 1993
Bacillus sp. **SPS-0**		60	8	6-7	75	–	99	0.7	Bataillon et al., 2000
Bacillus amyloliquefaciens		45	7	6.8-7	80	10.1	19.6	–	Breccia et al., 1998
Bacillus sp. **XE**		55	7	6	75	7.8	22	0.60	Debeire-Gosselin et al., 1992
Bacillus stearothermophilus **21**		55	7	7	60	4.8	39.5	3.80	Nanmori et al., 1990
Clostridium stercorarium	A	65	7	5.5-7	75	4.5	44	3.2	Berenger et al., 1985
	B			5.5-7	75	4.4	72	2.9	
	C			5.5-7	75	4.3	62	3.7	
Clostridium stercorarium **HX-1**	D	60	6-7	6.5	75	4.5	53	1.4	Sakka et al., 1991
Clostridium thermolacticum **TC 21**	A	65	6-7	6.5	80	4.4	39	0.40	Debeire-Gosselin et al., 1992
	B			6.5	80	4.5	66	0.53	
	C			6.5	80	4.6	65	0.48	
Streptomyces **T_7**		50	7	4.5-5.5	60	7.8	21.8	100	Keskar et al., 1989
Streptomyces sp. **B-12-2**	1a			6	55	7.5	26.4	5.80	Elgir et al., 1994
	1b			6	60	8.3	23.8	3.40	
	2			7	60	5.4	36.2	3	
	3			7	60	5	36.2	1.20	
	4			6	60	4.8			
Streptomyces thermoviolaceus **OPC520**	I	50	7	7	70	4.2	54	—	Tsujibo et al., 1992
	II			7	60	8	33	—	

Table 1 contd....

Source		Growth		Optimum		pI	MW [kD]	K_m [mg/ml]	References
		Temp [°C]	pH	Temp [°C]	pH				
Thermoanaerobacterium sp. **JW/SL-YS485**		60	6	6.2	80	4.3	350	3	Shao, et al., 1995
	I	55	6-7	7.8	75	4.2	36	2.50	Stutzenberg and Bodine 1992
				7.2		7.1	19	1.40	
				6.8		8.4	15	2.00	
Thermotoga sp. **Fj SS3 B1**				5.3	105	–	31	0.07	Ruttersmith et al., 1992
Thermotoga maritima **MSB8**	XynA								Winterhalter and Liebl, 1995
	XynB								
Thermotoga thermarum									Sunna et al., 1996
Thermophilic fungi									
*Talaromyces bysso-chlamydoides***YH-50**	X-a	50	6.2	5.5	75	4.3	76	N.D.	Yoshioka et al., 1981
	X-b-I			4.5	70	3.8	54	N.D.	
	X-b-II			5	70	4	45	N.D.	
Talaromyces emersonii	II	45	4.5	4.2	78	5.3	78.4	N.D.	Tuohy and Coughlan, 1993
CBS 814.70	III			3.5	67	4.2	58.2	N.D.	
Thermoascus aurantiacus **C436**		45	6	5.1	80	7.1	32	1.70	Tan et al., 1987

(a) Properties of thermophilic xylanases

Recent reviews have described characterization of xylanases from microbial systems [Kulkarni & Rao, 1996; Sunna & Antranikian, 1997; Warren, 1996]. The properties of some of the purified xylanases from extremophiles are described in Table 1. Xylanases from *Thermoanaerobacterium sp.* [Shao *et al.*, 1995] and *Thermotoga thermarum* [Sunna & Antranikian, 1996] showed higher molecular weights of 350 and 266 kDa respectively. Few bacterial and fungal xylanases show maximal activities at temperatures 60-80°C [Khasin *et al.*, 1993; McCarthy *et al.*, 1985]. The purified endoxylanases from various species belonging to the genus *Thermotoga* are optimally active at temperatures between 80 and 105°C [Simpson *et al.*, 1991; Sunna *et al.*, 1996; Winterhalter & Liebl, 1995]. Xylanase from *Dictyoglomus* sp. exhibited a half- life of 80 min at 90°C [Mathrani *et al.*, 1991]. *Clostridium stercorarium* xylanase exhibited a temperature optimum of 70°C and a half-life of 90 min at 80°C. The thermophilic fungi include *Thermoascus aurantiacus* [Khandke *et al.*, 1989], which produces a thermostable xylanase reported to be stable at 70°C for 24h, *Paecilomyces variota* [Krishnamurthy, 1989] and *Talaromyces byssochlamydoides* [Yoshioka *et al.*, 1981] with temperature optimum of 65-75°C at pH 5-6.5.

4. CLONING AND EXPRESSION

For commercial realization and

economic viability of xylanase production, it is necessary to identify organisms, which can hyperproduce the enzymes. Recombinant DNA techniques offer the means to enhance protein production. Genes for more than 50 xylanases have been cloned and sequenced [Schlacher *et al.*, 1996]. However, comparatively fewer genes of thermostable xylanases have been cloned.

(a) Heterologous cloning

Thermostable xylanase genes have been isolated from an assortment of genera and expressed in *E. coli.* The xylanase gene cloned from *Bacillus stearothermophilus* T-6 was of interest because the enzyme was optimally active at pH 9 and 65°C [Gat *et al.*, 1994]. Cloning and expression of a xylanase gene from the extreme thermophile *Dictyoglomus thermophilum* Rt46B.1 in *E. coli* has been reported [Gibbs *et al.*, 1995]. The recombinant enzyme was found to have an optimum temperature of 85°C. Walsh *et al.* 1998 cloned the thermophilic xylanase from the extreme thermophile *Thermotoga* in the yeast *Kluyveromyces lactis*. The xylanase enzyme was correctly processed and secreted by *K. lactis* transformants into the culture broth [Walsh *et al.* 1998]. The xylanase cloned from *Thermotoga maritima* showed an optimum temperature of 90°C at pH 5.5 and was stable upto 100°C [Chen *et al.*, 1997]. The recombinant xylanase from *Thermotoga neapolitana* was found to be stable at 90°C for 4 hours and a half-life of 2 hours at 100°C, and it had an optimum temperature of 102°C at pH 5.5 [Veoikodvorskaya *et al.*, 1997]. Hayashi et al. were the first to report the cloning of the xylanase component of the cellulosome from *Clostridium thermocellum* [Hayashi *et al.*, 1997]. Xyn C comprises of a single catalytic domain of family 10 in addition to the thermostabilizing domain and a dockerin domain. The optimum temperature of the enzyme was 80°C and the enzyme was stable at 70°C for 10 min in the absence of substrate. The bacterial genes encoding xylanases have been seen in close proximity in the genome to other genes encoding cellulase related functions. In *Bacillus stearothermophilus* the β-xylosidase gene (xylA) is located upstream of the xylanase gene (xynA) and the two genes are separated from each other by only two nucleotides. Nonetheless, the deletion analysis of the two genes indicates that they are independently transcribed [Baba *et al.*, 1994]. Similar linkage between xylanase and -xylosidase genes has been shown in the case of *Bacillus pumilus* and *Caldocellum saccharolyticum.* Helbers et al. have shown the high level expression of the thermostable xylanase from *Clostridium thermocellum* into transgenic tobacco plants [Helbers *et al.*, 1995]. The xylanase gene was introduced into the tobacco plant by an integration system with *Agarobacterium tumefaciens.*

(b) Homologous cloning

A number of heterologous proteins cannot be efficiently expressed in *E. coli*, due to the relatively abundant occurrence of rare codons in the cloned gene, the need for post synthetic modifications, toxicity of the coded protein to the host cells and susceptibility of the foreign protein to protease coded by *E. coli.* It is well known that higher expression levels are obtained in homologous host system. The homologous expression of the xylanase gene from an alkaliphilic, thermophilic *Bacillus* sp. has been demonstrated in a xyn$^-$ mutant *B. subtilis* A8 [Shendye & Rao, 1993], and *B. subtilis* MI 111 [Shendye *et al.*, 1994]. Enhanced production of the xylanases has also been reported by chromosomal gene integration in an alkalophilic, thermophilic *Bacillus* sp. [Shendye & Rao, 1993].

(c) Overexpression in *E. coli*

In order to make the cloning of a gene a commercial success it is important to overexpress the protein. Moreover, crystallographic studies which require copious amounts of pure protein, the overexpression of the cloned gene is the most feasible option to attain the goal. Luthi et al. have cloned and overexpressed a thermostable xylanase from *Caldocellum saccharolyticum* in *E. coli* [Luthi *et al.*, 1992]. High-level expression and single step purification of a thermostable xylanase from *Bacillus stearothermophilus* T-6 has been reported [Lapidot *et al.*, 1996]. The xylanase gene was cloned into T-7 polymerase expression vectors. The enzyme was found to constitute over 70% of the cell protein. Xylanase A from extremely thermophilic eubacterial strain Rt8BB.4 was overexpressed in *E. coli.* The xylanase activity from domain 2 was assosiated with a 36 kDa protein, which was stable at 70°C for at least 12 h at pH 7 [Dwivedi *et al.*, 1996].

5. THERMOSTABILIZING DOMAIN

The presence of repeated domains is common in bacterial xylanases, however little is known about their function. The thermostability of a thermostable xylanase from *Thermoanaerobacterium saccharolyticum* has been correlated to the N-terminal domains of the enzyme [46]. *Clostridium thermocellum* also contained a homologue of this domain that appeared to confer stability. There is a 28 % sequence homology between the thermostablilizing domains of xylanase Y from *C. thermocellum* and xylanase A from *T. saccharolyticum* [Fonles *et al.*, 1995]. The number and position of thermostablilizing domain are variable in respective enzymes; for example, a thermostabilizing domain occurs in the N terminal of *C. thermocellum* Xyn C, two occurs in the N terminus of *T. saccharolyticum* Xyn A, and one occurs in the middle of *C. thermocellum* Xyn Y [Hayashi *et al.*, 1997]. A similar thermostabilizing domain has been reported from a thermostable xylanase produced by a mesophilic *Bacillus* sp. strain NG-27 [Gupta *et al.*, 2000]. Thermostabilizing domains are found mainly in the thermophilic xylanase of family 10. Non-catalytic domains conferring thermostability have not yet been detected in family 11 xylanases. One of the reasons may be that these enzymes are inherently more thermolabile, unlike family 10 xylanases in which folding is such that it has been easy for them to evolve into a thermophilic enzyme. Although removal of these domains reduced the thermostability and optimal temperatures of *C. thermocellum* Xyn Y and *T. saccharolyticum* Xyn A, the interaction between catalytic domains and thermostabilizing domains responsible for thermostabilization remains to be studied.

6. SITE DIRECTED MUTAGENESIS

Initially enzyme active site residues were determined by selective chemical modification. Rapid developments in molecular biology techniques have made it possible for the individual amino acid to be substituted by site-specific mutagenesis. The highly conserved amino acid residues located at specific positions in xylanases are important in structure function analysis and hence are targeted for site directed mutagenesis (SDM). The conserved glutamate residues (Glu^{87} and Glu^{184}) have been shown to be present in the active site of xylanase A from *Schizophyllum commune* [Oku *et al.*, 1993]. The three dimensional structures of two bacterial xylanases from *Bacillus pumilus* and *B. circulans* and a fungal xylanse from *T. harzianum* have also revealed the presence of two completely conserved glutamic acid residues corresponding to Glu^{87} and Glu^{184} of xylanase A from *Schizophyllum commune*. SDM can also be used to improve the properties of the enzyme by replacing the key residues in the active site of the enzyme. The xylanase A gene from *Streptomyces lividans* was modified by SDM, selecting for mutations that improve the catalytic activity and thermostability of the enzyme [Moreau *et al.*, 1994]. The amino acids at positions 155, 156 and 173 were replaced either by amino acids having same polarity but a different side chain, or by amino acids having the same side chain with a different polarity.

All mutants synthesized a 43 kDa protein that reacted in immunoblots with specific anti-xylanase A antibodies. The F155Y, R156K, R156E and N173D mutant xylanases showed significantly higher activities than the wild type enzyme. An improvement in stability was obtained by replacing the arginine situated at position 156 of xylanase A by a glutamic acid residue (R156E). The mutation resulted in a modified enzyme with a temperature optimum that was 5°C higher than that of wild type. In fact, the same substitution occurs naturally in a xylanase produced by *Bacillus* sp. C-125 and in a xylanase produced by *Caldocellum saccharolyticum* [65]. Both xylanases have an optimum temperature of 70°C, which is the same as for R156E. It was observed that the mutation N173D resulted in an enzyme that had enhanced stability at 60°C. The substitution of Asn for Asp significantly increases the mean hydrophobicity. An increase in hydrophobicity in a region of a protein should reduce the number of water molecules having access to this locus or create a more compact protein as a result of new secondary structure interactions. This phenomenon could eventually increase protein stability as shown in the case of N173D xylanase. In *Bacillus circulans* mutant xylanase proteins, disulphide bridges conferred thermo protection as observed by the 15°C increase in thermostability [Wakarchuk, *et al.*, 1994].

7. CRYSTAL STRUCTURE

There are very few thermostable xylanases whose crystal structure has been determined. The crystal structure of the thermostable xylanase from *Thermomyces lanuginosus*, determined by single-crystal X-ray diffraction is similar to other family 11 xylanases. The structure of xylanase from *T. lanuginosus* is a compact, globular protein with two heavily twisted β-sheets. One sheet forms the "outer" surface of the enzyme and consists of five antiparallel strands while the other β-sheet consists of nine antiparallel strands. The most prominent feature of the structure is a long cleft spanning the whole molecule and containing the active site. The reasons for thermostablity of *T. lanuginosus* xylanase were analyzed by comparing its crystal structure with known structures of mesophilic family 11 xylanases. The thermostability of this xylanase is due to the existence of a disulphide bridge, which connects the C-terminus of the β-strand B9 (residue 110) with the N-terminus of the α-helix (residue 154). The presence of electrostatic interactions between ion pairs has also been implicated in the stabilization of the protein structure [Gruber *et al.*, 1994].

Thermoascus aurantiacus xylanase, belonging to family 10 of glycoside hydrolases ia a single polypeptide chain that assumes $(\alpha/\beta)_8$ TIM barrel fold. The presence of a disulphide bridge between C terminus of helix α7 and the N terminus of the β8 and the presence of more than ten salt bridges might be contributing to the thermostability and activity of this protein at elevated temperatures. Salt bridges between oppositely charged groups are an important aspect to protein structure stability and, when buried, contribute 3-5 kcal/mol in its stability. The presence of six short helices termed as "thermo helix" also contributes to the stability of the protein [Auerbach *et al.*, 1998]. The presence of a proline residue in the middle of a -helix may also contribute to the better packing in the protein and the formation of a more stable structure [Natesh *et al.*, 1999].

Xyn B from *Dictyoglomus thermophilum* Rt46B.1 is folded into a single domain based on two β-sheets that pack against each other. Xyn B is the most thermostable xylanase yet to be structurally characterized having a temperature optimum of 348 K. Several factors have been implicated for the thermostability of Xyn B. Firstly the solvent-accessible surface that is polar, is significantly higher than other xylanases. Xyn B has 83% polar surface when compared to the 78% polar surface area for *B. circulans* xylanase. An increased proportion of polar surface correlates strongly with increased thermostability [Vogt *et al.*, 1997]. Xyn B contains an extended B3-A5 connecting loop, which results in enhanced hydrogen bonding and results in higher thermostability.

The crystal structure of PVX, a thermostable, family 11 xylanase isolated from the thermophilic fungus *Paecilomyces varioti* Bainier has been determined by molecular replacement techniques using polyalanine coordinates of the *Thermomyces lanuginosus* xylanase. The structure has been determined at a high resolution (1.59 Å) and the final refined model has 194 amino acid residues and 128 water molecules. The structure belongs to an all β fold, with two curved β-sheets, forming the cylindrical active site cleft, and a lone α-helix, as present in other family 11 xylanases. The structure and the sequence of PVX were compared with the other available native structures of mesophilic and thermophilic xylanases belonging to family 11. The analysis provides the basis for the rationalization that the "hinge" region is made more compact in thermophiles by the addition of a disulphide bridge between Cys110 and Cys154 and a N-H...O hydrogen bond between Trp159 near the extremity of the lone α-helix and Trp138 on β-strand B8. Other factors which contribute to the thermostability of the enzyme are the associations involving buried water molecules, additional ion pairs and C-H...O and C-H...π type interactions dispersed in select areas [Kumar *et al.*, 2000].

8. CURRENT RESEARCH EFFORTS IN INDIA

Investigations on biotechnological, biochemical and molecular aspects of thermostable xylanases are being carried out in various laboratories in India. Archana and Satyanarayan have studied the production of a thermostable xylanase by thermophilic *Bacillus licheniformis* A99 in solid-state fermentation [Archana & Satyanarayana, 1997]. Dhillon *et al.* have studied the production, purification and characterization of a cellulase poor thermostable xylanase from *Bacillus circulans* AB16 [Dhillon *et al.* 2000]. The cloning, expression and sequence analysis of the gene encoding the thermostable endoxylanase from alkalophilic mesophilic *Bacillus* sp. strain NG-27 has been reported by [Gupta *et al.* 2000]. The crystal structure of a thermostable xylanase derived from *Thermoascus aurantiacus* has been studied at 1.8 Å resolution and its amino acid sequence has been proposed [Natesh *et al.*, 1999]. Whereas Kumar *et al.* have determined the tertiary structure of a family 11 xylanase from the thermophilic fungus *Paecilomyces varioti* Bainier at 1.59 Å resolution and proposed the amino acid sequence [Kumar *et al.*, 2000]. In our laboratory, an alkalophilic thermophilic *Bacillus* sp. (NCIM 59) was isolated from the hot springs of Vajreshwari (Thane, Bombay), which produced 50 IU/ml of cellulase-free xylanase at pH 10 and at 50°C in a simple medium containing wheat bran [Hinge *et al.*, 1989]. The practical application of xylanase from AT *Bacillus* in the biotreatment of bagasse pulp has been demonstrated [Kulkarni & Rao, 1996].

Two xylanases (Xylanase I and Xylanase II) were purified and characterized. The molecular weight of xylanase I and II were 35,000 and 15,800 respectively. The enzymes had similar temperature (50-60°C) and pH (Biely, 1985) optima. Both the enzymes were stable at 50°C for 96 h [Dey *et al.*, 1992]. Structure-function studies of Xyl I and Xyl II have revealed the presence essential tryptophan and carboxylic acid residues at the active site [Chauthaiwale & Rao, 1993, Chauthaiwale & Rao, 1994, Bieley, 1985]. The chemical modification studies in conjunction with the sequence data have shown that Trp-61in Xyl II is highly conserved and may play a crucial role in the interaction with the substrate [Nath & Rao, 1998]. Gene fragment encoding for the Xyl II has been cloned and expressed in *E. coli* and *Bacillus* [Shendye & Rao, 1993; Kulkarni *et al.*, 1995]. Chromosomal gene integration and amplification of xylanase gene fragment in AT *Bacillus* has resulted in a two-fold increase in the activity [Shendye & Rao, 1993]. The complete nucleotide sequence including the structural gene and the 5' and 3' flanking sequences of the xylanase gene has been determined [Kulkarni *et al.*, 1999]. Xyn II appeared to be a member of family G/11 of xylanases based on its molecular weight and basic pI (8.0). However, sequence homology revealed similar identity with families 10 and 11 of xylanases. The conserved triad (Val-Val-Xaa, where Xaa is Asn/Asp) was identified only in xylanases from alkaliphilic organisms. Our results implicate for the first time the concept of convergent evolution for Xyn II and provide a basis for research in evolutionary relationship among the xylanases from alkaliphilic and neutrophilic organisms. The genomic library of alkaliphilic thermophilic *Bacillus* was constructed in *E. coli.* A single recombinant (pATB 507 I) from the genomic library of alkaliphilic thermophilic *Bacillus* constructed in *E. coli* showed growth at 50°C. The thermophilic trait was transformable with 10^{-4} of expression amongst the p ATB 507 I transformants. Southern blot analysis revealed that a single locus (5.6 kb *Bam* HI fragment) was involved in the transfer of thermophilic trait [Shendye & Rao, 1996].

We are also working on a *Thermomonospora* sp., having optimum growth at pH 9 and temperature 50°C [George *et al.*, 2001]. The culture produced 125 IU/ml of xylanase when grown in shake flasks for 96 h [George *et al.*, 2001]. Xylanase I (Xyl I) produced by *Thermomonospora* sp. was purified to homogeneity and was classified as a family 10 xylanase based on its molecular weight (38,000 Da) and isoelectric point (4.1). K2d analysis showed that the secondary structure of Xyl I was made up of 38% α-helix and 10% β-sheet. The optimal temperature for the activity of Xyl I was 80°C. Xyl I was highly thermostable with half-lives of 86 min, 30 min and 15 min at 80, 90 and 100°C respectively. Investigations using *o*-phthalaldehyde (OPTA) as the chemical initiator for fluorescent chemoaffinity labeling and trinitrobenzene-sulphonic acid (TNBS) as chemical modifier have revealed the presence of a single lysine residue in the active site of Xyl I. Kinetic analysis suggested that lysine residue was present in the substrate-binding site [George *et al.*, 2001]. Similarly by using DEP as the chemical modifier a single histidine residue was identified in the active site of Xyl I. Kinetic studies indicated that the histidine residue was present in the catalytic site and may be involved in a network of hydrogen bonds, which are responsible for maintaining the ionization state of the two catalytic residues responsible for the hydrolysis of the β-1,4 glycosidic bond [George *et al.*, 2001].

9. FUTURE PROSPECTS

The long-term application of xylanase for the conversion of renewable biomass into liquid fuels is not yet economically feasible. In order to make the application of xylanases realistic the improvement of enzyme yields is of prime importance. The complete bioconversion of xylans to the sugar monomers has so far not been achieved; this is the main hurdle in the success of bioconversion process. The enzymatic cleavage of xylan to smaller oligosaccharides is itself a reversible reaction, hence a special process for the enzymatic cleavage of xylan needs to be designed. Such as the two-phase reactor in which the product can be continuously separated from the reactants. Cleaner biobleaching technology for the paper and pulp industry is currently concentrated in the developed countries, whereas renewable energy generation from agricultural waste has more relevance for the developing nations. In order to bring these ideas to reality a lot of work needs to be done. The commercially available xylanases do not meet the ideal criteria for application in paper industry i.e. optimum activity at pH 10 and temperature 90°C. Hence it is necessary to identify potent extremophilic xylanase producers, which are capable of producing enzymes active at extreme pH, and temperature. The crystal structures and sequence homology of few thermostable enzymes have been determined. Further studies on the gene sequence of thermostable enzymes will help to determine the basis of thermostability.

REFERENCES

Archana, A. and Satyanarayana, T. 1997. Xylanase production by thermophilic *Bacillus licheniformis* A 99 in solid state fermentation. *Enzyme Microbiol. Technol.* 21, 12-17.

Auerbach, G., Ostendrop, R., Prade, L., Korndörfer, I., Dams, T., Huber, R. and Jaenicke, R. 1998. Lactate dehydrogenase from the hyperthermophilic bacterium *Thermotoga maritima*: the crystal structure at 2.1 Å resolution reveals strategies for intrinsic protein stabilization. *Structure.* 6, 769-781.

Baba, T., Shinke, R. and Nanmori, T. 1994. Identification and characterization of clustered genes for thermostable xylan-degrading enzymes, beta-xylosidase and xylanase, of *Bacillus stearothermophilus* 21. *Appl. Environ. Microbiol.* 60, 2252-2258.

Bataillon, M., Cardinali, N., Castillon, N. and Duchiron, F. 2000. Purification and characterization of a moderately thermostable xylanase from *Bacillus* sp. SPS-0. *Enzyme Microb. Technol.* 26, 187-192.

Bérenger, J., Frixon, C., Creuzet, N. and Bigliardi, J. 1985. Production, purification and properties of thermostable xylanase from *Clostridium stercorarium.* Can. *J. Microbiol.* 31, 635-643.

Biely, P. 1985. Microbial xylanolytic systems. *Trends in Biotechnol.* 3, 286-290.

Blöchl, E., Rachel, R., Burgraff, S., Hafenbrandt, D., Jannasch, H.W. and Stetter, K.O. 1997. *Pyrolobus fumarii*, gen. and sp. nov., represents a novel group of archaea, extending the upper temperature limit for life to 113°C. *Extremophiles.* 1, 14-21.

Breccia, J.D., Siñeriz, F., Baigori, M.D., Guillermo, R.C. and Hatti-Kaul, R. 1998. Purification and characterization of a thermostable xylanase from *Bacillus amyloliquefaciens*. *Enzyme Microb. Technol.* 22, 42-49.

Chauthaiwale, J.V. and Rao, M. 1993. Specific interaction of guanidine hydrochloride with essential carboxyl group of xylanase from alkalothermophilic *Bacillus* sp. *Biochem. Biophys. Res. Commun.* 191, 922-927.

Chauthaiwale J. and Rao, M. 1994. Chemical modification of xylanase from alkalothermophilic *Bacillus*. *Biochim. Biophys. Acta.* 1204, 164-168

Chen, C., Adolphson, R., Dean, J.F.D., Eriksson, K.L., Adams, M.W.W. and Westpheling, J. 1997. Release of lignin from Kraft pulp by a hyperthermophilic xylanase from *Thermotoga maritima*. *Enzyme Microb. Technol.* 20, 39-45.

Coughlan, M.P. and Hazelwood, G.P. 1993. -1,4-D-xylan-degrading enzyme systems: biochemistry, molecular biology and applications. *Biotechnol. Appl. Biochem.* 17, 259-289.

Dahlberg, L., Holst, O., Kristjansson, J.K. 1993. Thermostable xylanolytic enzymes from *Rhodothermus marinus* grown on xylan. *Appl. Microbiol. Biotechnol.* 40, 63-68.

Debeire-Gosselin, M., Loonis, M., Samain, E., Debeire, P. 1992. Isoxylanases from the thermophile *Clostridium thermolacticum*. In: *Xylan and Xylanases* Prog. Biotechnol. vol 7 (Visser, J., Beldman, G., Kusters-van Someren, M. A., and Voragen, A. G.J., eds) pp 463-466. Elsevier, Amsterdam.

Dekker, R.F.H. and Richards, G.N. 1976. Hemicellulases: Their occurrence, purification properties and mode of action. *Adv. Carbohydr. Chem. Biochem.* 32, 277-352.

Deshpande, V.V., Hinge, J. and Rao, M. 1990. Chemical modification of xylanases: Evidence for essential tryptophan and cysteine residues at the active site. *Biochem. Biophys. Acta.* 1041, 172-177.

Dey, D., Hinge, J., Shendye, A., and Rao, M. 1992. Purification and properties of extracellular endoxylanases from alkalophilic thermophilic *Bacillus* sp. *Can. J. Microbiol.* 38, 436-442.

Dhillon, A., Gupta, J.K. and Khanna, S. 2000. Enhanced production, purification and characterization of a novel cellulase-poor thermostable, alkali-tolerant xylanase from *Bacillus circulans* AB 16. *Process Biochem.* 35, 849-856.

Dwivedi, P.P., Gibbs, M.D., Saul, D.J. and Bergquist, P.L. 1996. Cloning, sequencing and overexpression in *Escherichia coli* of a xylanase gene, *xynA* from the thermophilic *bacterium* Rt8B.4 genus *Caldicellulosiruptor*. *Appl. Microbiol. Biotechnol.* 45, 86-93.

Elegir, G., Szakács, G. and Jeffries, T.W. 1994. Purification, characterization, and substrate specificities of multiple xylanases from *Streptomyces* sp. strain B-12-2. *Appl. Environ. Microbiol.* 60, 2609-2515.

Fontes, C.M.G.A., Hazlewood, G.P., Morag, E., Hall, J., Hirst, B.A. and Gilbert, H.J. 1995. Evidence for a general role for non-catalytic thermostabilizing domains in xylanases from thermophilic bacteria. *Biochem. J.* 307, 151-158.

Gat, O., Lapidot, A., Alchana, T.I., Regueros, C. and Shoham, Y. 1994. Cloning and DNA sequence of the gene coding for *Bacillus stearothermophillus* T-6 xylanase. *Appl. Environ. Microbiol.* 60, 1889-1896.

George, S.P., Ahmad, A. and Rao, M.B. 2001. Studies on carboxymethyl cellulase produced by an alkalothermophilic actinomycete. *Bioresour. Technol.* 77, 171-175.

George, S.P., Ahmad, A. and Rao, M.B. 2001. A novel thermostable xylanase from *Thermomonospora* sp.: Influence of additives on thermostability. *Bioresour. Technol.* 78, 221-224.

George, S.P., Ahmad, A. and Rao, M.B. 2001. Involvement of a lysine residue in the active site of a thermostable xylanase from *Thermomonospora* sp. *Biochem. Biophys. Res. Commun.* 282, 48-54.

George, S.P. and Rao, M.B. 2001. Conformation and polarity of active site of xylanase I from *Thermomonospora* sp. as deduced by fluorescent chemoaffinity labeling: Site and significance of a histidine residue. *Eur. J. Biochem.* 268, 2881-2888.

Gibbs, M.D., Reeves, R.A. and Bergquist, P.K. 1995. Cloning, sequencing and expression of a xylanase gene from the extreme thermophilic *Dictyoglomus thermophilum* Rt 46B.1 and activity of the enzyme on fiber-bound substrate. *Appl. Environ. Microbiol.* 61, 4403-4408.

Gilbert, H.G., and Hazlewood, G.P. 1993. Bacterial cellulases and xylanases. *J. Gen. Microbiol.* 139, 187-194.

Gilead, S., and Shoham, Y. 1995. Purification and characterization of -L-arabinofuranosidase from *Bacillus stearothermophilus* T-6. *Appl. Environ. Microbiol.* 61, 170-174.

Goheen, D.W. 1981. Chemicals from wood and other biomass. Part II. The chemistry of conversion. *J. Chem. Educ.* 58, 544-547.

Gruber, K., Klintschar, G., Hayn, M., Schlacher, A., Steiner, W. and Kratky, C. 1998. Thermophilic xylanase from *Thermomyces lanuginosus*: high-resolution X-ray structure and modeling studies. *Biochemistry.* 37, 13475-13485.

Gupta, N., Reddy, V.S., Maiti, S. and Ghosh, A. 2000. Cloning, expression, and sequence analysis of the gene encoding the alkali-stable, thermostable endoxylanase from alkalophilic, mesophilic *Bacillus* sp. strain NG-27. *Appl. Environ. Microbiol.* 66, 2631-2635.

Hayashi, H., Tagaki, K.I., Fukumura, M., Kimura, T., Karita, S., Sakka, K. and Ohmiya, K. 1997. Sequence of xyn C and properties of Xyn C, a major component of the *Clostridiium thermocellum* cellulosome. *J. Bacteriol.* 179, 4246-4253.

Herbers, K., Wilke, I. and Sonnewald, U. 1995. A thermostable xylanase from *Clostridium thermocellum* expressed at high levels in the apoplast of transgenic tobacco has no detrimental effects and is easily purified. *Bio/Technology* 13, 63-66.

Hinge, J., Shendye, A., Srinivasan, M.C. and Rao. M. 1989. Process for preparing low molecular weight cellulase free xylanase from an alkalophilic thermophilic *Bacillus* sp. Patent application No. 796-DEL-89 dated 7.9.1989.

Keskar, S.S., Srinivasan, M.C., and Deshpande, V.V. 1989. Chemical modification of a xylanase from a thermotolerant *Streptomyces.* Evidence for essential tryptophan and cysteine residues at the active site. *Biochem. J.* 261, 49-55.

Khandke, K.M., Vithayathil, P.J. and Murthy, S.K. 1989. Purification of xylanase, -glucosidase, endocellulase, and exocellulase from a thermophilic fungus, *Thermoascus aurantiacus. Arch. Biochem. Biophys.* 274, 491-500.

Khasin, A., Alchanti, I. and Shoham, Y. 1993. Purification and characterization of a thermostable xylanase from *Bacillus stearothermophilus* T-6. *Appl. Environ. Microbiol.* 59, 1725-30.

Krishnamurthy, S. 1989. Purification and properties of xylanase and -glucosidase elaborated by the thermophilic fungus *Paecilomyces varioti* Bainier. PhD thesis, Indian Institute of Science, Bangalore, India.

Kulkarni N., Chautaiwale J. and Rao, M. 1995. Characterization of the recombinant xylanases in *Escherichia coli* from an alkaliphilic thermophilic *Bacillus* sp. NCIM 59. *Enzyme Microb. Technol.* 17, 972-976.

Kulkarni, N. and Rao, M. 1996. Application of xylanase from alkalophilic thermophilic *Bacillus* sp. NCIM 59 in biobleaching of bagasse pulp. *J. Biotechnol.* 51, 167-173.

Kulkarni N., Shendye, A., and Rao M. 1999. Molecular and biotechnological aspects of xylanases. *FEMS Microb. Rev.* 23, 411-456.

Kulkarni, N., Lakshmikumaran, M., and Rao, M. 1999. Xylanase II from alkaliphilic thermophilic Bacillus with a distinctly different structure from other xylanases: Evolutionary relationship to alkaliphilic xylanases. *Biochem. Biophys. Res. Commun.* 263, 640-645.

Kumar, P.R., Eswaramoorthy, S., Vithayathil, P.J. and Viswamitra, M.A. 2000. The tertiary structure at 1.59 Å resolution and the proposed amino acid sequence of a family-11 xylanase from the thermophilic fungus *Paecilomyces varioti* Bainier. *J. Mol. Biol.* 295, 581-593.

Lapidot, A., Mechaly, A. and Shoham, Y. 1996. Overexpression and single-step purification of a thermostable xylanase from *Bacillus stearothermophilus. J. Biotechnol.* 51, 259-264.

Lee, Y.E., Lowe, S.E., Henrissat, B. and Zikus, J.G. 1993. Characterization of the active site and thermostability regions of endoxylanase *from Thermoanaerobacterium saccharolyticum* B6A-RI. *J. Bacteriol.* 175, 5890-5898.

Lüthi, E., Bhana, J.M. and Bergquist, P.L. 1990. Xylanase from the extremely thermophilic bacterium *Caldocellum saccharolyticum:* over expression of the gene in *Escherichia coli* and characterization of the gene product. *Appl. Environ. Microbiol.* 56, 2677-2683.

Lüthi, E., Reif, K., Jasmat, N.B. and Bergquist, P.L. 1992. In vitro mutagenesis of a xylanase from the extreme thermophile *Caldocellum saccharolyticum. Appl. Microbiol. Biotechnol.* 36, 503-506.

Mathrani, I.M. and Ahring, B.K. 1991. Isolation and characterization of a strictly xylan-degrading *Dictyoglomus* from a man-made thermophilic anaerobic environment. *Arch. Microbiol.* 157, 13-17.

McCarthy, A.J., Peace, E. and Broda, P. 1985. Studies on the extracellular xylanase activity of some thermophilic actinomycetes. *Appl. Microbiol. Biotechnol.* 21, 238-244.

Moreau, A., Shareck, F., Kluepfel, D. and Morosoli, R. 1994. Increase in catalytic activity and thermostability of the xylanase A of *Streptomyces lividans* 1326 by site specific mutagenesis. *Enzyme Microb. Technol.* 16, 420-424.

Nanmori, T., Watanabe, T., Shinke, R., Kohno, A., and Kawamura, Y. 1990. Purification and properties of thermostable xylanase and -xylosidase produced by a newly isolated *Bacillus stearothermophilus* strain. *J. Bacteriol.* 172, 6669-6672.

Natesh, R., Bhanumoorthy, P., Vithayathil, P.J., Sekar, K., Ramakumar, S. and Viswamitra, M.A. 1999. Crystal structure at 1.8 Å resolution and proposed amino acid sequence of a thermostable xylanase from *Thermoascus aurantiacus. J. Mol. Biol.* 288, 999-1012.

Nath D. and Rao M. 1998. Structural and functional role of tryptophan in xylanase from an extremophilic *Bacillus*: Assessment of active site. *Biochem. Biophys. Res. Commun.* 249, 207-212.

Nicolson, H., Bechtel, W.J. and Mathews, B.W., 1988. Enhanced protein thermostability from designed mutations that interact with a-helix dipoles. *Nature* 336, 651-656.

Nielsen, P., Mathrani, I. M., and Ahring, B. K. 1993. *Thermoanaerobium acetigenum* spec. nov., a new anaerobic, extremely thermophilic, xylanolytic non-spore-forming bacterium isolated from an Icelandic hot spring. *Arch. Microbiol.* 159, 460-464.

Okazaki, W., Akiba, T., Horikoshi, K., Akahoshi, E. 1984. Production and properties of two types of xylanases from alkalophilic, thermophilic *Bacillus* sp. *Appl. Microbiol. Biotechnol.* 19, 335-340.

Oku, T., Roy, C., Watson, D.C., Wakarchuk, W., Campbell, R., Yaguchi, M., Jurasek, L., Paice, M.G. 1993. Amino acid sequence and thermostability of xylanase A from *Schizophyllum commune. FEBS Lett.* 334, 296-300.

Persidis, A. 1998. Extremophiles. *Nature Biotechnol.* 16, 593-594.

Ruttersmith, L.D., Daniel, R.M. and Simpson, H.D. 1992. Cellulolytic and hemicellulolytic enzymes functional above 100°C. *Ann. N. Y. Acad. Sci.* 672, 137-141.

Sakka, K., Maeda, Y., Hakamda, Y., Takahashi, N. and Shimada, K. 1991. Purification and some properties of xyalanase from *Clostridium stercorarium* strain HX-1. *Agric. Biol. Chem.* 55, 247-248.

Saul, D.J., Williams, L.C., Reeves, R.A., Gibbs, M.D. and Bergquist, P.L. 1995. Sequence and expression of a xylanase gene from the hyperthermophile *Thermotoga* sp. strain FjSS3-B.1 and characterization of the recombinant enzyme and its activity on kraft pulp. *Appl. Environ. Microbiol.* 61, 4110-4113.

Schlacher, A., Holzmann, K., Hayn, M., Steiner, W. and Schwab H. 1996. Cloning and characterization of the gene for the thermostable xylanase XynA from *Thermomyces lanuginosus. J. Biotechnol.* 49, 211-218.

Shao, W., DeBlois, S. and Wiegel, J. 1995. A high-molecular weight, cell-associated xylanase isolated from exponentially growing *Thermoanaerobacterium* sp. strain JW/SL-YS485. *Appl. Environ. Microbiol.* 61, 937-940.

Shareck, F., Roy, C., Yaguchi, M., Morosoli, R. and Kluepfel, D. 1991. Sequences of three genes specifying xylanases in *Streptomyces lividans. Gene* 107, 75-82.

Shendye, A. and Rao, M. 1993. Cloning and extracellular expression in *Eschericia coli* of xylanases from an alkaliphilic thermophilic *Bacillus* sp. NCIM 59. *FEMS Microbiol. Lett.* 108, 297-302.

Shendye, A. and Rao, M. 1993. Molecular cloning and expression of xylanases from an alkalophilic thermophilic *Bacillus* (NCIM 59) in *Bacillus subtilis* A8. *Enzyme Microb. Technol.* 15, 343-347.

Shendye, A. and Rao, M. 1993. Chromosomal gene integration and enhanced xylanase production in an alkalophilic thermophilic *Bacillus* sp. (NCIM 59). *Biochem. Biophys. Res. Commun.* 195, 770-784.

Shendye, A., Gaikaiwari, R. and Rao, M. 1994. Expression of the cloned xylanases from an alkalophilic, thermophilic *Bacillus* in *Bacillus subtilis. World J. Microbiol. Biotechnol.* 10, 414-416.

Shendye, A., and Rao, M. 1996. Genetic transfer of thermophilic trait for alkalophilic thermophilic Bacillus to *E. coli. Biochem. Biophys. Res. Commun.* 227, 386-392.

Simpson, H.D., Haufler, U.R. and Daniel, R.M. 1991. An extremely thermostable xylanase from the thermophilic eubacterium *Thermotoga. Biochem. J.* 277, 413-417.

Stutzenberger, F.J. and Bodine, A.B. 1992. Xylanase production by *Thermomonospora curvata. J. Appl. Bacteriol.* 72, 504-511.

Sunna, A., Puls, J. and Antranikian, G. 1996. Purification and characterization of two thermostable endo-1,4--D- xylanases from *Thermotoga thermarum. Biotechnol. Appl. Biochem.* 24, 177-185.

Sunna, A. and Antranikian, G. 1997. Xylanolytic enzymes from fungi and bacteria. *Crit. Rev. Biotechnol.* 17, 39-67.

Takahashi, H., Nakai, R. and Nakamura, S. 2000. Purification and partial characterization of a basic xylanase produced by thermoalkalophilic *Bacillus* sp. strain TAR-1. *Biosci. Biotechnol. Biochem.* 64, 887-890.

Tan, L.U.L., Mayers, P. and Saddler, J.N. 1987. Purification and characterization of a thermostable xylanase from a thermophilic fungus *Thermoascus aurantiacus. Can. J. Microbiol.* 33, 689-692.

Tsujibo, H., Miyamoto, K., Kuda, T., Minami, K., Sakamoto, T., Hasegawa, T. and Inamori, Y. 1992. Purification, properties, and partial amino acid sequences of thermostable xylanases from *Streptomyces thermoviolaceus* OPC-520. *Appl. Environ. Microbiol.* 58, 371-375.

Tuohy, M.G., and Coughlan, M.P. 1992. Production of thermostable xylan degrading enzymes by *Talaromyces emersonii. Bioresour. Technol.* 39, 131-137.

Uchino, F. and Fukuda, O. 1983. Taxonomic characteristics of a acidophilic strain of *Bacillus* producing thermophilic acidophilic amylase and thermostable xylanase. *Agric. Biol. Chem.* 47, 965-967.

Uffen, R.L. 1997. Xylan degradation: a glimpse at microbial diversity. *J. Ind. Microbiol. Biotechnol.* 19, 1-6.

Veoikodvorskaya, T.V., Volkov, I.Y., Vasilevko, V.T., Zverlov, V.V. and Piruzian, E.S. 1997. Purification and some properties of *Thermotoga neapolitana* thermostable xylanase B expressed in *E. coli* cells. *Biochemistry (Moscow)* 62, 66-70.

Vogt, G., Woell, S. and Argos, P. 1997. Protein thermal stability, hydrogen bonds, and ion pairs. *J. Mol. Biol.* 269, 631-643.

Wakarchuk, W.W., Sung, W.L., Campbell, R.L., Cunningham, A., Watson, D.C. and Yaguchi, M. 1994. Thermostabilization of the *Bacillus circulans* xylanase by the introduction of disulfide bonds. *Protein Eng.* 7, 1379-1386.

Walsh, D., Gibbs, M.D. and Bergquist, P.L. 1998, Expression and secretion of a xylanase from the extreme thermophile, *Thermotoga* strain FjSS3B.1, in *Kluyveromyces lactis Extremophiles* 2, 9-14.

Warren, R.A.J. 1996. Microbial hydrolysis of polysaccharides *Annu. Rev. Microbiol.* 50, 183-212.

Whistler, R.L. and Richards, E.L. 1970. *Hemicelluloses.* In: *The Carbohydrates.* pp. 447-469. Vol. 2a. Pigman, W. and Horton, D., Eds., Academic Press, New York.

Winterhalter, C. and Liebl, W. 1995. Two extremely thermostable xylanases of the hyperthermophilic bacterium *Thermotoga maritima* MSB8. *Appl. Environ. Microbiol.* 61, 1810-1815.

Wong, K.K.Y., Tan, L.U.L. and Saddler, J.N. 1988. Multiplicity of -1,4-xylanase in microorganism: Functions and applications. *Microbiol. Rev.* 52, 305-317.

Yoshioka, H., Nagato, N., Chavanich, S. Nilubol, N. and Hayashida 1981. Purification and properties of thermostable xylanases from *Talaromyces byssochlamydoides* YH-50. *Agric. Biol. Chem.* 45, 2425-2432.

CHAPTER - 4

POTENTIAL USE OF *AZOLLA* BIOFERTILIZER IN SUSTAINABLE RICE PRODUCTION

K. Kumar and S. Kannaiyan

Azolla Laboratory, Tamilnadu Agricultural University, Coimbatore - 641 003 (TN)

1. INTRODUCTION

Azolla is a green- floating water fern that floats in water and fixes atmospheric nitrogen in association with nitrogen fixing blue green alga *Anabaena azollae* (Moore, 1969). *Azolla* fronds consists of sporophyte with a floating rhizome and small overlapping bilobed leaves and roots. Rice growing areas in South East Asia and other third World countries have recently been evincing increased interest in the use of the symbiotic N_2 fixing water fern *Azolla* either as an alternate nitrogen source or as a supplement to commercial nitrogen fertilizers (Watanabe, 1982). *Azolla* is used as biofertilizer for wetland rice and it is known to contribute 40- 60 kg N ha^{-1} per rice crop (Kannaiyan, 1989a). The agronomic potential of *Azolla* is quite significant particularly for rice crop and it is widely used as biofertilizer for increasing rice yield (Watanabe *et al.*, 1981 and kannaiyan *et al.*, 1982). Rice crop response studies with *Azolla* biofertilizer in the People's Republic in China (FAO, 1977), Vietnam (Tran and Dao, 1973), Philippines (Watanabe, 1990), Thailand (Laddawan, 1985), Srilanka (Kulasooriya, 1985) and India (Kannaiyan, 1992a) have provided good evidence that *Azolla* incorporation into the soil as a green manure crop is one of the most effective ways of providing nitrogen sources for rice. The important factor in using *Azolla* as a biofertilizer for rice crop is its quick decomposition in soil and efficient availability of its nitrogen to rice. In tropical rice soil, the applied *Azolla* mineralizes rapidly and its nitrogen is made available to the rice crop in very short period (Kannaiyan, 1990).

Azolla- Anabaena Symbiosis

The eukaryotic water fern *Azolla* and the prokaryotic endosymbiont *Anabaena azollae* are actively involved in the symbiotic association and fix atmospheric nitrogen. The host plant *Azolla* provides the carbon source sucrose to the cyanobiont while the cyanobiont fix nitrogen and transfers to *Azolla* which in turn multiplies very rapidly. The cyanobiont is associated with dorsal lobes of the leaf from the onset of their development and is never in direct contact with the external environment and *A. azollae* is the only species found in symbiotic association with *Azolla* (Peters *et al.*, 1979). The growth and development of the *Azolla* and the *Anabaena* are synchronous. *Azolla* shoot tips are supported above the water surface by the stem axes and a small colony of *Anabaena* filaments is associated with every apical merits. The heterocyst of the symbiont *Anabaena* is the site of nitrogen fixation, while vegetative cells are involved in photosynthesis. The total nitrogen requirement of the *Azolla* fronds is supplied by the cyanobiont. The ammonia from N_2 fixation by the cyanobiont is released into the leaf cavity and absorbed by the frond's ammonia assimilating enzymes. Ammonia excretion by the symbionts of *Azolla* spp. was reported by Kannaiyan and Goyal (1988). Photoproduction of ammonia continuously by algal symbiont *A. azollae* was reported by Kannaiyan *et al.*, (1992).

2. GROWTH

The growth rate of *Azolla* is very rapid so much so that it doubles its weight in mere 2-3 days. Due to fast relative growth rate, the biomass production is heavy in a given time and hence the biomass is exploited as biofertilizer for rice. *Azolla* has the potential of maintaining an exponential growth rate under optimum conditions. The growth of *Azolla* is initially slow in the growth medium as well as in the rice field followed by fast growth (Kannaiyan, 1993b). The growth of *Azolla* is subjected to various environmental factors such as temperature, light intensity, season, humidity, wind, pH and salinity (Kannaiyan and Somporn, 1987; Latha *et al.*, 1992 and Rajagopal *et al.*, 1994). The response of *Azolla* to phosphorus fertilization is necessary for its growth and N_2 fixation (Kannaiyan *et al.*, 1981 and Singh and Singh, 1989). *Azolla* grown in phosphorus deficient solution had a reddish brown discoloration that spread from the centre of the frond to the tip of the body with reduction in size of the frond. Addition of phosphorus is most effective in stimulating the growth of *Azolla* (Singh, 1979). Kannaiyan (1985) reported higher biomass of *Azolla filiculoides* at 20 ppm of phosphorus but found that 5 - 10 ppm level was adequate for growth and multiplication. Singh (1977) recommended 4-6 kg P_2O_5 ha^{-1} per week for large scale multiplication of *A. pinnata* under field conditions. Kannaiyan *et al.*, (1982) have demonstrated the effectiveness of split application of phosphorus for *Azolla* growth and multiplication.

(a) Mass Production of *Azolla* Biomass

A simple *Azolla* nursery method for large scale multiplication of *Azolla microphylla* in the fields has been evolved for easy adoption by the farmers (Kannaiyan, 1982 and 1989b). The field selected for *Azolla* nursery must be thoroughly prepared and levelled uniformly. The field is divided into one cent plots (20 x 2 m) by providing suitable bunds and irrigation channels. Water is maintained at a depth of 10 cm. Ten kilos of fresh cattle dung mixed in 20 litres of water is sprinkled in each plot and *A. microphylla* SK-TNAU inoculum at 8 kg is inoculated into each plot. Super phosphate (100 g) is applied in three split doses at 4 days interval as top dressing fertilizer for *A. microphylla* SK-TNAU. For insect pest control, furadan (Carbofuran) granules at 100 g/plot is applied 7 days after inoculation. Fifteen days after inoculation, *A. microphylla* SK-TNAU is harvested and introduced into the main field as a source of primary inoculum. Fron one harvest,

100-150 kg fresh *Azolla* is obtained from each plot. Again *Azolla* is inoculated in the same plot and cattle dung, super phosphate, furadan granules are applied and harvested. Two cents (80 sq. m) nursery plots are required for raising *A. microphylla* SK-TNAU for inoculation to one hectare main field.

(b) Azolla Spore Production Technology

Sporulation in different *Azolla* species are not well understood and the factors responsible for sporulation and spore germination are not well defined. *Azolla microphylla* SK-TNAU a sporulating culture was selected for spore inoculum development. Fresh young fronds of *A. microphylla* SK-TNAU were inoculated in puddled and well prepared field under flooded condition. A basal nutrition of 10 kg fresh cattle dung was mixed in 20 litres of water and applied in 40 m^2. A week after inoculation of *A. microphylla* SK-TNAU the growth regulator GA_3 at 100 ppm was applied as foliar spray. Foliar spray with GA_3 induced the growth and stimulated the production of sporocarps. These sporulated fronds were harvested after four weeks and formed into a heap with flat surface. One hundred kilogram of sporulated fresh fronds of *A. microphylla* SK- TNAU formed a heap. The heap was then covered with a thin layer of clay soil slurry (20%) and allowed to undergo partial decomposition for a period of 21 days. Subsequently the heaps were removed and dried well under shade for a week period. The decomposed and well dried frond materials are called "Frond based spore inoculum" of *Azolla* (Kannaiyan, 1990). The viability of the spore inoculum was found to be good for 10-15 months (Shanmugaundaram and Kannaiyan, 1992).

Mahesh and Kannaiyan (1994) have investigated the influence of growth regulators viz., GA_3, NAA, 2, 4-D and BAP at various concentrations on the germination and fertilization of the megasporocarps of *A. microphylla* and found that GA_3 and BAP at 20 to 80 ppm stimulated the germination and fertilization of megasporocarps. The growth regulators also enhanced the survival of the sporelings and the heterocyst frequency of the algal symbiont in the sporelings.

3. INOCULATION OF FRESH BIOMASS OF *AZOLLA*

Fresh biomass of *Azolla* are broadcasted in the main field 7-10 days after transplanting rice. Inoculation of the fresh biomass of *A. microphylla* at 200 kg ha^{-1} could multiply faster and cover the rice fields as a green mat in 2-3 weeks periods with 15-25 t biomass accumulation. However, inoculation of even 50-100 kg biomass of *A. microphylla, A. pinnata, A. caroliniana* and *A. filiculoides* per hectare will multiply and establish in rice fields but it takes 4-5 weeks for full coverage in the transplanted rice. Under optimum conditions with cloudy days and low temperature in day/night cycle of 25±1°C/ 18±1°C, the inoculated *Azolla* spreads so fast in transplanted rice. *Azolla* technology is very efficient in terms of N_2 fixation and biomass accumulation during wet season due to the better environmental conditons prevailing during the second season rice in India particularly cloudly days coupled with low temperature favours its vegetative multiplication. The best suitable season for *Azolla* utilization as dual crop with rice is wet season. It can also be used for late dry season crop. Season is very critical for its multiplication and biomass accumulation. *A. microphylla* is known to survive even in summer season but its full potential as biofertilizer could be exploited only during second wet season rice crop in tropical conditions. Inoculation of *Azolla* in transplanted rice crop supports the growth and multiplication. Growing *Azolla* along with rice crop is called dual culture and it is not affecting the growth of rice crop in any way but is suppresses the weed growth in wet land rice (Kannaiyan *et al.*, 1984).

4. APPLICATION OF *AZOLLA* BIOFERTILIZER FOR RICE CROP

Srinivasan (1980) reported that *Azolla* as green manure could reduce the nitrogen input by 25-30% in lowland rice. *Azolla* incorporation as green manure increased the yield of rice by 9-38% (Singh 1977), 35.8% (Kannaiyan, 1981), 44% (Bartharkur and Talukdar, 1983). Subramani and Kannaiyan (1987) reported that application of gypsum at 5 t ha^{-1} with 60 kg N ha^{-1} as urea and *Azolla* inoculation as dual crop recorded the highest grain yield of rice. Jayanthi *et al.* (1994) have reported that incorporation of 10 t ha^{-1} fresh *Azolla* biomass along with 50 kg N ha^{-1} resulted in increased N uptake, available nitrogen in soil and rice yield. Gopalaswamy *et al.*, (1994) have investigated the potential of *Azolla* hybrids and reported that inoculation of *Azolla* hybrids significantly improved the soil organic carbon status besides increasing the rice yield (Table 1).

Table 1. Effect of incorporation of *Azolla* hybrids on the C:N ratio of rice soil

Azolla Hybrids	C:N ratio			
	Days after incorporation			
	0	15	30	45
Control	20.00	17.06	18.66	19.28
AH-C1	11.37	9.63	6.92	10.00
AH-C2	10.93	9.68	6.87	10.74
AH-C3	13.33	11.83	7.50	11.25
AH-C4	11.66	12.46	7.53	12.27
A. filiculoides	14.47	11.22	6.96	12.38
A. microphylla	16.11	11.18	6.09	14.00
Urea (30 kg N ha^{-1})	17.50	13.88	8.47	15.88
SEd	2.10	0.98	2.23	0.94
CD	4.37	2.04	4.64	1.95

The use of *Azolla* in rice field has considerably increased the soil microbial as well as soil enzymes activity (Kumar, 1994 and Kannaiyan, 1994) which in turn maintains the soil fertility. It is an important low cost biological input which plays a vital role in improving soil quality management in sustainable rice farming (Kannaiyan, 1992 a, b and 1993b).

In wetland rice ecosystem the productivity of rice could be increased by raising the efficiency of applied nitrogen either by natural means or by chemical fertilization. Latha *et al.*, (1988) have established that the inoculation of *Azolla* coupled with USG (Urea Super Granule) application have recorded significantly higher grain yield of rice than PU application. Thangaraju and Kannaiyan (1990) reported that *A. microphylla* and *A. filiculoides* coupled with USG application significantly increased the grain yield of rice. Kannaiyan (1992a) reported increased rice yield by the combined application of USG and *A. microphylla*. Shanmugasundaram and Kannaiyan (1994) have reported the inoculation of *Azolla* hybrid RS-SK-TNAU-1 with PU as well as USG recorded higher grain yield and N uptake in rice. The nitrogen fixing green manure plants viz., *Azolla* and *Sesbania rostrata* could be effectively utilized by incorporation with USG to increase the rice production in lowland rice ecosystem (Kumar *et al.*, 1995). The nitrogen use efficiency in lowland rice system coulb be increased substantially by adopting the integrated nutrient management system using nitrogen fixing green manures or *Azolla* biofertilizer.

Azolla inoculation has considerably increased the soil microbial as well as enzymes activity (Kumar, 1994 and Kannaiyan, 1994). Zhang Zhuang *et al.*, (1987) have reported that incorporation of 22.5-37.5 t fresh *Azolla* ha^{-1} biomass before planting increased the rice yield by 9.6 - 13.0 per cent. Patil and Mandhare (1991) reported that incorporation of 5 t *Azolla* ha^{-1} with 100 per cent recommended nitrogen level yielded 6.7 t ha^{-1} which is more or less equal to the incorporation of 5 t *Azolla* ha^{-1} with 75 per cent of the recommended dose of nitrogen. Alexander *et al.* (1992) have observed maximum rice yield increase with an output ratio recovery of 2.07 when 7.5 t fresh *Azolla* ha^{-1} biomass incorporated along with 90 kg N ha^{-1}. An increase in the grain yield of rice by the application of *Azolla* as green manure, dual crop or both with rice was equivalent to 30-40 kg N ha^{-1} and this has been well documented by several investigators in various countries such as China (Liu Chung Chu, 1987). The Philippines (Mabbayad, 1987; Watanabe, 1987), India (Singh, 1977; Kannaiyan, 1981), Thailand (Loudhapasitiporn and Kanareugsa, 1987), Sri Lanka (Kulasooriya *et al.*, 1987), USA (Talley *et al.*, 1977) and West Africa (Diara *et al.*, 1987). The use of *Azolla* in rice fieldhas considerably increased the soil microbial as well as enzymes activity (Kumar, 1994 and Kannaiyan 1994) which in turn maintained soil fertility (Table 2 & 3). It is an important low cost input, which plays a vital role in improving soil quality management in sustainable rice farming (Kannaiyan, 1992a; 1993b and 1995). Gopalaswamy *et al.*, (1994) have investigated the potential of *Azolla* hybrids and reported that inoculation of *Azolla* hybrids significantly improved the soil organic carbon besides increasing the rice yield. Dual cropping of *Azolla* is a most effective, ecofriendly practice which play a major role in sustainable rice farming (Kannaiyan, 1992a; 1993b and 1995). Though *Azolla Anabaena* symbiosis is the only plant - cyanobacterial symbiosis used as a rice biofertilizer for maintaining soil health and productivity, this system could still be improved physiologically by using the growth regulator gibberllic acid GA_3 (Gopalaswamy and Kannaiyan, 1996a). *Azolla* hybrids were found to undergo rapid decomposition and to increase the flood water ammoniacal N content (Gopalaswamy and Kannaiyan 1996b) which is very vital for the rice crop. By carefully screening of different *Azolla* cultures, *Azolla* hybrid AH-C2 was found to be saline tolerant (Gopalaswamy and Kannaiyan, 1996c). Rice production could be significantly increased with *Azolla* hybrids in

Table 2. Effect of incorporation of *A. microphylla* as green manure (GM)/ dual cropping (DC) and fertilizer N with neem cake on the activity of soil enzymes*

Treatments	Urease*	% increase over control	L-asparaginase **	% increase over control
Control	15.08	–	15.00	–
A. microphylla (GM) alone	18.43	22.21	19.92	32.82
A. microphylla (GM) + NC	20.20	33.95	22.82	52.16
A. microphylla (GM) + PU	20.56	36.33	23.24	54.96
A. microphylla (GM) + NC + PU	22.23	47.40	25.73	71.57
A. microphylla (DC)	17.95	19.02	18.93	26.21
A. microphylla (DC) + NC	20.00	32.62	22.70	51.36
A. microphylla (DC) + PU	20.04	32.88	23.04	53.63
A. microphylla (DC) + NC + PU	22.00	45.88	25.42	69.50
Prilled Urea (PU) alone	17.60	16.71	18.20	21.34
SED	0.73	–	1.02	–
CD	2.05	–	2.10	–

* - Observations recorded at the time of harvest
** -g of NH_3 released g^{-1} soil hr^{-1}
PU - Prilled urea at 60 kg N ha^{-1}
NC - Neem cake at 250 kg ha^{-1}
A. microphylla (DC) 500 kg ha^{-1}
A. microphylla (GM) 15 t ha^{-1}

Table 3. Effect of incorporation of *Azolla* mutants on the ammonical N contents of rice soil

Treatments	Ammonical nitrogen (μg^{-1} soil)									
	0th Day		7th Day		14th Day		21th Day			
	C	UC	C	UC	C	UC	C	UC	C	UC
Control	45.56	48.34	54.70	50.28	59.16	52.86	48.22	39.98	42.85	33.56
A. microphylla wild	52.10	47.90	126.42	102.14	115.48	110.42	82.59	71.36	68.19	54.38
A. microphylla wild + neem cake	49.50	44.19	98.05	75.67	142.56	117.86	107.12	92.60	84.68	69.77
A. microphylla - KK-SK-E_{20}	50.25	46.72	152.36	118.24	136.13	124.60	98.81	89.38	76.46	65.15
A. microphylla -KK-SK-E_{20} + neem cake	43.85	47.20	110.65	82.50	182.34	146.50	129.73	105.46	93.79	72.28
A. filiculoides wild	45.30	44.80	105.96	78.85	92.87	102.84	71.44	82.53	64.51	43.82
A. filiculoides wild + neem cake	53.15	41.62	95.34	66.43	119.32	108.65	94.35	79.61	75.37	61.74
A. filiculoides - KK-SK-C_5	42.20	45.58	136.21	108.25	100.44	86.31	81.69	67.42	69.85	54.27
A. filiculoides - KK-SK-C_5 + neem cake	46.60	47.52	101.76	87.60	157.55	129.74	106.24	94.83	88.16	63.48
Sed	2.50		6.94		7.30		4.99		8.81	
CD	5.25		14.59		15.34		10.48		18.50	

C: Cropped; UC : Uncropped

combination with 75 kg N ha^{-1} Urea Super Granule (Gopalaswamy and Kannaiyan, 1996d).

5. NEEM CAKE AND N USE EFFICIENCY

Neem cake is widely used as organic fertlizers for several crops. But it seems to have good nitrification inhibition capacity, due to the presence of alkaloids solanin, meliantrool, azadiractin, nimbin and nimbidin which possess a furan ring system responsible for inhibition (Jotwani and Srivastava, 1981). When neem cake is mixed with urea, the rate of inhibition increase with increasing proportion of neem cake. The nitrification inhibition period of neem cake is around 10-25 days. It retains ammonium N for longer period and higher NH_4+- -N contents are noticed in the soil with neem cake coated urea application (Subbiah and Kothandaraman, 1980).

The application of nitrification inhibitors along with urea in low land rice could inhibit nitrification and suppress the transformation of ammoniacal N and thereby accumulating, more ammoniacal N for effective utilization by rice crop (Manickam *et al.*, 1976; De Datta, 1986). The increase in rice yields due to the application of DCD mixed urea and neem cake mixed urea was reported (Sharma and Sinha, 1977; Simpson *et al.*, 1985). In the experiments conducted at Tamil Nadu Agricultural University, Coimbatore, the grain and straw yields of rice were enhanced significantly due to the application of neem cake and DCD. Of the two nitrification inhibitors, neem cake recorded maximum total N content in the plant, grain and straw yield when added during the incorporation of N_2 fixing green manure plants, *viz.*, *Sesbania rostrata* and *Azolla microphylla* compared to DCD. Among different treatments with different combinations of green manures, fertilizer N and nitrification inhibitors, *S. rostrata* incorporation (20 t ha^{-1}) with fertilizer N (75 kg ha^{-1}) and neem cake (250 kg ha^{-1}) registered the highest nitrogen accumulation, grain and straw yield in IR 64 rice. It has been apparently shown that both the nitrification inhibitors neem cake and DCD have influenced the grain yield of rice considerably. The beneficial effect on IR 64 rice, as a result of the addition of nitrification inhibitors during the incorporation of *A. microphylla* and *S. rostrata* was attributed to the significant biological and chemical changes brought by the inhibitors in rice soil which eventually influenced the availability of nitrogen to rice crop. Similar effect of nitrification inhibitors in increasing the yield of lowland rice and reducing the N loss have been well documented (Manickam *et al.*, 1976; Kamire and Sonar, 1979; De Datta *et al.*, 1983; Kumar *et al.*, 1992 and 1995). Kannaiyan (1996) more clearly elucidated the inhibitory action of neem cake on the N loss mechanism in rice soil eventually increasing the N use efficiency of *Azolla* biofertilizer and inorganic fertilizer N by rice crop (Fig. 1 and Table 4).

6. UREA SUPER GRANULES (USG) AND N USE EFFICIENCY

Deep placement of fertilizer in reduced soil has been considered the most efficient method to increase fertilizer N efficiency in lowland rice and is becoming increasingly relevant since the introduction of modified urea materials like urea super granules, can be placed in the soil by hand or with mechanical operator. The advantage of deep placement of urea super granules is that various losses like ammonia, volatilization, denitrification, run off and leaching can be minimised (Mikkelsen *et al.*, 1978; Savant *et al.*, 1983 and Kannaiyan, 1995). An additional advantage is that this mode of fertilization does not seem to interfere with dinitrogen fixation in flood water by *Azolla* and blue green algae (Roger *et al.*, 1980).

Several years of agronomic data from the International Network on Soil Fertility and Fertilizer Evaluation For Rice (INSF-FER) have shown that deep placement is generally more efficient than the traditional split application of urea (Craswell and De Datta, 1980; De Datta *et al.*, 1983). Sharma (1995) has documented the positive influence of USG on rice soil based on the field experiments conducted at different locations.

When the nitrogen fixing green manures like *Azolla* and *Sesbania rostrata*, biofertilizer like blue green algae are used, the application of neem cake coated urea and deep placement of USG proved better (Kannaiyan, 1993a). The application of different forms of urea along with *Azolla* inoculation has also been evaluated in the field experiments conducted during early and late wet seasons of 1984-85 at Tamil Nadu Agricultural University (Latha *et al.*, 1988). Significant increase in grain yield was obtained over 75 kg N ha^{-1} as prilled

Table 4. Economic analysis of the *Azolla* and neem cake package

Treatments	Grain yield (t ha^{-1})	Increase over 100 kg N ha^{-1}	Straw yield (t ha-1)	Increase over 100 kg N ha^{-1}	Additional cost over 100 kg N ha-1	Additional benefit over 100 kg N ha-1	MRR
Azolla (DC) (500 kg ha^{-1}) + PU 60 kg N ha^{-1}	6.24	0.17	8.36	0.41	140	697.50	4.98
Azolla (DC) (500 kg ha^{-1}) + Neem cake 250 kg	6.17	0.12	8.51	0.53	100	552.50	5.53
Azolla (DC) (500 kg ha^{-1}) + PU 60 kg N/ha + Neem cake 250 kg N/ha	7.93	1.86	10.01	2.03	890	7033.80	7.90
100 kg N/ha	6.07	–	7.98	–	–	–	–

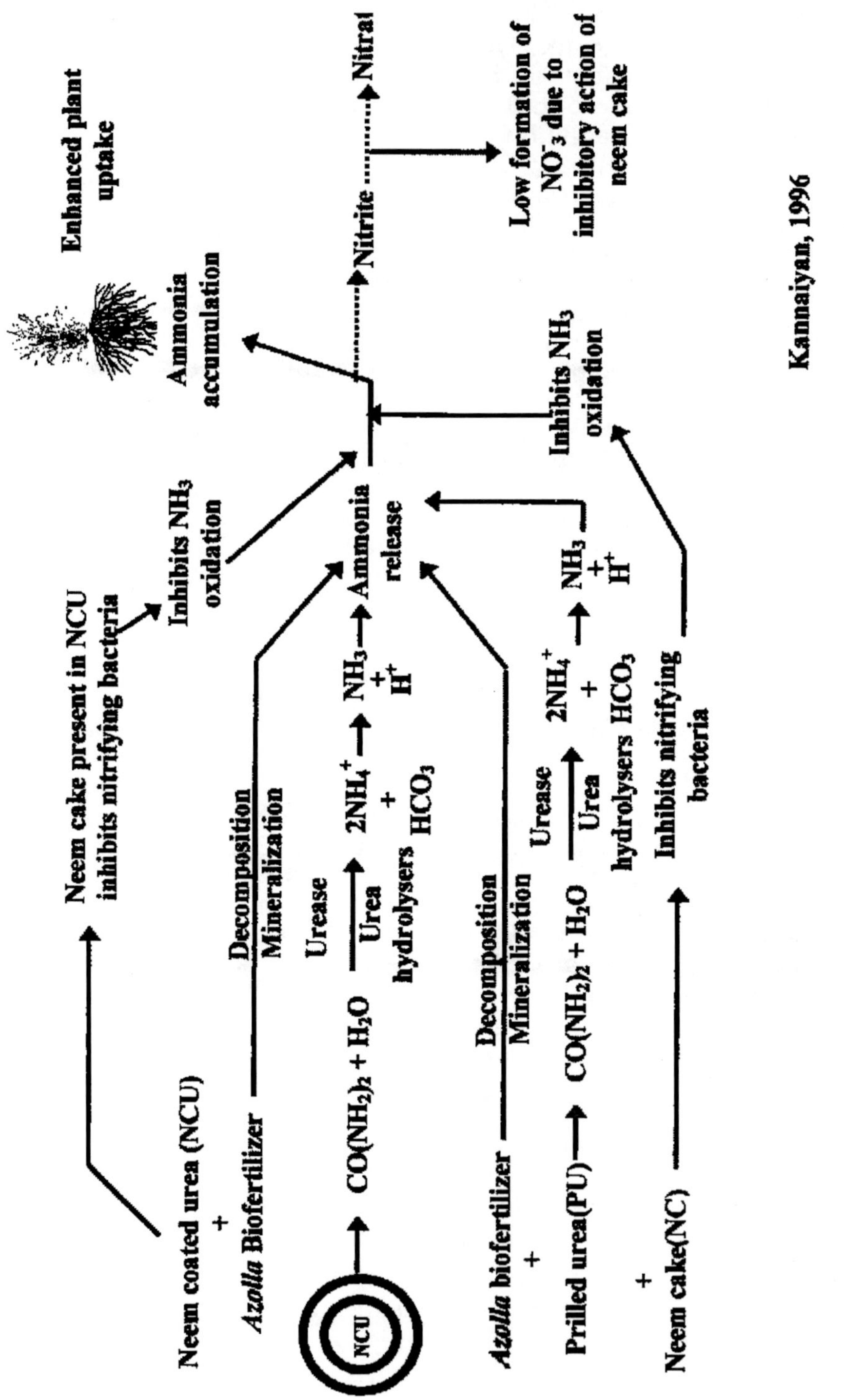

Fig. 1. Role of Neem cake on efficient utilization of Nitrogen in Rice crop

urea with *Azolla* as dual crop, due to the application of 75 kg N ha^{-1} as urea super granule or neem coated urea along with *Azolla* inoculation as dual crop in IR 50 rice during early wet season. The increased yield obtained to the continuous availability of nitrogen for rice crop (Table 5 & 6). From our results, it is evident that the nitrogen fixing green manure plants *viz., Azolla* and *Sesbania rostrata*, biofertilizer like blue green algae could be effectively utilized by incorporation with different forms of urea viz., neem coated urea, urea super granule etc., to increase the rice production under low land system. (Kumar *et al.*, 1995). The role of N_2 fixing manures and other associated systems which contributes nitrogen in rice field ecosystem and their impact on rice crop is given in Fig. 2.

Table 5. Effect of neem coated urea and *Azolla* inoculation on the grain yield of rice (ADT 36)

Treatments	Grain yield (Kg ha^{-1})	% increase over control
Control	2949	
Azolla as dual crop (DC)	3899	32.21
Neem cake 100 kg ha^{-1} + *Azolla* DC	4027	36.55
30 kg N ha^{-1} as prilled urea	3583	21.50
30 kg N ha^{-1} as Neem coated Urea (NCU)	3705	25.64
30 kg N ha^{-1} as NCU + *Azolla* DC	4144	40.52
60 kg N ha^{-1} as prilled urea	4133	40.15
60 kg N ha^{-1} as NCU	4333	46.93
60 kg N ha^{-1} as NCU + *Azolla* DC	4361	47.88
90 kg N ha^{-1} as prilled urea	4278	45.06
90 kg N ha^{-1} as NCU	4766	61.61
90 kg N ha^{-1} as NCU + *Azolla* DC	5149	74.60

Table 6. Influence of *Azolla* and *Sesbania* green manuring along with urea super granule application on the growth and yield of rice (IR 50)

Treatments	Grain yield (Kg ha^{-1})	Straw yield (Kg ha^{-1})
Untreated control	3700	6600
Prilled urea (100 kg N ha^{-1})	4567	8970
Urea super granule (USG) (100 kg N ha^{-1})	5100	11030
A. microphylla (20 t ha^{-1})	4500	7270
A. filiculoides (20 t ha^{-1})	4533	8130
S. rostrata (20 t ha^{-1})	5567	10700
S. aculeata (20 t ha^{-1})	5233	10700
USG + *A. microphylla*	6033	12400
USG + *A. filiculoides*	6267	12630
USG + *S. rostrata*	6633	14300
USG + *S. aculeata*	6267	13530

7. SOIL ENZYMES ACTIVITY

Soil enzymes or soil biocatalysts not only play a major role in the soil chemical and biochemical environment but also affect the rate at which nutrients become available to crop plants as well as to other forms of life in soil. Soil enzyme actvity is considered as an index of microbial activity as well as soil fertility (Burns, 1982). Soil enzymes such as urease and dehydrogenase activity were positively correlated with increasing organic carbon, total nitrogen and total phosphorus content (Barush and Mishra, 1984). Bolton *et al.* (1985) reported an increased activity of certain soil enzymes due to the application of green manures. Frankenberger and Tabatabai (1991) observed significant correlation of L-asparaginase activity with organic carbon and total nitrogen.

The incorporation of *Azolla* has resulted in the increased soil urease activity (Subramani and Kannaiyan, 1989; Thangaraju and Kannaiyan 1989 and Kannaiyan and Kalidurai, 1995). Thanikachalam *et al.*, (1984) have reported that the addition of neem cake, carbofuran,

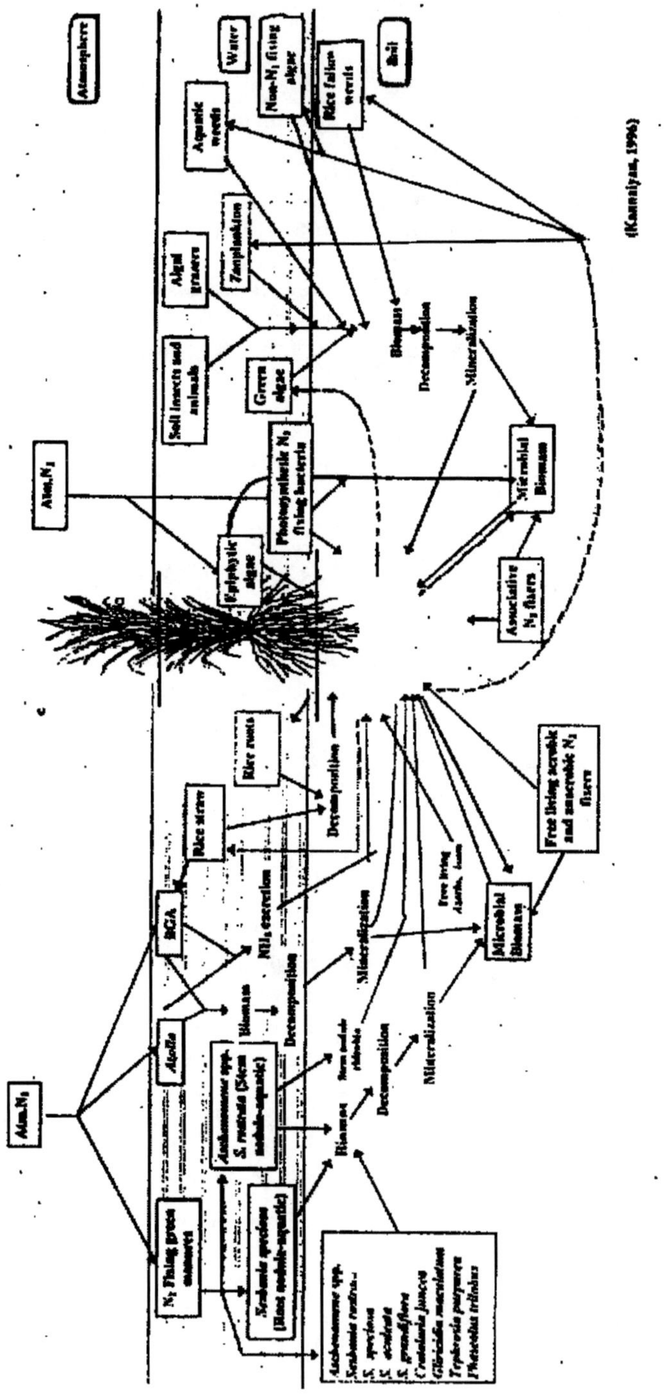

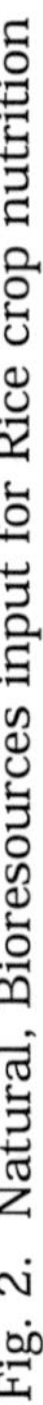

Fig. 2. Natural, Bioresources input for Rice crop nutrition

rice straw and *Azolla* increased the phosphatase activity in flooded soils. Significant increase in the activity of soil enzymes viz., dehydrogenase, phosphatase, cellulase and amylase due to incorporation of N_2 fixing green manures such as *S. rostrata, S. speciosa* and *A. microphylla* has been noticed (Kumar and Kannaiyan, 1992b; and Kannaiyan and Kalidurai, 1995).

8. POTENTIAL OF *AZOLLA*

It is a low cost biofertilizer which can contribute 40-60 kg N ha^{-1} per rice crop. It is symbiotically associated with a nitrogen fixing alga which helps the fern to fix molecular nitrogen. The relative growth rate is very high and the fern doubles its weight in 2-3 days. Because of its rapid growth *Azolla* produces a good biomass in a short periods. Within 20-30 days *Azolla* can multiply and produce biomass ranging from 15-24 t ha^{-1}. The nitrogen content is ranging from 3-5% on dry weight basis which is relatively higher than other green manure plants. Similarly potassium accumulation is between 3-6% on dry weight basis. *Azolla* can easily be grown along with rice in flood water and used as biofertilizer. *Azolla* can be incorporated easily when rice crop is in the field and it decomposes in the field rapidly and nitrogen is made available to rice plants. Since it produce spores i.e., microsporocarps and megasporocarps, the spores can with stand adverse conditons during summer when *Azolla* is not in the field and germinate and regenerate during cultivation cycle of rice crop. Therefore, once *Azolla* is introduced in a rice it can sustain and act as a renewable biofertilizer.

Besides nitrogen contribution to rice, *Azolla* also controls the weed flora in wet land rice which is an economic advantage for rice growers, the residual effect of *Azolla* is useful for subsequent crops grown after rice. In place where water scarcity is acute. Azolla can putforth roots in rice field and act as a mulch and conserve the moisture for 20-30 days. Besides its utilization as biofertilizer *Azolla* is also used as poultry and cattle feed since it has high protein and mineral content.

9. LIMITING FACTORS

Water is a pre-requisite for *Azolla* multiplication. *Azolla* cannot withstand desiccation. It cannot be used in upland rice. The initial *Azolla* inoculum requirement as fresh biomass is between 500-1000 kg ha^{-1}. Therefore fresh *Azolla* inoculum is to be maintained in nursery plots throughout the year. Storage value of fresh *Azolla* is also low and hence the fern in difficult to be transported to long distances. An irrigation net work and a net work for inoculum conservation, production and distribution are highly essential for *Azolla* utilization. Phosphorus is the most important element for *Azolla*. High temperature inhibits the growth. Temperature higher than $34 \pm 1^{\circ}C$ in day time is detrimental to *Azolla* growth. Pests likes pyralis and snails (*Limnaea acuminata*) affect *Azolla*. Black rot disease caused by *Rhizoctonia solani* is a serious one affecting *Azolla* in the multiplication plots.

10. CONCLUSION

Azolla is a potential nitrogen fixing aquatic fern for nitrogen contribution to rice crops under low cost rice production technology. It is renewable biofertilizer and the inoculated *Azolla* in a particular rice growing area survive for a long period through their spores and regenerate and spread naturally in rice growing areas. It is an ecological advantage for *Azolla* in wet land rice field ecosystem for their survival and subsequent development. The development of *Azolla* spore inoculants is under intensive field study and the success in this important area would eliminate the use of large fresh biomass for initial inoculation.

REFERENCES

Alexander, D., Sadanandan, N. & Karunakaran, K. 1992. Evaluation of *Azolla* as an organic matter for rice production. In : New Trends in Biotechnology (eds.) N.S. Subba Rao, C.Balagopalan and S.V. Ramakrishan, Oxford and IBH Publ. Co., New Delhki. pp. 255-262.

Baratharkur, H.B., & Talukdar, H. 1983. Use of *Azolla* and commercial nitrogen fertilizer in Jorhat, India. Inte. Rice Res. Newslett. 8(2) : 20.

Barush, M., & Mishra, R.R. 1984. Dehydrogenase and urease activities in rice field soils. Soil Biol. Biochem. 16 : 423-424.

Bolton, H. Jr., Elliot, L.F. Papendick, R.I. & Bezdicek, D.F. 1985. Soil microbial biomass and selected soil enzyme activities. Effect of fertilization and cropping practices. Soil Biol. Biochem. 17 : 297-302.

Burns, R.C. 1982. Enzyme activity in soil. Location and possible role in microbial ecology. Soil Biol. Biochem. 14 : 107-108.

Craswell, E.T., & De Datta, S.K. 1980. Recent developments in research on nitrogen fertilizers for rice. IRRI Research paper Series 49, pp. 11.

De Datta, S.K. 1986. Improving nitrogen fertilizer efficiency in lowland rice in tropical Asia. Fert. Res. 9 : 171-186.

De Datta, S.K., I.R.P. Fillery & E.I. Craswell. 1983. Results from recent studies on nitrogen fertilizer efficiency in wetland rice. Outlook on Agriculture. 12 : 125-134.

Diara, H.F., van Brandt. H. Diop., A.M. & Van Hove C. 1987. *Azolla* and its use in rice culture in West Africa. In : *Azolla* utilization. Proc. Workship on *Azolla* use, Fuzhou, Fujian, China, Inte. Rice Res. Inst., Los Banos, Philippines. pp. 147-152.

FAO. 1977. FAO soil Bull., 40 : 29-38.

Frankenberger, W.T. Jr. & Tabatabai, M.A. 1991. Factors affecting L-asparaginase activity in soils. Biol. Fertil. Soils. 11 : 1-5.

Gopalaswamy, G. & Kannaiyan, S. 1996a. Effect of growth regulator GA_3 on *Azolla*. Phykos 35 (1&2) : 183-188.

Gopalaswamy, G., & Kannaiyan, S. 1996b. Effect of inoculation of *Azolla* hybrids with fertilizer nitrogen on rice. 37th Ann. Conf. Assoc. Microbiologists India, p. 132 (Abstr.)

Gopalaswamy, G., & Kannaiyan, S. 1996c. Effect of different levels of phosphorus and sodium on *Azolla* biomass. 37th Ann. Conf. Assoc. Microbiologists, India. p. 132 (Abstr.)

Gopalaswamy, G., & Kannaiyan, S. 1996d. Inoculation of *Azolla* hybrids for rice. Paper presented in the X southern Regional Conference Microbial inoculants held during 10-11 Dec. at A.V.V.M. Sri Pushpam College, Poondi, Thanjavur.

Gopalaswamy, G., Anthoni Raj, S. Kempuchetty, N. & Abdul Karrem, A. 1994. Depression of soil microflora of continuous herbicide application. 34th Annual Conf. Assoc. Microbiologists India, Punjab, Agric. Univ., Ludhiana, India. p. 161 (Abstr.)

Jayanthi, M., Rajagopal, V. & Shantaram, M.V. 1994. Studies on conjuctive use of *Azolla* and fertilizer in rice cultivation, Natl. Seminar on *Azolla* and algal biofertilizers for rice, Tamil Nadu Agricultural University, Coimbatore, Tamil Nadu, India p. 23 (Abstr.)

Jotwani, M.G., & Srivastava, K.P. 1981. Neem insecticide of the future : Chemistry, toxicology and future strategy. Pesticide. 15 (2) : 12-14.

Kannaiyan, S. 1981. *Azolla* biofertilizer for rice. INSFFER Training Seminar, Int. Rice Res. Inst., Manila, Philippines p. 1- 11.

Kannaiyan, S. 1982. *Azolla* and rice. In : Multiplication and use of *Azolla* biofertilizer for rice production, Training pp 1-56. Tamil Nadu Agric. Univ., Coimbatore, India.

Kannaiyan, S. 1985. Potentially of *Azolla* biofertilizer for rice, In : Soil Biology (eds.) M.M. Mistra and K.K. Kapoor, Haryana Agric. Univ., Hisar, India pp. 265-259.

Kannaiyan, S. 1989a. Inte. Rice Res. News Let. 14(5) : 21.

Kannaiyan, S. 1989b. *Azolla* propagation. In : Organic recycling in Asia and Pacific. FAO, Bangkok, Thailand, RAPA Bulletin 5 : 5-7.

Kannaiyan, S. 1990. Paper presented at the symposium on Biotechnology and International Trade. 1992 and Beyond.

Kannaiyan, S. 1992a. Potentiality of *Azolla* as biofertilizer in sustainable rice farming. In : Biodiversity Implications for global food security (eds.) M.S. Swaminathan and S. Jana, MSS Res. Foundation, Madras pp. 221 - 233.

Kannaiyan, S. 1992b. *Azolla*, biofertilizer technology for rice crop. Tamil Nadu Agric. Univ. Coimbatore, Tamil Nadu, India, p. 62.

Kannaiyan, S. 1993a. Nitrogen contribution by *Azolla* to rice crop. Proc. Indian Natl. Science. Acad., 59 (5 & 6) : 21-31.

Kannaiyan, S. 1993b. Biofertilizers for rice. Tamil Nadu Agric. Univ., Coimbatore, Tamil Nadu, India p. 60.

Kannaiyan, S. 1994. *Azolla* Biotechnology. Biotechnology in India. Spl. Vol. 15 - 24.

Kulasooriya, S. A. 1985. Tech. Document, Inte. Atomic Energy Agency, Vuenna, p. 129.

Kannaiyan, S. 1995. Biofertilizers for rice status and scope. In : Rice Management Biotechnology (ed) S. Kannaiyan Assoc. Pub. Co., New Delhi, pp. 237-265.

Kannaiyan, S., & Somporn, C. 1987. Studies on sporulation, biomass production and nitrogen fixing potential of eleven cultures of the aquatic fern *Azolla*. Indian J. Microbiol. 27 : 22-25.

Kannaiyan, S., & Goyal S.S. 1988. Effect of light intensity and salt stress on the growth and ammonia excretion by algal symbionts of *Azolla*. Paper presented in 29th Annu. Microbial. Conf. Haryana Agri. Univ. Hissar, Haryana India.

Kannaiyan, S., & Kalidurai, M. 1995. Nitrogen fixing potential of stem nodulating *Sesbania rostrata* and its effect on rice yield. In : Rice Management Biotechnology (ed.) S. Kannaiyan, Assoc. Pub. Co., New Delhi, pp. 173-195.

Kannaiyan, S., Thangaraju, M. & Oblisami, G. 1981. Mineral nutrition of *Azolla*. Tamil Nadu Agric. Univ. Newslett. 8 : 2.

Kannaiyan, S. Thangaraju, M. & Oblisami, G., 1982. Inte. Workshop on practical application of *Azolla* for rice Production Univ. Pureto rico Hagagez Campus, Pureto Rico.

Kannaiyan, S., Thangaraju, M. & Oblisami, G. 1984. Effect of *Azolla* application on grain yield of rice. Agric. Res. J. Kerala 22 : 183-185.

Kannaiyan, S., Sopko, B. Rao K.K. & Hall, D.O. 1992. Ammonia excreting by the algal symbiont *Anabaena azollae* (AS-DS). In : BNF and Biogas Technology (eds) S. Kannaiyan, K. Ramaswamy, K. Kumar & K. Ilamurugu, Tamil Nadu Agri. Univ. Coimbatore, Tamil Nadu, India. pp. 12-15.

Kamire, K.K., & Sonar, K.R. 1979. Mineralization of urea- N with 'Neem' (*Azadirachta indica*) and 'Karanj' (*Pongamia glabrarent*) cakes. J. Maharashtra Agric. Univ. 4 : 10-13.

Kulasooriya, S.A., Hirmburegama, W.K. & Abeysekara S.W. 1987. Use of *Azolla* in Sri Lanka. In : *Azolla* Utilization. Proc. Workshop on *Azolla* use. Fuzhou, Fujian, China, Inte. Rice Res. Inte., Los Banos, Phillipines, pp. 131-140.

Kumar, K. 1994. Studies on the development of *Azolla* mutants and their nitrogen fixing potential, microbial decomposition and biofertilizer value for rice. Ph.D. Thesis, Tamil Nadu Agrl. Univ., Coimbatore, Tamil Nadu, India, p. 323.

Kumar K. & Kannaiyan, S. 1992. Changes in the activity of soil enzymes during decomposition of N fixing green manures. 33rd Annual Conference of the Association of Microbiologists of India. Goa University, Goa, (Abstr.) P. 57.

Kumar, K., Gopalaswamy, G. & Kannaiyan, S. 1992. Effect of incorporation of nitrogen fixing plants *Azolla microphylla* and *Sesbania rostrata* with nitrification inhibitors neem cake and dicyandiamide. In : Biological Nitrogen Fixation and Biogas Technology (eds.) S. Kannaiyan, K. Ramaswamy, K. Ilamurugu and K. Kumar, Tamil Nadu Agric. Univ. Coimbatore, Tamil Nadu, India pp. 49-53.

Kumar, K. Thangaraju, M. & Kannaiyan, S. 1995. Use of nitrification inhibitors and nitrogen use efficiency in low land rice soil. In : Rice Management Biotechnology (ed.) S. Kannaiyan, Assoc. Publ. Co., New Delhi, pp. 101-100.

Laddavan, L. 1985. Inte. Workshop on *Azolla* use, Fujian Acad. Sci. Fujian, China.

Latha, K.R., Kannaiyan, S. & Subramanian, S. 1988. Uptake of nitrogen in different growth phases of rice through *Azolla* biofertilizer. Acta Agron. Hungarica. 37 (3-4) : 239-243.

Latha, K.R., Subramanian, S. Thangaraju, G.S. & Kannaiyan, S. 1992. Response of rice to *Azolla* inoculation and fertilizer nitrogen. Agric. Res. J. Kerala. 30 : 21-27.

Liu Chung Chu. 1987. Re-evaluation of *Azolla* utilization in agricultural production. In : *Azolla* utilization Proc. on the Workshop on *Azolla* use. Fuzhou, Fujian, China, Inte. Rice Res. Instt., Philippines, pp. 67-76.

Loudhapasitiporn, L. & Kanareugsa, C. 1987. *Azolla* use in Thailand. In : *Azolla* utilization Proc. on the workshop on *Azolla* use. Fuzhou, Fujian, China, Inte. Rice Res. Instt., Philippines, pp. 119-122.

Mabbayad, B.B. 1987. The *Azolla* program of the Philippines. In : *Azolla* Utilization Proc. on the Workshop on *Azolla* use. Fuzhou, Fujiamn, China, Inte. Rice Res. Instt., Philippines, pp. 101-108.

Mahesh, G., & Kannaiyan, S. 1994. Influence of certain growth regulators on the germination and fertilization of sporocarps *Azolla microphylla*. Paper presented in the National Seminar on *Azolla* and Algal biofertilizer for rice. Tamil Nadu Agric. Univ., Coimbatore, India, 1-3, March, 1994 (Abstr.) p. 16.

Manickam, T.S., Selvaseelan, A. & Krishnamurthy, K.K. 1976. Nitrification of urea treated with nitrification inhibitors, coated fertilizers and slow acting fertilizers. Fert. News. 21(9) : 64-65.

Mikkelsen, D.S., De Datta, S.K. & Obcemae, W.N. 1978. Ammonia volatilization losses from flooded rice soils. Soil Sci. Soc. Am.J., 442 : 725-730.

Moore, A.W. 1969. *Azolla* biology and agronomic significance. Bot. Rev. 35 : 17-35.

Patil, P.L., & Mandhare, V.K. 1991. Effect of direct incorporation of *Azolla* with nitrogen levels on yield of rice. 31st Annu. Conf. Assoc. Microbiol. India, Tamil Nadu Agric. Univ., Coimbatore, Tamil Nadu, India, p. 135 (Abstr.)

Peters, G.A., Mayne, B.C. Ray, T.B. & Toia, R.E. 1979. Physiology and biochemistry of the *Azolla Anabaena* symbiosis. In : Symbiosis. In : Nitrogen and Rice, pp. 325-344. Inte. Rice. Res. Inst. Los Banos, Laguna. Philippines.

Rajagopal, V., Jayanthi, M. & Shantaram, M.V. 1994. Growth studies on *Azolla* under Telangana conditions of Andra Pradesh. Natl. Seminar on *Azolla* and algal Biofertilizer for rice, Tamil Nadu Agric. Univ., Coimbatore, Tamil Nadu, India, p. 15 (Abstr.)

Roger, P.A., Kulasooriya, S.A. Third A.C. & Craswell, E.T. 1980. Deep placement of nitrogen fertilizer application, compatible with algal nitrogen fixation in wetland rice soils. Plant and soil. 57 : 137-142.

Savant, N.K., Craswell, E.T., & Kiamond, R.E. 1983. Use of urea super granules for wetland rice : A Review fert. News. 28 (8) : 27-35.

Shanmugasundaram, R. & Kannaiyan, S. 1992. Effect of growth regulator and fungicide combination on germination, fertilization and viability of the dried sporocarps of *Azolla microphylla*. Phykos. 37 (1 & 2) : 61-67.

Shanmugasundaram, R., & Kannaiyan, S. 1994. Studies on the development of *Azolla* and algal biofertilizer for rice. Tamil Nadu Agric. Univ., Coimbatore, India (Abstr.) 0.17.

Sharma, S.K. 1995. Urea Super Granules - Its performance and prospects for rice production. In : Rice Management Biotechnology. (ed.) S. Kannaiyan, Assoc. Publ. Co., New Delhi, pp. 117-134.

Sharma, N.N. & Sinha, U.P. 1977. Neem cake blending of urea for raising efficiency of applied nitrogen to wheat. Indian. J. Agron. 22 : 106-107.

Simpson, J.R., Freney, J.R. Muirhead, W.A. & Leuning, R. 1985. Effect of phenyl phosphorodiamidate and dicyandiamide on nitrogen loss from flooded rice. Soil. Sci. Soc. Am. J. 49 : 1426-1431.

Singh, P.K. 1977. Multiplication and utilization of fern *Azolla* containing nitrogen fixing algal symbiont, a green manure in rice cultivation.Ilrizo. 26 : 125-137.

Singh, P.K. 1979. Use of *Azolla* in rice production in India. In : Nitrogen and rice, Inte. Rice Res. Instt., Manila, Philippines. p. 407-418.

Singh, A.L. & Singh, P.K. 1989. N_2 - fixation in Indian rice field (*Azolla* and Blue green algae). Agrobotanical Publisher, Bikaner, India. P. 236.

Srinivasan, G. 1980. *Azolla* manuring and grain yield of rice. Inte. Rice Res. Newslett. 6 : 25.

Subbiah, S., & Kothandaraman, G.V. 1980. Effect of neem cake blending of urea on ammonium and nitrate N content in soil under field conditions. J. Indian. Soc. Soil Sci. 28 : 136-137.

Subramani, S. & Kannaiyan, S. 1989. Effect of the water fern *Azolla* on enzyme activity in soil under rice cultivation. Mikrobiologiya, 58 (1) : 118-121.

Subramani, S., & Kannaiyan S. 1987. Effect of incorporation of *Azolla* biofertilizer on the grain yield of rice. J. Agron. And Crop Sci. 159 : 308-311.

Talley, S.N., Talley, B.J. & Rains D.W. 1977. Nitrogen fixation by *Azolla* in rice fields. In : Genetic engineering for nitrogen fixation (ed.) Alexander Hollaender. Plenum Publishing Company, New York, USA, pp. 250-282.

Thangaraju, M., & Kannaiyan, S. 1989. Studies on the urease activity status of soil by incorporating certain *Azolla* cultures. 29th Ann. Conf. Assoc. Microbiologists of India, HAU, Hissar (Abstr.) P. 87.

Thangaraju, M. & Kannaiyan, S. 1990. Use of nitrification inhibitors with *Azolla* biofertilizer and fertilizer on nitrogen use efficiency in rice. In : Biotechnology of Biofertilizers for Rice Crop (ed.) S. Kannaiyan, Tamil Nadu, India pp. 131-140.

Thanikachalam, A., Rajakannu, K. & Kannaiyan, S. 1984. Effect of neem cake, carbofuran and *Azolla* application on phosphatase activity in soil. 25th Ann. Conf. Assoc. Microbiologists of India, Pant Nagar, (Abstr.) p. 7.

Watanabe, I., Ke-Zhi, B. Berja, N.S. Espinas, C.R. Ito, O. & subudhi, B.P.R. 1981. Inte. Rice Res. Inst. Research Paprt Sceries, No. 69, Phillippines.

Watanabe, I. 1982. Microbiology of tropical soils and plant productivity (eds.) Y.R. Dommegues and H.G. Oiem, Hague : Martinus Nijhoff Dr. W. Jank Publishers. Pp. 169-185.

Watanabe, I. 1987. Fertility evaluation for rice. In : *Azolla* Utilization. Proc. on the workshop on *Azolla* use. Fuzhou, Fujian, China, Inte. Rice Res. Instt., Philippines, pp. 197-208.

Watanabe, I. 1990. Int. Symp. on rice res., New Frontiers. Rajendra Nagar, Hyderabad, India, p - 1-18.

Zhang, Zhuang-Ta, Ke Yu-Si, Ling De-Quan, Duan Bing-Yuan & Liu Xi-Lian. 1987. Utilization of *Azolla* in agricultural production in Guandong Province, China. In : *Azolla* Utiliztion. Proc. on the Workshop on *Azolla* use. Fuzhou, Fujian, China, Inte. Rice. Res. Instt., Philippines, pp. 141-145.

CHAPTER – 5

PHOSPHATE SOLUBILIZING ACTIVITY OF MICROBES AND THEIR ROLE AS BIOFERTILIZER

Jyoti Saxena and Vinay Sharma

Department of Bioscience and Biotechnology, Banasthali Vidyapith,
P.O. Banasthali Vidyapith, Raj. - 304 022

1. INTRODUCTION

Phosphorus occupies a critical position, both in plants and in the biology of the soil. It is a macronutrient required for plant growth. Pierre (1938) referred to it as the 'master key' element in crop production. Phosphorus cycle is mainly sedimentary cycle as no appreciable amounts of gaseous phosphorus compounds are involved, thereby restricting it to the lithosphere and hydrosphere only. The availability of soluble phosphate from weathering of apatite containing rocks may initially have been the rate-determining factor in early life development. In most ecological systems the phosphate content is the limiting factor for growth.

Phosphorus is not found free in nature and almost always occurs in fully oxidised state as phosphate. The element is widely distributed as phosphate in soils, rocks, oceans, living cells and in most other materials (Fig. 1). Though most soils contain insoluble inorganic phosphates or organic phosphorus, these are of no use to crops unless either solubilized or mineralized. Around 95.99% of total phosphates are insoluble which are not directly available as phosphorus to the crops.

Generally plants take up their phosphorus not only as H_2PO_4 but also as HPO_4^{-2} in small quantities (Beever and Burns, 1980). Unfortunately, when P is applied to the soil as fertilizer, it quickly combines with other chemicals, forming compounds, which are not available to plants. Only about 25% of water soluble phosphate is taken up by plants in the year of its application and the remaining portion is converted into insoluble forms due to interaction between added soil phosphates and soil constituents (Bhattacharyya and Jain, 1996).

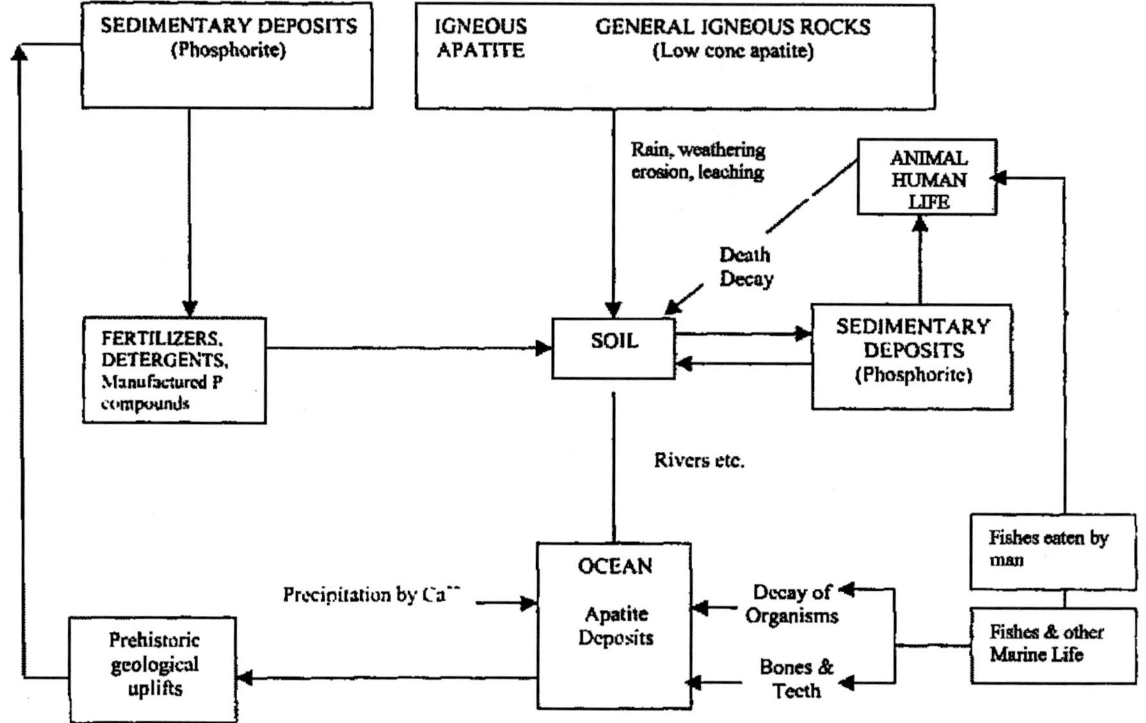

Figure 1. Natural and Artificial Cycles of Phosphorus

Many soil microorganisms have the ability to solubilize mineral phosphates in pure culture conditions by the secretion of extracellular organic compounds (Duff and Webley, 1959; Bajpai and Sundara Rao, 1971; Banik and Dey, 1982; Gaur, 1987). These phosphate solubilizing microbes (PSMs) are believed to play an important role in supplementing P to plants, thus allowing the sustainable use of phosphate fertilizers. Out of several PSMs known, bacteria and fungi are the most important ones residing in the root region of plants. These efficient microorganisms have shown consistency in their capacity to solubilize fixed soil phosphorus from different sources (Gaur, 1972). These properly selected PSMs along with low-grade rock phosphates can be used quite economically in place of chemically synthesized costly fertilizers as rock phosphate has been reported to be efficiently solubilized by soil microorganisms.

Phosphorus is estimated to be in insufficient amounts in most of the Indian Soils as available P. According to one of the recent compilations in 1994 based on soil tests for available P in Indian soils, 43.9% of areas are in the low category, 48.8% are in the medium and only 1.9% has high P status (Hasan, 1994).

In view of the above facts there is an increasing realization in India and throughout the world to explore the possibilities of efficiently utilizing the less available P source and PSMs offer a viable alternative. Their use as biofertilizer is environment friendly and serves as low cost agricultural input.

2. DISTRIBUTION OF PHOSPHORUS IN NATURE

More than 200 different phosphate minerals are known but only those in the apatite group occur in sufficient abundance and concentration to serve as commercial sources of the element. These are distributed in rocks in different parts of the globe. The commonest apatite deposits mainly consist of fluorapatite [$Ca_{10}(PO_4)_6F_2$], chlorapatite [$Ca_{10}(PO_4)_6C_{12}$] and hydroxyapatite [$Ca_{10}(PO_4)_6(OH)_2$].

Apatatic phosphate rocks occur mostly as sedimentary deposits, termed as phosphorite. The largest and most important phosphorite deposits are found in Morocco, U.S.A. (Florida and N. Carolina), Kazakhstan, China and Tunisia. It also occurs as igneous phosphate rock but less commonly. This form is highly crystalline and much purer than sedimentary phosphorite. Commercially important igneous rock formations of crystalline fluorapatite are found in the Kola Peninsula of USSR, the Transvaal and Brazil but this account for only about 15% of the world total of mined rock. At present, Morocco is the world's largest exporter and probably has the largest phosphorite reserves, which are economically workable with present technology. Present rate of world usage exceeds 100 million tons per annum with about 90% of this being put to fertilizer use. Total known apatite reserves have been estimated to last, at this rate of consumption, for a period of at least 1000 years. A second source of phosphate is guano, a natural deposit formed from decaying bones and excreta from fish-eating birds. Guano deposit are found in Chile, Peru and Seychelles but it accounts for less than 10% of the world's phosphate production. The third and the minor source of phosphorus is basic slag, a waste product from blast furnaces operating on iron ores.

3. PHOSPHORUS SOLUBILIZING MICROORGANISMS

A fairly large volume of information has accumulated over the years on the characteristics of phosphate solubilizing microbes (PSMs). These studies have been conducted under laboratory as well as under green house conditions in pot culture and field experiments. There are several reports available about the isolation of various PSMs from the rhizosphere, non-rhizosphere, rhizoplane and also from other environments such as rock phosphate deposits, marine environments etc. (Rakade and Patil, 1992; Thakkar *et al.*, 1994).

Many bacteria, cyanobacteria and fungi present in soil are reported to solubilize phosphate from inorganic phosphatic compounds and make then available to plants. In Indian alluvial soils, 14.3%, 35% and 30% of the total bacterial, actinomycetes and fungal population, respectively possessed the ability of solubilizing $Ca_3(PO_4)_2$ (Banik and Dey, 1982). Further, a significant positive correlation between the number of the above phosphate solubilizers, the total number of soil microorganisms and level of total phosphate in soil, which showed a linear relationship with each other, was also observed. The PSMs isolated include fungi from the genus *Aspergillus*, *Penicillium* and *Chaetomium*, bacteria from the genus *Bacillus*, *Micrococcus* and *Arthrobacter* and actinomycetes from the genus *Streptomyces* (Banik and Dey, 1982; 1983). *Trichoderma viride* has also been described recently which proved to be more efficient than *Bacillus megatherium* and *A. awamori*. The other two species of *Trichoderma*, *T. harzianum* and *T. psedokoningii* also solubilize to a lesser extent (Anusuya and Jayarajan, 1998). *A. candidus*, *A. fumigatus*, *A niger*, *A awamori* and *A. japonicus* from the genus *Aspergillus* and *P.*

simlicissimum and *P. aurartiogriseum* from the genus *Penicillium* solubilize phosphate very efficiently (Gaur and Sachar, 1980; Banik and Dey, 1982; Singal *et al.*, 1994).

Narsian *et al.*, (1994) isolated 27 Aspergilli, 7 Penicillia and 1 *Rhizopus* sp. from rhizosphere of 24 crop plants, compost and garden soils in India. Nahas (1996) isolated and tested 31 bacteria and 11 fungi from soils of Brazil, out of which 8 bacteria viz., 2 strains of *Bacillus subtilis*, 3 strains each of *B. licheniformis, B. macerans* and *B. megatherium, Pseudomonas* sp. and 8 fungi viz, *Aspergillus niger, A. ochraceous, Eupenicillium shearli, Penicillium implicatum, P. minioluteum, P. viridicatum* and *P. purpurogenum* showed the activity, *Pseudomonas* and *P. purpurogenum* being the most efficient phosphate solubilizers. *Thiobacillus ferroxidans* was proved to be very efficient rock phosphate solubilizer (Mahapatra *et al.*, 1985). Mba (1996) isolated nine strains of *Streptosporangium* from casts of tropical earthworm, which has high RP solubilizing activity. *Tolypocladium geodes* were found first time from Scottish soils by Jones *et al.*, (1991), which showed good capacity as iron phosphate solubilizer.

4. FACTOR DETERMINING THE PHOSPHATE SOLUBILIZATION

(a) Type of insoluble phosphates

The type of insoluble phosphate present in soil may also affect the solubilization capacity of different microorganisms. Apatite is the most common type of insoluble phosphate found in nature. Inorganic P also occurs as poorly soluble Ca, Al or Fe phosphates viz., tricalcium phosphate (TCP) [$Ca_3(PO_4)_2$], dicalcium phosphate (DCP) [$CaHPO_4$, $2H_2O$], rock phosphate (RP) (Ca phosphate), variscite [$AlPO_4$, $2H_2O$] and $FePO_4$. Calcium phosphate are dominant in neutral to alkaline soil while iron and aluminium phosphates occur in acidic soil. These have to be mineralised to be in assimilable forms before plants can take them. Bardiya and Gaur (1974) studied RP solubilization by some very efficient strains of *Aspergillus, Penicillium* and *Bacillus.* All of these showed lesser extent of activity as compared to the relative of PSMs using different phosphates. This finding indicates that solubilization of phosphates varies greatly with the nature of phosphate and the organisms. Banik and Dey (1982) reported the highest solubilization from $Ca_3(PO_4)_2$ followed by $AlPO_4$ and $FePO_4$ by *A. candidus, A. fumigatus, Bacillus megatherium* and *Streptomyces* species. *Aspergillus japonica, A. foetidus* and *Thiobacillus ferroxidans* were better solubilizers where low-grade rock phosphate were mixed with pyrites (Mahapatra *et al.* 1985; Singal *et al.*, 1994). Illmer *et al.*, (1995b) reported solubilization of hardly soluble $AlPO_4$ with *Pseudomonas* sp., *Penicillium simplisissimum* and *P. aurantiogriseum.* In hyroponic studies with these microorganisms, the plant *Lepidium* and $AlPO_4$ led to an increase in dry weight and Al-content of plants.

Three forms of most commonly found insoluble phosphates viz., DCP, TCP and hydroxyapative were chosen to see the effect of solubilization by 2 strains of *Aspergillus niger* and *Penicillium claviformis* and *P. aurantiogriseum.* DCP was least efficiently solubilized by all the strains whereas TCP was solubilized maximally (Basu, 1999). According to Narisian *et al.* (1994) among the various phosphates tested for solubilization by *Aspergillus* and *Penicillium* sp, TCP supported the highest activity; DCP was the next best for all the fungi, followed by $AlPO_4$ and $FePO_4$. Rock phosphate was solubilized to a lesser extent than TCP but was superior to bone meal as a phosphate source. Arora and Gaur (1979) tested a range of bacteria and fungi for their ability to solubilize TCP, DCP, hydroxyapatite and RP. Species of *Pseudomonas, Bacillus, Escherchia, Aspergillus* and *Penicillium* effectively solubilized TCP and hydroxyapatite. Rock phosphate was more resistant to solubilization, relatively small amounts being solubilized by most species. Fungi were more effective than bacteria in general. The studies on different types of RPs show that Jordan RP solubilized more than Indian RP (Singh *et al.*, 1984). Six strains of *A. awamori* were compared for their phosphate (AP), and four types of RP. Four strains preferred TCP while 2 strains AP. Different RPs were utilized in the following decreasing order : Sonarai RP Udaipur RP Hirapur RP (Narsian *et al.*, 1993). The effect of soluble phosphate concentration on the solubilization of fluorapatite by *A. niger* was studied by Nahas and Assis (1992). Acid phosphatase production and fluorapatite solubilization were decreased when the concentration of soluble phosphate was enhanced. It appears that the process of solubilization by fungi is governed by external phosphate levels as also observed for phosphatases.

Besides the type of phosphates, their concentration also plays an important role. PS activity with TCP decreased with the rise in its concentration (Narsian *et al.*, (1993). Gaur and Sachar (1980) studied optimum amount of RP to be applied for maximum PO_4^{-3} dissolution. The results showed that lower the quantity of PO_4^{-3} supplied, greater was the conversion percentage. The treatment receiving lowest quantities of phosphate remained more acidic than treatments receiving higher amount of RP.

(b) Time of incubation

Time of incubation also affects the phosphate solubilization in *in vitro* conditions. Narsian *et al.*, (1994) reported that *A. aculeatus* solubilized 94% TCP on 4th day of incubation which was far more than that of *A. niger* which solubilized 49.6% on 7th day. TCP solubilization data also showed that the highest phosphate solubilization (93%) occurred on the 2nd day, after which the values fluctuated but showing invariably greater activity on 4th and 6th day compared to those on 1st, 3rd and 5th day. The drop in PS activity on 3rd day may be due to the availability of soluble form of phosphate that has a reducing effect on phosphate solubilization (Darmval *et al.*, 1989). The pH of the medium fell from 7 to 4.6 after 24 hours and remained acidic thereafter. Similar observations were made by Illmer and Schinner (1992) with the species of *Pseudomonas* and *Penicillium* in presence of DCP. The growth increased up to 4th day after which there was a slight decrease and then it remained unaltered during the later period of incubation. Several studies on PS activity of Aspergilli and Penicillia showed that *Penicillium* spp. released phosphate early from RP than *Aspergillus* spp. (Chhonkar and Subba Rao, 1967; Agnihotri, 1970; Asea *et al.*, 1989). *Bradyrhizobium* and *Rhizobium* spp. have been reported to show a sigmoid increase in solubilized phosphate and concomitant decrease in phosphate solubilization of medium from 0 to 6 days of inoculation as seen in *A. niger* Basu (1999) reported that *A. niger* strains solubilized DCP, TCP and hydroxyapatite very efficiently for the 1st 6 days, and then the rate of solubilization decreased and finally attained an almost constant value-giving rise to sigmoid curve. The *Penicillium* spp. did not give this smooth sigmoid graph. They gave rather zigzag curves having a number of peaks showing a few increases and decreases in soluble phosphate.

(c) Morphology and growth characteristics

Solubilization of RP appeared to be related with their growth type in broth. *Bacillus firmus* strains producing an orthodox mucoid growth were found to be better solubilizing agents than the strains producing a turbid and mucoid growth.

Also, PS activity of *A. fumigatus* and *A. candidus* was intensified with sporulation, which was later found to be true for *Streptomyces* sp. This may be due to greater demand of phosphorus during sporulation (Banik and Dey, 1981, 1982).

(d) Type of energy sources

Carbon source for the active proliferation and organic acid production by microorganisms is an important factor in phosphate dissolution. The effect of varying dosage of glucose was investigated on RP solubilization by *A. awamori*. It increased due to addition of greater quantities of energy source. Maximum solubilization was brought about 3% glucose (Gaur and Sachar, 1980).

Roos and Luckner (1984) also showed that *P. cyclopium* solubilized phosphate in presence of glucose in medium. Gaur *et al.*, (1979) studied the effect of varying degree of concentration of glucose on RP solubilization by *A. awamori*. The PS activity of fungus *A. aculeatus* in TCP was seen in presence of 11 carbon compounds individually and the results indicate that the replacement of glucose invariably resulted in a decreased PS activity which occurred in the following order : glucose > mannitol > sucrose = xylose > glycine > maltose > sorbitol > arabitol > arabinose > fructose > galactose > lactose. Goel (2000) reported that *A. awamori* preferred sucrose followed by glucose, maltose and mannitol whereas *A. aculeatus* preferred maltose followed by glucose, sucrose and mannitol.

(e) Type of metal ion

There are some reports, which show that metal ions also seem to have effect on PS activity of microbes. Sayer *et al.* (1995) observed the activity of *P. simplicissimum* and *A. niger* in presence of eight insoluble metal compounds. Almost one-third of the natural isolates screened solubilized at least one of ZnO, $Zn_3(PO_4)_2$ and $CO_3(PO_4)_2$ with five strains being capable of solubilizing all the three. Cobalt phosphate was the most toxic of the three compounds, zinc phosphate was the least toxic and most resistant to solubilization while zinc oxide was the most readily solubilized compound. Effect of Mn^{+2}, Al^{+3} and Fe^{+3} was reported by Basu (1999). The activity of *A. niger* decreased in presence of Mn^{+2} and Fe^{+3} whereas Al^{+3} did not have any significant effect on it. According to Gaur and Sachar (1980) Al^{+3}, Fe^{+3} and Ca^{+2} may be inhibitory to the growth and activity of fungi. *Pseudomonas* and *Bacillus* spp. showed better solubilization of TCP up to 0.5% NaCl. *B. circulans* solubilized 86% of TCP at this level (Gaind and Gaur, 1999).

(f) Type of nitrogen sources

Asea *et al.*, (1988) attributed higher phosphorus concentration in plant amended with NH_4 to better development of roots and acidification of rhizosphere. Halder *et al.*, (1992) showed phosphate solubilization by strains of *Rhizobium* and *Bradyrhizobium* in medium containing KNO_3 as sole source of nitrogen. *R. leguminosarum* bv. *viceae* caused highest solubilization and the value was 360 μg P/ml of culture filtrate. Phosphate solubilization by a strain of *Rhizobium* in medium containing KNO_3 increased linearly till about 24 hr whereas in presence of $(NH_4)_2SO_4$ it progressed linearly till 36 hr. In general the replacement of $(NH_4)_2SO_4$ in the medium with KNO_3 caused a decreased solubilization of phosphate from hydroxyapatite by the strains of *Rhizobium* and *Bradyrhizobium*. The difference may be attributed to the employment of different mechanisms for the generation of acidity in the cultures. Presence of NH_4 in the growth medium of *P. cyclopium* was reported to result in the development of inorganic acid following an operation of NH_4/H^+ exchange mechanisms in presence of NH_4Cl (Roos and Luckner, 1984). Effect of different nitrogen sources on PS activity from TCP by different fungal strains was studied by Gaur (1990). Of all the compounds tested, $(NH_4)_2SO_4$ found to be the most superior. Gaur (1990) has reported NH_4NO_3 and $NaNO_3$ as best nitrogen sources for *A. awamori* and *P. digitatum*.

(g) Plant growth regulators and vitamin production

Bacteria isolated from rhizosphere soils are known to produce growth-regulating substances (Brown, 1972), and some of them are capable of dissolving phosphate (Barea *et al.*, 1976). For example, introduction of phosphate dissolving bacteria resulted in improvements in the establishment of *Azotobacter* in rhizosphere and in mycorrhizal development, due to production of growth regulating substances (Azcon *et al.*, 1976; Barea *et al.*, 1978). The culture supernatant fluids of 50 phosphate dissolving bacteria isolated from rhizospheres of crop plants were examined for IAA, gibberellins and cytokinins but only 27 strains could solubilize RP. Twenty bacteria synthesized all 3 types of plant hormones, 43 produced IAA, 29 gibberellins and 45 cytokinin like substances.

Studies were conducted to determine the relationship of vitamin production and ability to dissolve bicalcium phosphate in bacteria isolated from rhizosphere, rhizoplane and control soil. Production of vitamin B_{12}, riboflavin, niacin, pantothenate and biotin was determined. All isolates producing one or more vitamins solubilized phosphates. Production of vitamins B_{12}, riboflavin and niacin by rhizosphere isolates and of riboflavin by rhizoplane isolates was also correlated with ability to solubilize phosphate (Baya *et al.*, 1981).

5. MECHANISMS OF PHOSPHATE SOLUBILIZATION

Results of various experiments indicate that there must be several solubilization mechanisms regarding phosphate mobilization (Fig. 2). A fall in pH during the growth of PSM in liquid medium containing insoluble phosphate has often been reported (Gerretson, 1948; Swaby and Sperber, 1959). Several scientists attribute solubilization of inorganic P sources solely to the production and release of organic acids (Sperber, 1957; Sachar and Gaur, 1984; Halder *et al.*, 1992). PSMs have been found to produce various acids viz., citrate (Cunningham and Kuiack, 1992), oxalate (Lapeyrie *et al.*, 1991), Ketogluconate (Kucey, 1988), lactate and succinate (Kucey, 1983), gluconate (Eckhardt, 1979) and a few other acids (Berthalin *et al.*, 1991; Yadav and Singh, 1991). Seven different acids viz., malic, glyoxalic, succinic, formic, citric, tartaric and L-ketobutyric were identified during phosphate solubilization by *Pseudomonas* sp and succinic, citric and tartaric acids by *A. awamori* and *P. digitatum* by Bardiya and Gaur (1974). *A. niger* produced citrate, oxalate and gluconate, *A. candidus* citric and tartaric acid whereas *Bacillus firmus* formed succinic acid and also 2-Ketogluconic acid in lesser amounts. A fall in pH during the growth of PSM in liquid medium containing insoluble phosphate has often been reported (Gerreston, 1998). Organic acids act as good chelators of divalent cations of Ca^{++} accompanying release of phosphates from insoluble phosphatic compounds by microoranisms. However, there are several reports, which contradict the necessity of production of organic acids during phosphate solubilization (Banik and Dey, 1983). This finding shows that there should be some more pathways for phosphate solubilization in microbes.

Illmer and Schinner (1992) worked on optimum carbon concentration of P solubilization. They also reported that varying Mg and Na concentration in medium did not affect PS activity of

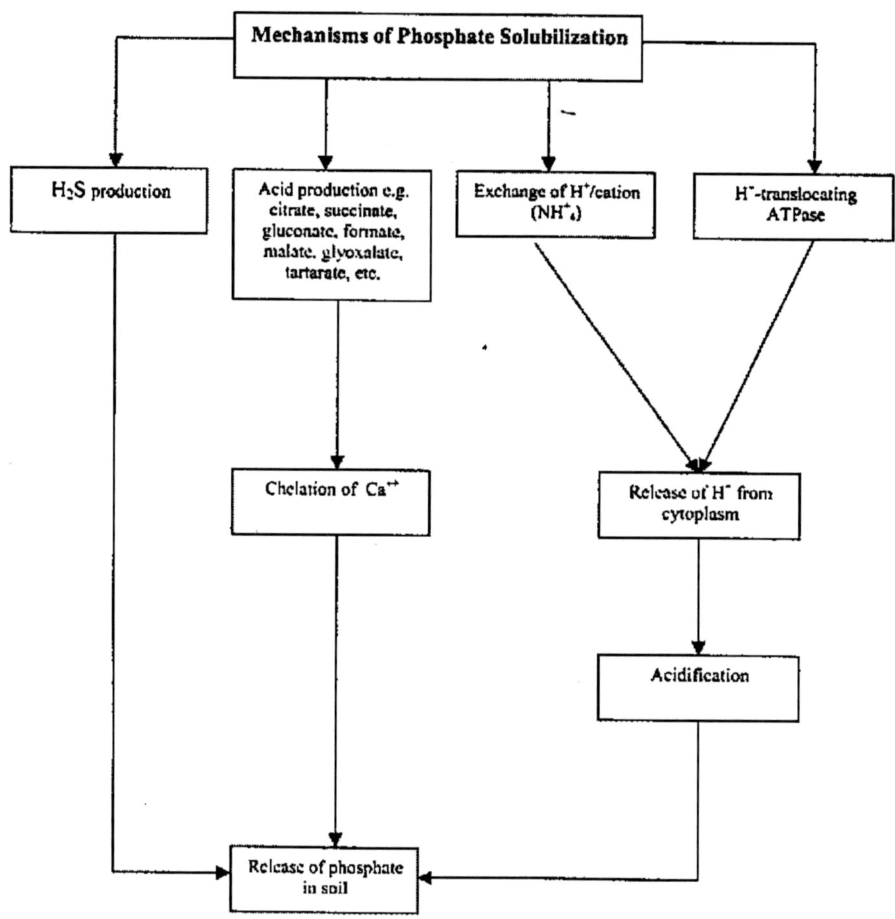

Figure 2. Various Mechanisms of Phosphate Solubilization

Penicillium and *Pseudomonas* spp. from $CaHPO_4$, however, K played a key role in P uptake.

Earlier, Rose (1957) demonstrated that *A. niger.* could bring $FePO_4$ into soil by producing H_2S gas. Sperber (1958) suggested that H_2S production by soil bacteria could be a factor in releasing phosphate from ferric phosphate. Thus inorganic phosphate solubilization by microbial means has been attributed to the process of acidification, chelation, H_2S production and exchange reaction in the growth environment. But according to Asea *et al.* (1988) organic acids may play the important role although this is not the only possibility for phosphorus solubilization, as is confirmed by the weak or even lacking linear correlation between pH and the amount of solubilization phosphate.

H^+ excretion is another possible mechanism to explain calcium phosphate solubilization accompanied by a pH decrease (Illmer and Schinner, 1995a). According to this theory protons may originate from NH_4 assimilation - an often observed phenomenon (Taha *et al.*, 1969; Banik and Dey, 1983; Kucey, 1983;

Dighton and Boddy, 1989; Parks *et al.*, 1990) - or from respiratory H_2CO_3 production (Jurinak *et al.*, 1986). H^+ from respiration cannot be entirely responsible for phosphate solubilization. Nevertheless, respiratory acidification may play a role as a starting aid. Contrary to this, H^+ excretion accompanying NH_4 assimilation, in principle, can account for total phosphate solubilization. In this mechanism the release of H^+ from the cytoplasm to the outer surface takes place. This may happen in exchange for cation (especially NH_4) uptake or with the help of H^+ translocating ATPase, which is located in the plasmalemma and uses the energy of ATP hydrolysis (Beever and Burns, 1980). Following this hypothesis, rock phosphate would be solubilized directly at the cell surface. H_2PO_4 or HPO_4^{-2} depending upon pH could rapidly be absorbed. Roos and Luckner (1984) observed that an isolate of *Penicillium cylopium* on NH_4^+/H^+ exchange mechanism in presence of NH_4Cl and glucose in the medium resulting in the generation of inorganic acid which caused reduction of medium pH.

One more possibility is the alteration of solubility product by acid producing chelate with Ca^{+2} (Kucey, 1983). Beever and Burns (1980) described a similar possibility i.e. internal or external accumulation of Ca^{+2}. Especially, oxalate would be suited to bind Ca^{+2}. Indeed calcium oxalate crystals are knwon to occur in or on fungal hyphae. Accumulation of Ca^{+2} by internal oxalate (or other acid) could help to explain why some strains have a high ability for RP solubilization without acidification of the medium.

The precipitation of $CaCO_3$ during phosphate solubilization was reported by Perez *et al.*, (1983). Direct oxidation pathway for glucose dissimilation forms the metabolic basis for the strong mineral phosphate solubilizing phenotypes. The direct oxidation of glucose results in the production of gluconic acid and often 2-ketogluconic acid in the periplasmic space which in turn results in acidification of the region adjacent to the cell. Calcium phosphates in soil are dissolved by their acidic proteins. It is proposed that extracellular acidification via direct oxidation pathway may play an ecological role in some environments including soil where calcium phosphates provide a significant role of unavailable mineral phosphate.

6. PHOSPHATE SOLUBILIZING MICROBES AS BIOFERTILIZER

In recent years, use of microbial inoculants as a source of biofertilizers has become a hope for most of the countries, as far as economical and environmental view points are concerned.

Penicillium bilaji has been shown to increase yields and phosphorus uptake by canola (*Brassica napus* L.) in beans and wheat in green house and field experiments (Kucey, 1987, 1988). This species is believed to enhance phosphate availability to crop roots by solubilization of inorganic phosphate minerals which are subsequently released into the soil solution and are available for plant uptake. Addition of 20 mg P/kg soil as Florida RP and inoculation with *P. bilaji* resulted in plant straw yields by canola nearly equivalent to that resulting from the addition of 10 mg P/kg soil in the form of commercial monoammonium phosphate (MAP) alone.

Field experiments conducted with rice by Vendan and Subramanian (1999) for 2 years with phosphobacteria showed a three-fold increase in the availability of phosphorus in soil, as it solubilizes the unavailable phosphates. The phosphobacteria applied plots recorded 42.5 kg/ hectare of available phosphorus content, whereas the uninoculated control registered only 14 kg/ hectare. Application of phosphobacteria through seed treatment (3 packets/hectare), seedling dipping (5 packets/hectare) and soil application (10 packets/hectare) with 50% of recommended phosphatic fertilizer (25 kg/ hectare) resulted in the higher yield of 6.3 tons/hectare, which was 37% increase over the control. Application of phosphobacteria could help to save 50% of phosphatic fertilizer besides getting higher solubilization of phosphorus and yield of rice.

Seed inoculation with *Bacillus megatherium* var. *Phosphaticum* and *Pseudomonas* sp. has been found to increase the yield of various crops (Sundara Rao, 1965; Taha *et al.*, 1969). Kavimandau and Gaur (1971) inoculated the maize seeds with *Pseudomonas* and sown in the soil treated with hydroxyapatite tagged with P^{32}, as an insoluble phosphate source at the rate of 80 kg P_2O_5/hectare. Data revealed significant increase in dry matter yield and phosphorus uptake by plants due to seed inoculation with *Pseudomonas*. Kundu *et al.*, (2000) found *Pseudomonas* and *Bacillus* as promising strains under pot house conditions with maize, sorghum, pearlmillet, cotton and cowpea. Their number also improved in the rhizosphere region. Gupta *et al.*, 2000) reported positive role of microbes in terms of reducing 15 kg P_2O_5 besides improving crop yield.

Gaur and Ostwal (1972) showed that *Bacillus polymyxa* significantly increased the grain and straw yields of wheat and the phosphorus uptake by grains. The increases were comparable to that obtained with superphosphate treatment. The results also depicted that inoculation with cultures was more effective when RP was applied to soil. In general, the available phosphorus content of soils increased in presence of farmyard manure, rice straw or any other form of phosphate which shows that due to shortage of energy source, the activity of PSMs could not be fully exploited (Gaur, 1972; Banik and Dey, 1982). The grain yield and phosphate uptake was substantially augmented by inoculation with phosphate dissolving microorganisms, both with and without organic matter. The increase in grain yield without added organic manure ranged from 68 to 115% over the control due to inoculation of various strains of *Pseudomonas, Schwanniomyces, A. awamori* and *P. digitatum.* Farmyard manure also proved a good source and agent for increasing phosphate availability of soils. Applications of 1% of it significantly enhanced the yield of wheat crop (Gaur, 1972).

There are reports, which show that PSMs along with RP and superphosphate may improve the quality of nitrogenous biofertilizers also. Role of nitrogen fixing bacteria and PSMs as biofertilizers was enlisted with regard to the current situation of production and use of groundnut in sandy loam soils with phosphate solubilizing bacteria. The phosphate uptake increased from 88.94 and 155.27 mg/plant to 114.63 and 207.56 mg/plant, respectively (Manjunath and Devi, 1990). A pot culture experiment with Delhi soil was carried out to see the effect of lime and rock phosphate pelleted seed on the nodulation of soybean inoculated with *Rhizobium japonicum.* The number of nodules/plant and dry weight of the plants increased with both time and rock phosphate pelleting (Iswaran and Jauhri, 1969).

Compost is an ideal material for improving the microbial activity of soil and for maintenance of soil humic substances. The attempts were made to assess the effect of addition of RP, inoculation of PSMs and nitrogen fixing bacteria together in improving the manurial value of compost. It was found that additon of RP with mixed inoculum of *Azotobacter chroococcum* and *A. awamori* gave the maximum increase of 31.9% in nitrogen content (Poincelot, 1975; Sadasivam *et al.*, 1981).

Apart from phosphate solubilizers, some mycorrhiza can also help in mobilization of phosphorus. Mycorrhiza is a distinct morphological structure, which develops as a result of mutualistic symbiosis between some specific root-inhabiting fungi and plant roots. Plants, which suffer from nutrient scarcity, especially phosphorus and nitrogen, develop mycorrhiza. In recent years, the significant of use of artificially produced inoculum of mycorrhizal fungi has increased due to its multiferous role in plant growth and yield and resistance against climatic stresses and pests. Vesicular-arbuscular mycorrhiza (VAM) is the most important phosphate mobilisers. *Glomus, Gigaspora* and, *Acaulospora* are worth mentioning VAMs for phosphorus mobilization. Inoculation of chickpea and soybean with *Glomus* spp. had significant positive effect on yield; phosphorus uptake and reduction of dry root rot and collar rot disease (Bhattacharyya and Jain, 1996). In recent years studies were done using host plants like maize, wheat, beans, cowpea and several grasses (Bhattacharyya and Jain, 1996; Kucey, 1987). Institute for Mycorrhizal Research and Development, USA and Athens and Abbort Laboratory, USA have developed a mycelial inoculum of *Pisolithus tinctorius* in a vermiculite-peat moss substrate with a trade name 'Myco-Rhiz' which is commercially available in large quantities. In 1982, about 1.5 million pine seedlings were produced with Myco-Rhiz in the USA. During 1930-40 phosphate solubilizing organisms were marketed as 'Phosphobacteria' in USSR. In Czechoslovakia, the same was marketed in the name of 'Fosfo 24'. Indian Agriculture Research Institute (IARI) also sold the biofertilizer as 'IARI Microphos Culture'. Now several commercial organizations in India are producing phosphorus biofertilizers and according to an estimate a saving of approximately of US $ 92300 in foreign exchange on the import of phosphatic fertilizers can be done (Bhattacharyya and Jain, 1996).

REFERENCES

Agnihotri, V.P. 1970. Solubilization of insoluble phosphate by some fungi isolated from nursery seed beds. Can. J. Microbiol. 16: 877-880.

Anusuya, D. and Jayarajan, R. 1998. Solubilization of phosphorus by *Trichoderma viride*. Curr. Sci. 74 (5) : 464-465.

Arora, D. and Gaur, A.C. 1979. Microbial solubilization of different inorganic phosphates. Ind. J. Exp. Biol. 17 (11) : 1258- 1261.

Asea, P.E.A., Kucey, R.M.N. and Stewart, J.W.B. 1988. Inorganic phosphate solubilization by two penicillium species in solution culture and soil. Soil Biol. Biochem. 20 : 459-464.

Azcon, R., Barea, J.M. and Hayman, D.S. 1976. Utilization of rock phosphate in alkaline soils by plants inoculated with mycorrhizal fungi and phosphate solubilizing bacteria. Soil Biol. Biochem. 8 : 135-138.

Bajpai, P.D. and Sundara Rao, W.V.B. 1971. Phosphate solubilizing bacteria. Soil. Sci. Pl. Nutr. 17 : 41-53.

Banik, S. and Dey, B.K. 1981. Phosphate solubilizing microorganisms of lateritic soil II. Effect of inoculation of some TCP solubilizing microorganisms on available phosphorus content of soil. Zentrlbl. Bakteriol II Abt. 136 : 487-492.

Banki, S. and Dey, B.K. 1982. Available phosphate content of an alluvial soil as influenced by inoculation of some isolated phosphate solubilizing microorganisms. Plant and Soil. 69 : 353- 364.

Banik, S. and Dey, B.K. 1983. Phosphate solubilizing potentiality of the microorganisms capable of utilizing $AlPO_4$ as sole phosphate source. Zentrlbl. Bakteriol. 138 (1) : 17-23.

Bardiya, S. and Gaur, A.C. 1974. Isolation and screening of microorganisms dissolving low grade rock phosphate. Folia Microbiol. 19 : 386-389.

Barea, J.M., Navarro, E. and Montoya, E. 1976. Production of plant growth regulators by rhizosphere phosphate-solubilizing bacteria. J. App. Bacteriol. 40 : 129-134.

Barea, J.M., Ocampo, J.A., Azcon, R., Olivares, J. Montoya, E. 1978. Effects of ecological factors on the establishment of *Azotobacter* in the rhizosphere. Ecological Bulletin (Stockholm) 26 : 325-330.

Basu, P. 1999. Phosphate solubilization by different strain of Aspergilli and Penicillia. M.Sc. dissertation thesis, Banasthali, Vidyapith, pp. 64.

Baya, M.A., Boehhing, R.S. and Ramos-Cormenzana, A. 1981. Vitamin production in relation to phosphate solubilization by bacteria. Soil Biol. Biochem. 13 : 527-531.

Beever, R.E. and Burns, D.J.W. 1980. Phosphorus uptake, storage and utilization by fungi. Advances in Bot. Res. 8 : 127-219.

Bertholin, J., Leyval, C., Laheurte, F. and De Gludice, P. 1991. Involvements of roots and rhizosphere microflora in the chemical weathering of soil minerals. In plant root growth : an ecological perspective (D. Atkinson, Ed.). Special Publication Series of the British Ecological Society, No. 10, Blackwell Scientific, Oxford.

Bhattacharyya, P. and Jain, R.K. 1996. Phosphorus biofertilizers. Kisan, World. 49-50.

Brown, M.E. 1972. Plant growth substances produced by microorganisms of soil and rhizosphere. J. App. Bacteriol. 35 : 443.

Chhonkar, P.K. and Subba Rao, N.S. 1967. Phosphate solubilization by fungi associated with legume root nodule. Can. J. Microbiol. 13 : 749-752.

Cunningham, J.E. and Kuiach, C. 1992. Production of citric and oxalic acid and solubilization of calcium phosphate by *Penicillium bilaji* J. Appl. and Environ. Microbiol. 58 : 1451-1458.

Darmval, N.S., Singh, R.B. and Rai, R. 1989. Isolation of phosphate solubilizers from different sources. Curr. Sci. 58 : 570-571.

Dighton, J. and Boddy, L. 1989. Role of fungi in nitrogen, phosphorus and sulfur cycling in temperate forest ecosystems. In Nitrogen, Phosphorus and Sulfur utilization by fungi (L. Boddy, R. Marchant and D. Read, Eds.), Cambridge Univ. Press. Cambridge.

Duff. R.B. and Wesley, D.M. 1959. 2-Ketogluconic acid as a natural chelator produced by soil bacteria. Chem. Industry. 1376- 1377.

Eckhardt, F.E.W. 1979. Uber die einwirkung hetertropher mikroorganismen auf die zersetzung silikatischer minerale. Zeetschrift fur Pflanzenernahrung und Bodenkunde. 142 : 434-445.

Gaur, A.C. 1972. Role of phosphate solubilizing microorganisms and organic matter in soil productivity. Symp. Soil productivity. 259-268.

Gaur, A.C. 1987. Organic manures and biofertilizers. Div. of Microbiology, IARI, New Delhi, pp. 46.

Gaur, A.C. 1990. Phosphate solubilizing microorganisms as biofertilizers, Omega Scientific Publishers, New Delhi, pp 26-29.

Gaur. A.C. and Ostwal, K.P. 1972. Influence of phosphate dissolving Bacilli on yield and phosphate uptake of wheat crop. Ind. J. Exp. Biol. 10 (5) : 393-394.

Gaur, A.C. and Sachar, S. 1980. Effect on rock phosphate and glucose concentration on phosphate solubilization by *Aspergillus awamori*. Curr. Sci. 49 : 553-554.

Gaind, S. and Gaur, A.C. 1999. Microbial phosphate solubilization as influenced by sodium chloride. Ind. J. Exp. Biol. 37 : 209- 210.

Gerrtson, F.C. 1948. The influnence of microorganisms on the phosphate intake by the plant. Plant and Soil. 1 : 51-81.

Goel, D. 2000. Phosphate solubilization by different species of *Aspergillus* in different conditions. M.Sc. dissertation thesis, Banasthali Vidyapith, pp. 31.

Gupta, R.P., Brar, A.S., Hunjan, R.K. and Pandhar, M.S. 2000. Development of PSM biofertilizers for cotton. Abt. 41st annual conference of association of microbiologists of India. 105.

Halder, A.K., Banerjee, A., Misra, A.K. and Chakraborty, P.K. 1992. Role of NH_4 or NO_3 on release of soluble phosphate from hydroxyapatite by Rhizobium and Bradyrhizobium. J. Basic Microbiol. 32 (5) : 325-330.

Hasan, R. 1994. Phosphorus research in India (ed. Dev. A.). Proceedings. IARI Pub.

Illmer, P. and Schinner, S. 1992. Solubilization of insoluble phosphates by microorganisms isolated from forest soils. Biol. Biochem. 24 (4) 389-395.

Illmer, P. and Schinner, S. 1995a. Solubilization of inorganic calcium phosphates-Solubilization mechanisms. Soil Biol. Biochem. 27 (3) 257-263.

Illmer, P., Barbato, A. and Schinner, F. 1995b. Solubilization of hardly-soluble $AlPO_4$ with P-solubilizing microorganisms. Soil Biol. Biochem. 27 (3) : 265-270.

Iswaran, V. and Jauhri, K.S. 1969. Effect of pelleting soybean seed with lime and rock-phosphate on its nodulation in Delhi soil. The Mysore J. Ag. Sci. 3 (4) : 1-2.

Jones, D., Smith, B.F.L., M.J. and Goodman, B.A. 1991. Phosphate solubilizing fungi in a Scottish upland soil. Mycol. Res. 95 (9) : 1090- 1093.

Jurinak, J.J., Dudley, L.M., Allen, M.F. and Knight, W.G. 1986. The role of calcium oxalate in the availability of phosphorus in soils of semiarid regions : a thermodynamic study. Soil Sci. 142 : 255-261.

Kavimandan, S.K. and Gaur, A.C. 1971. Effect of seed inoculation with *Pseudomonas* sp. on phosphate uptake and yield of maize. Curr. Sci. 40 (16) : 439-440.

Kucey, R.M.N. 1983. Phosphate solubilizing bacteria and fungi in various cultivated and virgin Alberta soils. Can. J. Soil. Sci. 63 : 671-678.

Kucey, R.M.N. 1987. Increased phosphorus uptake by wheat and field beans inoculated with a phosphorus - Solubilizing *Penicillium bilaji* strain and with vesicular-arbuscular mycorrhizal fungi. J. Appl. Environ. Microbiol. 53 : 2699-2703.

Kucey, R.M.N. 1988. Effect of *Penicillium bilaji* on the solubility and uptake of phosphorus and micronutrients from soil by wheat. Can. J. Soil. Sci. 68 : 261-270.

Kucey, R.M.N. and Leggett, M.E. 1989. Increased yields and phosphorus uptake by westar canola (*Brassica napus* L.) inoculated with a phosphate solubilizing isolate of *Penicillium bilaji*. Can. J. Soil Sci. : 69 : 425-432.

Kundu, B.S., Gera, R., Sharma, N. and Lakshminarayana, K. 2000. Specificity of phosphate solubilizing bacteria with host-plants.

Lapeyrie, F. Ranger, J. and Varelles, D. 1991. Phosphate solubilizing activity of ectomycorrhizal fungi *in vitro*. Can. J. Prot. 69 : 342-346.

Mahapatra, S.S.R., Mishra, A.K. and Chandra, D. 1985. Solubilization of rock phosphate by *Thiobacillus ferrooxidans*. Curr. Sci. 54 (5) : 235-237.

Manjunath, H. and Devi, L.S. 1990. Effect of phosphate solubilizing bacteria on phosphate availability to groundnut from rock phosphate. Curr. Res. Univ. Agr. Sci. 19 : 56-57.

Mba, C.C. 1996. Rock phosphate solubilizing *Streptosporangium* isolates from casts of tropical earthworm. Resources, Conservation and Recycling. 17 : 211-217.

Nahas, E. 1996. Factors determining rock phosphate solubilization by microorganisms isolated from soil. World J. Microbiol. and Biotech. 12 : 367-572.

Nahas, E. and Assis, L.C. 1992. Phosphate rock solubilization by *Aspergillus niger* in different types of vinasse. Pesquisa Agropecuiria Brasileira 27 : 325-331.

Narsian, V. Thakkar, J. and Patel, H.H. 1993. Solubilization of natural rock phosphates and pure insoluble inorganic phosphates by *Aspergillus awamori*. Ind. J. Exp. Biol. 31(9) : 747-749.

Narsian, V. Thakkar, J. and Patel, H.H. 1994. Isolation and screening of phosphate solubilizing fungi. Ind. J. Microbiol. 32 (2) : 113-118.

Narsian, V. Thakkar, J. and Patel. H.H. 1995. Mineral phosphate solubilization by *Aspergillus aculeatus*. Ind. J. Exp. Biol. 33 (2) : 91-93.

Parks, E.J., Olson, G.J., Brinckman, F.E. and Baldi, F. 1990. Characterization by high performance liquid chromatography (HPLC) of solubilization of phosphorus in iron ore by a fungus. J. Ind. Microbiol. 5 : 183-190.

Perez, G.I., Rivadeneyra, M.A., Salmeron, V., and Cormenza, A. 1983. Calcium carbonate participation during phosphate solubilization. Anales de Edafologiacy-Agrobiologia. 42 : 2237- 2242.

Pierre, W.H. 1938. Phosphorus deficiency and soil fertility. Year Book of Agr., USDA, Washington, D.C. 377-396.

Poincelot, R.P. 1975. The biochemistry and methodology of composting. Conn. Agric. Exper. Sta. Bull. 754. pp. 18.

Rokade, S.M. and Patil. P.L. 1992. Phosphate solubilizing microorganisms : a review. Ind. J. Maharashtra Agric. Univ. 17 (3) : 458-465.

Roos, W. and Luckner, M. 1984. Relationship between proton extrusion and fluxes of NH_4 and organic acids in *P. cyclopium*. J. Gen. Microbiol. 130 : 1007-1009.

Rose, R.E. 1957. Techniques for determining the effect of microorganisms on insoluble inorganic phosphate. Nz. J. Sci. Tech. 38 : 773-780.

Sadasivam, K.K., Mathur, R.S., Magu, S.P. and Gaur, A.C. 1981. Enrichment of compost by *Azotobacter* and phosphate solubilization. Zbl. Bakt. 11 : Abt. 628-630.

Salih, H.M., Yahya, A.M., Abdul Rahim, A.M. and Unam, B.H. 1989. Availability of phosphorus in a calcareous soil treated with rock phosphate or superphosphate as affected by phosphate dissolving fungi. Plant and soil. 120 : 181-185.

Sayer, J.A., Raggett, S.L. and Gadd, G.M. 1995. Solubilization of insoluble metal compounds by soil fungi : Development of a screening method for solubilizing activity and metal tolerance. Mycological Res. 99 : 987-993.

Singal, R., Gupta, R. and Saxena, R.K. 1994. Rock phosphate solubilization under alkaline condition by *Aspergillus japonicus* and *A. foetidus*. Folia Microbiol. 39 (1) : 33 - 36.

Singh, H.P., Pareek, R.P. and Singh, T.A. 1984. Solubilization of rock phosphates by phosphate solubilizers in broth. Curr. Sci. 53 (22) : 1212-1213.

Sperber, J.I. 1957. Solubilization of mineral phosphates by soil bacteria. Nature. 180 : 994-995.

Sperber, J.I. 1958. Solution of apatite by soil microorganisms producing organic acids. Aus. J. Agri. Res. 9 : 782-787.

Sundara Rao, W.V.P. 1965. Bacterial fertilizers. In : Hand book of manures and Fertilizers. ICAR Pub. Ch. 9. pp. 186.

Swaby, R.J. and Sperber, J. 1959. Nutrition of legumes. Ed. E.G. Hallsworth. Butterworths, London.

Taha, S.M., Mahmood, S.A.Z., Halim E.L. - Damaty, A.H. and Hafez, A.E. 1969. Activity of phosphate dissolving bacteria in Egyptian soil. Pl. Soil. 31 : 149-160.

Thakkar, J., Narsian, V. and Patel, H.H. 1994. Isolation and screening of phosphate solubilizing fungi. Ind. J. Microbiol. 34 (2) : 113-118.

Vendan, R.T. and Subramanian, M. 1999. Phosphate bacteria boost rice yield. The Hind. Sci. Tech.

Yadav, K. and Singh, T. 1991. Phosphorus solubilization by microbial isolate from a Cacifluvent. J. Ind. Soc. Soil Sci. 39 : 89-93.

CHAPTER - 6

FRANKIA — A PROMISING NITROGEN FIXING ACTINOMYCETES

Mahadeva Swamy, M.N. Sreenivasa and M.K. Naik

College of Agriculture, Raichur, Karnataka

1. INTRODUCTION

Today the scenario is, we have less than 70 million hectares of forest area, out of its total land area of 328.05 million hectares, which works out to about 22 per cent of total geographical area and per capita forest works out to be 0.13 hectare. There is a heavy demand for forest products all over the world. To increase the crop productivity, nitrogen is most essential nutrient, but is found in low levels in soil. Certain non-leguminous trees and woody shrubs form root nodules and fix atmospheric nitrogen under natural condition. The organism responsible and found in the nodules, is an actinomycete, identified as *Frankia* sp. This symbiotic association between actinomycete and the roots of non-leguminous plants is termed as "*Actinorrhiza*". Actinorrhizal plants are usually involved in the early successional stages of plant communities at low nutrient sites, and are found colonizing disturbed and marginal habitats. Actinorrhizas have been reported in 23 genera of woody plants belonging to eight families (Table 1) (Bond 1976; Baker, 1984 and Lechevalier, 1984).

Actinorrhizal plants are of economic importance, particularly for forestry and land reclamation. *Casuarina*, which is widely distributed in tropical regions, is reported to be the world's best firewood in terms of K cal/unit mass (Anonymous 1980). *Casuarina* is a valuable tree because firewood is an essential commodity in developing countries (Anonymous, 1984). In addition, *Casuarina* fixes atmospheric nitrogen, (Dommergus, 1963).

The nitrogen fixed ranges from 5-260 Kg N/ha/year and biomass produced due to actinorrhizal association is taken advantage in several applications in agroforestry. Tree breeding, timber production, soil stabilization and reclamation, studding of other crops and parked vehicles, wind breaks, amenity planting, organic matter

Table 1. Important *Actinorhizal* host-plant Genera

Genus	Family	Nodulated species Identified	Reference
Alnus	Betulaceae	48	Baker 1982, Baker 1984.
Casuarina	Casuarinaceae	32	Diem 1987.
Coriaria	Caoririaceae	16	Mirza 1991.
Datisca	Datiscaceae	02	Chaudary *et al* 1987.
Elaeagnus	Elaeagnaceae	39	Baker 1979.
Hippophae	Elaeagnaceae	02	Gauthier *et al.*, 1981, Faure *et al.*, 1983.
Myrica	Myricaceae	29	Bloom *et al.*, 1989, Burgraff 1981.
Ceanothus	Rhamnaceae	31	Lechevalier *et al.*, 1984.
Colletia	Rhamnaceae	03	Akkermans *et al.*, 1984.
Comptonia	Myricaceae	01	Callaham *et al.*, 1978.
Cowaria	Rosaceae	02	Baker 1984
Dryas	Rosaceae	02	Baker 1984
Purshia	Rosaceae	02	Baker 1984

accumulation, nutrient recycling and reduction on soil erosion are important applications.

Endophytic existence involves root hair infection, intercellular ramification of mycelium, nodule formation and nitrogen fixation by specialized spherical structures called vesicles. Reproductive structure like sporangia are also formed in several host nodules. They can be planted following fire, volcanic eruption and logging. Some excellent reviews of current and potential uses of such plants and their management in different parts of the world have recently been produced by Benoit and Berry (1990) and Dawson (1990).

In the world where the cost of nitrogenous fertiliser is escalating and the problem of deforestation is very great, the potential of actinorrhizal plants must be brought to the notice of land managers. According to Benoit and Berry (1990), the adaptive ability of actinorrhizal plants can greatly reduce the cost of revegetation and soil stabilization plantings. In addition, plant nitrogen needs are met which would otherwise have to be met by the addition of nitrogenous fertilizer, can alleviate the problem of the contamination of downstream waters subsequent to the often indiscriminate use of chemical fertilizer.

2. BIOLOGY

The first reproducible studies on the growth of *Frankia* in pure culture have only been reported (Callaham et al., 1978; Quispel and Tak, 1978). In the root nodules, *Frankia* may produce sporangia SP (+) or may not. The SP (-) strains are claimed to be more infective (Houwers and Akkermans 1981) but may be less efficient in N_2- Fixation (Van Dijk 1978). Normally nitrogen fixing structures are found as vesicle clusters and reproductive structures as sporangia in the root nodules of most hosts excepts in *Casuarina*. However, isolates from *Casuarina* do produce vesicles and sporangia on laboratory media (Tyson and Silver, 1979). A rapid loss of biomass of *Frankia* isolates from *Chilean* shrubs at the end of log phase of growth on laboratory media have been attributed to autolysis of actinomycete mycelium due to protolytic enzymes and aminopeptidase activities.

Frankia strains usually are coded with initials of the host plants name from which the strain was isolated and a proposal for a more universal coding system has been arrived at by Lechevalier (1983) where in she presents as many as 10-13 informations with each *Frankia* strain while cataloging. Reddell and Bowen (1985) provided evidence for the occurrence of more than one *Frankia* strains in single nodules of Casuarinaceae. An attempt by Raj et al. (1984) to obtain a mutant of *Frankia* strain Cc 1.17 capable of growing in N-free medium without vesicle formation, by treating with introsoguanidine, resulted in a number of mutants that were producing less number of vesicles. However, the original strain Cc. 1.17 was not able to produce vesicles at 35°C on medium containing NH_4 ions, although it produced a large number of them at 25°C and a few at 30°C. They also found that pre shaken cultures of *Frankia* could tolerate and reduce C_2H_2 at higher O_2 concentration compared to tolerance by stationary cultures.

Immuno-gold labelling of ultrathin cryosections of *Frankia* sp. Cc 1.17 showed specific labelling of nitrogenase in the vesicles (Meesters *et al.*, 1987).

2.1. Isolation of *Frankia* from root nodules of *Actinorrhizal* host

For a long time, the major obstacle to research on this symbiosis was the inability of the researchers to reinfect host plants with isolates from actinorrhizal nodules. Callaham and his associates in 1978 at Harvard forest succeeded for the first time in isolating the pure culture of *Frankia* that could be demonstrated to reinfect the host plant. Root nodules were surface sterilized, macerated using enzymes and then plated on yeast extract medium to obtain the actinomycete. Since this first confirmed success in isolating *Frankia* from *Comptonia peregrina*, numerous other successful isolations from species of *Ceanothus*, *Comptonia*, *Casuarina*, *Elaeagnus*, *Myrica* and *Purshia* have been reported (Baker, 1982).

Over the past 10 years a range of techniques have been used to isolate *Frankia*, from actinorrhizal root nodules including nodule dissection techniques (Burgraaf, 1984; Diem 1983), serial dilution of crushed nodule (Diem, 1982; Gautheir *et al.*, 1981), sucrose density fractionation of crushed nodule suspensions (Baker and Torrey, 1979) and filtration technique for purification of *Frankia* vesicle clusters from root nodules (Benson, 1982; Burgraaf, 1984).

To be successful the isolation procedures must accomplish the following criteria :

1. Removal of toxic phenolic compound from root nodule homogenate.
2. The endophyte separation from contaminating microorganisms.
3. The endophyte separation from contaminating plant tissue (Blom 1982)

Isolation of *Frankia* from older species i.e., *Alnus crispa* by using osmium tetroxide treatment and isolation method was given by Norman (1982). Li and Ding (1986) isolated an infective strain of *Frankia* from root nodule of *Casuarina cunninghamiana* by nodule piece method. *Frankia* strains were isolated from actinorrhizal nodules of *Alnus, Casuarina, Datisca, Elaegnus* and *Hippophae* (Chaudary and Sajjad Mirza 1987) by using double layer of agar system consisting of an upper layer of one percent agar which included the endophyte and lower layer of 1.5 per cent agar which contained the nutrients.

2.2. Morphological and cultural characteristics of *Frankia*

The isolated actinomycete is filamentous, grows slowly in pure culture on yeast extract medium, produces sporogenous bodies which form large number of ovoid spores. (Callaham *et al.*, 1978). *Frankia* growing on isolation plates are slow growers; aerobic, gram-positive actinomyetes; bear septate filaments and vegetative sporangia containing non-motile spores; may be pigmented, fix nitrogen *via* specialized *Frankia* vesicles; possess a multilamellar envelope exterior to the cellwall; contain meso-diaminopimelic acid and cross react with anti *Frankia* sera (Baker 1984). Some of the strains of *Frankia* could produce pigments, viz., blood red, tan brown, yellow, brick red etc., in culturing media (Baker and Torrey, 1979, Benson and Hanna, 1982; David, 1982 and Zhongze and Torrey, 1985).

Several studies on *Frankia* have revealed that the organism has three main types of structures as found both *in vivo* as well as defined medium viz. :

1. Procaryotic, highly branched, septate hyphae of 0.3 to 1.0 μm diameter.
2. Inter calarily or terminally positioned sporangia, which consists of sporangiospores of size 1-2 μ m.
3. Vesicles which are the sites of nitrogen fixation, about 4-6 μm size (Baker and Torry 1979; Lalonde, 1979; Necomb *et al.*, 1979; Zhang *et al.*, 1984)

Parsons *et al.* (1987) found that *Frankia* produces vesicles with outer walls thickened by multiple lipid-like monolayers. Lamont *et al.* (1988) investigated the composition of the envelope of vesicles of *Frankia*. The results obtained were consistent with the view that the envelope of *Frankia* vesicles consist largely of lipid and serves as a barrier to diffusion of oxygen. Schultz and Benson (1989) studied the ability of nitrogenase-containing *Frankia* vesicles to regrow vegetative hyphae. They found that the vesicles exhibited hyphal outgrowth when incubated in certain defined media. They observed that single or multiple hyphal extension grew out from the vesicles.

2.3. Ultrastructural studies of *Actinorrhizal* root nodules

In axenic root nodules of *Alnus crispa* the endophyte was present in three forms: the hypha, branched or unbranched measured between 0.3mm to 1.0mm, the vesicle (3 mm-5mm), which was septate and bacteria like cells (0.5 mm to 1.5mm). The structure was comparable to that in normal nodule (Lalonde and Fortin, 1973). The nodules of one year old nursery grown seedlings of *Alnus cordata* varied in number and size. The symbiont consisted of hyphae with complete septa of 0.5-0.7 mm in diameter and vesicles of 3.0 mm diameter showing the same characteristics as the Actinomycetales, although a definite classification could not be made. The nodules of *Elaeagnus angustifolia* formed more frequently in regions of the root lacking root hairs.

The entry point of the hyphae appeared to be between the epidermal cells of the root with subsequent hyphal growth occurring in the intercellular spaces between cortical cells. Berry *et al.* (1986) studied the structure of *Frankia*, its mode of penetration into the cell wall of *Alnus rubra* roots and changes in root hair structure at the infection site.

3. INFECTION, NODULATION AND NITROGEN FIXATION

About 30 species of Casuarinaceae are known to be nodulated. Nitrogen fixation occurred at soil temperature above 20°C (Reddell and Bowen 1985) and nitrogen fixation was impaired beyond 0.15 mg

NaCl/g soil (Bowen and Reddell, 1986). In culture, vesicles are generally spherical but within the host their shape varies and can be spherical, pear-shaped or club-shaped depending on host species (Gardner 1976). Vesicle shape appears to be host determined. The same *Frankia* strain are capable of forming differently shaped vesicles within different hosts. No vesicles are seen to be produced within *Casuarina* nodules although isolates from these nodules produce spherical vesicles in culture. The vesicles as the site of nitrogen fixation has been verified by immunolocalisation of the enzyme nitrogenase both in culture by Meesters *et al.* (1987).

It has been suggested that nitrogenase is synthesized only in vesicles as there has been correlation between vesicle production and C_2 H_2 reduction (Tjepkema *et al.,* 1980). However, some strains produce vesicles in a medium containing combined nitrogen but apparently with no nitrogenase activity (Akkermans *et al.,* 1984), suggesting that the relationship is not absolute. For the majority of Actinorrhizal genera endophytic vesicle wall is thought to constitute a barrier to oxygen diffusion thus providing protection for the nitrogenase within. The non-formation of vesicles by *Frankia* in *Casuarina* may thus reflect the low PO_2 within the infected cells of these nodule tissues. [Silvester *et al.* (1990) and Gallon (1992)].

Although several *Frankia* strains seem to be promiscuous, other have marked symbiotic specialization in host selection. The root hair deformation caused by *Pseudomonas cepacia* (helper bacteria) and the associated cell wall histological changes (fluorochrome acridine orange) in the host *Alnus rubra* is apparently due to high molecular weight compound (filtrate studies) and the process is host specific. The deformation helps in successful nodulation of *Alnus rubra* by *Frankia* (Berry and Torrey, 1983).

Aldrich-Blake (1932) collected nodules from India which were 12 cm in diameter. Nodulation in nursery seedlings is possible by adding soil or leafy litter collected from the immediate vicinity of effectively nodulated plants.

3.1 Effect of *Frankia* on *Actinorrhizal* plants

Inoculation with crushed nodules containing *Frankia* increased the nodulation and dry matter production of *Casuarina equisetifolia* by 70 and 42 per cent, respectively (T. Bhat *et al.*, 1987), older seedlings inoculated with *Frankia* at planting yielded better shoot height, nodule number and shoot dry weight compared to the ones inoculated 4 weeks later (Stower and Smith, 1985). It has been estimated that *Casuarina equisetifolia* seedlings inoculated with *Frankia* strain ORS_021001 fixed about 40 to 60 Kg nitrogen per hectare per year at normal densities of 10,000 trees per hectare (Gauthier *et al.*, 1985). Zang Zhang Ze and Torrey (1985) showed that the *Frankia* strain HEP11 nodulates effectively, seedlings of *Casuarina equisetifolia* and *Casuarina cunninghamiana* forming nodules with vertically growing nodule roots which are effective in acetylene reduction. It has been observed that over a three year period, *Alnus rubra* and *Alnus glutinosa* returned 10 to 12 Kg nitrogen per hetare to the ecosystem (Wheeler *et al.*, 1986).

The response of *Casuarina equisetifolia* to inoculation with *Frankia* strain ORS 021001 was studied. The inoculated and uninoculated trees were randomly sampled after 1, 2, and 3 years of transplantation. When biomass was expressed as dry weight, increase due to inoculation were similar at the three sampling dates, 45, 36, and 40 per cent, respectively, Nitrogen fixation, estimated using the different methods reached 2.48, 12.25 and 13.44%, nitrogen fixed annually per tree. The ecological function of ineffective *Frankia* strains and their behaviour in comparision with effective *Frankia* strains was indicated by an increase in plant height and total dry weight when inoculated with effective *Frankia* strains and simultaneous inoculation with a constant amount of an ineffective strain (Sougoufara *et al.*, 1989). The *Casuarina cristata* seedlings inoculated with nodule organisms from *Casuarina gluca* formed an average of 118 nodules per seedling whereas those inoculated with nodule organisms from *C. cristata* formed 2.7 nodules/seedlings. But fixation was higher in the latter, which was indicated by higher shoot nitrogen content. Mahadeva Swamy *et al.* (1998) revealed nodule crush inoculum was most efficient in promoting plant growth and nitrogen fixation in *Casuarina equisetifolia* seedlings.

3.2 *Mycorrhizae* and *Actinorrhizal* plants

In legumes, nodules appear to resist *Vesicular arbuscular* mycorrhizae fungal colonization but this does not seem to be the case in actinorrhizal nodules, since *Vesicular arbuscular* mycorrhizae association with nodules has been described in *Alnus* sp. and *Caenothus* sp. (Rose and Youngberg 1981). Apart from the enhancement of phosphorus uptake by *Vesicular arbuscular* mycorrhizae fungi in Actinorrhizal plants. They also increased nitrogenase activity of nodules and consequent improvement of nitrogen nutrition of plants.

The root system of *Caenothus vehutinus* was colonised by *Glomus gerdemanili* and a nitrogen fixing actinomycete *Ceanothus* seedlings inoculated with both these microsymbionts showed increase in total dry weight, nodule number and nodule weight, levels of nitrogen, calcium and phosphorus and increase in nitrogenase activity.

Mahadeva Swamy and Sreenivasa (2000) showed that the mycorrhizal spores and percentage of *Vesicular arbuscular* mycorrhizal colonisation were maximum in *Frankia* plus *Glomus fasciculatum* inoculated *Casuarina* seedlings. These seedlings also had highest population of bacteria, fungi and actinomycetes in their rhizosphere suggesting synergistic or additive effect of *Frankia* and *Glomus fasciculatum*.

4. ECOLOGY

Usually *Frankia* occupies two types of ecological niches, the root nodule and soil. The nature and extent of their free living existence are still not clear. The presence of soil borne microorganisms capable of forming root nodules on plants other than legumes has been known. (Quispel 1990) *Frankia* were obligate symbionts and it establishes on the basis of host range. Many observations show *Frankia* strains to be present in soils well outside the normal geographic range of appropriate host plants (Benecke 1969). Survival and persistence of soil have generally been studied by using host plants as bioassay agents for *Frankia* (Smolander 1990). The capacity of different soils to induce nodulation in *Alnus incana* was studied, by inoculating grey older seedlings growing in liquid culture with 20 fold serial dilutions. The nodulation capacity of the soil under *Alnus* varied between 101 and 105 and between 0 and 3.1 x 10^3 units (nu) cm^{-3} of soil (Smolander *et al.* 1987) and was at least as high as that under older or even higher. An average of 69% of the nodules belonged to the spore negative group (Smolander *et al.*, 1990). Among factors controlling *Frankia* soil population is the pH. For most plants there is a narrow pH optimum near neutrality at which nodulation is optimal, and this coincides with the optimurn growth PH of *Frankia* cells. The number of nodulation units in soil as measured by most probable number nodulation tests on *Alnus* sp. is strogly correlated with soil pH upto pH 8.0 but not with any other common soil property.

5. APPLICATION

In China, *Casuarina* is used for house construction and also for large size boats and furniture. The timber yield is reported to be $4m^3$ ha/year (FAO, 1978) *Casuarina* planted over 1 m ha at the coastal shelter belt (length 3000km, 0.5-5km width) in the Guangdong provinance in China has increased the yields of agricultural crops from 1.4 to 4.8 t/ha. The belt acts as wind break and stabilizes the sand and soil beneath. Inoculation of seedlings with *Frankia* doubled the wood production in 44 months at Craford (south Australia) and increased the height by 50-60 per cent 13-14 months after transplanting at Zimbabwe; and enhanced wood volume. By 34 to 95 percent with application at Gympie (Queensland) campared to unionculated control plants and *Frankia* entrapped in alginate was used as incoculum. Our *Frankia* ce series of strains (Mahadeva Swamy and Raj J., 1998) isolated from *Casuarina equisetifolia* have been forming vesicles in laboratory media and are capable of reducing acetylene in N-free medium. These strains upon inoculation to plants in pot culture have enhanced nodulation and biomass of seedlings.

Dual inoculation of *Frankia* sp. and *Glomus fasciculatum* resulted in higher biomass and sporecount and nodulation contents of *Casuarina equisetifolia* seedlings, compared to single inoculum formed by either of them. Presence of hyphae within nodular tissue indicated endophytic nature of *Frankia* (Mahadeva-swamy and Sreenivasa 2000).

6. CONCLUSION

Studies on *Frankia* really took off in 1978, when Callaham and coworkers successfully isolated it from the plant *Comptonia* (Callaham *et al.*, 1978). The number of *Frankia* strains has grown several fold since then today endophytes from different generas are known.

Although the information on Frankia is not fully satisfactory, a fair amount is now available about the biology, isolation, ecology, host preference, infection, nodulation and application. Work is also needed on basic side to establish the role of vesicles in binding nitrogen, possibly through detection of essential components of enzyme nitrogenous in vesicles. Or 15_N incorporation into vesicles. There exists many opportunities for studies on the actinorrhizal symbiosis. However, the research areas holding the most potential future, include studies on genetics and molecular aspects of host-plant infection and symbiotic interactions. *Frankia* -plant interaction studies in identification of genes involved formation would help in laying the foundation for unifying the hypothesis to explain its symbiosis.

Further, research and surveys are required for selection of highly salt tolerant actinorrhizal plants and *Frankia* genotypes, Which would help in ameliorating the salt affected soil. The major problem encountered in research on *Frankia* was the longlasting problem of isolation from root nodules. However, a range of techniques have been developed in the past decade to isolate *Frankia* from actinorrhizal root nodules thus, enabling for furthering the cause of research in this exciting field. Since the cost of nitrogenous fertilizers is escalating and the problem of deforestation is at its peak, the issues concerning potential application of *Frankia* and their commercial exploitation must be addressed thoroughly.

REFERENCES

Akkermans, A. D. L., Hafeez, F., Roelfsen, W., Chaudhary, A. H. & Baas, R. 1984. Ultrastructure and nitrogenase activity of *Frankia* grown in pure culture and in actinorrhizae of *Alnus, Colletia*, and *Datisca* spp, p. 311-319. In C. Veeger and W. E. Newton (ed.), Advances in Nitrogen Fixation Research. Nyhoff/Junk Publishers, The Hague, The Netherlands.

Aldrich-Blake, R. N. 1932. On the fixation of atmospheric nitrogen by bacteria living symbiotically in root nodules of *Casuarina equisetifolia* Oxford for mem., 14: 20-21.

Anonymous, 1980. Firewood Crops, National Academy of Sciences, Press Washington, D. C.

Anonymous, 1984. Casuarinas- Nitrogen fixing trees for adverse sites. U. S.National Research Council, National Academy Press, Washington, D. C.

Baker, D., Torrey, J. G., & Kidd, G. H. 1979. Isolation by sucrose density fractionation and cultivation *in vitro* of actinomycetes from nitrogen. Nature. 281:76-78

Baker, 1982. A cumulative listing of isolated *Frankia*, the symbiotic nitrogen fixing actinomycetes, Actinomycetes. 17: 35-42.

Baker, D. & O'Keefe, D. 1984. A modified sucrose franctionation procedure for the isolation of *Frankia* from actinorrhizal root nodules and soil samples. Plant soil. 78:23 - 28.

Baker, D. D. 1987. Relationships among pure cultured strains of *Frankia* based on host specificity. Physiol. plant. 70:245-248.

Baker, D., & D. O' Keefe. 1984. A modified sucrose fractionation procedure for the isolation of *Frankia* from actionorrhizal root nodules and soil samples. Plant Soil 78:23-28.

Baker, D. 1982. A cumulative listing of isolated *Frankia*, the symbiotic, nitrogen-fixing actinomycetes. Actinomycetes 17:35-42.

Baker, D. 1984. *Frankia*; new light on an actinomycete symbiont In: Biological Biochemical and Biomedical aspects of Actinomycetes, Academic Press, Orlando, pp. 563-74.

Baker, D. 1988. Modification of catalog numbering system for *Frankia* strains and revision of strain numbers for DDB Frankia collection. Actionmycetes 20:85-88.

Baker, D., Torrey, J.G., & Kidd, G.H. 1979. Isolation by sucrose density fractionation and cultivation *in vitro* of actinomycetes from nitrogen - fixing root nodules. Nature (London). 281: 76-78.

Baker, D., W. Newcomb, & Torrey, J.G. 1980. Characterization of an ineffective actionorrhizal microsymbiont, *Frankia* sp. Eull. Can. J. Microbiol. 26:1072-1089

Benecke, U. 1969. Symbionts of alder nodules in New Zealand. Plant soil 30:145-149.

Benoit, L.F. & Berry, A.M. 1990. Methods for production and use of actinorrhizal plants in forestry, low maintenance and revegetation. In : Schwintzer C.R., Tjepkema J.D (eds). The biology of *Frankia* and actinorrhizal plants. Academic Press. London, pp 281-298.

Benson, D. R. & Hanna, D. 1982. *Frankia* diversity in an older stand as estimated by sodium dedeoyl sulphate polyacrylamide gel electrophoresis of whole cell proteins. Can. J. Bot. 16:2929-2933.

Benson, D.R. & Hannah, D. 1982. *Frankia* diversity in an alder stand as estimated by sodium dodecyl sulphate polyacrylamide gel electrophoresis of whole cell proteins. Can J. Bot. 16 : 2929 - 2933.

Berry, A. M. & Torrey, J. G. 1983. Root hair deformation in the infection process of *Alnus rubra. Can. J. Microbiol.* 61: 2863-2976.

Berry, A. M., Mclntyre, L. & McCully, M.E. 1986. Fine structure of root hair infection leading to nodulation in the *Frankia-Alnus* symbiosis. Can. J. Bot.. 64:292-305.

Blom, J. 1982. Carbon and nitrogen source requirement of *Frankia* strains. FEMS Microbial Letter. 13 : 5-55.

Bloom, R.A., Lechevalier, M.P. & Tate, R.L. 1989. Physiological, chemical, morphological, and plant infectivity characteristics of *Frankia* isolates from *Myrica pennsylvanica*: coorelation to DNA restriction patterns. Appl. Environ. Microbiol. 55: 2161 - 2166.

Bond, G. 1976. The results of the IBP survey of root nodule formation in non-leguminous angiosperms. In: Symbiotic nitrogen fixation in plants. Cambridge Univ. Press, England, PP.443-444.

Burggraaf, A. J. P. & Shiption, W.A. 1983. Studies on the growth of *Frankia* isolates in relation to infectivity and nitrogen fixation (acetylence reduction). Can. J. Bot. 61:2774-2782.

Burggraaf, A. J. P., Quispel, A., Tak, T. & Valstar., J. 1981. Methods of isolation, and cultivation of *Frankia* species from actinorhizas. Plant soil. 61:157-168.

Burgraaf, A. J. P. 1984. Isolation, cultivation and characterization of *Frankia* strains from Actinorrhizal root nodules, Ph.D. thesis, University of Leiden, The Netherlands.

Callaham, D., Delredici, P. D. & Torrey, J. G. 1978. Isolation and cultivation *in vitro* of the actinomycete causing root nodulation in Comptonia Science. 199: 899-902.

Chaudhary, A. H. & Sajjad Mirza, M. 1987. Isolation and characterization of *Frankia* from nodules of actinorrhizal plants of Pakistan. Physiol. Plant. 70:255-258.

David, R. B. 1982. Isolation of *Frankia* strains from alder actinorrhizal root nodules Appl. Environ. Microbiol., 44:461-465.

Dawson, J.O. 1986. Actinorrhizal and agriculture, Outlook on Agri. 15(4):202-08.

Dawson, J.O. 1990. Interactions among actinorrhizal and associated plant species. In: Schwintzer C. R. & Tjepkema J.A. (eds). The biology of *Frankia* and Actinorrhizal plants Academic Press, London. pp 299-316.

Diem H.G. & Dommergues, Y.R. 1990. Current and potential uses and management of Casuarinaceae in the tropics and subtropics. In: Schwintzer R & Tjepkema JD (eds). The biology of *Frankia* and actinorrhizal plants. Academic Press, London, pp 317-342.

Diem, H. G., Gauthier, D. & Dommurgnues, Y. R. 1982. Isolation of *Frankia* from nodules of *Casuarina equisetifolia* Can J. Microbial. 28: 526 - 530.

Diem, H. G., Gauthier, D. & Dommergues, Y.R. 1983. An effective strain of *Frankia* from *Casuarina* sp. Can. J. Bot. 61:2815-2821.

Dommergues, Y. R. 1963. Evaluation due to N_2 fixation de L' azote dans unsolreboise en filao (Nitrogen - Fixation rate of a dune soil affected with *Casuarina equisetifolia*). Agrochemical. 105: 179-187.

FAO. 1978. China, Forestry support for Agril Forestry paper, 12 FAO Rome.

Faure-Raynaud, M. & Moiroud, A. 1983. Symbiose *Frankia.* Culture in vitro de frankia sp. sur differentes souces de carbone et d' zazote. C. R. Acad. Sci. Paris ser. III. 296 : 757-760.

Gallon. J. R. 1992. Reconciling the incompatible: N_2 fixation and O_2. New Phytol 122 : 571-609.

Gardner, I.C. 1976. Ultrastructural studies of non-leguminous root nodules. In: Nutman, P.S. (ed) Symbiotic nitrogen fixation in plants. University Press Cambridge, pp 485-495.

Gauthier, D., Diem, H. G. & Dommergues, Y. 1981. Infectivite et effectivite de souches de Frankia isolates de nodules de *Casuarina equisetifolia* et d' Hippophae rhamnoides. C. R. Acad. Sc. Paris Ser. III 293:489-491.

Gauthier, D., Diem, H. G., Dommergues, Y. R. & Gutry, F. 1985. Assessment of N_2 Fixation by *Casuarina equisetifolia* inoculated with *Frankia* 021001 15N method, Soil Biology Biochem. 173 (3) : 375.

Gauthier, D., Diem, H.G. & Dommergues, Y. 1981. *In vitro* nitrogen fixation by from *Casuarina* nodules. Appl. Environ Microbial. 41 : 306 - 308.

Houwers, A. & Akkermans, A.D.L. 1981. Influence of inoculation on yield of *Alnus glutinosa* in the Netherlands, Plant soil. 61 : 189 - 202.

Lalonde, M. 1979. A simple and rapid method for the isolation, cultivation *in vitro* and characterization of *Frankia* strains from *Alnus* root nodules. In: Symbiotic nitrogen fixation. Wheeler and Perry, D. A., Oregon state univ. press, corvallis, oregon pp. 480.

Lalonde, M. & Fortin., J.A. 1973. Microscopic Photomiqnedes nodules racinaires exeniquesd *Alnus Crispa* var. moths. Can, J. Microbiol. 19: 1115-1118.

Lamont, H. C., Silvester, W.B. & Torrey, J. G. 1988. Nile red flurosence demonstrates lipid in envelope of vesicle from N_2 fixing culture of *Frankia* Can J Microbid., 34 (5) : 656-660

Lechevalier, M. P. & Ruan, J. S. 1984. Physiology and chemical diversity of *Frankia* spp. isolated from nodules of *Comptonia peregrina* and *Ceanothus americans*. Plant Soil. 778 : 15 - 22.

Lechevalier, M.P. Baker, D. & Horriere, F. 1983. Physiology, chemistry, serology, and infectivity of two *Frankia* isolates from *Alnus incana* subsp. *rugosa*. Can J. Bot. 61:2826-2833.

Lechevalier, M. P. & J. S. Ruan 1984. Physiology and chemical diversity of *Frankia* spp. Isolated from nodules of *Comptonia peregnna* (L.) Coult. and *Ceanothus americanus* L. Plant soil. 78:15-22.

Lechevalier, M. P. & Lechevalier, H. A. 1984. Taxonomy of *Frankia* p. 575-582. In L. Ortiz- Ortiz, L. F. Bojalil, and V. Yakoleff (ed.) Biological, biochemical, and biomedical aspects of Actinomycetes. Academic Press, Inc., New York.

Li. Z. G. & Ding. J. A. 1986. Isolation of *Frankia* sp. fsccol from root nodules of *Casuarina cunninghamiana* and its cultivation *in vitro* and nodulation conditions, Acta microbial. sin. 26 (41) : 295-301.

Mahadeva Swamy & Raj, J. 1998. Effect of dual inoculation with *Frankia* and *Pseudomonas cepacia* on growth and nodulation of *Casuarina equisitifolia*, 39th AMI conference. Mangalore. P. 249. (Abstr).

Mahadeva Swamy & Sreenivasa, M. N. 2000. Response of *Casuarina* for dual inoculation with *Frankia* sp and VAM fungus, *Glomus fasciculatum* 41st AMI, Conference., Jaipur. P.116 (Abstr.).

Meesters, T.M., Tvan Genesen & Akkermans, A.D.L. 1985. Growth acetylene reduction activity and localisation of nitrogen in relation to vesicle formation in *Frankia* strains Cc. 17 and Cpi. 2, Arch. Microbial. 143 : 137 - 142.

Meesters, T.M., Vaa Vhet, W.M. & Akkermans, A.D.L. 1987. Nitrogenase is restricted to the vesicles in *Frankia* strain EANI. Physiol plant. 70:267-271.

Mirza, M. S. D., Hahn & Akkermans, A. D. L. Isolation and characterization of *Frankia* strains from *Coriaria nepalensis*. Syst. Appp Microbiol. (in press).

Mirza, M. S., Janse, J.D., Haga, D. & Akkermans, A.D.L. 1991. Identification of a typical *Frankia* strains by fatty acid analysis, FEMS Microbial. Lett. 83:91-98.

Newcomb, W., Challham, D., Torrey J. G. & Peterson, R. L. 1979. Morphogenesis and fine structures of the actinomycetes endophyte of Nitrogen fixing root nodules of *Comptonia peregrina* Bot. Gaz., R10 (Suppl.): 522-24.

Norman, P. & Lalonde, M. 1982. Evaluation of *Frankia* strains isolated from Can. J. Microbial. 28:1133-1142.

Parsons, R., silvester, W. B., Harrio, S., Grujlevs 1987. *Frankia* vesicles provide inducible and absolute oxygen protection for nitrogenase Pl. physiol. 83(4): 728-731.

Pommer, E. H. 1959. Uber die Isolierung des Endophyten ausden Wnrzethnollchen *Alnus glutinosa* Gaertn. und uber erfol. Greuche reinfektions. Ber. Disth. Bot. Ges. 72 :138-150.

Quispel, A. & Tak, T. 1978. Studies on the growth of the endophyte of *Alnus glutinosa* (L.) Gaertn. in nutrient solutions. New Phytol. 81:587-600.

Quispel, A. 1990. Discoveries, discussions, and trends in research on actinorrhizal root nodule symbioses before 1978, p 15-33. In Schwintzer, C. R. & C. R. Tjepkema, C. R. (ed). The biology of *Frankia* and actinorrhizal plants. Academic Press, Inc., New York.

Raj. J., Meesters, T. M. & Akkermans, A. D. L. 1984. Studies on growth vesicles formation and nitrogen fixation (C_2H_2 reduction) by *Frankia* SP. CC1.17. Post Doctoral Report, Agril. Univ. of Wageningen, The Netherlands. PP. 39.

Reddell & Bowen, G. D. L. 1985. *Frankia* sources affects growth, nodulation and nitrogen fixation in *Casuarina* sps. New Phytol. 100 : 115.

Rose, S. L. & Youngberg C.T. 1981. Tripartite associations in snowbrush (*Ceanothus velutinus*): effect of vesicular - arbuscular mycorrhiza on growth, nodulation and nitrogen fixation. Can. J. Bot. 59 : 34-39.

Schultz, N. A. & Benson, D. R. 1989. Development potential of *Frankia* vesicles J. Bacterial. 171(12) : 6873-6877.

Silvester, W. B. & Winship, L. J. 1990. Transient responses of nitrogenase to acetylene and oxygen by actinorrhizal nodules and cultured *Frankia*. Plant physiol. 92: 480-486.

Smolander, A. & Sarsa, M. 1990. *Frankia* strains in soil under *Betula pendula*: behaviour in soil and in pure culture. Plant soil 122:129-136.

Smolander, A. 1990. *Frankia* populations in soils under different tree species with special emphasis on soils under *Betula pendula*, Plant soil. 121 : 1 - 10.

Smo lander, A. & Sundman. V. 1987. *Frankia* in aicd soils of forests devoid of actinorrhizal plants. *Physiol. Plant.* 70: 297-303.

Songoufora, B. Duhonx, E. & Dommergues, Y. R. 1989. Choice of criteria of selection in nitrogen fixing tree *Casuarina equisetifolia*. Acta. Oecol. 10(2): 215-224.

Stower, M. D. & Smith, J. E. 1985. Inoculation and production of container -grown red alder seedlings. Plant soil, 87 : 153 - 60

Tirumaleshwara Bhat, P., Raj. J. & Rai. P.V. 1987. A comparative study on biology of carbon metabolism and nitrogen fixation of Cc2 and Cp strains of *Frankia* sp. and effects of strain Ce2 on *Casuarina equisetifolia* 27th Conference of AMI (Abstr.)

Tjepkema, J. D., Ormerod, W. &. Torrey, J.G. 1980.Vesicle formation and acetylene reduction activity in *Frankia* sp. Cpll cultured in defined nutrient media. Nature. 287: 633-635.

Tyson, J. H. & Silver, W. S. 1979. Relationship of ultrastructure to acetylene reduction (nitrogen fixation) in root nodules of *Casuarina*. Bot. Gaz. 140 (Suppl.): S44 - S48.

Van Dijk, C. 1978, Spore formation and endophyte diversity in root nodules of *Alnus glutinosa* (L.) Vill. New Phytol. 81:601-615.

Wheeler, C. T., Hooker, J. E., Crowe, A. & Berrie, A. M. M. 1986. The improvement and utilization in Forestry of nitrogen fixation by actinorrhizal plants with special reference to *Alnus* in Scotland. Plant Soil. 21 : 393 - 406.

Zhang Z., Lopez, M. F. & Torrey J. G. 1984. A comparison of cultural characteristics and infectivity of *Frankia* isolated from root nodules of *Casuarina* species, Plant soil. 71:79-80 .

Zhong-Zhoghze & Torrey, J. G. 1985. Studies of an effective strain of *Frankia* and *Allocasuarina* of Casuarinaceae. Plant Soil. 87 : 1-16.

CHAPTER – 7

ENDOPHYTIC ASSOCIATION OF *Azorhizobium caulinodans* FOR NODULATION AND NITROGEN FIXATION IN CEREAL CROPS

S. Kannaiyan and K. Kumar

Azolla Laboratory, Tamilnadu Agricultural University, Coimbatore - 641 003 (TN)

1. INTRODUCTION

All the plants either legumes or non-legumes require, nitrogen the major nutrient, to form the green pigment chlorophyll and also to synthesize proteins and other organic compounds within the plant system for their growth and sustainability. In general, legumes such as soybean, groundnut, cowpea, mung bean, redgram, peas, chickpea, clovers, lucerne etc. are infected by the root nodule bacterium - *Rhizobium sp.*, and produce nodules and fix nitrogen from the atmospheric air. Legume nodules contain millions of *Rhizobium*, which convert nitrogen from the air into a form available to the plant and this simple biological process is known as "Biological Nitrogen Fixation" - BNF (Earl and Ausubel, 1993). BNF by legume - rhizobial symbiosis is still by far the best as far as nitrogen fixation and contribution is concerned. But non - legumes such as rice, wheat, corn, *Sorghum*, cumbu, sugarcane and grasses *etc.* do not possess the nodules in their system. Extending legume type BNF to non-legume crops would make them less dependant on external N supply (Kannaiyan 1993a, Kannaiyan, 1995).

Rice which is a staple food crop of the world entirely depends on nitrogen for its nutrition, crop growth and yield. It is grown in flooded condition in many rice growing countries in Asia (Kannaiyan, 1990; Kannaiyan, 1993b; Kannaiyan *et al.*, 1993). It is likely that higher and more consistent nitrogen fertility in wetland soil could be attributed to biological nitrogen fixation (BNF). BNF is an important process in rice farming system because it is an inexpensive source of nitrogen for rice which would not only provide economic benefit to small farmers but also improve the long term soil fertility in rice soil eco-system (Kannaiyan,

1994; Kannaiyan *et al.*, 1997). Also, nitrogen fertilization of non-leguminous crops such as maize is one of the most expensive inputs in agriculture. Eliminating nitrogen fertilization of maize will reduce the dependence on fossil fuels. Thus, the development of a symbiosis between diazotrophic bacteria and rice and maize would be of economic importance (Kannaiyan and Kumar, 2000; Kannaiyan *et al.*, 2000).

Nitrogen fixing bacteria in association with rice and maize are well known but their contribution to N supply is minor. Their productivity is severely limited by available N. Yields beyond our national average will not be attainable without chemical N fertilizers on innovative forms of BNF. In upland conditions where the BNF option is difficult to exploit and where other factors such as phosphorus deficiency intervene, the highest sustainable yields are unlikely to exceed 20 t/ha./ There are several strategies like improving efficient association between cereal roots and diazotroph, induction of N_2 fixing nodules in cereal roots and transfer of N_2 fixation capacity to rice genome for providing additional N inputs (Bennett and Ladha, 1992; Kannaiyan, 1998). Rice plant root associated nitrogen fixation involves complex interactons among the cultivars, bacterial population and the environment. Several strategies have been envisaged to achieve this goal, including the transfer of bacterial genes from free-living diazotrophs to non-legume crops (Triplett, 1996) and the induction of paranodules by adding 2, 4- D, NAA or cell wall degrading enzymes (Kennedy, *et al.*, 1997; Al-Mallah *et al.*, 1989; Amutha and Kannaiyan, 2000). Recently, it has been investigated at Tamil Nadu Agricultural University (TNAU) and found that the symbiotic nitrogen fixing bacterium *Azorhizobium caulinodans* isolated from the stem nodules of *Sesbania rostrata* could colonize and produce nodular structures in rice and maize roots (Kannaiyan, 2000).

2. ENDOPHYTIC NITROGEN FIXATION

"Endophytic nitrogen fixation is a process where the nitrogen fixing bacteria enter the plant roots through cracks or fissures in the lateral roots, colonize the inter-cellular spaces (cortex, within few cells of epidermis, inner cortex to epidermis, vascular tissues and xylem), where without differentiating into bacteroids fix atmospheric nitrogen micro-aerobically, with the help of the enzyme "nitrogenase" (Kennedy *et al.*, 1997). In the earlier days, it was thought that only by inducing nodular structures to provide a place for the organism specifically, BNF can be extended to cereals (Arora, 1954). But now recent researches have vividly brought down the difficulties and increased the scope for N_2 - fixation in cereals by means of endophytic nitrogen fixation (Kannaiyan, 2000; Stoltzfus *et al.*, 1997).

The extension of biological nitrogen fixation to non-legume crops such as rice, maize *etc.* would be of enormous economical and environmental impact. The discovery that bacteria of the genera *Rhizobium* and *Bradyrhizobium* can form effective nitrogen nodules on the non-legume *Parasponia* (Davey *et al.*, 1997) provided added impetus to this approach. Many of these nodular structures have to be artifically induced by adding 2, 4-D (Kennedy *et al.*, 1990; Tchan *et al.*, 1991; Buvana and Kannaiyan 2000) or cell wall degrading enzymes (Al-Mallah *et al.* 1989; Tchan *et al.*, 1991; Cocking *et al.*, 1990; Amutha and Kannaiyan, 2000). Cocking *et al*, (1993) reported that *Rhizobium* of *Parasponia* and *Aeschynomene* produce nodule like structure on the emerging lateral roots of rice and maize without any enzyme treatment. Kennedy and Tchan (1992) studied various approaches for the induction of nodule like strucutres on the roots of rice and other cereals.

Zeman *et al.*, (1992) reported that the exogenous application of the synthetic auxin 2, 4-D to rice induces modified root outgrowths (MROs) which result from the induction of meristem. Reports have been made that certain rhizobia can enter maize and rice at low frequency (Cocking *et al.*, 1994). Christiansen-Weniger (1996) reported nodule like structure formation in rice roots treated with 2, 4-D and inoculated with *A. caulinodans*. In maize, Christiansen -Weniger (1988) reported para-nodulation by *Azospirillum brasilense*. Recently it has been discovered that the endophytic diazotrophs *Acetobacter diazotrophicus* and *Herbaspirillum* spp. could contribute substantially to the growth of sugarcane *via* nitrogen fixation and without the formation of any nodulelike structures (Boddey *et al.*, 1995). It is suggested to have a thorough knowledge about the plant genes involved in nodulation and symbiotic nitrogen fixation and the function of their gene products before the development of true nodulation of cereals (de Brujin *et al.*, 1995; Kennedy *et al.*, 1997; Webster *et al.* 1997). Now, opinion has consequently changed and the emphasis is to try to establish stable endophytic associations between diazotrophic bacteria and non-legume crops.

Two major approaches are currently being undertaken to try to extend biological nitrogen fixation to non-legume crops. The first is to look for natural endophytic diazotrophs of non-legume crops and to assess such endophytic nitrogen fixation. This is being exploited with sugarcane (Boddey *et al.*, 1995) and other crops such as rice and wheat (Sabry *et al.*, 1997). The second approach is to determine whether the rhizobia can internally colonize and ultimately fix nitrogen in non-legumes. The first step is to study the possibility of intercellular colonization of diazotrophs with non-legumes including entry, spread and the mechanism involved. A sensitive and quantifiable system is needed for this approach. Reports of inter-cellular colonization by diazotrophs in non-legumes have been made. Reinhold-Hurek *et al.* (1993) isolated *Azoarcus* sp. BH-72 from Kallar grass and reported their endophytic association with rice and other grasses. Microscopic observation of colonization is greatly facilitated by using a genetically modified bacterial strain tagged with a reporter gene (Webster *et al.*, 1997). The efficacy of endophytic colonization by diazotrophs *Azospirillum* and *Azorhizobium* tagged with *lac* Z marker genes in rice was compared. It was noticed that the colonization by diazotrophs was mainly through crack entry. In the case of *Azospirillum* the bacteria entered the lateral cracks and their invasion was not so deep and very much limited to endodermis and cortex, whereas, *Azorhizobium* exhibited deep colonization. They first got entry through lateral cracks and traverse intercellularly and colonized xylem of rice var. ADT 36 (Gopalaswamy *et al.*, 1998).

3. PRE-REQUISITES FOR AN ENDOPHYTIC RELATIONSHIP

There are several prerequisites for the establishment of an endophytic situation (Quispel, 1991) :

(i) Hostile reactions from the host, like hypersensitive reactions and production of phytoalexins have to be prevented.

(ii) The bacteria must find an inter or intracellular niche, where sufficiently reducing conditions are provided and the PO_2 is regulated by diffusion barriers and/or leghaemoglobins.

(iii) Membrane systems around the endosymbiont of microbial and host origin must be adapted in such a way that a bi-directional transport of substances is possible.

4. PARA-NODULAR BACTERIA

Bacteria growing within the paranodule displayed variable morphology, differing in size and cell contents. Electron transparent regions within some bacterial cells of these sections may have accumulations of Poly-B-hydroxybutyrate. Considerable amounts of indeterminate material were found together with these bacteria. This was surprising, since the bacteria were extracellular in the middle lamellar space or in rare cases, present in dead cells *Azorhizobium caulinodans* ORS 571 induces the formation of root and stem nodules on the tropical legume *Sesbania rostrata, Azorhizobium* (that can use nitrogen while free living) cells are Gram negative, small rods, 0.5 to 0.6 µm by 1.5 to 2.5 µm and motile. The cells have peritrichous flagella on solid medium and one lateral flagellum in liquid medium. Colonies on agar are circular and have a creamy color and obligately aerobic in nature. It fixes atmospheric N_2 under aerobic conditions and grow well on N_2 with vitamins present in a nitrogen free medium (Dreyfus *et al.*, 1988). Organic acids such as lactate and succinate are the favorite carbon substrates for both NH_4 and N_2 dependent growth. Gebhardt *et al.* (1984) noticed the higher tolerance of *Azorhizobium* to oxygen upto 9µ moles. Balasubramani and Kannaiyan (1992a) found maltose, trehalose and sucrose were found to be the best carbon sources.

This bacterium contains essential nodulation genes as that of *nod ABC* of rhizobia (Goethals *et al.*, 1989). Hence it is possible to elucidate the *nod* genes in the biosynthesis of the Nod factor induced by host plant exudate. In view of reports of associative N_2 fixation by *Azorhizobium* in rice fields, it will be interesting to study this behaviour and identify functions that contribute to the interactions between *Azorhizobium* and rice plants. The induction of paranodules and the invasion of rice and maize by *Azorhizobia* either spontaneously or after the treatment of the roots with cell wall degrading enzymes have been reported (Cocking *et al.*, 1990). The nodule like structures induced by 2, 4-D can be inhabited by diazotrophs (Nie, 1983). It is possible that 2, 4-D has at least two effects :

1. An auxin-like activity controlling development of nodular structure;
2. An effect enabling increased frequency of infection of plant root cells and colonization of nodular structures with certain bacteria.

The choice of the bacterial strain, *A. caulinodans - ORS 571* was based on three characteristics (de Bruijin, *et al.*, 1995). Firstly as it is unlikely that there will be bacteroid differentiation in non-legume-plants, ORS 571 has the unusual ability among the rhizobia to fix nitrogen in free-living conditions without differentiation into bacteroids. Secondly, as it is also unlikely that non-legumes will be able to reproduce the sophisticated oxygen barriers which exist in legume nodules to protect the nitrogenase enzyme complex from inhibitory oxygen levels, ORS 571 can tolerate up to 12 µM dissolved oxygen while fixing nitrogen in culture. Thirdly ORS 571 forms nodules on its host plant following crack entry infection i.e. inter-cellulary between adjacent cells.

In TNAU, much work has been made on the isolation and characterization of efficient isolates of *Azorhizobium caulinodans* from both *S. rostrata* and rice soil previously cropped with *S. rostrata.* In earlier times, stem nodules in *S. rostrata* has attracted much attention in terms of biomass-production and nitrogen contribution to wet land rice (Kannaiyan, *et al.*, 1988; Kannaiyan and Kalidurai 1989a & b, 1994; Balasubramani and Kannaiyan, 1991, 1992; Kalidurai and Kannaiyan,

1991c; Sundaravarathan and Kannaiyan 1998; Kannaiyan *et al.*, 1998). Kalidurai and Kannaiyan (1991a) have reported that both stem and root isolates of *A. caulinodans* could tolerate combined nitrogen at 2.5 ppm and fix N_2. *S. rostrata* callus inoculated with *A. caulinodans* showed nitrogenase activity under *In vitro* conditons (Balasubramani, 1990; Chitra, 1992; Amutha, 1997). The epiphytic occurrence of *A. caulinodans* in *S. rostrata* plant parts including leaves, stem, flowers, pistil was reported (Balasubramani and Kannaiyan, 1991 a, b; Balasubramani *et al.*, 1992 a, b). Interestingly the bacterium *A. caulinodans* was isolated from the mouth parts of ants, mealy, bug, beetles (Kannaiyan, 1997).

The movement of *A. caulinodans* within *S. rostrata* was studied and established by isolation from different plant parts of inoculated *S. rostrata* (Chitra and Kannaiyan, 1994). They also reported the presence of *A. caulinodans* both in the rice rhizoplane region and in rice soil.

5. MODE OF ENTRY AND COLONIZATION

The invasion of root cells by endophytic bacteria and their colonization have attracted the interest of many workers. The bacteria are attracted by certain root exudates. In cereals like rice and maize, since the exopolysaccharide production and the production of flavonoids like naringenin, luteoline and daidzein are low, by supplying them artificially at very low concentrations invasion can be enhanced (Kennedy *et al.*, 1997; Sabry *et al.* 1997). The bacterium searches for cracks or fissures through which it can enter the roots. The bacteria enters through cracks in the lateral roots. It moves through the intercellular spaces and finds a site of low redox potential better suited for N_2 - fixation. A large number of bacteria were also found associated with the para-nodules (Kennedy *et al.*, 1997).

6. PARA-NODULES

Para-nodules are nodular structures produced by modification from the lateral roots of many plant species due to the stimulating effect of auxins such as 2, 4-D (Tchan and Kennedy, 1989). These structures may be inhabited by *Azospirilla* or *Azorhizobia* but quite dissimilar to legume nodules (Kennedy and Tchan, 1992).

In case of Azospirilla, the bacteria colonize inter-cellularly almost exclusively, usually in the basal zone of the para-nodules where plant cells appear loosely packed. The colonization is similar with *Azorhizobium* also and there is no evidence of intra-cellular colonization of cereals as occurs in legume nodules (Nie, 1983). However, recently it has been shown that *Azorhizobium caulinodans* colonizes the xylem vessels of rice (Gopalaswamy *et al.*, 2000) as a potential endophyte.

7. STRUCTURE OF PARA-NODULES

Para-nodular structures were formed when seedlings were treated with 2, 4-D. The para-nodules were spheroidal in shape, and less than a millimeter in diameter. The frequency of para nodulation on the young roots was 1-4 nodules per cm of root length. Sections of para-nodules show the presence of central vascular tissue, connected to the root style. Several layers of cortical cells, usually without recognizable cytoplasmic contents, surrounded this rudimentary stele. This is in contrast to the organisation of vascular tissue in legume nodules, in which stele surrounds the infected cells. In early stages, root-cap like structures were found on the para-nodules. In later stages, outer cells in the region showed swelling and sloughed off frequently. Occasionally bacteria were observed inside the outer 'empty' cortical cells but never in large numbers. In contrast, extensive bacterial infection was confined to the inter-cellular spaces. Serial sectioning of para-nodules showed that the bacteria and associated material were present in large, irregular channels open to the exterior. Examination of roots by phase-contract and light microscopy indicated that extensive colonization of the root surface and root hair by the endophytic bacteria. Electron microscopical observation of the paranodules revealed the presence of large numbers of bacteria in inter-cellular spaces. These irregular spaces or channels were developed by splitting of the middle lamella, during bacterial colonization. These channels were surrounded by shrunken and almost empty cortical cells. Only cytoplasmic debris was observed in these empty cells.

8. EFFECT OF GROWTH REGULATORS AND CELL WALL DEGRADING ENZYMES ON INDUCTION OF PARA-NODULES IN RICE AND MAIZE

The inoculation of the bacterium, *A. caulinodans* with growth regulators produces lateral rootlets formation in cereal crops. During the development of lateral root emergence, a minute way is formed on both sides of the lateral root by which the bacterium, *A. caulinodans* could colonize well and invade into the roots of cereal crop through crack entry and gradually reach to the inner region of the rice root tissues. Seven days old rice seedlings of ADT36, when inoculated with the symbiotic nitrogen fixing bacterium, *A. caulinodans*, along with the growth regulators such as 2, 4-D, NAA, Kinetin could induce the formation of nodular structures. Similar observations with the growth regulators such as 2, 4-D, NAA, Kinetin could induce the formation of root nodular structures. Root nodular structures were produced with *Sorghum*, maize, ragi, cumbu and sugarcane. Among the cereal crops tested maize crop responded well for the inoculation of *A. caulinodans* and prominent nodular structures have been noticed. Buvana and Kannaiyan (2000) studied the effect of growth regulator NAA and cell wall degrading enzymes (cellulase and pectinase) on induction of para nodules in rice and maize by *A. caulinodans*. Inoculation of *A. caulinodans* with NAA and cell wall degrading enzyme mixture (cellulase + pectinase) decreased the root growth of seedlings compared to inoculation of *A. caulinodans* alone. Inoculation of *A. caulinodans* and cell wall degrading enzyme mixture developed lateral rootlets which were short and thickened. The number of nodule like structures and total nitrogen content was maximum with inoculation of *A. caulinodans* strains in combination with cell wall degrading enzyme mixture and NAA (Table 1, 2). Inoculation of *A. caulinodans* was found to increase the root and shoot growth, total biomass and lateral rootlets of rice seedlings over uninoculated control. The growth regulator kinetin was used at 3.0, 5.0 and 7.0 ppm concentration and the levels above 3.0 ppm was found to be inhibitory to the seedlings by recording decreased root and shoot growth, total biomass and lateral rootlets. The inoculation of *A. caulinodans* could nullify the adverse effect of kinetin to some extent compared to the incorporation of growth regulator treatment alone. Increased concentrations of kinetin with inoculation of *A. caulinodans* increased the number of

Table 1. Effect of NAA and inoculation of *A. caulinodans* strains with enzyme mixture (cellulase + pectinase) on the number of lateral rootlets (LR) and the number of nodule like structures (N) of rice (ADT 36)

Treatments	RB-SK-1		RB-SK-2		RB-SK-3		RB-SK-4		RB-SK-5		RB-SK-6		RB-SK-7		RB-SK-8		RB-SK-9		RB-SK-10	
	LR	N	LR	N	LR	N	LR	N	LR	N	LR	N	LR	N	LR	N	LR	N	LR	N
Uninoculated	64.67	–	65.67	–	63.00	–	67.67	–	65.33	–	64.67	–	65.00	–	66.00	–	62.33	–	65.33	–
A. caulinodons (A.c.) alone	86.00	–	85.33	–	83.33	–	87.00	–	85.00	–	87.33	–	84.33	–	86.00	–	81.67	–	84.33	–
NAA (3.0 ppm)	55.67	–	53.00	–	57.33	–	51.33	–	52.00	–	54.00	–	54.67	–	56.67	–	57.33	–	57.33	–
NAA + *A.c*	81.00	32.00	82.00	34.67	84.33	–	82.67	32.33	82.33	34.00	84.00	31.67	84.33	33.00	85.00	–	80.6	34.33	82.00	34.23
NAA + *A.c* + Cellulase	57.33	24.33	58.67	23.00	59.33	–	57.00	25.00	56.67	25.67	60.67	23.33	57.67	27.33	60.00	–	60.33	24.67	61.67	23.33
NAA + *A.c.* + Cellulase + Pectinase	77.67	35.33	78.67	32.00	76.00	–	76.67	34.33	75.00	36.00	79.67	33.33	77.00	35.33	80.67	–	74.67	36.00	74.00	35.33
NAA + *A.c* + Pectinase	58.00	23.33	56.67	18.33	57.67	–	56.67	21.33	55.33	21.33	56.33	22.33	56.66	24.67	57.67	–	57.00	24.00	57.33	22.00
A.c. + Cellulase	65.67	–	68.33	–	68.33	–	70.00	–	64.00	–	68.33	–	70.67	–	70.67	–	66.00	–	65.67	–
A.c. + Pectinase	65.33	–	64.67	–	68.67	–	65.00	–	63.00	–	68.33	–	66.33	–	65.67	–	65.00	–	64.00	–

	LR		N	
	SEd	CD	SEd	CD
Treatment	0.34	0.66	0.61	1.25
Culture	0.35	0.70	0.99	1.98
Treatment x Culture	1.06	2.09	1.99	3.95

Table 2. Effect of NAA and inoculation of *A. caulinodans* strains with enzyme mixture (cellulose + pectinase) on the total nitrogen content of rice (ADT 36)

Treatments	Total nitrogen content (%)									
	RB-SK -1	RB-SK -2	RB-SK -3	RB-SK -4	RB-SK -5	RB-SK -6	RB-SK -7	RB-SK -8	RB-SK -9	RB-SK -10
Uninoculated	0.217	0.220	0.213	0.222	0.223	0.226	0.217	0.213	0.224	0.223
A. caulinodans (A.c) alone	0.268	0.318	0.305	0.318	0.317	0.317	0.295	0.296	0.312	0.310
NAA (3.0 ppm)	0.188	0.181	0.175	0.176	0.182	0.180	0.178	0.182	0.180	0.180
NAA + A.c	0.364	0.373	0.375	0.374	0.378	0.365	0.365	0.344	0.377	0.363
NAA + *A.c* + Cellulose	0.330	0.349	0.356	0.343	0.359	0.348	0.335	0.325	0.353	0.324
NAA + *A.c.* + Cellulose + Pectinase	0.447	0.456	0.454	0.457	0.459	0.451	0.446	0.349	0.456	0.441
NAA + *A.c.* + Pectinase	0.313	0.335	0.325	0.336	0.335	0.327	0.320	0.325	0.339	0.318
A.c. + Cellulose	0.224	0.242	0.242	0.248	0.239	0.233	0.226	0.231	0.236	0.214
A.c. + Pectinase	0.224	0.239	0.235	0.237	0.232	0.226	0.229	0.228	0.231	0.221

	SEd	CD
Treatment	0.002	0.004
Culture	0.002	0.004
Treatment x Culture	0.006	0.012

Table 3. Effect of Kinetin and inoculation of *A. caulinodans* ZB-SK-5 on the number of nodule like structure of rice

Treatment	Para-nodules (Numbers)			
	ADT-36	Purple puttu	ASD-16	TKM-9
Uninoculated	—	—	—	—
A. caulinodans (A.c) alone	—	—	—	—
Kinetin (3.0 ppm)	—	—	—	—
Kinetin (3.0 ppm) + A.c	20.67	22.67	20.33	25.33
Kinetin (5.0 ppm)	—	—	—	—
Kinetin (5.0 ppm) + A.c.	31.67	31.67	27.00	35.30
Kinetin (7.0 ppm)	—	—	—	—
Kinetin (7.0 ppm) + A.c.	46.00	46.67	48.00	50.30
CD (P = 0.05)	3.705	4.819	4.909	7.005

nodule like structures in the roots of young rice seedlings. The number of nodule like structure were relatively more in rice variety TKM-9 compared to the rice variety ADT-36 (Table 3). It is interesting to note that the increased concentrations of kinetin with inoculation of *A. caulinodans* increased the number of nodule like structures in rice seedlings. However, there is no nodule like structure formation in rice roots when *A. caulinodans* and growth regulator kinetin were treated in the seedlings separately (Amutha and Kannaiyan, 2000).

9. ROLE OF FLAVONOIDS ON INDUCTION OF PARANODULES

Flavonoids are a class of phenolics secreted by plant roots as part of the regulator system for colonization of *A. caulinodans*. Four families of flavonoids such as flavones, flavonols, flavonones and iso-flavonones have been shown to have nod-gene inducing activity. Naringenin- a specific flavonoid is known to stimulate the colonization of *A. caulinodans* in rice roots. It is a signal molecule for attracting the population of *A. caulinodans*. But rice root did not produce the flavonoid Naringenin. The flavonoid Naringenin is considered as a potential signal molecule to activate the colonization of *A. caulinodans* in rice roots and, therefore, a specific gene could be introduced into rice plants for synthesis of Naringenin. Flavonoids also enhance the growth rate of certain rhizobia and phenolic compounds in general could induce expression of a variety of genes in plant associated bacteria. The flavonoids naringenin and diadzein, at low concentration were found to significantly stimulate the frequency of lateral root cracks (LRC) and inter-cellular colonization by *A. caulinodans* in non-legumes by acting as chemical stimulants (Gough *et al.*, 1996). Flavonoids have been shown to accumulate in many plants during infection. However, there are few reports in cereals crops showing such accumulation. Naringenin was detected in rice leaves exposed to UV irradiation. Kumar *et al.*, (1996) while studying the effect of flavonoid naringenin on the endophytic colonization by *A. caulinodans* in rice, reported favourable stimulation of lateral root formation by naringenin in ADT36 rice. They also reported that the main mode of entry of *A. caulinodans* in rice roots is by crack entry through the emerging lateral roots. The effect of naringenin (3.0 ppm), enzyme mixture and *A. caulinodans* on induction of para nodules in rice and maize was reported (Senthil Kumar, 2000). It was found that addition of naringenin enhanced the short thickened lateral root formation. The seedlings treated with narigenin, enzyme mixture and *A. caulinodans* recorded maximum number of lateral rootlets, rhizoplane population, nodule like structures and nitrogenase activity. Lateral root development of rice and colonization of lateral root cracks by bacteria were shown to be stimulated by the flavonoid naringenin. The effect of flavonoid on root morphology of different rice varieties was studied and found that flavonoid naringenin at 5 x 10^{-5} M concentration significantly stimulated the lateral root formation in rice varieties (Gopalaswamy *et al.*, 1998).

10. ENDOPHYTIC COLONIZATION OF XYLEM OF RICE ROOTS BY *A. CAULINODANS*

The diazotroph, *Azorhizobium caulinodans* - ORS 571 colonizes the xylem of rice roots (Gopalaswamy *et al.*, 2000). Colonization of the xylem of rice by *A. caulinodans*. ORS 571 is of interest because it might provide a suitable nutritional and environmental niche for nitrogen fixation. Rice seedlings (ADT 36, CO43, CR 1009 and the American cv Lemont) were growing aseptically in the presence of flavonoid naringenin inoculated with ORS 571 (pXLGD4), carrying the *lacZ* reporter gene. By using the bacterial strains carrying a marker gene, it is now possible to detect both visually and enzymatically the internal colonization of inoculated diazotrophs. Generally the colonization was more in older roots than younger roots. None of the rice varieties showed positive ARA due to inoculation and this might be due to the lack of carbon sources in the tube grown experiment. Microscopic analysis of sections of inoculated rice roots showed the colonization of xylem rice roots by *A. caulinodans*. So far endophytic colonization study using the marker strain of *A. caulinodans* was confined to rice only among cereals. Once the bacterium is established in xylem with adequate colonization it might be capable of fixing nitrogen. Thus, colonization of xylem of rice roots would create an excellent condition for the bacterium to enter into the entire system of rice plant including the leaf system which would be a boon to full fill the requirements of nitrogen through biological nitrogen fixation.

11. ECOLOGICAL AND ECONOMIC ASPECTS

Here it has to be considered whether under field conditions, these non-legumes show the properties that would be expected on the basis of green house experiments. As it happens in the case of legumes, it is now being realized that these endophytes also aid in improving soil properties (Dobereiner *et al.*, 1995). The crop residues and stubbles in a major route for the transfer of fixed nitrogen to the soil. The decomposition of dead roots and nodules, and a possible excretion of nitrogen have also to be considered (Rinaudo *et al.*, 1983). It is thought that these organisms have an ability to act as pioneers in nitrogen poor soils and subsequently raise the level of soil nitrogen (Jain and Patriquinn, 1984).

In summary, there is reasonable evidence that this endophytic nitrogen fixation can attain a magnitude which is of definite ecologfical significance and economic advantage. Additionally, it should be noted that these endophytes constitute a group of species which can tolerate adverse conditions of growth such as low temperature, or waterlogged and acid soils. Thus the group is in a position to fulfill a role of nitrogen providers under very varied circumstances and conditions.

12. POTENTIAL FOR FUTURE RESEARCH

Extending endophytic nitrogen fixation to all cereal crops of economic significance may be tried by the following two methods:

(a) Genetic Modification of Host Plant

- To avoid hypersensitive reactions of the host plant.
- To avoid production of phyto-alexins.
- To increase host compatibility to accommodate endophytic nitrogen fixers.
- To provide areas of low redox potential for the establishment of endophytes by means of diffusion barriers or leghaemoglobins.
- To make the membrane systems around the endosymbiont facilitate bi-directional transport of substances.
- To study *nif*-gene transfer.

(b) Genetic Modification of the Micro-organisms

- To induce production of auxins such as 2, 4-D, NAA etc.
- To produce cellulases and pectinases, which can degrade the cell walls.
- To make the endophytes more tolerant to oxygen concentrations.
- To increase the ability of endophytes to colonize both intra and inter-cellularly.
- To increase the span of microbial life in vegetative phase.
- To make the microorganisms use the host materials as energy source but without any phyto-pathogenic effect.
- To make endophytic microorganisms produce sufficient amounts of superoxide dismutase.

- Selection of promising strains possessing high nitrogenase activity.
- To screen germplasm for diazotrophic colonization.
- To investigate root morphologenesis in relation to invasion.
- To study in detail the interactive responses between flavonoids and nod-factor.
- To investigate the metabolic programming of cereals for attachment of micro-organism and protection of nitrogenase from oxygen inactivation.

13. CONCLUSION

The ambitious goal of achieving nitrogen fixation in cereals naturally arouses controversy. However, despite the biological obstacles, there are strong and obvious grounds for achieving it based on environmental and economic considerations (Cocking *et al.*, 1990). This goal is also towards a more organically based agriculture (Kalidurai and Kannaiyan, 1991b). The progress made in this field of study since 1992 suggests that the research is on schedule to deliver positive outcomes in the medium term of 5-10 years.

Currently there are challenges with respect to factors such as whether the high numbers of bacteria found endophytically in seedlings (10^8 cells g^{-1} fresh weight roots) can be maintained in larger plants and whether nitrogenase activity will be adequate with the O_2 concentrations found in soil. Overcoming these obstacles will require significant genetic improvements of both host plant and bacteria. These improvements can be hastened by selections made in plants for superior genotypes favouring endophytic colonization of the plant.

REFERENCES

Al-Mallah, M.K., Davey, M.R. & Cocking, E.C. 1987. Formation of nodular structure on rice seedlings by rhizobia. J. Exp. Bot. 40 : 473-478.

Amutha, K. 1997. Studies on the inoculation of *Azorhizobium caulinodans* on induction of paranodules and nitrogen fixation in rice. M.Sc. Ag. Thesis, Tamil Nadu Agric. Univ., Coimbatore, Tamil Nadu, India, p. 147.

Amutha, K. & Kannaiyan, S. 2000. Effect of kinetin and inoculation of *Azorhizobium caulinodans* on induction of lateral rootlets and para nodules in rice. J. Microb. World. 2 (1) : 1-7.

Arora, N. 1954. Morphological development of the root and stem nodules of *Aechynomene indica.* J. Phytomorphology. 4 : 211-216.

Balasubramani, G. 1990. Studies on nitrogen fixation in stem nodulating *Sesbania rostrata* by *Azorhizobium caulinodans.* M.Sc. (Ag.) Thesis, Tamil Nadu Agric. Univ., Coimbatore, Tamil Nadu, India, p. 128.

Balasubramani, G. & Kannaiyan, S. 1991a. Epiphytic occurrence of *Azorhizobium caulinodans* in leaves and flowers of *Sesbania rostrata.* IRRI INSURF Netwatch. 1 : 4-5.

Balasubramani, G. & Kannaiyan, S. 1991b. Natural spread of *A. caulinodans* in stem nodulating *S. rostrata.* INSURF IRRI Netwatch. 1 : 5-6.

Balasubramani, G. & Kannaiyan, S. 1992a. *In vitro* screening of media for nodulation by stem and root azorhizobial isolates of *Sesbania rostrata.* India J. Microbiol. 32 92) : 211-212.

Balasubramani, G. & Kannaiyan, S. 1992b. Influence of waterlogging condition on stem nodule formation in *Sesbania rostrata* IRRI Netwatch 2 (3) : 6-7.

Balasubramani, G., Krishnaveni, S. & Kannaiyan, S. 1992a. Inhibition of *Azorhizobium* and *Rhizobium* by seed Leachates. World J. Microbiol. Biotechnol. 8 (5) : 542-543.

Balasubramani, G., Krishnaveni, S. & Kannaiyan, S. 1992b. *Sesbania* species Leachates and the growth of *Azorhizobium* and *Rhizobium*. In : Organic recycling in Asia and the Pacific, FAO RAPA Bull. : 8 : 7-8.

Bennett, J., & Ladha, J.K. 1992. Feasibility of nodulation and nitrogen fixation in rice. In : Nodulation and Nitrogen Fixation in Rice. Potential and Prospects (eds.) G.S. Khush and J. Bennett. Inte. Rice. Res. Inst. Philippines, pp. 1-14.

Boddey, R.M., de Oliveria, O.C., Urquiaga, S. Reis, V.M. de Olivares, F.L. Baldani, V.L.D. & Dobereiner, J. 1995. Biological nitrogen fixation associated with sugar cane and rice : Contributions and prospects for improvement. Plant Soil. 174 : 195-209.

Buvana, R. & Kannaiyan, S. 2000. Studies on the effect of NAA and enzyme mixture on induction of nodule like structures in rice by inoculation with *Azorhizobium caulinodans*. Paper presented in 41st Annual conf. of AMI, November 25-27, 2000, Birla Institute of Scientific Research, Jaipur, Rajasthan, India. p. 99.

Chitra, C. 1992. Studies on seed coating with certain carbon sources on the survival of *Azorhizobium caulinodans* in seeds and foliar inoculation in *Sesbania rostrata* M.Sc. (Ag.) Thesis. Tamil Nadu Agricultural University, Coimbatore, Tamil Nadu, India, p. 230.

Chitra, C. & Kannaiyan, S. 1994. Application of stem nodulating *Sesbania rostrata* for rice production. In : Rice management Biotechnology (ed.) S. Kannaiyan, Assoc. Publ. Co., New Delhi, India, pp. 217-236.

Christiansen-Weniger, C. 1996. Endophytic establishment of *Azorhizobium caulinodans* through auxin-induced root tumors of rice (*Oryza sativa* L.) Biol. Fertil. Soils. 21 : 293- 302.

Christiansen-Weniger, C. 1988. Influence of plant growth substances on growth and nitrogenase activity from *Azospirillum brasilense*, In : *Azospirillum* IV : Genetics, physiology, ecology (ed.) W. Klingmuller. Springer Verlag, Berlin, Heidelberg, pp. 141-149.

Cocking, E.C., Al-Mallah, M.K. Benson, E. & Davey, M.R. 1990. Nodulation and nitrogen fixation of non-legumes by rhizobia. In : Nitrogen fixation : achievements and objectives (eds.) P.M. Gresshoff, L.E. Roth, G. Stacey and W.E. Newton, Chapman and Hall, New York, pp. 813-820.

Cocking, E.C., Davey, M.R. Kothari, S.L. Srivastava, Y.X. Jing, J. Ridge, R.W. & Rolfe, B.G. 1993. Altering the specificity control of the interaction between rhizobia and plants. Symbiosis. 14 : 123-130.

Cocking, E.C., Kothari, S.L. Batchelor, C.A. Jain, S. Webster, G. Jones, J. Jotham, J. & Davey, M.R. 1994. Nodulation of non-legume crops : a new look. Agro-food-Industry Hi-Tech. 1 : 21-24.

Davey, M.R., Webster, G. Manders, G. Ringrose, F.L. Power, J.B. & Cocking, E.C. 1997. Effective nodulation of micro-propagated shoots of the non-legume *Parasponia andersonii* by *Bradyrhizobium* J. Exp. Bot. 44 : 863-867.

de Bruijin, F.J., Jing, Y. & Dazzo, F.B. 1995. Potential and pitfalls of trying to extend symbiotic interactions of nitrogen fixing organisms to presently non-nodulated plants, such as rice. Plant Soil. 174 : 225-240.

Dobereiner, J., Baldani, V.L.D., & Reis, V.M. 1995. Endophytic occurrence of diazotrophicbacteria in non-leguminous crops. In : NATO ASI series, vol. G37 (*Azospirillum* VI and related microorganisms) (eds.) I. Fendrik, M. del Gallo, J. Vanderleyden and M. de Zamaroczy, Berlin, Heidelberg : Springer-Verlag. pp. 3- 14.

Dreyfus, B.L., J. Garcia, L. & Gillis, M. 1988. Characterization of *Azorhizobium caulinodans* gen. nov. sp. nov., a stem nodulating nitrogen fixing bacterium isolated from *Sesbania rostrata*. Inte. J. Syst. Bacteriol. 38 (1) : 89-98.

Earl. C.D, & Ausubel, F.M. 1993. The genetic engineering of nitrogen fixation. Nutr. Rev. 41 : 1-6.

Gebhardt, C., Turner, C.L., Gibson, A.H. Dreyfus, B.L. & Bergerson, F.J. 1984. Nitrogen fixing growthin continuous culture of a strain of *Rhizobium* spp. isolated from stem nodule of *Sesbania rostrata* J. Gen. Microbiol. 130 : 843-848.

Goethals, K., Gao, M. Tomepke, K. Van Montagu, M. & Holsters, M. 1989. Common nod ABC genes in Nod locus I of *Azorhizobium caulinodans* nucleotide sequence and plant inducible expressions. Mol. Gen. Genet. 219 : 289-298.

Gopalaswamy, G., Kannaiyan, S. O'Callaghan, K.J. Davey, M.R. & Cocking E.C. 1998. Colonization of xylem of rice (*Oryza sativa*) by *Azorhizobium caulinodans* ORS - 571. Paper presented in IIIrd European N_2 Fixation Conference, Lunteren, Netherlands.

Gopalaswamy, G., Kannaiyan, S. O'Callaghan, K.J. Davey, M.R. & Cocking, E.C. 2000. The xylem of rice (*Oryza sativa*) is colonized by *Azorhizobium caulinodans*. Proc. R. Soc. Lond. 267 : 103-107.

Gough, C., Webster, G. Vasse, J. Galera, C., Batchelor, C O'Callaghan K.J. & M. Davey, 1996. Specific flavonoids stimulate intercellular colonization of non-legumes by *Azorhizobium caulinodans*. In Biology of plant - microbe interactions (eds.) G. Stacey, B. Mulin and P.M. Gresshoff, Proc. 8th Internatl Cong. Molec. Plant- Microbe interactions. Knoxville, USA, 14-19 July. pp. 409-415.

Jain, D. & Patriquinn, D. 1984. Root hair deformation, bacterial attachment and plant growth in wheat - *Azospirillum* interaction. Appl. Environ. Microbiol. 48 : 1208-1213.

Kalidurai, M. & Kannaiyan, S. 1991a. *Sesbania* as a biofertilizer for rice. Bioresource Technol. 36 (2) : 141-145.

Kalidurai, M. & Kannaiyan, S. 1991b. Effect of *Sesbania rostrata* on N_2 uptake and yield of lowland rice. J. Agron. Crop Sci. 163 : 284-288.

Kalidurai, M. & Kannaiyan, S. 1991c. Studies on the biomass production, nodulation and nitrogen fixation in stem nodulating *Sesbania rostrata*. In : Biological nitrogen fixation associated with rice production (ed.) S.K. Dutta and C. Sloger, Oxford & IBH publishing Co. Pvt. Ltd., New Delhi, pp. 291-305.

Kannaiyan, S. 1990. Biotechnology of Biofertilizers for rice. Tamil Nadu Agri. Univ. Coimbatore, Tamil Nadu, India, p. 225.

Kannaiyan, S. 1993a. Nitrogen contribution by *Azolla* to rice crop. Proc. Indian Natn. Sci. Acad. B. 59 (3 &4) : 309-314.

Kannaiyan, S. 1993b. Biofertilizers for Rice. Tamil Nadu Agricultural University Press, Coimbatore, Tamil Nadu, India, pp. 225.

Kannaiyan, S. 1995. Biofertilizers for rice - Status and Scope. In : Rice Management Biotechnology, (ed.) S. Kannaiyan, Assoc. Pub. Co., New Delhi, India, pp. 237-266.

Kannaiyan, S. 1997. The International momentum for using biofertilizers. The Indian Experience. In : Biological Nitrogen Fixation : The Global Challenge and Future needs (eds.) I. R. Kennedy and E.C. Cocking. The Rockfeller Foundation Bellagio Conference Centre, Italy, pp. 69-70.

Kannaiyan, S. 1998. Induction of root nodule structures in rice for nitrogen fixation. CADRI Newsletter : 11 (11): 1-3.

Kannaiyan, S. 2000. Induction of nodulation and endophytic nitrogen fixation in rice and maize. Paper presented in the Indian Science Congress, 88th Session, January 3-7, 2001, IARI, New Delhi, India.

Kannaiyan, S. & Kalidurai, M. 1989a. Studies on the nitrogen fixing potential of stem nodulating *Sesbania rostrata*. In : *In vitro* approaches for plant Breeding (ed.) J.P. Varghese, Ashram Press, Manganam, Kottayam, Kerala, India. p. 91- 98.

Kannaiyan, S. & Kalidurai, M. 1989b. Nitrogen fixing potential of stem nodulating *Sesbania rostrata* and *in vitro* approaches to plant breeding. Proc. Natl. Sem. Mol. Biol. Biotech. Kottayam, Kerala, India, pp. 75-84.

Kannaiyan, S. & Kalidurai, M. 1995. Nitrogen fixing potential of stem nodulating *Sesbania rostrata* and its effects on rice yield. In : Rice Management Biotechnology, (ed.) S. Kannaiyan, Assoc. Pub. Co., New Delhi, India, pp. 173-198.

Kannaiyan, S. & Kumar, K. 2000. Endophytic N_2 fixation in cereals crops. Paper presented in the International Conference on Biotechnology and Biodiversity, November 14-16, 2000, Tribhnvan University, Kathmandu, Nepal.

Kannaiyan, S., Thangaraju, M. Subramani, P. & Udayasuriyan, V. 1988. Studies on the nodulation, biomass production and nitrogen fixation in stem nodulating *Sesbania rostrata*. Int. Conf. Plant. Sci. New Delhi (Abst.)

Kannaiyan, S., Aruna, S.J. Merina Prem Kumari. S. & Hall, D.O. 1997. Immobilized cyanobacteria as a biofertilizer for rice crops. J. crops. J. App. Phycol. 9 : 167-174.

Kannaiyan, S., Kumar, K. & G. Gopalaswamy, 2000. Endophytic N_2 fixation in cereal crops. Paper presented in the 8th International symposium on Nitrogen fixation with Non-legumes, University of Sydney, Sydney, Australia.

Kannaiyan, S., Palaniappan , SP. & Sundaravaradhan, S. 1998. Nitrogen Fixing green manures for sustainable rice production. Inte. Rice. Res. Inst. Philippines and Tamil Nadu Agri. Univ. Coimbatore, Tamil Nadu, India.

Kannaiyan, S., Rao, K.K. & Hall, D.O. 1993. Immobilization of *Anabaena azollae* from *Azolla filiculoides* in polyvinyl foam for ammonia production in a photobioreactor system. World J. Microbiol. Biotechnol. 9 : 160-165.

Kennedy, I.R. & Y.T. Tchan, 1992. Biological nitrogen fixation in non-leguminous field crops : recent advances. Plant Soil. 141 : 93-118.

Kennedy, I.R., Zeman, A. Tchan, Y.T. New, P.B. Sriskandarajah, S. & Nie, Y.F. 1990. Biological nitrogen fixation and prospects for yield increases in wheat. Transactions, 14th international Congress Soil Science III. 146-151.

Kennedy, I.R., Pereg-erk, L. Wood, C. Deaker, R. Gilchrest, K. & Katupitiya, S. 1997. Biological nitrogen fixation in non- leguminous field crops : facilitating the evolution of an effective association between *Azospirillum* and wheat. Plant Soil. 194 : 65-79.

Kumar, K. Kannaiyan, S. & Cocking, E.C 1996. Studies on the colonization of *Azorhizobium caulinodans* in rice roots and their role in symbiotic nitrogen fixation. Paper presented in the 37th Annual Conference of Microbiologists of India, IIT, Chennai, 4-6, December, 1996 p. 118 (Abstr.)

Morris, R.A., Meelu, O.P. Dreyfus, B. & Dommergues, Y.R. 1983. Influence of *Sesbania rostrata* green manure on the nitrogen content of rice crop and soil. Soil Biol. Biochem. 15 : 111-113.

Nie, Y.E. 1983. Studies on induced nodulation in plants lacking root nodules. Nature. 6 : 326-360.

Quispel, A. 1991. A critical evaluation of the prospects for nitrogen fixation with non-legumes. Plant Soil. 137 : 1-11.

Reinhold-Hurek, B., Hurek, B.T. Gillis, M. Hoste, B. Vancanneyt, M. Kersters, K. & Deley, J. 1993. *Azoarcus*, gen. nov. Nitrogen - fixing Proteobacteria associated with roots of Kallar grass (*Leptochloa fusca* (L.) Benth) and description of two species, *Azoarcus indigens* sp. nov. and *Azoarcus communis* sp. nov. Int. J. Syst. Bacteriol. 43 : 574-584.

Rinaudo, G., Dreyfus, B. & Dommergues, Y.R. 1983. Influence of *Sesbania rostrata* green manure on the nitrogen content of rice crop and soil. Soil Biol. Biochem. 15 ; 111-113.

Senthil Kumar, M. 2000. Studies on the induction of paranodules by the stem nodulating *Azorhizobium caulinodans* and its nitrogen fixing activity in rice and maize. M.Sc. (Ag.) Thesis, Tamil Nadu Agricultural University, Coimbatore, Tamil Nadu, India.

Stoltzfus, J.R., Malarvizhi, Sr. R. Ladha, J.K. & de Bruijin, F.J. 1997. Isolation of Endophytic bacteria from rice and assessment of their potential for supplying rice with biologically fixed nitrogen plant and Soil. 194 : 25-36.

Sundaravarathan, S. & Kannaiyan, S. 1999. Mineralization of nitrogen fixing water fern *Azolla* and stem nodulating *Sesbania rostrata* with fertilizder N coupled with neem cake and their effect on rice. In : Vistas of rice research (ed) Dr. M. Subramanian, TRRI, Aduthurai, pp. 410-431.

Tchan, Y.T. & Kennedy, I.R. 1989. Possible N_2 fixing root nodules induced in non-legumes. Agricultural Science (AIAS Melbourne New Series) 2 : 57-59.

Tchan, Y.T. Zeman, A.M.M. & Kennedy, I.R. 1991. Nitrogen fixation in para-nodules of wheat roots by introduced free-living diazotrophs. Plant Soil. 137 : 43-47.

Triplett, E.W., 1996. Diazotrophic endophytes : Progress for nitrogen fixation in monocots. Plant Soil. 86 : 29-38.

Webster, G., Gough, C. Vasse, J. Batchelor, C.A. O'Callaghan, K.J. Kotharri, S.L. Davey, M.R. Denarie, J. & Cocking, E.C. 1997. Interactions of rhizobia with rice and wheat. Plant Soil. 194 : 115-122.

Zeman, A.M., Tchan, Y.T., Elmerich, C. & Kennedy, I.R. 1992. Nitrogenase activity in wheat seedlings bearing para-nodules induced by 2, 4-dichlorophenoxyacetic acid (2, 4-D) and inoculated with *Azospirillum*. Research in Microbiology. 143 : 847-855.

CHAPTER – 8

EXTRACTION OF METALS FROM ORES USING BACTERIA — A REVIEW

N. Anuradha, N.S. Sashidhar and H. Polasa

Department of Microbiology, Osmania University, Hyderabad - 500 007, India.

1. INTRODUCTION

The ever increasing demand for metals and fast depletion of known and accessible high grade ores coupled with limited ore resources has compelled to utilize the low grade ores. Extraction of metals from these low grade ores using conventional technologies is not economically viable. This gave the impetus to search for an appropriate economically feasible technology to recover greater amounts of metals from the lower grade ores. It is in this context that bacterial leaching and the role of the microorganisms in altering minerals and metals in solution has assumed a role of considerable importance. The bacterial leaching process is not only economical as compared to the conventional methods but is also essentially pollution free.

The enthusiasm of the microbiologists working on the development of these new "bio-mining" techniques is matched by a need in the mineral industry to find alternatives to conventional methods of mining, ore processing and waste-water treatment. The continued depletion of high-grade mineral resources, the resulting tendency for mining to be extended deeper underground, the growing awareness of environmental problems associated with the smelting of sulfide minerals and the burning of sulfur-rich fossil fuel and the rising cost of the energy required in the conventional recovery methods has drawn the attention of the mining companies towards the biological process.

The earliest recorded mining activity, which can be attributed to bacteria, involved the recovery of copper from mine waters of Rio Tinto, Spain in 1670. The presence of bacteria in the leaching fluid was not confirmed in that operation until 1962. However, in 1922 Rudolf *et al.*, reported the microbiological leaching of metal sulfides by an unidentified autotrophic bacterium. The economic importance of the extraction of metals from low-grade sulfide-bearing ores was thus suggested. In 1947, Colmer and Hinkel first isolated the microorganism, *Thiobacillus ferrooxidans*, from an acid drainage source, which was responsible for the oxidation of sulfide minerals.

The ability of the microorganisms to attack and dissolve mineral deposits has had a major impact on the mining industry. Biological processes are generally less energy-intensive and less polluting than most non-biological ones, and so the role of biological technology in mining, ore processing and waste-water treatment is likely to become increasingly important. As a result of microbial metabolism, insoluble metal deposits (usually metal sulfides or oxides) are converted to soluble metal sulfates. The solubilization process is termed bioleaching and occurs naturally wherever suitable conditions are found for the growth of the ubiquitous bioleaching bacteria. While the metal content of an ore is a major factor in the cost of metal recovery by conventional mining techniques, it is of less importance in bioleaching technology. Bacterial activity can be as effective in extracting all the metal from 0.3% ore as from a 0.03% ore. In addition, it is often possible to carry out bioleaching of deep or low-grade deposits *in situ*, thereby saving the cost of bringing vast tonnage's of ore and waste rock to the surface.

Bioleaching is also a useful alternative to conventional procedures in treating recalcitrant ores. For example, traditional methods for the treatment of recalcitrant gold-bearing ores involves processes that (when compared to bioleaching processes) consume large quantities of energy.

Bioleaching technology also has potential environmental benefits. The uncontrolled leaching of metals and acid from mine dumps and tailing dams (a result of the activity of the naturally occurring bioleaching bacteria) is a long-standing problem. Controlled leaching and recovery of metals from waste dumps can result in the recovery of valuable metals as well as protecting the environment from this source of pollution (Douglas E. Rawlings, 1995).

Hence, the advantages of bioleaching over the conventional metal mining procedures can be summarized as:

- Low energy needs – ores need not be dried as with pyrometallurgy and the maximum process temperatures can be reduced from 1500°C to less than 100°C. This could reap a huge energy saving for ores, especially wet ones like lateritic nickel wherein 50% of the total energy used to extract metallic nickel is consumed in the initial ore drying process.
- No SO_2 pollution – pyrometallurgy and SO_2 emissions are eliminated by the total use of hydrometallurgy.
- Lower acid usage – the acid-making ability of the bacteria reduces the need for purchased acid make up in leaching.
- Lower operating costs – data on Cu are as low as 77-88 cents/kg for a modern leaching plant, about half of the $1.60 to $1.80/kg for conventional pyrometallurgy (Burger, 1985).
- Lean ores and waste products – metals can be recovered from sources technically or economically untouchable by conventional hydro - or pyrometallurgy.
- The action of bacteria imparts easier separation of waste Fe precipitates (jarosite) from leachants. This may be due to a re-association of Fe and SO_4^{2-} in bioleach solutions, nucleating a more crystalline jarosite phase.

Today bacteria are being deliberately exploited to recover millions of pounds of copper from billions of tons of low-grade ore. Copper obtained in this way accounts for more than 10% of the total U. S. production (Corale L. Brierley, 1982). In recent years bacterial leaching has also been applied to the recovery of another non-ferrous metal: Uranium.

2. GENERAL MECHANISM OF METAL SOLUBILIZATION

Metals can be dissolved from insoluble minerals directly by the metabolism of microorganisms or indirectly by the products of their metabolism. The bacterial leaching of metal sulfides can be considered as a biochemical oxidation process, which is catalyzed by microorganisms. A generalized reaction is often used to express the biological oxidation of a mineral sulfide involved in leaching.

$$MS + 2O_2 \xrightarrow{\text{Microorganisms}} MSO_4$$

where M is ideally a bivalent metal.

The leaching process is the end result of the bacteria acting upon the mineral sulfide, which serves as an energy source in the presence of other nutrients. Microbiological leaching processes involve complex interactions between biological (microorganisms), aqueous (nutrients in the medium) and solid mineral phases. The biological reactions involved in extractive metallurgy are primarily oxidation of sulfur or mineral sulfides. Generally metal sulfides are insoluble in aqueous media while the sulfates are soluble. Hence, is the metal dissolution process. Many metals can be leached from their corresponding sulfides (Lundgren and Silver, 1977). The main leaching mechanisms involved are direct leaching, indirect leaching and galvanic conversion:

Based on the electrochemical experiments, numerous investigators have demonstrated that under sterile conditions the elemental sulfur produced during the leaching process, forms a protective layer on the mineral surface, thus slowing down the oxidation process. The microorganisms remove this protective reaction layer by oxidation. Bacteria also increase the initial leaching rate by directly attacking the mineral particle. Also, presence of small concentrations of soluble iron associated with copper ores, help in the generation of Fe^{3+} by bacteria which being a strong oxidizer helps indirectly in leaching of the metal.

(a) Direct bacterial leaching

In direct bacterial leaching, there is physical contact between the bacterial cell and the mineral sulfide surface, and the oxidation to sulfate takes place via several enzymatically-catalyzed reactions.

In this process, pyrite is oxidized to iron (III) sulfate according to the following reactions:

$$4FeS_2 + 14O_2 + 4H_2O \xrightarrow{\text{Bacteria}} 4H_2SO_4 + 4FeSO_4 \quad ......(1)$$

$$2H_2SO_4 + 4FeSO_4 + O_2 \xrightarrow{\text{Bacteria}} 2Fe_2(SO_4)_3 + 2H_2O \quad ...(2)$$

The direct bacterial oxidation of pyrite is best summarized by the reaction:

$$4FeS_2 + 15O_2 + 2H_2O \xrightarrow{\text{Bacteria}} 2Fe_2(SO_4)_3 + 2H_2SO_4 \quad ..(3)$$

Investigations by Torma (1971) have shown that the following non-iron metal sulfides can be oxidized by *Thiobacillus ferrooxidans* in direct interaction: covellite (CuS), chalcocite (CuS_2), sphalerite (ZnS), galena (PbS), molybdenite (MoS_2), stibnite (Sb_2S_2), cobaltite (CoS) and millerite (NiS).

Therefore, direct bacterial leaching can be described according to the following reaction:

$$MeS + 2O_2 \xrightarrow{\text{Bacteria}} MeSO_4 \quad .... (4)$$

where MeS is the metal sulfide.

There is some evidence that the bacteria have to be in intimate contact with the mineral surface. The mechanism of attachment and the initiation of metal solubilization are not completely understood. Obviously the bacteria do not attach to the whole mineral surface but prefer specific sites of crystal imperfection, and metal solubilization is due to electrochemical interactions.

(b) Indirect bacterial leaching

In indirect bacterial leaching, the bacteria generate a lixiviant which chemically oxidizes the sulfide mineral which in acid solution is ferric iron. The metal solubilization can be described according to the following reaction:

$$MeS + Fe_2(SO_4)_3 \longrightarrow MeSO_4 + 2\ FeSO_4 + S^o \quad ....(5)$$

To maintain enough amount of iron in solution the chemical oxidation of metal sulfides must occur in an acid environment below pH 5.0. The ferrous iron arising in this reaction can be reoxidized to ferric iron by *Thiobacillus ferrooxidans* or *Leptospirillum ferrooxidans* and as such can take part in the oxidation process again. In indirect leaching, the bacteria do not need physical contact with the mineral surface. They only have a catalytic function because they accelerate the reoxidation of the ferrous iron, which takes place very slowly in the absence of bacteria. As shown by Lacey and Lawson (1970), the pH of the medium plays an important role in indirect leaching in the range of pH 2-3 bacterial oxidation of ferrous iron is about $10^5 - 10^6$ times faster than the chemical oxidation of the ferrous iron.

The sulfur arising simultaneously (equation, 5) may be oxidized to sulfuric acid by *Thiobacillus ferrooxidans*, but oxidation by *T. thiooxidans* which frequently occurs together with *T. ferrooxidans* is much faster:

$$2S^o + 3O_2 + 2H_2O \xrightarrow{\text{Bacteria}} 2H_2SO_4 \quad ....(6)$$

The role of *T. ferrooxidans* in bioleaching obviously consists in creating favorable acid conditions for the growth of ferrous iron oxidizing bacteria such as *T. ferrooxidans* and *L. ferrooxidans*. A well-known example of indirect leaching process is the extraction of uranium from ores, where insoluble tetravalent uranium is oxidized to the water-soluble hexavalent stage of uranium:

$$U^{IV}O_2 + Fe_2(SO_4)_3\ U^{VI} \longrightarrow O_2SO_4 + 2\ FeSO_4 \quad ...(7)$$

The lixiviant may be generated by *T. ferrooxidans* by the oxidation of pyrite that is very often associated with uranium ore. Besides the indirect leaching of uranium, there is some evidence that *T. ferrooxidans* can oxidize U^{IV} to U^{VI} enzymatically and uses some of the energy of this reaction for the assimilation of CO_2 (Tuovinen, 1982).

Hence, bioleaching is based on the interaction of biological and chemical oxidation processes. In nature and in technical applications both mechanisms direct and the indirect leaching would undoubtedly occur in concert (Bosecker, 1997).

(c) Galvanic conversion

A third leaching mechanism is termed galvanic conversion. Physical contact between two dissimilar metal sulfide phases immersed in an electrolyte (dilute sulfuric acid/ferric sulfate solution) creates a galvanic cell. For example, in the $CuFeS_2$/ FeS_2 couple, chalcopyrite has the lower electrical rest potential and becomes the anode while pyrite behaves as a cathode. As a result, the chalcopyrite phase undergoes rapid dissolution while the pyrite phase remains essentially unaffected (i.e., is passivated). *Thiobacillus ferrooxidans* may accelerate the reaction by continuously oxidizing the film of elemental sulfur that would otherwise form a physical barrier to diffusion of copper and iron salts away from the reacting phase. The contribution of galvanic conversion leaching in the environment is unknown (Hutchins *et al.*, 1986).

3. BIOLEACHING OF ORES

A large variety of metal containing ores may be leached through the bacterially assisted oxidation of insoluble metal sulfide to soluble metal sulfates. Copper, Uranium and more recently gold-bearing arsenopyrite are the ores that are leached in the greatest tonnage's, although this technology can also be applied to ores containing sulfides of zinc, lead, cobalt, nickel, bismuth and antimony.

(a) Copper ores

At least 15% of world production or 1.1 x 10^6 tons of copper are bioleached annually (Ehrlich and Brierley, 1990). It is estimated that in 1984, copper valued at $350 million and representing 10% of U.S. copper production was recovered by bioleaching. The use of bioleaching to recover this valuable reserve is growing as remaining ore bodies are depleted and conventional extraction technology becomes uneconomic. According to the 1995 data, Chile was the world's major copper producing country with copper reserves of 1.5 x 10^8 tons of which 4.7 x 10^7 (32%) are contained in ores of a grade too low for conventional technology to be profitable (Douglas E. Rawlings, 1995).

Chemolithotrophic bacteria such as *T. ferrooxidans* attack and oxidize a variety of copper sulfide ores and convert insoluble copper sulfide to soluble copper sulfate.

$$4\,CuFeS_2 + 17\,O_2 + 2\,H_2SO_4 \rightarrow 4\,CuSO_4 + 2\,Fe_2(SO_4)_3 + 2\,H_2O$$
(chalcopyrite)

$$CuS + 2\,O_2 \xrightarrow{T.\ ferrooxidans} CuSO_4$$
(covellite)

$$2\,Cu_2S + O_2\ 2\,H_2SO_4 \xrightarrow{Bacteria} 2\,CuS + 2\,CuSO_4 + 2\,H_2O$$
(chalcocite)

(b) Lead and Zinc sulfides

These sulfides are found frequently associated in varying proportions and these metals are separated from the minerals by selective flotation. However, the application of this procedure to lean ores or complex sulfide ores containing copper in the form of salts and pyrite is difficult. One approach to the problem may be represented by bioleaching:

In the case of zinc sulfides:

$$ZnS + 4H_2O \longrightarrow Zn^{2+} + SO_4^{2-} + 8H^+ + 8\bar{e}$$

and

$$ZnS + 4Fe_2(SO_4)_3 + 4H_2O \longrightarrow ZnSO_4 + 8FeSO_4 + 4\,H_2SO_4$$

Oxidation of ZnS is, however, markedly enhance by *T. ferrooxidans*.

In case of lead sulfides :

Galena (lead ore, PbS) oxidizes chemically according to the following equation:

$$PbS + H_2SO_4 + \tfrac{1}{2}\,O_2 \longrightarrow PbSO_4 + H_2O + S^o \;.\; (1)$$

and in the presence of iron the oxidation is expressed as:

$$PbS + Fe_2(SO_4)_3 \longrightarrow PbSO_4 + 2\,FeSO_4 + S^o \quad (2)$$

In the presence of bacteria:

$$PbS + 2O_2 \xrightarrow{Bacteria} PbSO_4 \ldots\ldots\ldots\ldots (3)$$

Biological oxidation can be direct as shown in equation (3), and / or indirect as in equation (2) with the microorganisms oxidizing elemental sulfur and ferrous iron producing sulfuric acid and Fe^{3+} which help in metal solubilization.

(c) Cobalt and Nickel sulfides

Cobalt occurs in the form of cobaltite (CoAsS), Linnaeite [Co_3S_4(Fe,Cu)], carrollite ($CuS.Co_3S_3$) while nickel occurs in the form of millerite (NiS), ullmannite (NiSbS), pentlandite (Fe, Ni_9S_8) etc.

The following equation has been

proposed for the bacterial mediated solubilization of cobalt by Torma *et al* (1971):

$$CoS + 2O_2 \xrightarrow{Bacteria} CoSO_4$$

Biologically mediated nickel sulfide oxidation proceeds according to the equation:

$$NiS + 2O_2 \xrightarrow{Bacteria} NiSO_4$$

In case of nickel containing pentlandite:

$$(Ni, Fe)_9S_8 + 17^5/_8\ O_2 + 3^1/_4\ H_2SO_4 \xrightarrow{Bacteria}$$

$$4^1/_2\ NiSO_4 + 2^1/_4\ Fe_2(SO_4)_3 + 3^1/_4\ H_2O$$

(d) Uranium ores

Uranium does not occur as sulfides but as insoluble oxides. Uranium minerals are often associated with metal sulfides such as iron pyrite, pyrrhotite and chalcopyrite. In the case of metal sulfides, the production of sulfuric and ferric sulfate is the mechanism by which *Thiobacillus ferrooxidans* indirectly accelerates the leaching of uranium ores involving the following chemical reactions (Mathur, *et al.*, 1992):

$$2\ FeS_2 + 7^1/_2\ O_2 + H_2O \longrightarrow 2FeSO_4 + H_2SO_4$$

$$2\ FeSO_4 + ½\ O_2 + H_2SO_4 \xrightarrow{T.\ ferrooxidans} Fe_2(SO_4)_3 + H_2O$$

$$FeS_2 + Fe_2(SO_4)_3 \longrightarrow 3\ FeSO_4 + 2\ S$$

$$2\ S + 3\ O_2 + H_2O \xrightarrow{Bacteria} 2\ H_2SO_4$$

$$\underset{(Insoluble)}{UO_2 + 2\ Fe^{3+}} \longrightarrow \underset{(Soluble)}{UO_{22+} + 2\ Fe2+}$$

4. MICROORGANISMS INVOLVED IN LEACHING

Most commercial bioleaching operations are carried out with a consortium of highly acidophilic, chemolithotrophic bacteria which may include *Thiobacillus ferrooxidans*, *Thiobacillus thiooxidans*, *Leptospirillum ferrooxidans* and acidophilic heterotrophic bacteria belonging to the genus *Acidophilium* (Harrison, A. P., 1984). Although, *T. ferrooxidans* is considered to be the most important member of the consortium, mixed cultures of bacteria is often more efficient at ore decomposition than *T. ferrooxidans* alone. *T. ferrooxidans* and *T. thiooxidans* are more effective in leaching certain ores than either organism alone. Similarly, *L. ferrooxidans* and *T. organoparus* can degrade pyrite (FeS_2) and chalcopyrite ($CuFeS_2$).

Thiobacilli that attack some sulfide minerals and certain soluble sulfur compounds under neutral conditions are often found in sulfide ore deposits and in other habitats where sulfur is available. *Thiobacilli* of this type may be responsible for the initial increase in the acidity that establishes an environment conducive to the growth of the more acidophilic leaching bacteria.

Most of the above bacteria are mesophilic and at temperatures greater than 45°C, several other types of bacteria including the moderately thermophilic "Thiobacillus TH" strains and various *Sulfolobus* species dominate (Norris, P. R., 1990). At even higher temperatures, extremely thermophilic archaebacteria such as *Sulfolobus* and *Acidianus* species are found. These groups of bacteria are also able to rapidly attack a variety of ores.

At temperatures between 60 and 75°C and under neutral conditions the filamentous bacterium *Thermothrix thioparus* oxidizes sulfhydryl ions (HS^-), sulfite ions $(SO_3)^-$, and thiosulfate ions. There is an increasing evidence of the wide spread existence of *Thermothrix* species and similar filamentous, sulfur-oxidizing bacteria in thermal springs and near volcanic fissures. Their existence in sulfur-bearing springs suggests they could prepare such environments for the more acidophilic species.

Among the most interesting of the leaching microorganisms are the moderately thermophilic and acidophilic bacteria designated TH (for thermophilic and Thiobacillus-like). Various strains of TH bacteria can grow on ferrous iron or on minerals such as pyrite, chalcopyrite, covellite (CuS) and pentlandite [$(Fe, Ni)_9S_8$]. An organic supplement to the mineral diet is apparently required by these organisms.

One of the most important micro-organisms implicated in bioleaching is the extremely thermophilic and acidophilic species of the *Sulfolobus*. These bacteria are usually found in the hot springs and volcanic fissures at temperatures that can exceed 60°C. Some strains of *Sulfolobus* bacteria have different structures from that of most bacteria. *Sulfolobus acidocaldarius* and *S. brierleyi* oxidize sulfur and iron for energy relying on either carbon dioxide or simple organic compounds for carbon. Minerals that resist most microorganisms, such as chalcopyrite and molybdenite (MoS_2), are readily attacked by *Sulfolobus*, and the resulting metals are not toxic to the organism. Molybdenum, which is extremely toxic even to the metal-tolerant *Thiobacilli* is readily endured by *S. brierleyi* in concentrations as high as 750mg/L. *Sulfolobus acidocaldarius* can oxidize Fe^{2+} at temperatures as high as 80°C – 85°C. This organism was shown to oxidize Fe^{2+} in acid hot springs having total iron concentration of 1–200 ppm, of which about 50% is ferrous (Agate, Personal Communication).

An acid-tolerant *Metallogenium* has been reported from mesoacidic, iron-bearing groundwater's. It is filamentous organism consisting of branching filaments usually encrusted with iron. The organism tolerates pH range from 3.5 to 6.8, with an optimum at 4.1 (Agate, Personal Communication).

The environment found at different microsites within the ore body is different, exhibiting climatic as well as mineralogical variations, and this determines the establishment of different microbial population. Considering the fact that different microbial populations able to tolerate different pH, temperature and metal concentrations thrive and aid in bioleaching of a wide variety of metals from the ores, commercial bioleaching operations are often carried out with mixed cultures of bacteria.

5. PHYSIOLOGY AND BIOENERGETICS OF *Thiobacillus ferrooxidans*

Thiobacillus ferrooxidans based on the 16S rRNA sequence, with the exception of strain m-1, can be classified into the - subdivision of Proteobacteria (Lane *et al.*, 1992).

The genome of *T. ferrooxidans* is GC-rich. The presence of a wide spectrum of GC mol% (55-65) and 7 DNA homology groups assigned by Harrison, Jr. (1982) indicates the existence of a high level of genomic diversity amongst the strains of *T. ferrooxidans*.

T. ferrooxidans is a Gram negative, rod–shaped bacterium, has a cell size in the range of 0.3 to 0.5 µm long and a physiology, which is well suited for growth in an inorganic mining environment. It obtains its carbon by fixing atmospheric carbon dioxide and is obligately autotrophic (Kelly, *et al.*, 1989). Not only is the bacterium autotrophic, but also, it is capable of fixing atmospheric nitrogen (Mackintosh, 1978). Diazotrophy is therefore likely to be a general property of these bacteria.

These are chemotrophs and the energy is obtained by the oxidation of either ferrous ion to ferric or reduced – sulfur compounds to sulfuric acids. The bacterium is acidophilic and can grow in the pH range of 1.5 to 6.0, the optimum for maximum growth rate being between 2.0 and 2.5. Though, *T. ferrooxidans* survives in low pH environment, its internal pH is close to neutrality resulting in the generation of a pH gradient across the cytoplasmic membrane. It survives in the temperature range 2°C to 40°C, but does best in the range of 28°C to 35°C.

It grows best in an aerobic environment with oxygen as an electron acceptor. In the absence of oxygen, reduced sulfur compounds or formate is used as an electron donor and ferric ions can serve as the electron acceptor.

The components of the electron transfer system, which are coupled to the oxidation of Fe^{2+} by *Thiobacillus ferrooxidans* are expected to function in the periplasmic space of the bacterium. This follows from the fact that the Fe^{2+} oxidizing enzyme shows optimal activity around pH 4.0 *in vitro* and all the components are more stable at pH 3.0 – 4.0 than the counterparts found in bacteria that grow at neutral pH. These enzymes and proteins seem to be bound to the outer side of the plasma membrane. Several electron transfer components have been purified which participate in the oxidation of Fe^{2+} by the *T. ferrooxidans*. On the basis of the interactions between the isolated components, the electron transfer coupled to the oxidation of Fe^{2+} by the bacterium seems to occur through the system:

Fe^{2+} → Fe(II) – cytochrome *c* oxioreductase (an Fe/S protein)

→cytochrome *c*-552 →cytochrome *c* oxidase (*aa*$_3$ type) → O_2

Rusticyanin

[Schematic representation of electron transfer coupled to the oxidation of Fe^{2+} by *Thiobacillus ferrooxidans*. Redrawn from: Molecular aspects of the electron transfer system which participates in the oxidation of ferrous ion by *T. ferrooxidans* (Tateo Yamanaka and Yoshihiro Fukumori, FEMS Microbiology Reviews, 1995 (17))]

The electron transfer from Fe^{2+} to O_2 in *T. ferrooxidans* is thought to occur as follows:

The electrons from Fe^{2+} ions are first accepted by Fe(II) – cytochrome *c* oxidoreductase. This enzyme donates electrons directly to cytochrome *c*-552 and the resulting ferroocytochrome *c*-552 is oxidized by *aa*$_3$-type cytochrome *c* oxidase. Rusticyanin is also reduced in the presence of cytochrome *c*-552, and the reduced form of rusticyanin is also oxidized by the oxidase, hence, the copper protein expands the electron pathway form cytochrome *c*-552 to the oxidase. Although Fe(II)-cytochrome *c*-reductase is destroyed when it is contacted with cytochrome *c*-552 *in vitro*, the presence of rusticyanin protects the enzyme from the destruction. Therefore, rusticyanin is necessary for the electron transfer system not only to expand the electron pathway to cytochrome *c* oxidase, but also to donate electrons to cytochrome *c*-552 from Fe(II)-cytochrome *c* oxidoreductase. As ferrocytochrome *c*-552 is rapidly oxidized by the oxidase, it can also act as the electron donor for the oxidase (T. Yamanaka., 1995).

There are four main features of the life style and characteristics of *T. ferrooxidans*, which are relevant to bioenergetic considerations:

- The small amount of energy available from the growth substrates (Fe^{2+}, O_2)
- The bacterium's chemolithotrophy and therein the requirement to fix its own CO_2 and if necessary, to fix its own N_2
- The bacterium is an obligate acidophile
- The cell membrane contains an unusually high specific content of electron – transport chain components.

Thiobacillus ferrooxidans grows optimally on Fe^{2+} at pH 2.0 with an electron – transport chain mediating between the two half – reactions of the overall reaction:

$$2\ Fe^{2+} \rightarrow 2\ Fe^{3+} + 2\ e^-$$

$$2\ e^- + \frac{1}{2} O_2 + 2\ H^+ \rightarrow H_2O$$

$$\text{Sum: } 2\ Fe^{2+} + \frac{1}{2} O_2 + 2\ H^+ \rightarrow 2\ Fe^{3+} + H_2O$$

Thus, the overall reaction consumed O_2, H^+ and Fe^{2+} yielding H_2O, Fe^{3+} and energy for CO_2 fixation and cell growth (W. John Ingledew, 1981).

The pair of electrons transferred yields enough energy to ensure the formation of ATP when coupled to oxidative phosphorylation.

$$ADP + P_i \rightarrow ATP$$

Ingledew *et al.*, (1977) proposed a mechanism for the generation of proton electrochemical gradients in acidophilic *Thiobacillus ferrooxidans* for the Fe^{2+} oxidation.

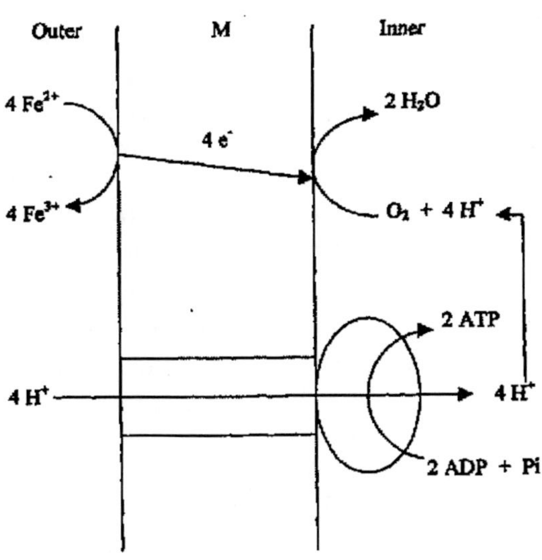

Fig: Mechanism for the generation of proton electrochemical gradients in acidophilic Thiobacilli. Model of Ingledew *et al* for Fe^{2+} oxidation by *Thiobacillus ferrooxidans*. Redrawn from: Energy Conservation in Acidophilic Bacteria, John G. Cobley and John C. Cox (Microbiological Reviews, Dec. 1983).

Ingledew and his associates calculated the membrane potential to be –10mV and the electrochemical proton activity difference to be 256mV; they also estimated that this value is adequate to support the formation of ATP.

However, when compared to the oxidation of glucose that of ferrous iron does not yield much energy. Lees *et al* (1969) determined the free energy yield from ferrous iron oxidation to be close to 6.5kCal mol^{-1} which is barely enough to synthesize 1mol of ATP. The same authors also approximated that 18.5mol of Fe^{2+} would have to be oxidized to assimilate 1mol of carbon.

The CO_2 fixation reactions in *T. ferrooxidans* are coupled to the production of energy from Fe^{2+} oxidation. The pathway for CO_2 assimilation is a reductive Pentose Phosphate cycle. Under certain growth situations, the organism can be utilize compounds other than CO_2 for its carbon requirement. Although a number of laboratory studies have shown the adaptability of the iron grown cultures to the use of organic substrates for energy and carbon, little is known about the organisms capacity for heterotrophy / mixotrophy in nature (Lundgren and Silver, 1959).

6. FACTORS INFLUENCING MICROBIAL ACTIVITY

From all the published accounts and our experiments of microbial leaching it can be concluded that the following factors are the important in leaching with the help of bacteria.

- An available source of substrate
- A supply of carbon-dioxide which is the sole source of carbohydrate for the *T. ferrooxidans*
- A supply of oxygen for the oxidation process

- A supply of essential nutrients including nitrogen
- Water as the carrier of the nutrients, source of trace elements and as a solvent for the solvated metal
- An acidic medium

Like all other life process, the leaching of ores by microbes is influenced by environmental factors. These include :

(a) pH

Hydrogen ions play an important role in the energy supply mechanism of most of the acidophiles including *T. ferrooxidans*. The microbe-mediated oxidation of ferrous iron and metal sulfides involves electron transfer and movement of hydrogen ions. Hence, hydrogen ion concentration (pH) of the liquid media where biohydrometallurgical processes are carried out has a definite effect on microbial life as well as on the related hydrometallurgical process. Also, as discussed earlier, the indirect leaching of metals by is dependent on pH. Therefore, optimum pH ranges and intervals within which the microorganisms are active can be established. pH ranges have to be optimized in order to prevent the detrimental precipitation of jarosite-like compounds, which might interfere in the leaching activity of the microorganisms.

(b) Temperature

Each microbial species is characterized by a well-defined temperature range, which the microbes can tolerate, and by optimal growth temperature at which growth and reproduction rates are highest. Hence, biohydrometallurgical process effectiveness is temperature dependent. The bioleaching process is affected by two competing factors: First, the enhancement in the reaction rate usually brought about by an increase in temperature and the decrease in microbial activity as the temperature increases and approaches upper tolerance limit of the microorganism. The reaction rate and its temperature dependence are typical for each mineral dissolution and it also depends on the microbial strain (March *et al.*, 1983). Hence, the lower the temperature range of the microorganism involved the greater the competition with in the above two mentioned factors. For different strains of *T. ferrooxidans*, the optimum range is 25°C to 45°C (Marchlewitz, B., Swhartz, W., 1961). The optimum values for ferrous iron oxidation were found in the range of 28°C to 32°C, whereas for metal sulfide leaching, the optimum temperature is 35°C; microbial mediated oxidation ceases between 55°C to 65°C and at higher temperatures only chemical oxidation can be observed (Bryner, L. C., *et al.*, 1967).

However, in case of thermophiles such as *Sulfolobus* significant mineral sulfide bioleaching was reported at 65°C and 70°C (Norris, P. R., and Parrot, L., 1986).

Microorganisms can also survive at very low temperatures but both growth rates and metabolic activities pronouncedly decrease than at the optimal temperature. However, bioleaching oxidation process being an exothermic reaction, generates lot of heat during the leaching process. This high temperature is again inhibitory to mesophilic microbial activity.

Hence, temperature should be suitably controlled in bioleaching process, since exceedingly high temperature increases occur due to the heat generated by oxidation reactions. Moreover, establishment of optimum temperature range at which the microorganisms can show maximal metal leaching is also very important.

(c) Pulp density

The pulp density expressed as Mass of solids / Volume of medium x 100 may have a significant effect on the extraction of metals by the bacteria. Available reports implicate that the extraction rates increase linearly upto 15% pulp density but decrease with a further increase in the pulp densities. At lower pulp the extraction rates are lower since the rate of growth of the organisms is limited by the availability of the required energy source. On the other hand, at higher pulp densities, decrease in bacterial activity may be due to the interference of increased number of solid particles with the effective mass transfer of carbon-dioxide and oxygen required for maintaining bacterial activity (Dave, S. R., *et al.*, 1981).

(d) Period of incubation

When a bacterial species is introduced into a new environment, it usually enters a lag phase during which time it gets itself acclimatized to the surroundings. This also applies to *T. ferrooxidans* species introduced into leaching medium containing the flotation concentrate and it might reflect on the bioleaching rates of the bacterium. Hence, establishment of optimum period of incubation such that optimal leaching of the metals is achieved by the bacterium is very important.

(e) Ammonium sulfate concentration

Mackintosh, M. E., (1978), reported the ability of *T. ferrooxidans* to fix atmospheric nitrogen but ammonium nitrogen added as ammonium sulfate to the medium is usually used as suitable source of nitrogen for the microbe. Torma, *et al* (1970), provided evidence that ammonium concentration controls the yield and hence the percent metal leached by the microorganism. However, ammonium sulfate in excess retards the leaching by the microorganism (Lazaroff, N., Sigull, W., and Wadderman, A., 1982) and hence fixation of an optimum range for the microorganism is extremely important.

(f) Effect of Ferrous sulfate

Increasing concentrations of heavy metal ions such as Cu^{2+}, Ni^{2+}, and also Fe^{2+} and Fe^{3+} in a bioleaching system can inhibit bacterial activity. In general, microorganisms are sensitive to nutritional requirements in their growth phase and starvation brings about changes in the nature of bacterial cell surface and their hydrophobicity. This eventually affects their growth characteristics. Ferrous sulfate is the principal nutrient for the bioleaching microorganism *T. ferrooxidans*. This microbe derives its energy during the oxidation of Fe (II) to Fe (III). So, the availability of ferrous sulfate in the leaching system affects the growth of the microbe. The concentration of ferrous sulfate in the leaching system also affects metal solubilization from the mineral particle, since, the ferric iron generated, by the oxidation of ferrous iron by the bacteria, being a strong oxidizing agent attacks the mineral particle and helps in metal release into the leachate. Hence, presence of ferrous sulfate in the leaching medium, in optimum concentrations, might affect the percentage of the metal finally solubilized into the medium.

(g) Inoculum size

When a limited number of living microbes, say a few hundred thousand individuals belonging to the same strain – called an "inoculum" – are introduced into a liquid medium of essential nutrients at suitable temperature and pH, the organisms will grow.

The process of cell growth involves the uptake of matter needed for biosynthesis and the conversion of compounds for energy supply from the cells environment as well as release of metabolic end products. In the case of pyrite bioleaching by *T. ferrooxidans* or *L. ferrooxidans* or *T. thiooxidans*, CO_2, N, P, K and Ca are some of the compounds or elements used for biosynthesis whereas the oxidation of iron from ferrous to ferric satisfies the energy requirements. The rates of these processes vary considerably during growth and the concentrations of nutrients and metabolites may in turn pronouncedly affect microbial growth.

McGoran *et al* (1969) reported that in case of bioleaching of zinc, the rates of leaching of the metal and of the oxidation of iron are related to bacterial growth. With a gradual increase in the initial dose of the inoculum, the metal extracted also increases correspondingly, but at higher dosage of the inoculum the leaching rates drastically decreased. Hence, establishment of an optimum range for the inoculum dosage is required to achieve optimal metal leaching rates.

7. TOXICITY AND TOLERANCE TO HEAVY METALS

T. ferrooxidans exhibits unusually high level of resistance to most of the heavy metal cations. However, it is extremely sensitive to most of the inorganic anions.

Heavy metals differ from each other with respect to biological requirement and the resistance mechanisms. This forms the basis for classifying them into 2 groups. In group I are included the metal ions which have no biological role to play and are very toxic. Mercury, cadmium and lead fall into this group. The other group includes the metals which are required for cellular growth and maintenance but are toxic when in excess – eg. Cu, Zn and Ni. The homeostatic mechanism operating for these 2 classes of metal ions are different. A resistance mechanism for the first group of metals should just prevent the metal ion from either entering the cell or if it enters, contact with its cellular targets should be prevented. In the case of Cu, Zn and Ni-group of metals, the cell should detoxify the excess, while simultaneously maintaining an optimal essential level.

The concentration of soluble metal sulfates keeps building up in the mining environments where active bioleaching processes occur. Most of the heavy metal ions are capable of undergoing redox transitions due to their multiple valency states and may thus interfere with the cellular systems of the biomining microorganisms.

Acidophiles, in general, are particularly tolerant to high concentration of metal ions as compared to other microorganisms. The metal tolerance levels of *T. ferrooxidans* are high and it possesses a remarkable capacity to increase its tolerance from sub-toxic to toxic level of cations. Starting from 4mM Cu^{2+} concentration, *E. coli* can be adapted to tolerate 20mM Cu^{2+} concentrations while *T. ferrooxidans* can be adapted to withstand 400mM Cu^{2+} concentrations (Tetaz and Luke, 1983). Adaptation of *T. ferrooxidans* has resulted in the development of strains which are capable of growing in the presence of various metal ions at following concentrations: Zn^{2+} (120 g/L); Ni^{2+} (72 g/L); Co^{2+} (30 g/L); Cu^{2+}(55 g/L); Fe^{2+} (160 g/L); and U_3O_8 (12 g/L). In contrast, Ag^{+}, Hg^{+} and Mo^{2+} are inhibitory at very low levels of 0.1 ppm, 2 ppm and 5 ppm respectively (Norris and Kelly, 1977). Tuovinen *et al* (1971) reported that the pre-adaptation of *T. ferrooxidans* to a metal, for example Cu^{2+}, results in acquired tolerance to related metal ions, Zn^{2+} and Ni^{2+} much more easily than an un-adapted culture.

T. ferrooxidans possesses remarkable capacity to increase its level of tolerance when exposed to step-wise increase of metal ion concentration (Tuovinen, *et al.*, 1971). Tolerance may be the result of effective competition by the H^{+} ions, at low pH conditions, for the negatively targeted sites at the cell surface (Norris and Kelly, 1978).

Determinants of heavy metal resistance have been frequently found on the plasmids in many Gram-negative bacteria (Foster, 1983; Silver, 1992). But in some cases, the resistance determinants have been localized on the chromosome as well (Silver and Mishra, 1988).

Resistance to metal ions is brought about by any of the 4 mechanisms:

- The most common mechanism involves efflux of the specific cation or anion out of the cell. Examples include, *czc* determinant of *Alcaligens eutrophus* which encodes proteins required for efflux of cadmium, zinc and cobalt (Nies and Silver, 1989)
- Detoxification achieved by changing the valency state of the ion or by introducing a modification. This mechanism operates for specific metal ions such as mercury where reduction of Hg^{2+} to Hg^{o} produces a non-toxic, volatile product, which diffuses out of the cell. Since other metals are not volatile in their elemental state and would get reoxidized, this method of detoxification is not usually found.
- Sequestration of specific mineral-ion binding components. Examples include three small proteins in *Psuedomonas putida* involved in cadmium resistance (Higham *et al.*, 1984)
- Modification of the transport system by the chromosomally encoded cadmium system in cadmium resistance system in *Bacillus subtilis* (Ladagga *et al.*, 1985).

Inorganic anions are generally growth inhibitory in *T. ferrooxidans*. The only acceptable anion for cell growth is SO_4^{2-}. It is dominant anion of the growth environment and is also essential for iron oxidation, as seen by investigations with whole cells and with cell envelopes (Tuovinen, 1971). Anionic requirement can be specific or non-specific. The specific requirement is satisfied by SO_4^{2-}. Dugan and Lundgren (1967), postulated that sulfate along with phosphate forms complexes with ferrous which facilitate ferrous oxidation and electron transport into the cell. Non-specific anionic requirement involves stabilization of Fe_2+ and Fe^{3+} aquated complexes in acid solution. Though, this again is mainly brought about by SO_4^{2-} anion, PO_4^{3-}, AsO_4^{3-} and TeO_4^{2-} can partially replace SO_4^{2-} (Lazaroff, 1977, 1983). NO_3^{2-} and Cl^- are tolerated at low concentrations, but are inhibitory at high levels. With gradual stepwise adaptation *T. ferrooxidans* can be made resistant to both anions.

8. FLOTATION AND BIOLEACHING

The beneficiation of base-metal sulfide ores is generally based on the selective production of lead, zinc and copper flotation concentrates from which the respective metals are extracted by metallurgical process. The gradual extraction and/or impoverishment of available mineral resources will probably lead to the search for more advanced solutions to the problem of the beneficiation of some refractory ores, the treatment of which by selective flotation yields poor technical and economical results.

The present techniques of sulfide ore beneficiation are almost always based on selective flotation processes of the run-of-mine ore, aimed at producing single-metal final concentrates ready for further metallurgical treatment. These processes represent the best solution to the problems that are involved in the beneficiation of the easier ores, since they enable high recoveries to be obtained economically. However, these processes are grossly inadequate for the beneficiation of lean ores. Two factors contribute to the detriment of its overall technico-economic balance: the poor quality of the concentrates obtainable, which greatly reduces their commercial value, and the lower flotation recovery, which combined with metallurgical recovery, in its turn marred by the quality of the concentrates, produces a relatively low overall recovery. Increasing world demand and supply difficulties, however, especially felt in industrialized countries, emphasize the urgent need for an effective solution to the problem of the exploitation of these far from ideal resources (lean ores).

T. ferrooxidans is an organism characterized by its ability to accelerate kinetics of inorganic sulfur and ferrous iron oxidation. Metallic sulfide bioleaching can be defined as a process of biochemical oxidation catalyzed by a living organism. In the classical chemical processes the main advantages of direct leaching of the run-of -mine ores over the flotation are, basically, the high metal recoveries and savings at the grinding stage, chiefly whenever the liberation of commercial minerals, essential for the subsequent selective flotation step necessitates very fine grinding. Various drawbacks include: presence of Ca or Mg among gangue minerals, their solubilization may involve the consumption of excessive amounts of reagents utilized for the attack, and the large amounts of solids to be leached necessitates a high energy consumption for agitation in stirred vessels and cause metal losses.

Bioleaching processes, in comparison with the classical processes, present the basic advantage of considerably lowered reagent consumption, thus cutting overall costs: however, for a satisfactory metal solubilization rate fairly fine grinding of the ore is required before the attack.

Thus, bioleaching combined with flotation processes, offer a great promise in the benefaction of sulfide minerals. Bulk flotation of the non-ferrous metals may be carried out on a run-of-mine ore resulting in the production of a bulk concentrate with only traces of Ca or Mg. The metals that constitute the bulk flotation concentrate are recovered from the pregnant liquor produced by its bioleaching (Giovanni Rossi, 1980).

On a final note we would like to add briefly the results of the experiments carried out in our laboratory using two strains of *Thiobacillus ferrooxidans*. These two strains of bacteria exhibited a selectivity to leach nickel from the sulfide concentrate of a uranium ore. Initially, these organisms could leach only 7.88% of nickel present in the concentrate. However, optimization of various physical, chemical and biological factors increased the nickel leaching ability of these organisms to 65%. Further both the strains were serially adapted to grow at higher concentrations of copper and nickel. They were also adapted to grow in the presence of the sulfide concentrate alone. These metal tolerant strains could leach nickel upto 90.99%, which was highly significant. Although the copper content in this batch of flotation concentrate was higher than nickel (13.5% Cu and 11% Ni), the bioleaching was lesser than (9%) than nickel by these two *T. ferrooxidans* strains indicating that these two strains are specific for nickel bioleaching.

REFERENCES

Agate, A. D. Iron oxidizing bacteria (Personal Communication).

Bosecker, K. 1997. Bioleaching: Metal solubilization by microorganisms. FEMS Microbiol. Rev. 20, 591-604.

Brierley, C. L. 1982. Microbiological mining. Scientific American. p. 42-51.

Cobley, G. John and John C. Cox. 1983. Energy conservation in acidophilic bacteria. Microbiological reviews. Dec., p. 579-595.

Dave, S. R., Natarajan, K. A, Bhat, J. V. 1981. Effects of physico-chemical parameters on the bacterial leaching of sphalerite concentrates. Transactions of the Indian Institute of Metals. Vol. 34, No. 2.

Duncan, D. W., Walden, C. C., Trussel, P. C. 1967. Recent advances in the microbiological leaching of sulfides. Tans. Soc. Min. Eng., A. I. M. E., 238, 122.

Giovanni Rossi 1980. Benefaction of a complex sulfide ore and concentrate leaching. In: Biohydrometallurgy. Giovanni Rossi *(ed.)*, McGraw Hill Publications, Hamburg, Germany.

Harrison, A. P., Jr. 1982. Genomic diversity among strains of *Thiobacillus ferrooxidans*, and genomic comparison with *Thiobacillus thiooxidans*. Arch. Microbiol. 131: 68-76.

Harrison, A. P. 1984. The acidophilic Thiobacilli and other acidophilic bacteria that have their habitat. Ann. Rev. Microbiol., 38: 265-292.

Hutchins, S. R., Davidson, M. S., Brierley, J. A., Brierley, C. L. 1986. Microorganisms in reclamation of metals. Ann. Rev. Microbiol., 40: 311-336.

Ingledew, W. J., Cox, J. C., Halling, P. J. 1977. A proposed mechanism for energy conservation during Fe^{2+} oxidation by *Thiobacillus ferrooxidans*. Chemiosmotic coupling to net H^+ influx. FEMS Microbiol. Lett., 2: 193-197.

Ingledew, W. J. 1982. The bioenergetics of an acidophilic chemolithotroph. Biochemica et Biophysica Acta, 683, 89-117.

Kelly, D. P., Harrison, A. P. 1989. Genus *Thiobacillus Beijernik*. P: 1842-1858. In: J. T. Stanley, M. P. Byrant, N. Pfennig, and J. G. Holt *(eds.)*. Bergey's Manual of Systematic Bacteriology, Vol. 3. The Williams & Wilkins Co., Baltimore.

Lacey, D. T., Lawson, F. 1970. Kinetics of the liquid-phase oxidation of acid ferrous sulfate by the bacterium *Thiobacillus ferrooxidans*. Biotechnol. Bioeng., 12, 29-50.

Less, H., Kwok, S. C., Suzuki. 1969. The thermodynamics of iron oxidation by the Ferrobacilli. Can. J. Microbiol. 15, 43-46.

Lundgren, D. G., Silverman, M. P. 1959. Studies on chemoautotrophic iron bacterium *Ferrobacillus ferrooxidans*. 1. An improved medium and a harvesting procedure for securing high cell yields. J. Bacteriol. 77: 642-647.

Mackintosh, M. E. 1978. Nitrogen fixation by *Thiobacillus ferrooxidans*. J. Gen. Microbiol., 34, 263-283.

March, R. M., Norris, P., Leroux, N. W. 1983. Growth and mineral oxidation studies with *Sulfolobus*. In: Progress in Biohydrometallurgy. Rossi, G. and Torma, A. E. *(eds.)*, Assoc. Miner. Sarda. Iglesias., p.71.

Mathur, A. K., Sivananda, S. R., Dwivedy, K. K. Mathur, S. K. 1992. Bacterial leaching of Uranium ore from a Quartz-Chlorite schist host rock and its effects on sulfides. Exploration and Research for atomic minerals, Vol. 5, p. 105-111.

Norris, P. R. 1990. Acidophilic bacteria and their activity in mineral sulfide oxidation. P. 3-22. In: H. L. Ehrilich and C. L. Brierley *(eds.)*. Microbial mineral recovery. McGraw-Hill books Co., New York.

Rawlings, D., Kusano, T. 1994. Molecular genetics of *Thiobacillus ferrooxidans*. Microbiol. Rev. Vol. 58. 39-55.

Rawlings, D. E., Simon Silver 1995. Mining with microbes. Biotechnology. Vol. 13. 775-778.

Tateo Yamanaka and Yoshihiro Fukumori 1995. Molecular aspects of the electron transfer system, which participates in the oxidation of ferrous iron by *Thiobacillus ferrooxidans*. FEMS Microbiol. Rev. 17, 401-413.

Torma, A.E. 1971. Microbiological oxidation of synthetic Cobalt, Nickel and Zinc sulfides by *Thiobacillus ferrooxidans*. Rev. Can. Biol., 30, 209-216.

Tetaz, T.J., Luke, R.K. 1983. Plasmid-controlled resistance to copper in *Escherichia coli.* J. Bacteriol. 154 (3): 1263-1268.

Tuovinen, O.H., Niemela, S. A., Gyllenberg, H.G. 1971. Effect of mineral nutrients and organic substances on the development of *T. ferrooxidans*. Biotechnol. Bioeng., 13: 517.

Tuovinen, O.H., Niemela, S.A., Gyllenberg, H.G. 1971. Tolerance of *Thiobacillus ferrooxidans* to some metals. Antonie van Leuwenhoek, 37: 489-496.

CHAPTER – 9

BIOLOGICAL REMEDIES FOR GROUNDWATER POLLUTION

P. Kaushik and I.K. Sharma***

* Department of Botany and Microbiology, Gurukul Kangri University, Haridwar- 249 404
** Central Ground Water Board, State Unit Office, Kangra, H.P.

1. INTRODUCTION

Water is the most important ecological factor for the development of plants, animals and life of human being. The earth is a large ecosystem which can be divided into small units on the basis of their morphology, physiology climate, habitat and plant communities. Development of any ecosystem depends on either surface water sources or ground water sources. Due to scarcity and pollution of surface water sources, dependency of human being is shifted towards ground water. But now the situation has arrived that ground water is also over exploited and water table is declining in some parts of the country. The quality of ground water is being polluted by municipal, industrial and agricultural wastes. This had lead to the ecological disturbance and the human life is also effected drastically. The biological pollution in ground water may be detected, prevented and controlled by biosensors and bioremediation. The objective of the present study was to review the physico-chemical and biological parameters of ground water quality for various purposes. The principal sources and causes for ground water pollution have also been evaluated. Attempts are made to find out certain organisms/bacteria which can be utilised to detect the biologically active pollutants and micro-organisms to get rid of water pollution.

2. MATERIALS AND METHODS

(a) Quality of Ground Water

The quality required of a ground water supply depends on its purpose; thus, needs for drinking water, industrial water and irrigation water vary widely.

(b) Quality of Drinking Water

The physico-Chemical standard for drinking water suggested by WHO are given in Table 1.

Table 1. The World Health Organization drinking water standards.

Characteristic	Limit of general acceptability (mg/l)	Allowable limit (mg/l)
Total solids	500	1500
Colour (H)	5	50
Turbidity	5	25
Chloride	200	600
Iron	0.3	1
Manganese	0.1	0.5
Copper	1.0	1.5
Zinc	5	15
Calcium	75	200
Magnesium	50	150
Magnesium & Sodium Sulphate	500	1000
Phenols	0.001	0.002
Fluoride	1.0	1.7
Nitrate	45	100
Sulphate	–	200
PH	7-8	
Toxic Substances		
Lead		0.1
Arsenic		0.05
Selenium		0.01
Chromium		0.05
Cadmium		0.01
Cyanide		0.05

(c) Biological Standard

Bacteriological analysis is important for detecting biological pollution of ground water. Most pathogenic bacteris in ground water are indigenous to the intestinal tract of animal and humans, but isolating them from natural water is difficult in the laboratory. Because bacteria of the coliform group are relatively easy to isolate and identify, standard tests to determine their presence or absence in a water are taken as a direct indication of the safety of the water for drinking purpose. Coliform test results are reported as the most probable number (MPN) of coliform group organisms in a given volume of water. By analysis of a number of separate portions of a water sample, the MPN is computed from probability tables for this purpose.

Table 2. Bacteriological standards of World Health Organization

Substance	Examined	Maximum permissible limit
Standard portions	10 ml	Not more than 10% in one month
Standard portions	100 ml	Not more than 60% in one month shall show coliforms.

(d) Quality of Water for Industrial Uses

Recommended quality limits for water used for various industries, compiled from numerous sources are given in Table 3.

(e) Water Quality for Irrigation

Salt is the most impotant constituent of irrigation water which harm plant growth physically by limiting the uptake of water through modification of Osmotic pressure. Effect of salt on soil is cause of changes in soil structure, permeability and certain indirect effects on plant growth.

Boron is necessary in very small quantities for normal growth of all plants, but in large concentration it becomes toxic. The quality classification of water for irrigation are given in Table 4.

3. POLLUTION OF GROUND WATER

Ground water pollution may be defined as the artificialy induced degradation of natural ground water quality. The principal sources and causes of ground water pollution are to be divided into four categories, municipal, industrial, agriculture and miscellaneous.

Table 3. Range in recommended limiting concentrations for industrial process waters (units are mg/l, except as noted) (after Amer, Water Works Assoc)

Use	Turbidity , Units	Color Units	Toxic and oder Threshold	Dissolved solids	Hardness as $CaCO_3$	Alkalinity $CaCO_3$	pH, Units	Chlorides as Cl
Air-conditioning	–	–	Low	–	–	–	–	–
Baking Boller feed	10	10	None- low	–	–	–	–	–
Brewing	0-10	0-10	None-low	500-1,500	–	75.00'	6.5-7.0'	6.0-100
Carbonated beverages	1-2	5-10	None-low	850	200-250	50-100	–	250
Confectionary	–	–	Low	50-100	Soft	–	>7.0	–
Dairy	–	None	None	500	100	–	–	30
Drinking, Food canning and freezing	5	15	J. Inoflona lve	500	–	–	–	250
Food canning and freezing	1-10	–	None-low	850	–	30-250	>7.5	–
Food equipment washing	1	8-20	None	850	10	–	–	250
Food processing general	1-10	5-10	Low	850	10-250	30-250	–	–
Ice manufacture	5	5	Low	170-1,300	–	–	–	–
Laundary	–	–	–	–	0.50	00	0.0-0.0	–
Paper and pulp fina	10	5	–	200	100	75	–	–
Paper, groundwood	50	30	–	500	200	150	–	75
Paper, kraft, bleached	40	25	–	300	100	75	–	0.2
Paper, kraft, unbleached	100	100	–	500	200	150	–	1.0
Paper, soda and sulfate pulps	25	5	–	250	100	75	–	0.1
Rayon and secinin fiber pulp production	5	5	–	100	5	50-75	–	0.05
Rayon manufacture	0.3	–	–	–	55	–	7.8-8.3	0.0
Sugar	–	–	–	Law	Low	–	–	0.1
Tanning	20	10-100	–	–	50-500	130	0.0-0.0	0.1-0.2
Textile	0.3-25	0-70	–	–	0-50	–	–	0.1-0.2

Table 3 contd...

Table 3 contd...

Use	Sulfates as SO_4	Iron as Fe	Manganese as Mn	Iron plus Manganese	Hydrogen sulfide	Florides as F	Other requirements
Air-conditioning	–	0.5	0.5	0.5	–	–	Not corrosive or slime promoting
Baking Boller feed	–	0.2	0.2	0.2	0.2	–	Potable if sicam is used for food preparation
Brewing	–	0.1	0.1	0.1	0.2	1.0	Potable, numerous other requirements
Carbonated beverages	250	0.1-0.2	0.1-0.4	0.1-0.4	0-0.2	0.2-1.0	Potable : COD 1.5: organic aller. Infinitesimal: algae and protozoa none
Confectionary	–	0.2	0.2	0.2	0.2	–	Potable
Dairy	00	0.1-0.3	–	–	–	–	Potable; NO_3 N, 5.5 : NO_3 N.0; NH_3, N.trace only COD as $KMnO_4$, 12
Drinking, Food canning and freezing	250	0.3	–	–	–	14-2.4	Potable
Food canning and freezing	–	0.2	0.2-0.3	0.2-0.3	1.0	1.0	Potable; free from saprophytic organisms NaCl, 1.000-1.500; NO_3, N, 2.0 : NH_3. N.0.4
Food equipment washing	–	–	0.1	0.1	–	1.0	Potable: organic matter. Infinitesimal
Food processing general	–	0.2	0.2-0.3	0.2-0.3	–	1.0	Potable
Ice manufacture	–	0.2	0.2	0.2	–	1	Potable : SIO_2.10
Laundary	–	0.2	0.2-1.0	0.2-1.0	–	–	
Paper and pulp fina	–	0.05	–	–	–	–	Soluble SIO_2.20, free CO_2.10 : Residual Cl_2.2
Paper, ground wood	–	0.1	–	–	–	–	Soluble SIO_2.20, free CO_2.10
Paper, kraft, bleached	–	–	0.1	–	–	–	Soluble SIO_2.50, free CO_2.10
Paper, kraft, unbleached	–	–	0.5	–	–	–	Soluble SIO_2.100, free CO_2.10
Paper, soda and sulfate pulps	–	–	0.05	–	–	–	Soluble SIO_2.200, free CO_2.10
Rayon and secinin fiber pulp production	–	–	0.03	0.05	–	–	Al_2O.0:50, SI.25: Cu.5
Rayon manufacture	7.8-8.3	–	0.0	0.0	–	–	–
Sugar	–	20	–	–	–	–	Ca, 20:Mg. 10:bicarbonate, as $CaCO_3$.100 alerile, no saprophytic organisms
Tanning	0.0-0.0	–	0.1-0.2	0.2	–	–	Dicarbonic hardness
Textile	–	100	0.05-1.0	0.2-1.0	–	–	Low : COD a; heavy metals none : Ca.10:Mg. 5 : bicarbonate as $CaCO_3$. 200

Table 4. Quality classification of water for irrigation

Water Class	Precent Sodium	Specific Conductance, μS/cm	Boron, mg/l		
			Sensitive Crops	Semitolerant Crops	Tolerant Crops
Excellent	<20	<250	<0.33	<0.67	<1.00
Good	20-40	250-750	0.33-0.67	0.67-1.33	1.00-2.00
Permissibile	40-60	750-2000	0.67-1.00	1.33-2.00	2.00-3.00
Doubtful	60-80	2000-3000	1.00-1.25	2.00-2.50	3.00-3.75
Unsuitable	>80	>3000	>1.25	>2.50	>3.75

(a) Municipal Sources and Causes

(i) Sewer leakage

Leakage of sewage into the ground is of common occurrence, especially from old sewers. Leakage may result from defective sewer pipe, ruptures from heavy load, seismic activity etc., and this sewage some times pollute the ground water due to infiltration in soil. It introduces high concentration of BOD, COD, nitrate, organic, chemicals, and possibly bacteria into ground water.

(ii) Liquid wastes

Wastes water in an urban area may originate from domestic uses that introduce bacteria, viruses and inorganic or organic chemicals into ground water. Shallow wells are widely employed to place surface runoff and sometimes untreated municipal waste water under ground in fresh water aquifers. Such disposal wells have been criticized from a health stand point because of the potential for pollutants to be released directly into an aquifer. The problem is most critical where disposal wells are near pumping wells and where the beneficial effects passing through time grained materials may be absent, such as in basalt and limestone aquifers.

(iii) Solid wastes

The land disposal of solid wastes creates an important source of ground water pollution. A landfill may be defined as any land area serving as a depository of urban or municipal, solid waste. Leachate of landfills creates problem of pollution of ground water where high rainfall and shallour water tables occur. Important pollutants frequently found in leachate include BOD, COD, iron, manganese, chloride, nitrate, hardness and trace elements.

Hardness, alkalinity, and total dissolved solids are often increased, while generation of gases such as methane, carbon dioxide, ammonia and hydrogen sulfide, are further by products of landfills.

(b) Agricultural Sources and Causes

(i) Irrigation return flows

Approximately one half to thirds of the water applied for irrigation of crops is consumed by evapotranspiration; the remainder, termed irrigation return flow drains to surface channels or joins the underlying ground water. Irrigation increases the salinity of irrigation return flow from these to ten times that of the applied water.

The degradation results from the addition of salt by dissolution during the irrigation process, from salts added as fertilizers or soil amendments, and from the concentration of salts by evapotranspiration. Principal actions include calcium, magnesium and sodium; major anions include bicarbonate, sulfate, chloride and nitrate. Because irrigation is the primary use for water in arid and semiarid region, irrigation return flow can be the major cause of ground water pollution in such regions.

(ii) Animals' wastes

Where animals are confined within a limited area, as for beef or milk production, large amounts of wastes are deposited on the ground. Thus for the 120 to 150 days that a beef animal remains in a feedlot, it will produce over a half ton of manure on a dry weight basis. With thousands of animals in a single feedlot, the natural assimilative capacity of the soil can become overtaxed. Storm runoff in contact with the manure carries highly concentrated pollutants to surface and subsurface waters. Animal wastes may transport salts., organic loads, and bacteria into the soil. Nitrate-nitrogen is the most important persistent pollutant that may reach the water table.

(iii) Fertilizers and soil amendments

When fertilizers are applied to agricultural land, a portion usually leaches through the soils and to the water table. The primary fertilizers are compounds of nitrogen, phosphorus and potassium; and phosphate and potassium fertilizers are readily absorbed on soil particles and seldom constitute a pollution problem. But nitrogen in solution is only partially used by plants or absorbed by the soils, and it is the primary fertilizer pollutant.

Soil amendments are applied to irrigated lands to alter the physical or chemical properties of the soil. Lime, gypsum and sulfur are widely used for this purpose, substantial amounts of these soil amendments may eventually leach to the groundwater, thereby increasing its salinity.

(iv) Pesticides

Pesticides can be significant in agricultural areas as a diffuse source of ground water pollution. The presence of these materials in ground water even in minute concentrations can have serious consequences in relation to the potability of the water. The impact of pesticide residue, however, depends on rainfall or irrigation rates and soil characteristics. Most pesticides are relatively insoluble in water, while others are readily absorbed by soil particles or are subject to microbial degradation.

(c) Industrial Sources and Causes

(i) Liquid wastes

The water in industrial plants are used for cooling, sanitation and manufacturing and processing. Cooling water that is softened before use to inhibit scale formation produces waste waters with salts and heat as important pollutants.

Ground water pollution can occur where industrial waste waters are discharges into pits, ponds or lagoons, thereby enabling the wastes to migrate down to the water table of ground water.

(ii) Tank and pipeline leakage

Underground storage and transmission of a wide variety of fuels and chemicals are common practices for industrial and commercial installations. These tanks and pipelines are subject to structural failures so, subsequent leakage becomes a source of ground water pollution. Petroleum and petroleum products are responsible for much of the pollution.

(iii) Mining activities

Mines can produce a variety of ground water pollution problems. Pollution depends on the material being extracted and the milling process : coal, phosphate and uranium are major contributors, metallic ores for production of iron, copper, zinc and lead are also importand, stone, sand and ground quarries although numerous are chemically much less important.

(d) Miscellaneous Sources and Causes

(i) Septic tanks

The most numerous and widely distributed potential sources of ground water pollution are septic tanks. It is constructed in individual houses for household wastewater treatment, and in commercial establishments, hospitals, industries etc. where community sewer systems are not available, a septic tank is a water tight basin but some times seepage occurs due to old and defaultic construction and sewage infilter to ground water. Due to it bacteria, viruses and nitrogen added into ground water create severe ground water pollution.

(ii) Salt water intrusion

Polluted surface water bodies that contribute to ground water recharge becomes source of ground water pollution. Many municipal water supply wells are located adjacent to rivers to ensure adequate flows, they can serve as important ground water pollution mechanism.

4. RESULTS AND DISCUSSION

Biological remedies of polluted waters of the rivers dates back to vedic period. The vedic scholars suggested how some plants can help us to get rid of micro organisms that are produced in a polluted river. The 3rd *mantra* under the *sukta* 37 of *kand* 4 of the Atharvaveda means guggulu (*Commiphora mukul*), peelu (*Salvadora oleoides*), naldi or mansi (*Nardostachys jatamansi*) aukshagandha and pramandini botanically not certain are the five drugs that repel microbes (in, *Kaushik*, 1988). Besides the aforesaid five drugs arjun *Terminalia arjuna* and Karkari (probably *Pistacia integerrima*) also help to get rid of pathogens.

(a) Biosensors

Biosensors are the organisms/bacteria which can be utilised to detect biologically active pollutant. Kaushik (1996) proposed the role of *Streptococcus faecalis* (Streptococci) and *Bifidobacterium bifidus* which survive in open waters for a short time; the intentional discharge of untreated sewage water into the Ganga at Haridwar by any negligent staff of sewage treatment plant can be significantly indicated. He also emphasized the need to use *vibrio fisheri, Photobacterium, Xanthomonas, Bradyrhizobium, Lactococcus* and *Escherichia coli* as biosensors.

(b) Bioremediation

The bioremediation is the use of microorganisms to get rid of the water pollutants.

Some bacteria can use pollutants as source of energy while others can degrade them into less harmful substances by producing enzymes. In this way toxins can be removed from underground wells, chemical spills, toxic waste sides and oil spills. *Pseudomonas* and *Bacillus* can be exploited for bioremediation. The bacteria which are capable to degrade petroleum products can be isolated and applied to oil spills. The oil being rich in carbon and sulphur but low in nitrogen and phosphorus which are essential for growth of microorganisms. Therefore, oil soluble fertilizer containing nutrient factor can be spread on oil spills with the suitable bacteria to metabolise the oils efficiently. Since the oil requires oxygen for its metabolisation. In case when oil spills on land enough tilling of the soil should be done with maximum irrigation.

The coal mine waste rich in sulphur content due to iron sulphide can be improved by oxidizing sulphide to sulphate. The sulphate, however enter water stream that lowers pH of the waters harming the aquatic organisms. The low pH also lead to the formation of insoluble iron hydroxide that forms the yellow precipitates which can be often seen on the surface of such water.

(c) Removal of nutrient by biological method

Nebel (1981) reported that a number of plants can be used to remove nutrients from water. The plants such as water hyacinths and water lilies can be used for live stock feed. The cattails (*Typha*) and 'reeds' (*Saccharum*) may be used in weaving straw-mats and baskets. Such plants can be anaerobically decomposed to produced methane or burn directly as a fuel. The eutrophication may be prevented in nutrient enriched water body. That leads excessive growth of algae and than depletion of dissolved oxygen as dead algae is consumed by decomposers. There are several other aquatic plants besides above said angiosperms as *Lemna, Wolffia, Utricularia* and Pteridophytes. (*Marsilea, Azolla* and *Salvinia*).

(d) Biological leaching of Uranium and Copper

Thiobacillus ferroxidants can be used in the recovery of unprofitable grade of uranium and copper ores. It can be applied to maximized the uranium and copper contents in subsurface soil water where these are found in beyond permissible limit.

Present study reveals that biosensors and bioremediation play effective role to detect, control and in prevention of water pollution. *Streptococcus faecalis, Bifidobacterium bifidus, Vibrio fisheri, Photobacterium, Xanthomonas, Bradyrhizobium, lactococcus* and *Escherichia coli* are suggested for use as biosensors.

Pseudomonas and *Bacillus* may be used for bioremediation to get rid of water pollutants. Eutrophication may be prevented in nutrient enriched water body by the use of several aquatic plants like *Lemna, Wolbbia, Utricularia* and Pteridophytes. The quality of water which is not suitable for drinking purpose may be used for irrigation and industrial purpose. Water quality required for different type of industries are different (Table 3). The crops for growth should be selected according to available water quality for irrigation. Necessary arrangements and precautions are must for control and prevention of water pollutions.

ACKNOWLEDGEMENT

The authors are highly grateful to Dr. S.K. Sharma, Member (SAM), Central Ground Water Board and Dr. D.K. Chadha Chairman, Central Ground Water Board, Ministry of Water Resource, Govt. of India for providing critical comments and valuable suggestions. The authors are also thankful to Dr. S.C. Dhiamn Scientist 'D' Central Ground Water Board, Kangra (H.P.) for providing kind cooperation during the studies.

REFERENCES

Allen, MJ and EE Geldreich, 1975. Bacteriological Criteria for ground water quality. Ground Water 13 (1) 45-51.

Anonymous 1981. Standard Methods for the Examination of Water and Waste Water. APHA, Washington, ppI - XL VII + 1-1134.

Anonymous 1998 Dec. 12, Bacteria can be used to combat the manace of pollution, Amer Ujala (In Hindi p-2).

Davis SN and RJM De Wiest 1966. Hydroecology, John Wiley and Sons, New York.

Devlin RM and FH Witham 1986. Plant Physiology 4th ed. CBS Publishers and Distributors, Delhi.

Johansen, DA 1940. Plant Microtechnique McGraw Hill Book Co. Inc. New York and London P-491.

Kaushik, P. 1996. Introductory Microbiology Emkay Publications Delhi IIX + 344.

Kaushik P 1988. Indigenous Medicinal Plants Including Microbes and Fungi. Today and Tomorrow Printers & Publishers New Delhi P VII + 243.

Nebel BJ 1981. Environmental Science : The way The world works, Printice-Hall Inc. New Jursey.

Todd DK 1980. Ground Water Hydrology, John Wiley & Sons New York, 2nd Edn.

Walton C 1970. Ground Water Resource Evaluation, McGraw - Hill, New York.

CHAPTER – 10

OPTIMIZATION OF FACTORS AFFECTING NICKEL BIOLEACHING BY TWO STRAINS OF *Thiobacillus ferrooxidans*

N. Anuradha, A.K. Mathur**, N.S. Sashidhar* and H. Polasa**

*Dept. of Microbiology, Osmania University, Hyderabad
**Atomic Minerals Division for Exploration and Research BARC, Hyderabad.

1. INTRODUCTION

Microorganisms play a predominant role in the solubilization, transport, and deposition of metals and minerals in the environment. Microbial mining technology presents an economic alternative for today's minerals and waste water treatment industries, at a time when high grade minerals resources are being depleted, energy costs are increasing and adverse environmental effects are becoming more apparent as a result of conventional technologies [Hutchins *et al*, 1986]. Bioleaching of lean grade ores and their concentrates are being used as a possible alternative to the conventional chemical leaching methods [Lundgren *et al.*, 1986]. The bacterium most studied in context of its application in mineral processing is *Thiobacillus ferrooxidans*. It is an aerobic, autotrophic, acidophillic, microorganism, which requires sulfur or ferrous iron as energy for its growth and maintenance.

Apart from the various factors that tend to affect the efficacy of the microorganisms involved in the leaching process, temperature, pH, pulp density and inoculum size are the most important factors affecting the adaptability and the ability of the microbes to leach metals from the ores.

In the present study, an attempt was made to optimize the above parameters for leaching nickel from an sulfide concentrate obtained from a uranium ore by two strains of *Thiobacillus ferrooxidans*.

In our present study, we used the sulfide concentrate obtained from a uranium ore and proceeded to optimize the conditions required by two strains of *T. ferrooxidans* to primarily leach nickel and copper. The flotation concentrate contains substantial amounts of copper (13-16%) and nickel (10-12%). The concentrate is generally used to recover copper as metal. The recovery is affected due to the high percentage of the nickel in the concentrate, as it is very difficult to separate nickel from copper by chemical process. Therefore an economically viable process would be selective bioleaching of nickel from the ore concentrate would be highly desirable. Preliminary experiments in our laboratory showed that the bacterial strains available with us showed a selectivity to leach nickel from the flotation concentrate. However, the percentage of nickel leached was extremely low (8.08% by MCMB–203 and 7.226% by MCMB – 231). Experiments were designed to optimize the conditions of leaching so as to obtain maximal percentage of nickel into the leach liquor.

2. MATERIALS AND METHODS

(a) Bacterial strains and Maintenance

The two strains of *Thiobacillus ferrooxidans*, MCMB-203 and MCMB-231 used in the present study were obtained from Agarkar Research Institute, Pune, India. These cultures were maintained in $M9K^+$ liquid medium [Silverman and Lundgren, 1959] at 30°C on a gyratory shaker with periodical transferring to fresh medium.

(b) Sulfide concentrate

The sulfide concentrate used in the experiments was recovered by flotation technique from an uranium ore at Uranium Corporation of India Ltd., (UCIL), Jaduguda, Bihar, India and it was kindly provided by Atomic Minerals Division for Exploration and Research (AMD-ER), Hyderabad. This batch of sulfide concentrate predominantly consisted of 25.46% Fe, 13.52% Cu, 10.96% Ni and 4.12% of Mo. The particle size was 250μm.

(c) Bioleaching procedure

In carrying out the bioleaching experiments, varying amounts of sulfide concentrate was first autoclaved for 15 min at 15 lbs pressure and then transferred aseptically into 90ml of $M9K^-$ medium (lacking $FeSO_4$), which was pre-autoclaved for 15min at 15 lbs. pH of this leaching medium was adjusted to 2.5 with 0.1N H_2SO_4. All the experiments were carried out in 250ml conical flasks. The flasks were incubated on a gyratory shaker at 150rpm and 30°C temperature. The pH of the leaching medium was monitored every 24 hours and adjusted to pH 2.5 with 0.1N H_2SO_4 until it stabilized at 2.5 which usually took 7-8 days of incubation on the shaker. Inoculation of bacteria into the leaching media was done only after pH was stabilized.

Bacterial inoculum was prepared by first centrifuging the cultures (in log phase) at 7000 rpm at 4°C temperature. The cell pellet was washed with 0.1N H_2SO_4 to remove any traces of jarosite (iron precipitate). This cell pellet was then suspended in sterilized M9K medium and the cell density of the cultures was fixed by measuring $O.D_{590}$ to 2.8 [Silverman & Lundgren, 1959]. These cell suspensions were used as inoculum in all experiments. After the initial pH stabilization of the leaching medium at 2.5, a 10% (v/v) inoculum of each strain was made into the flasks containing the above leaching medium. The inoculated flasks as well as the uninoculated control flasks (containing 10g of the sulfide concentrate in 100ml of $M9K^-$ medium) were incubated on a gyratory shaker at 150 rpm and 30°C temperature. At the end of 28 days of experimentation (excluding the time taken for initial pH stabilization), the contents of all the flasks were filtered through Whatman No. 1 filter paper. The leach liquors were then analyzed by atomic absorption spectroscopy for the estimation of Cu and Ni leached.

The experimental protocol followed to study the effect of varying pH on the bioleaching ability of MCMB-203 and MCMB-231 was similar in all respects as described above, except that the initial pH of the leaching media was stabilized at 1.5, 1.8. 2.0, 2.3, 2.5 and 3.0 using 0.1N H_2SO_4.

To study the effect of temperature on bioleaching, similar protocol of bioleaching described in the previous section was followed. The initial pH of the leaching media was stabilized at 2.3 and the flasks were incubated on gyratory shaker at 25°C, 28°C, 30°C, 33°C, 35°C and 40°C.

In studying the effect of pulp density on bioleaching, varying amounts of the sulfide concentrate was added to the leaching media to carry out the bioleaching experiments. A 5% (w/v), 10% (w/v), 20% (w/v), 40% (w/v) and 60% (w/v) pulp densities have been used. The bioleaching protocol was similar as described above. The pH and temperature were maintained at 2.3 and 28°C respectively in the subsequent experiments.

To study the effect of varying bacterial inoculum sizes on bioleaching, different volumes of the inoculum, 5% (v/v), 10% (v/v), 20% (v/v) and 30% (v/v), was added to the leaching system. The optimum pulp density of 10% was used in the subsequent experiments.

3. RESULTS

The effect of varying pH in the range of 1.5 to 3.0 was studied on the efficacy of MCMB-203 and MCMB-231 to leach nickel and copper using floatation concentrate. The results presented in Table 1 indicate that percentage of nickel and copper leached from the flotation concentrate by the two strains of *Thiobacillus ferrooxidans* MCMB-203 and MCMB-231 increased from pH 1.5 to pH 2.3. The maximum leaching at optimum pH was 31.05% amd 37.50% for MCMB – 203 and MCMB – 231 respectively. There was, however, a sharp decline in the percent nickel and copper leached at pH 2.5 and 3.0.

Experiments were conducted at six different temperatures ranging from 25°C to 40°C keeping the optimum pH 2.3. The results are presented in Table 2. Maximum percentage of nickel and copper was leached at 28°C. The percentage of nickel leached from the sulfide concentrate declined with an increase in the temperature above 28°C.

Among the various other factors effecting the bioleaching process, pulp density, which is the ratio of solid to liquid, is an important factor. In the present study, 10% pulp density was found to be optimum for bioleaching Nickel (54.962% and 57.515% for MCMB – 203 and MCMB – 231 respectively) and copper (5.052% and 5. 751% for MCMB – 203 and MCMB – 231 respectively). Further increase or decrease in pulp density decreased the percentage of metals leached by both microorganisms [Table 3].

Table 1. Effect of pH on Ni & Cu bioleaching by MCMB-203 and MCMB-231

	MCMB-203		MCMB-231	
pH	% Cu leached	% Ni leached	% Cu leached	% Ni leached
1.5	0.34	14.58	0.781	18.92
1.8	0.97	18.92	1.139	22.62
2.0	1.14	20.04	1.650	25.72
2.3	2.24	31.05	4.049	37.50
2.5	0.44	8.08	0.858	7.226
3.0	0.36	4.96	0.57	4.53

Note: The % of copper and nickel leached were calculated from the amount of the respective metals bioleached into the leach liquors by *Thiobacillus ferrooxidans* strains MCMB-203 and MCMB-231 at varying pH.

The effect of inoculum size on bioleaching has been carried out. It was found that 10% inoculum size was optimum for the bioleaching process [Table 4] of Nickel (67.0% and 66.00% for MCMB - 203 and MCMB - 231 respectively) and copper (5.04% and 4.89% for MCMB - 203 and MCMB - 231 respectively).

4. DISCUSSION

The use of microorganisms to leach metals from ores though is a very old process has been an area of intense research in the recent past at a stage when the high grade ores are fast depleting. Bioleaching is a better alternative to chemical process in leaching metals from

Table 2. Effect of temperature on Ni & Cu bioleaching by MCMB-203 and MCMB-231

	% Copper leached			% Nickel leached		
Temperature	Control	MCMB-203	MCMB-231	Control	MCMB-203	MCMB-231
25°C	0.97	3.98	4.97	2.72	19.49	22.40
28°C	1.14	4.09	5.32	2.9	39.14	45.28
30°C	1.23	4.11	5.54	3.54	33.63	35.33
33°C	0.86	3.55	4.12	3.97	18.44	21.50
35°C	0.74	2.64	2.98	3.94	13.62	16.96
40°C	0.42	1.52	1.76	4.05	4.52	3.90

Note: The % of copper and nickel leached were calculated from the amount of the respective metals bioleached into the leach liquors by *Thiobacillus ferrooxidans* strains MCMB-203 and MCMB-231 at varying temperature. The pH was 2.3.

Table 3. Effect of Pulp density on Ni & Cu bioleaching by MCMB-203 and MCMB-231

Pulp	% Copper leached		% Nickel leached	
Density (w/v)	MCMB-203	MCMB-231	MCMB-203	MCMB-231
5	4.754	4.904	45.264	47.810
10	5.052	5.751	54.962	57.515
20	1.997	1.658	32.464	30.520
40	1.374	1.263	25.219	27.327
60	1.152	1.196	18.575	20.864

Note: The % of copper and nickel leached were calculated from the amount of the respective metals bioleached into the leach liquors by *Thiobacillus ferrooxidans* strains MCMB-203 and MCMB-231 at varying pulp densities. The pH and temperature were maintained at 2.3 and 28°C respectively.

Table 4. Effect of Inoculum size on Ni & Cu bioleaching by MCMB-203 and MCMB-231

Pulp	% Copper leached		% Nickel leached	
Inoculum size (v/v)	MCMB-203	MCMB-231	MCMB-203	MCMB-231
Control	1.77	1.77	5.24	5.24
5%	2.46	2.54	55.00	58.00
10%	5.04	4.89	67.00	66.00
20%	3.88	3.67	52.00	54.00
30%	1.97	2.01	45.00	42.00

Note: The % of copper and nickel leached were calculated from the amount of the respective metals bioleached into the leach liquors by *Thiobacillus ferrooxidans* strains MCMB-203 and MCMB-231 at varying inoculum size. The optimal pH, temperature and pulp density were maintained.

lean ores, thus being economical at the same time without any adverse effect on environment.

In the present study, the various physical parameters effecting the bioleaching efficiency of *Thiobacillus ferroxidans* have been studied and optimized.

The optimum pH at which both the strains exhibited maximum bioleaching efficiency was 2.3 (Table 1). Though it has been reported by Silverman [1967] that *T. ferrooxidans* was active in the pH range of 1.5-5.5. At pH values above 2.5, a strong precipitation of iron compounds (jarosite) occurs which cover the mineral particles (Vian *et al*, 1986) which in turn adversely affects the leaching process by preventing the oxidizing action of ferric ions. Also, some of the bacterial cells might have been trapped in the iron precipitates formed at pH above 2.3 consequently decreasing the number of active bacterial cells available for attacking the ore particle.

The reaction rate and its temperature dependence are typical for each mineral dissolution and it also depends on the microbial strain being employed in the bioleaching process. Modak *et al.* [1996], suggested that the activity of *T. ferrooxidans* is inhibited at temperatures higher than 35°C where inactivation and ultimately thermal death of the organism might occur. Hence, temperature is another factor which influences both the bioleaching efficiency and the growth of *T. ferrooxidans* in the leaching media. The effect of temperature on the percent copper leached was, however, not very significant since there was no major difference in the percent copper leached in the temperature range of 25°C to 40°C. However, the optimum temperature for bioleaching of nickel by MCMB-203 and MCMB-231 was found to be 28°C (Table 2).

The pulp has a significant effect on the extraction of metals by the bacteria. The optimal pulp density at which a bioleaching microbe can show maximum metal leaching ability is unique to each *T. ferrooxidans* strain. Available reports indicate that the extraction rates increase linearly upto 15% pulp density but decrease with a further increase in the pulp densities. Bosecker [1986], suggested that an increase in the pulp density involves an increase in total particle surface area, which will accelerate the leaching process. Hence, leaching experiments were carried out at pulp densities varying from 5 g to 60 g of flotation concentrate in the leaching medium. In our experiments we found that, there was an initial increase in the amount of nickel bioleached with an increase in the pulp density upto 10% (Table 3). Above 10%, a gradual and drastic decline was noted in the bioleaching efficiency. A similar trend was shown by both the organisms in case of copper leaching. Increase in the pulp density above optimum levels will effect the mass transfer of O_2 and CO_2 resulting in the reduction of leaching efficiency of bacteria.

The rate of an enzymatically - catalyzed reaction is affected by the concentration of the catalyst, which in the bioleaching reactions are the bacterial cell suspensions. Hence, we studied the effect of inoculum/culture size on the efficiency of nickel and copper leaching from the flotation concentrate by MCMB-203 and MCMB-231. Inoculum size, either lower or higher than 10% (v/v) of culture as inoculum showed a decrease in the amount of nickel and copper leached from the floatation concentrate (Table 4). At lower inoculum sizes, there are too few cells free in the aqueous phase to facilitate higher metal solubilization. Contrary to this, at higher inoculum sizes, an increase in bacterial population results in competition for survival and only the metabolically active and potent cells would survive annihilating the weaker species resulting in decreased leaching rates [Vian et al, 1986].

Further studies are in progress to increase the bioleaching efficiency of these strains by adapting them to flotation concentrate and to nickel and copper which are the major constituents of the leaching system.

REFERENCES

Bosecker, K. 1986. Leaching of lateritic nickel ores with heterotrophic microorganisms. In: R. W. Lawrence, R. M. R. Branion, H. G. Ebner (Eds.) Fundamental and applied biohydrometallurgy. Elsevier Publ.

Hutchins, S. R., Davidson, M. S., Brierley, J. R. and Brierley, C. L. 1986. Microorganisms in the reclamation of metals. Ann. Rev. Microbiol. 40 : 311-336.

Lundgren, D. G., Valkova, M. and Reed, R. 1986. Chemical reactions important in the bioleaching and bioaccumulation. In: Ehrlich, H. L., Holmes, D. S (Eds.): Workshop on biotechnology for the mining, metal refining and fossil fuel processing industries. Biotechnol. Bioeng. Symp., No. 16 : 7.

Modak, J. M., Natarajan, K. A. and Sanghamitra Mukhopadhyay 1996. Development of temprature tolerant strians of *T. ferrooxidans* to improve bioleaching kinetics. Hydrometallurgy. 42 : 51-61.

Silverman, M. N. and Lundgren, D.G. 1959. Studies on chemolithotrophic iron bacterium *Ferroobacillus ferrooxidans*. 1. An improved medium and a harvesting procedure for securing high cell yields. J. Bacteriol., 77 : 642-647.

Silverman, M.P. 1967. J. Bacteriol. 94 : 1046-1051.

Tuovinen, O.H. 1993. Bioleaching and Mineral biotechnology. Current opinion in Biotechnology, 4 : 344-355.

Vian, M., Creo, Y. C., Dalmistri, C., Gionni, A., Palazzolo, P. and Levi, G. 1986. *Thiobacillus ferrooxidans* selection in continuous culture. In: R. W. Lawrence, M. R. Branian, H. G. Huber (Eds.) Fundamental and Applied Biohydrometallurgy. Elsevier publication.

CHAPTER – 11

DEVELOPMENT OF NEW ANTIMICROBIAL AGENTS — ISSUES AND OPTIONS

C.K.M. Tripathi and Vinod Bihari

Fermentation Technology Division, Central Drug Research Institute, Chattar Manzil Palace, Lucknow-226 001, India

1. INTRODUCTION

The discovery of novel pharmaceuticals or products, through screening microbial secondary metabolites is an expanding activity throughout the pharmaceutical industry and the one that has been extremely fruitful. There is widespread acceptance that microorganisms are a virtually unlimited source of novel structures with many potential therapeutic applications. The structural novelty and diversity of microbial secondary metabolites make them a logical source for exploitation not only for molecules that exhibit antibiosis, but also molecules that interact with receptors, signal transduction or biosynthetic pathways which represent potentially important drug targets (Yarbrough *et al.*, 1993). More recently, the integration into microbial screening of pharmacology and molecular biology have focused efforts on novel molecular targets and the development of new screening technologies. This has altered the approach to screening microbial metabolites, thus bringing new pupeatives into what historically has been the domain of microbiologist and natural chemist. By expanding the targets and ability to detect novel activities the perspective of screening has drastically expanded the scope of developing new antibacterial agents (Gould, 1999).

It is desirable to introduce as much metabolite diversity into a screening programme as is possible. To achieve metabolite structural diversity, it is generally agreed that a diverse and novel repertoire of microbes is desirable. Treatment of soil and selective pressures are routinely applied in the initial isolation and incubation procedures to favour the growth of desired organisms (Williams *et al.*, 1989). Freshly obtained wild organisms yield greater microbial metabolite diversity than isolates held in culture collections. Metabolic diversity and production are functions of the biosynthetic capabilities of the organism and the fermentation conditions which allow for the expression of this capacity. Fermentation parameters can be manipulated to encourage the production of diverse secondary metabolites (Omura, 1992).

The issues related with the development of new antimicrobial agents have to advance in multiple directions. Clearer understanding of the mechanisms involved in bringing about multidrug resistant organisms, target modeling giving new scope for antimicrobial design based on genomics and proteomics and judicious application of the antimicrobial agents should be synchronized (Austin *et al.*, 1999).

2. FACTORS CONTRIBUTING TO THE RESISTANCE FOR ANTIMICROBIAL AGENTS

Drug resistance in bacterium was likely to have been present, to some extent, prior to the modern antibiotic era. However, the maintenance and spread of resistance detriments has increased in parallel with the use of antimicrobials(Livermore, 1995). The use of antimicrobials is directed towards invading organisms rather than creating alterations in host physiology. Antibiotic use in an individual patient may result in changes in microbial ecology that has the potential to affect other. Although not proven, considerable evidence suggests that antibiotic use is casually linked to the selection of resistance in bacteria. However, the exact quantitative relationship between antibiotic use and resistance is unknown (Sprart, 1994). The problem of antibiotic resistance is complex and involves interaction of multiple factors relating to the particular organism, the drug being used as well as the environment. At any given time, the rate of resistance within a health care setting is a function of, first, the rate of introduction into the environment from the community including long term care facilities, second amplification by antibiotic use, and third, the effectiveness of infection control efforts. Resistance is introduced and spread in health care settings by a variety of mechanisms, with selection pressure by antibiotic use as arguably the most important factor to maintain its pressure. The epidemiology of the spread of resistant organisms has been well described for many bacteria, but the epidemiology of antibiotic use as its specific contribution to amplification and maintenance of resistance is poorly understood.

Four processes contribute to the accumulation of resistance : first, species with inherent resistance and advantages; secondly, resistant mutants of hitherto susceptible strains are selected; thirdly, transferable genes disseminate among bacterial strains, carried by plasmids, transposons and integrons; forthly some resistant strains achieve epidemic spread among patients, hospitals and countries; and last inactivation of the antimicrobial agents by the pathogens (Livermore and Dudley, 2000). The relative importance of these processes varies with the pathogen and setting. It is now clearly established that acquisition of a gene from exogenous source, conferring resistance by any means, is the most common solution adopted by bacteria to escape antimicrobial activity. However, several antibiotics were known to be of bacterial origin, implying the existence of protective mechanisms in producer organism and thus a potential source for resistance genes. There are several means for bacteria to acquire and transfer exogenous DNA but conjugation is the main process (Monroe and Polk, 2000). The first significant component in the process of antibiotic resistance flux in microbial populations is gene culture by way of transposons, which can move from replicon to replicon and disseminate through conjugation. The crucial question is how do transposons acquire the resistance genes. There are two types of transposons : composite transposons, found in both Gram positive and Gram negative species, and transposons harbouring an integron, which are mainly observed in Enterobacteriaceae and Pseudomonads. With this system, bacteria are capable of stockpiling exogenous genetic loci to establish an appreciable antimicrobial armamentorium, and multi resistant integron harbouring upto different cassettes.

Within past several years we learned that genes for multidrug efflux are wide-spread in the prokaryotic world. In Gram negative bacteria they always seem to be present on chromosomes and create intrinsic resistance to antibiotics, dyes and detergents. Increased expression of these pumps do increase MICs of many agents. Such an increase in MIC may be only marginal in organisms with high outer memebrane permeability such as *E. coli*. The genes for multidrug efflux are present on plasmids in *S. aureus*. Increased efflux mutations are caused in these genes. One factor favouring selection of such mutations is the weak inhibitory activity of the selective agents. It has been suggested that the drug that are less susceptible to conventional resistance mechanisms are, paradoxically, more likely to select for multidrug efflux mutations. Indeed, fluro-quinolones, a synthetic agent, often select efflux type mutation in *P. aeruginosa* in a clinical setting (Naikaido, 1998). Our increased understanding of the mechanism of this process suggests that it may be possible to design drugs that are less efficiently pumped out. Combined therapy with a drug that is not a substrate of ubiquitous efflux system would be another possibility. Either inhibiting the efflux pumps or increasing outer membrane permeability in Gram negatives would be another excellent solution.

One way that bacteria resist antimicrobial agents involves 'drug sensors', proteins that bind a range of structurally unrelated antimicrobial drugs and induce expression of multidrug resistance pumps. The basis for drug recognision is now becoming clear from the recently determined crystal structure of drug sensor. The drug sensors are providing us with useful insight into the functions of multidrug resistance pumps and will undoubtedly be very useful in searching for natural multidrug resistance pump substrate. The drug sensors share an important feature with the larger transporters they regulate, which is remarkable ability to discriminate between drug and self molecular understanding how the small regulators do this will be very useful in helping to decipher the working of the pumps (Lewisk, 1999; Lee *et al.*, 2000; Renau *et al.*, 1999).

3. RATIONALE APPROACHES TO DESIGN NEW DRUGS BASED ON UNDERSTANDING THE BIOCHEMICAL BASIS OF BACTERIAL RESISTANCE

3.1 Analogs of existing antibiotics stable to enzymatic inactivation

Enzymatic inactivation of antibiotics is an important mechanism of bacterial resistance to β-lactam antibiotics, chloramphenicol, aminoglycosides and macrolides. To encounter this problem, enzyme stable analogs have been developed and marketed, most notably with the β-lactam e.g. the isoxasolyl penicillins, imipenem and merophenem and aminoglycosides e.g. amikacin (MacKenzie *et al.*, 1997).

3.2 Inhibition of bacterial enzymes that inactivate antibiotics

The bacterial enzymes that degrade or modify antibiotics are themselves potential targets for drug design, leading to combination products which contain an antibiotic with a specific inhibitor. Clavulanic acid was the prototype molecule that established the value of the strategy. It exhibits weak antibacterial activity, but binds with high affinity and essentially irreversibly to many bacterial β-lactamases. Other β-lactamase inhibitors have also been introduced e.g. sulbactam, tazobactam (Hill *et al.*, 1998).

3.3 Analogs of existing antibiotics not recognized by bacterial efflux pump

In some bacteria resistance to the macrolides, quinolones, and tetracyclines is mediated by efflux pumps that prevent antibiotic accumulation by the organism. The synthesis of the analogs of these antibiotic classes that are not recognized by efflux pumps yet retain antibacterial activity provides in principle a solution that can be used to circumvent transport mediated resistance (Smith, 1999).

3.4 Inhibition of efflux pumps that remove antibiotics from the cell

The efflux systems that remove antibiotics from bacteria are themselves potential targets for drug action that would lead to the products containing an antibiotic and an inhibitor which prevents efflux of the drug from the cell. Analogs of existing antibiotics that bind to modified target sites in resistant bacteria will provide alternative antibacterial agents.

Modification of the target site for antibiotic action results in resistance to a number of antibiotics including tetracyclines, β-lactams and glycopeptide glycylcyclines, new carbapenem analogs. N-alkyl-substituted glycopeptides have successfully shown to overcome the antibiotic resistance because of the side modification.

4. IDENTIFICATION OF NEW TARGETS FOR ANTIMICROBIAL ACTION

An alternative approach to the problem of emerging resistance to current antibiotics is to seek structurally novel antibiotics that inhibit new molecular targets (Xia *et al.*, 1999). Such agents are unlikely to be susceptible to existing mechanisms of resistance because of their structural novelty and unique mode of action.

4.1 New molecular targets involved in bacterial growth

Nucleic acids, proteins and cell wall synthesis are already the targets of existing antimicrobial agents. In the search of new microbial targets, a number of other processes that would affect bacterial growth have been suggested. Inhibition of lipoprotein and lipopolysaccharide biosynthesis, various enzymes of protein synthesis, metabolic transport are the ones tried. Many of these pathways have been validated as candidate targets for antibiotic action.

4.2 New molecular targets involved in bacterial infection

Recent interest in the mechanisms by which pathogenic bacteria cause disease has raised the possibility of designing new agents that act against gene products expressed primarily or exclusively during infection. An anticipated advantage of developing such agents will be the likely absence of preexisting resistance mechanisms.

4.3 Bacterial virulence as a target for antimicrobial therapy

Before discussing bacterial virulence as a target for antimicrobial chemotherapy, it is necessary to define virulence and pathogenicity. We will define here a pathogen as any organism that has the capacity to cause disease in a host at any time. What is probably more useful in terms of targetting virulence for antimicrobial chemotherapy is to distinguish between those genes that are specially expressed during the disease process and those that are not. In order to understand virulence factor expression of a number of techniques that analyse bacterial gene expression *in vivo* have been developed (Allsop, 1998; Hansel *et al.*, 1995). *In vivo* expression technology (IVET) allows the identification of genes specifically induced in host tissues. This approach uses the host to enrich for genes that are expressed in host tissues during the pathogenesis of infection and can be applied in principle to any pathogen. Infection of mice with *Salmonella typhimurium* has been investigated with this technique (Rosamond 2000; Handfield and Levesque, 1999). The idea is to insert certain indicator genes in various places in the bacterial chromosome and then look for bacteria that express the indicator genes only during infection. Those bacteria must have the indicator genes associated with a virulence gene. Drugs directed against products of virulence genes might arrest the infective process and would also hit previously unknown targets. Using IVET techniques, the investigators have found that one gene expressed during mouse infection is a gene previously known to break down fatty acids that bacteria meet outside their own cells. It is suggested that during an infection, the bacteria might be trying to destroy fatty acids of the host (Mahan *et al.*, 1993; Lee & Camilli, 2000).

Yet another target in bacterial virulence is the sensing and intracellular signaling system that pathogens use to sense hostile environment in the human host and turn certain genes in response. Signature tagged mutagenesis (STM) allows for an assessment of which genes are essential, or atleast useful, for viability *in vivo*. Other methods seek to determine which messenger RNAs are transcribed (or not) during infection. Substractive hybridization has been used successfully to assess difference in gene transcription in culture and in the host. A relatively new technology, transcriptional profiling, enables the investigators to analyse the entire mRNA content of a bacterial culture under different conditions.

Another method for determining the virulence factor of a particular bacterium is to analyse the sequence of the entire genome (Tang *et al.*, 1998). Of the bacterium genome completely sequenced to date, at least 60 are from pathogenic ones. Identification of the homologs to known virulence factor genes in a given bacterium could suggest that they play a role in pathogenic process. Genomic sequencing has led to the discovery that virulence factor often reside on specific regions of the chromosome (pathogenic island) whose genetic content differ from the rest of the chromosome and which can resemble mobile elements. The presence of these islands in one srain and not another has been shown to dictate the virulence capability of the bacterium harbouring it.

4.4 Establishment of pathogen in host as a target for antimicrobial therapy

It is advocated that targetting those processes that occur early in infection, before the solid establishment of a bacterial consortium, represent the most viable area for chemotherapy that does not directly kill or inhibit the growth of bacterium. Certain proteins have been found to be secreted by the pathogens prior to the adhesion to the host. Identification and inhibition of these specific proteins and the presence of these inhibitors in the host will deter the infection process.

5. GENOMICS : USEFUL TOOL FOR DEVELOPING NEW ANTI-MICROBIALS

From the beginning, the identification and development of an antimicrobial agent has been proven to be a lengthy and multifaceted process requiring the interaction of many scientific disciplines (Barry *et al.*, 2000). Today, the utilization of genomic information and technology, and its potential for identifying novel antimicrobial agents is the latest concept. Nevertheless, genomics has already revolutionized the discovery process for antimicrobial by accelerating target identification and evolution, assay development, mechanisms of action studies and follow up support for medicinal chemistry programmes. Till now new class of antimicrobial agents have been mostly identified from natural products or their derivatives. With the target directed approach however, lead-chemistry follow up has to address the issues of cellular accumulation, efflux, spectrum and other aspects of achieving drug like characteristics. Cellular accumulation and efflux issues or crucial factors that were previously circumvented by the whole cell assays have probably favoured the discovery of natural products for antimicrobial activity. Natural products are typical complex molecules that evolve with cell penetration properties that allows their release from the producing organism as well as entry into surrounding microbes. Genomic information has illustrated the prevalence of efflux pumps in all organism which are significant in mediating both innate and acquired resistance to many antimicrobial agents. Some companies have even embarked on programmes to find inhibitors of these pumps that could enhance the activity of or prevent the development of resistance to, current antimicrobial therapies. Genomic information should enhance our understanding of the selectivity and role of these pumps in both cellular process and in drug efflux. Chemical structure activity relationship (SAR) may then be assessed co-ordinately with various efflux mechanisms or cell permeability parameters. Genomics should also address cellular uptake issues either by allowing SAR to determine what cell wall property in preventing drug or perhaps by identifying novel essential enzymes that can be used to design prodrug version of antimicrobials. Prodrug that are activated by essential enzymes could be used to improve cellular uptake and accumulation parameters or reduce host toxicity. The current need to find new class of antimicrobial agents is driving the efforts to find chemistry that is active against unique targets. The successful application of new paradigms to address cellular accumulation issues may be required to discover the novel drugs of the genomics era (Weinstock *et al.*, 2000; Stranss and Falkows, 1997).

5.1 Genomics for target identification and validation

As primary tenet of genomics-based drug discovery is that sequence data base searching can begin to separate an organism genes into several broad classification to dictate how inclusive or exclusive gene sets can be for the cellular functions. Valid drug targets are, as the first criteron, those genes that encode protein (or structural/catalytic RNA molecules) required for the cells to grow. Many different processes have been described for performing an analysis to identify the genes essential for growth. Classical genetic approaches such as screening for temperature sensitive mutations and transposon mutagenesis have been applied on a large scale (Cialperin *et al.*, 1999; Alksne and Projar, 2000; Watanabe *et al.*, 1999; LeClerq *et al.*, 1988; Black and Hare, 2000).

Knowing the complete genome of microbial pathogens will accelerate the commercial development of novel pharmaceuticals to treat pathogenic microbes. By simultaneously analyzing the microbial genome and the host genome it will become possible to define those genes that are uniquely important for microbial survival in a given environmental context and those that are redundant or superfluous in that context. It will also be possible to learn why a given pathogen is virulent in the context of a specific host, when toxic cytokines are activated within a host, whether a given pathogen has evolved proteins with molecular mimicry capable to frustrating host immunity or inducing autoimmunity, and how a pathogen is able to survive in a state of latency impervious to the hosts immune system (Frieden *et al.,* 1993; Wanzel and Wong, 1999; Guillemot, 1999; White *et al.*, 2000). The sum total of this information will make it possible to define medications or vaccines that induce specific and effective immunity against the pathogen while minimizing unwanted or toxic side effects in the host. This will revitalize the entire private sector antibiotic and vaccine industry (Klugman and Feldman, 1999; Chopra *et al.*, 1997).

REFERENCES

Alksne LE, Projan, SJ 2000. Bacterial virulence as a target for antimicrobial chemotheraphy. Curr Opin Biotechnol. 11, 625-636.

Allsop AE 1998. New Antibiotic discovery, novel screens novel targets and impact of microbial genomics. Curr Opin Microbiol. 1, 530-534.

Austin DJ, Kristinsson KG, Anderson RM 1999. The relationship between the volume of antimicrobial consumption in human communities and the frequency of resistance. Proc Natl Acad Sci USA. 96, 1152-1156.

Barry CE III, Slayden RA, Sampson AE, Lee RE 2000. Use of genomics and combinatorial chemistry in the development of new antimycobacterial drugs. Biochem Pharmacol. 59, 221-231.

Black T, Hare R 2000. Will genomics revolutionize antimicrobial drug discovery. Curr Opin Microbiol. 3, 522-527.

Chopra I, Hodgson J, Metcalf B, Poste G 1997. The search for antimicrobial agents effective against bacteria resistant to multiple antibiotics. Antimicrob Agents Chemother. 41, 497-503.

Cialperin MY, Koonin EV 1999 Searching for drug targets in microbial genomics. Curr Opin Biotechnol. 10, 571-578.

Frieden TR, Munsiff SS, Low DE, Willey BM, Williams G, Faur Y, Eisner W, Warren S, Kreiswirth B 1993. Emergence of vancomycin resistant enterococci in New York City. Lancet. 342, 76-79.

Gould IM 1999. A review of the role of antibiotic policies in the control of antibiotic resistance. J Antimirob Chemother. 43, 459-465.

Guillemot D 1999. Antibiotic use in humans and bacterial resistance. Curr Opin Microbiol. 2, 494-498.

Handfield M, Levesque RC 1999. Strategies for isolation of *in vivo* expressed genes from bacteria. FEMS Microbiol Rew. 23, 69-91.

Hansel M, Shea JE, Gleeson C, Jones MD, Dalton E, Holden DW 1995. Simultaneous identification of bacterial virulence genes by negative selection. Science. 269, 400-403.

Hill DA, Herford T, Barratt D 1998. Antibiotic usage and methicillin resistant *Staphylococcus aureus* : an analysis of casuality. J Antimicrob Chemother. 42, 676-677.

Klugman KP, Feldman C 1999. Penicillin and cephalosporin resistant *Streptococcus pneumoniae*. Emerging treatment for an emerging problem. Drugs. 58, 1-4.

LeClerq R, Derlot E, Duval J, Courvalin P 1988. Plasmid mediated resistance to vancomycin teicoplanin in *Enterococcus faecium*. N Engl J Med. 1, 157-161.

Lee A, Mao W, Warren MS, Mistry A, Hoshino K, Okumura R, Ishida H, Lomovskaya O 2000. Interplay between efflux pumps may provide either additive or multiplicative effects on drug resistance. J Bacteriol. 182, 3142-3150.

Lee SH, Camilli A 2000. Novel approaches to monitor bacterial gene expression in infected tissue and host. Curr Opin Microbiol. 3, 97-101.

Lewisk 1999. Multidrug resistance : Versatile drug sensors of bacterial cells. Curr Biol. 9, R 403-R 407.

Livermore DM, Dudley MN 2000. Antimicrobials : better use, better drugs, or both? Curr Opin Microbiol. 3, 487-488.

Livermore DM 1995. lactamases in laboratory and clinical resistance. Clin Microbiol Rev. 8, 557-584.

MacKenzie FM, Forbes KJ, Dorai-John T, Amyes SG, Gould IM 1997. Emergence of a carbapenem resistant *Klebsiella pneumoniae*. Lancet. 350, 783-786.

Mahan MJ, Slauch JM, Hanna PC, Camilli A, Tobias JW, Waldor MK, Mekalanos JJ 1993. Selection of bacterial genes that are specially induced in host tissues : the hunt for virulence factor. Infect Agents Dis. 2, 263-268.

Monroe S, Polk R 2000. Antimicrobal use and bacterial resistance. Curr Opin Microbiol. 3, 496-501.

Naikaido H 1998. Multiple antibiotic resistance and efflux. Curr Opin Microbiol. 1, 516-523.

Omura S 1992. Trends in the search for bioactive microbial metabolites. J Indust Microbiol. 10, 135-156.

Renau TE, Leger R, Flamme EM, Sangalang J, She MW, Yen R, Gannon CL, Griffith D, Chamberland S, Lomovskaya O *et al.* 1999. Inhibitors of efflux pumps in *Pseudomonas aeruginosa* potentiate the activity of the fluroquinolone antibacterial levoploxacin. Med Chem. 42, 4928-4931.

Rosamond J, Allsop A 2000. Harnessing the power of the genome in the search of new antibiotics. Science. 287, 1973-1976.

Smith DW 1999. Decreased antimicrobial resistance after changes in antibiotic use. Pharmacotherapy. 19, 129-132.

Sprart BG 1994. Resistance to antibiotics mediated by target alterations. Science. 264, 388-393.

Strauss EJ, Falkows 1997. Microbial pathogenesis : genomics and beyond. Science. 276, 707-712.

Tang CM, Hood DW, Moxon ER 1998. Microbial genome sequencing and pathogenesis. Curr Opin Microbiol. 1, 12-16.

Wanzel RP, Wong MT 1999. Editorial response : managing antibiotic use – impact of infection control. Clin Infect Dis. 28, 1126-1127.

Watanabe K, Lyobe S, Inoue M, Mitsuhashi S 1991. Transferable imipenem resistance in *Pseudomonas aeruginosa*. Antimicrob Agents Chemother. 35, 147-151.

Weinstock GM, Smajs D, Hardham J, Norris SJ 2000. From microbial genome sequence to application. Res Microbiol. 151, 151-158.

White RL, Friedrich LV, Mihm LB, Bosso JA 2000. Assessment of the relationship between antimicrobial usage and susceptibility : differences between hospital and specific patient care areas. Clin Infect Dis. 31, 16-23.

Williams DH, Stone MJ, Hauck PR, Pahman SK 1989. Why are secondary metabolites (natural products) synthesized. J Nat Prod. 52, 1189-1208.

Xia M, Lunsford RD, McDevitt D, Lordanescus S 1999. Rapid method for the identification of essential genes in *Staphylococcus aureus*. Plasmid. 42, 144-149.

Yarbrough GG, Taylor DP, Rowlands RT, Crawford MS, Lausre LL 1993. Screening microbial metabolites for new drugs – theoretical and practical issues. J Antibiot. 46, 535-544.

CHAPTER – 12

BIOTECHNOLOGICAL APPROACHES IN MANAGEMENT OF SOILBORNE DISEASES

M.K. Naik, Mahadevaswamy and Arun Sataraddi

College of Agriculture, Raichur – 584 101

1. INTRODUCTION

There are numerous strategies available for management of soil borne plant diseases including the use of resistant varieties, crop rotation, tillage practices, organic amendment, solarization, fungicidal application and biocontrol etc., although none have been completely effective except in few instances. Challenges continue for scientists to develop innovative and eco-friendly technologies for management of soilborne pathogens, many of which may rely on traditional method of disease control while others may rely heavily on modern biotechnological approaches.

The biotechnological developments in ecology and management of soil borne pathogens have been mostly from the use of molecular tools to study the biocontrol mechanism and pathogenesis. Few applications of recombinant technology have so far been incorporated into management strategies against soilborne disease. This review would describe examples and prescribe strategies as to how biotechnological approaches have been exploited and can be used directly or indirectly to manage soil borne diseases.

The most important biotechnological research areas so far explored is the use of disease resistant transgenic plants that express microbial genes involved in the suppression of root infection. Next important research area is the use of transgenic microorganisms as biocontrol agents in which one microbe expresses genes from another micro-organisms whose products are inhibitory to soilborne pathogens. Lastly, the use of nucleic acid hybridization technologies for detection, identification, monitoring of soil borne pathogens and introduced biocontrol agents is another area being explored recently.

2. MANAGEMENT OF RESISTANT GENES

The agricultural development is based on high yielding varieties and improved tillage practices, water management, crop protection and optimum use of fertilizers. In various crops few cultivars involving particular generally top plasma have completely replaced the local land races and there is a phenomenal decrease in genetic variability. The substitution of a few crop varieties on a large geographical tract constitutes a major crop protection hazard. Management is based on the philosophy of maintaining the loss below the economic injury level either by reducing the amount of spread of inoculum or reducing the rate of spread of the disease.

There are a number of schemes available inorder to manage resistant genes both in conventional and molecular approaches. Among the conventional approaches, varietal diversification, multilines, variety mixtures, advance warning of vulnerable genes, gene deployment, mixed cropping, gene pyramiding resistant gene cycling, and durable resistance have been adopted in some crops against diseases.

Some novel approaches are now finding their application in conventional breeding methodology. Most of these innovative approaches show tendency to make use of in-vitro techniques.

2.1. Tissue Culture and Embryo Rescue

It is only during the past few years that some tissue culture techniques have proved a valuable to conventional method of plant breeding. Embryo rescue has also shown a tremendous potential for modifying the host to make it resistant to soil borne diseases. Resorting to embryo rescue has incorporated Disease resistant genes. Read and Collins (1978) transferred resistance to black shank diseases caused by *Phytophthora parasitica* from *Nicotiana nesopnila* and *N. reponda* to tobacco.

2.2 Somatic Cell Hybridization

The protoplast fusion products in many cases have resulted in the formation of hybrids. Grosser *et al.* (1988) obtained a somatic hybrid between *Poncirus* and *Citrus poncirus* is resistant to foot rot caused by *Phytophthora parasitica*.

2.3 Somaclonal Variation

The cell culture itself can generate stable heritable variation, including resistance genes, against soil borne disease. Hortman *et al.* (1984) selected cell lines in Lucerne resistant to the toxin produced by *Fusarium oxysporum* and later regenerated plants. Somaclones of Lucerne which were resistant to wilt caused by *Verticillium* species has been identified. (Latunde Dada and Lucas 1983). It has been demonstrated in several cases that disease resistant somaclone can be obtained and that technique holds potential for developing genetic strains resistant to some soil borne pathogens.

2.4 Genetic Engineering

Phenomenal achievements have been made in molecular biology and genetics in developing a novel technology now called as genetic engineering. It has now become possible to isolate or construct desirable genes if their products of nucleotide sequence is known with suitable promoter, the gene constructs can be incorporate into any alien host using direct and indirect D.N.A transfer methods. Several novel methods including Ti plasmid has been used as a gene vector for insertion genes. Electroporation, microinjection and microprojective bombardment are the other direct gene transfer techniques being used in genetic engineering.

3. TRANSGENIC PLANTS

Breeding plants for resistance to soil borne diseases through conventional methods has met with mixed results. Use of resistant varieties has been effective in more specialized pathogens, such as vascular wilt pathogens like *Fusarium* and *Verticillium* but for many of the unspecialized pathogens, such as Pythium. *Sclerotium, Rhizoctonia* and *Gaeumannomyes*. It has not been possible to obtain resistance through conventional breeding.

Transgenic plants offer a promising means for developing resistant plants for managing soil borne disease. Much of the research in transgenic plants has gone into development of plants resistance herbicides or viral and bacterial foliar pathogens. For example, Broglie *et al.* (1992) transformed tobacco and Canola plants with endo-chitinase for their susceptibility to seedling and root diseases caused by *Rhizoctonia solani.* The use of promoter region 35 D camv transcript resulted in progeny with a lower percentage of *R. solani* induced seedling mortality than in non- transformed plants.

Similarly, Montanelli and Nascari (1991) transformed potato with *Agrobacterium rhizogens* carrying application gene for cercopin biosynthesis form lepidopterns insect. *Hyalaphara ceropia.* Cercopin is antibacterial compound and highly active against many bacterial pathogens (Jaynes et al., 1987). Fresh tissue extracted taken from transgenic regenerants expressing cercopin protein was inhibitory to *Pseudomonas solanacearum.*

The lack of available genes encoding pathogen-suppressive traits and the present state of our knowledge on ecology and epidemiologist of soil borne pathogens are the reasons for not many transgenic plants against soil borne pathogens. Our major limitation to the development of useful transgenic plants for soil borne disease management is our limited understanding about the mechanism of pathogenesis and biological control. In order to develop transgenic plants,. Useful genes for soil borne disease to develop transgenic plants, useful genes for plant transformation must be available. Currently, few pathogen suppressive genes from plant, animal or microbial pathogenesis have been identified and cloned. Future efforts in molecular aspects of pathogenesis by soil borne fungal pathogens and in mechanisms of biocontrol are likely to facilitate the isolation of useful genes and subsequent development of transgenic plants for management of soil borne diseases.

4. TRANSGENIC MICROORGANISMS

These are likely to have their greatest utility as biocontrol agents of soil borne pathogens in which specific strains are constructed to express genes from another micro organisms whose products are inhibitory to soil borne pathogens. The limitations expressed about transgenic plants also apply to the development of transgenic microorganisms. Yet, there are examples, which illustrate the usefulness of their strategy. For example, the gene coding the detoxification activity was expressed (Utsumi *et al.,* 1988). When excised tomato plans were treated with cells of transformed *E. coli* strain and placed in a fusaric acid solution, the damage from fusaric acid as greatly reduced in treated plants as compared with untreated control plants. Similar, non-pathogenic, FA detoxifying content of *Pseudomonas solanacearum* was capable of reducing symptom development on plants inoculated with *Fusarium oxyspo- rum* f. sp. *lycopersici* (Tbyoda *et al.,* 1998).

The enteric bacterium *Serratia marcescens* known not only for its chitinase production but also a bioagent against soil borne pathogens (Ordentlich *et al.*, 1988. Sneh *et al.*, 1985). Chitinolytic enzymes produced from *S. marcescens* is inhibitory to chitin containing soilborne fungi. The chitinase genes have been cloned (Fuchs *et al.*, 1986. Jones *et al.*, 1986. Shapira *et al.*, 1989. Sundheim *et al.*, 1988). And used for the development of transgenic microorganisms. Two cosmid clones from a library of *S. marcescens* strain BJL 200 DNA have now been mobilized into *Pseudomonas fluorescens* strain to inhibit germ tube growth of *Fusarium oxysporum* f. sp. *conglutinans* and *F. oxysporum* f. sp. *redolens* and hyphal growth *R. solani* and *Magnaporthe grisea* as compared with wild type of strains. Further, treating radish seeds with the chitinase positive transformation of *P. fluorescens* NRRL B-15135 significantly reduced disease caused by *F. oxysporum* f. sp. *redolens* over wild type of strain.

Similarly, application chitinase gene from *S. marcescens* designated as chiA, was cloned (Jones *et al.*, 1986) and expressed in *E. coli* strain A 2097 (Sharpia *et al.*, 1989). Partially purified chitinase preparation was effective in bursting hyphal tips of *Sclerotium rolfsii* and in reducing bean seedling infection by *S. rolfsii* and *R. solani* by up to 62 per cent. When bean seedlings were drenched with viable cells of *E. coli* expressing the chi A gene disease incidence was significantly reduced as compared with untreated control. The expression of chitinate genes in transgenic bacteria is currently the best example of the development of genetically engineered biological control agents for the suppression of soil borne diseases.

4.1 Transgenic Bacteria

So far, the best documented success of a biocontrol agent against soilborne pathogen has been the use of *Agrobacterium radiobacter* strain K-84, which has been used successfully for over 20 years to control crown gall in various plants all over the world (Moore and Warren, 1979). However, when failure of biocontrol was noticed (Ryder and Jones, 1990), due to the fact that strain K-84 contains the plasmid pAg K-84 which carries genes that encode agrocin –84 biosynthesis as well as genes encoding agrocin-84 resistance (Ryder *et al.*, 1987). It is also observed that pAg K-84 is readily modifiable into other Agro bacteria, including virulent strains carrying pTi. That means the virulent strains of *A. tumefaciens* carrying pTi and sensitive to agrocin-84, can acquire pAg K-84 and thus become agroncin-84 resistant (Ellis *et al.*, 1979). Thus it is possible that K-84 could be rendered ineffective as a biological control agent if pTi from virulent *A. tumefaciens* strains could be transferred to *A. radiobacter* strains K-84.

Thereafter, Jones *et al.* (1988) developed a strategy to construct transfer deficient deletion mutant in which part of the transfer region of pAg K-84 was deleted. This strain was constructed by deleting 2.8 kg of 3.9 kb transfer region of pAg K-84. The resultant strain was designated as sd K-1026 which possessed all the invitro phenotypic traits of the parental K– 84 with the exception of its inability to transfer pAg K-84. furthermore, K-1026 was as effective as K-84. in protecting almond seedlings in the field (Jones and Kerr 1989). Strain K 1026 was registered in 1988 in Australia under the name "Nogall" as peat based formulation for the control of crown gall on stone fruits, nuts and roses and has been used successfully as replacement for the old strain K-84. To date, *A. radiobarter* strain K-1026 is the genetically engineered microbe registered for the biological control of plant disease.

5. MOLECULAR MECHANISM OF PATHOGENESIS AND BIOCONTROL

Molecular genetic approaches have become powerful tools in studying the roles of specific microbial traits in plant pathogenesis and biocontrol process with soilborne pathogens. It is now possible to directly address question regarding microbial metabolites and gene products in complex process such as pathogenesis and biological control. The recombinant DNA techniques aid the investigators to experimentally isolate DNA sequences that regulate specific phenotypes and thus can assess the role of these genes and their corresponding gene products in pathogenesis and biocontrol. The knowledge of these processes will inevitably lead to eventual application for manage- ment of soil borne diseases.

5.1 Molecular genetics of pathogenesis in Soilborne pathogen

Detailed knowledge of molecular genetics of pathogenesis of soilborne pathogen has mostly come from studies on bacterial pathogens. Most comprehensive understanding infection processes of soilborne bacterial pathogens is with *Agrobacterium tumefaciens* (Kado, 1991). There is considerable research base with *Pseudomonas solanacearum* (Bouchr *et al.*, 1992) and soft rotting species of *Erwinia* (Daniels *et al.*, 1988, Kotoujansky 1987). The methods of genetic transformation of filamentous fungi have not been so well developed (Leong and Holden, 1989)unlike bacterial pathogens where processes of gene transfer and manipulation are well under stood. However, transformation systems have been developed for some filamentous fungal pathogens including soilborne pathogens like *Fusarium oxysporum* (Kistler and Benny, 1988), *Gaeumannomyces graminis varticitici* (Henson *et al.*, 1988) and *Nectria haematococca* (*Fusarium solani* f. sp. *pisi*). (Dickman and Kolattukudy 1987, Tugeon *et al.*, 1987).

Work of Van Etten and his colleagues is the best example to show how these recombinant – DNA technologies have been used to identify genes encoding pathogenicity traits in fungal soilborne pathogens (Van Etten *et al.*, 1989). Early studies demonstrated that the detoxification of the pea phytoalexin, pisatin, could be correlated with pathogenesis in pea pathogens. *Fusarium solani* sp. *pisi* (telemorph *Nectria heamtococca*) was choosen as a model fungus to study the detail significantly of piston detoxification. The detoxification of pisatin, could be correlated with athogenesis in pea pathogens. *Fusarium solani* f sp. *pisatin* though the enzymatic action of cytochrome P–450 manoxygenase enzyme, pisatin demethylase (Matthews and Van Etten 1983). Isolates of *N. haematoccoca* able to demethlate pisatin (Pda') were also able to infect pea plants whereas those that could not demethylate pisatin (pda') were non-pathogenic (Van Etten *et al.*, 1980).

The development of efficient transformation system for *N. haematococca* enabled weltring *et al.*, 1988 to transform *Aspergillus nidulans*. Application pda fungus with a genomic library of *N. haematococca*. The transformants were selected on a minimal medium containing tryptophan and screened for pisatin dimethylane (pda) activity. The gene encoding pisatin demethylase was isolated and found to be expressed consistently in *A. nidulans*. However, research in this area is warranted to know the genes that are linked to pathogenicity and virulence in other soilborne fungal pathogens which will provide important information of the development of management strategies against soilborne disease.

5.2 Molecular Mechanism of Biological Control of Soilborne Pathogens

Our knowledge of biocontrol process mediated by bacteria is somewhat more advanced than that for fungal systems mainly because the basis for recombinant

DNA technology has come from bacterial biocontrol processes is that majority of the studies have been confined to fluorescent *Pseudomonas* species particularly comprising only 5 to 10 strains of *P. fluorescens*. The studies have focused on antibiotic biosynthesis or siderophore biosynthesis against only three soilborne pathogens. Apart from *P. fluorescens*, the best studies on biocontrol agent are strains of *S. marcescens*. A broader diversity of biocontrol agents, against a wider array of soilborne pathogens must be investigated before application suitable collection of biocontrol genes can be assembled and biocontrol process can be developed. Then it is possible to manipulate the biocontrol behavior consistently.

Similar to bacterial biocontrol agents, research on molecular mechanisms of fungal biocontrol processes has been confined to *Trichoderma* and *Gliocladium* generas. Genetic recombination through conventional breeding strategies is highly problematic since they do not readily form a sexual stage (Papavizas, 1985). Similar to the situation with pathogenic soilborne fungi one of the major limiting factors in genetic manipulation of fungal biocontrol agents is the limited availability of adequate genetic manipulation of fungal biocontrol agent is the limited availability of adequate gene transfer technologies. Transformation systems have only recently been described for *Trichoderma* (Goldman *et al.*, 1990, Herrera Estrella *et al.*, 1990, Pentilla *et al.*, 1987, Ulhoa *et al.*, 1992) and *Gliocladium* (Ossana and Mischeke, 1990, Thomas and Kennerly 1989). The protoplast fusion has become application common method for genetic recombination particularly in strains of *Trichoderma harzianum* (Peer and Chet 1990, Stasz *et al.*, 1988, Stasz, 1990) due to the absence for appropriate transformation systems for biocontrol fungi, Peer and Chet (1990) showed the fusion of protoplast from relatively non-suppressive auxotrophic strains of *T. harzianum* and obtained protrophic progency which were highly effective in suppressing *Rhizoctonia* damping off in cotton. They were also more effective than the parental strains in over growing colonies of *S. rolfissi*, *R. solani* and *P. aphanidermatum*.

6. DETECTION AND IDENTIFICATION USING RECOMBINANT DNA TECHNIQUES

The range of techniques available for identification and monitoring of soilborne pathogens has expanded over the past decade including immunological assays and most recently nucleic acid hybridization and related nucleic acid chemistry techniques to identify and track soilborne plant pathogens in soil and plant tissues. However, nucleic acid techniques are having improved sensitivity and accuracy and thus are regarded important once in detection. Accuracy, recession and speed are the critical factors in all diagnostic procedures, Conventional methods of culturing and microscopic analysis are time consuming in disease management. Many soil borne pathogens multiply and proliferate without producing host symptoms and in such cases their detection and monitoring are vital in preventing population reaching economic threshold levels (Naik and Sen 1999). Although the economic threshold levels have been determined in few of the soilborne pathogens such as *Fusarium udum* on pigeonpea at field level (Naik 1983, Reddy 1996, Naik *et al.*, 1997) but is essential to have the field data for many soilborne pathogens for conformity of these improved and sensitive techniques.

Several factors affected hybridization between an experimentally derived probe and the chromosomal or plasmid DNA of the organism in question under detection. Important factors include specific reaction conditions such as temperature and osmolarity of reaction solutions, the match between the nucleotide sequence of the probe and its target nucleic acid, the absolute and relative abundances of the probe and target sequences and amplification of target sequences.

Probes consist of application cloned or chemically synthesized DNA sequence identical or very similar to application sequence found in the target organism. The development of probes for the detection of different pathovers or strains requires considerable research to assure that the use of probes will hybridize differentially to the pathovers or strains of interest. Te probe must be specified enough to make application positive identification to the pathovers or strains of interest. The probe must be specified enough to make a positive identification of a particular strain and rule out other. There are several new tools, such as polymerase chain reaction (PCR) amplification and non-radioactive labelling of probes which enhance sensitivity or softy of these methods. The development of PCR technology increases the detection sensitivity or organism present in low population levels and greatly simplifies or accelerates the routine detection of organisms that are not easily cultured.

Nucleic acid hybridization approach for soilborne pathogens identification and monitoring have been generated from research data on other aspects of soil microbial ecology, biotechnology risk assessment and to a lesser extent, research with foliar pathogens. Although techniques are rapidly being adapted in detection and monitoring of soilborne pathogens, yet nucleic acid based diagnostic kits are not readily available for general use in detection of soilborne pathogens.

6.1 Probes and specificity

The nucleic acid hybridization technique distinguishes among genetically distinct but morphologically similar micro-organism and thus achieves highest level of specificity. Because of high sensitivity, nucleic acid methodology makes it a potentially a useful tool to detect extremely low population of pathogens or associated microbes. It also offers the possibility of rapidly detected pathogens which are impossible or difficult to culture.

Many variations of nucleic acid hybridization methodologies have been employed to detect and identify microbial populations from water, soil and the rhizosphere. Ford and Olson (1988) and Miller and Martin (1988) have reviewed these. Many of them are original devised or modified to detect organisms that had been genetically engineered and the interest in such work was to determine the fate of genetically engineered or modified organism. Among them colony hybridization (Sambrook *et al.* 1989) are dot blot analysis (Kafatos *et at.* 1979) have been used. This method is useful when one seeks a simple positive or negative answer to find out whether a colony carries the target gene.

Selective medium plating techniques have been widely used for enumeration of soil borne pathogens such as *Fusarium* (Naik 1990, Naik & Sen 1999, Naik *et al.* 1997, Naik *et al.* 1997b) *Rhizoctonia, Sclerotium, Pythium* and *Phytophthora* species. In combination with selective media technique, it is possible to monitor the dynamic of the soil microbial community and determine the effects of the addition of an organism on that are perhaps in interpretation as well as population sampling procedures. Many a times soil borne diseases are the result of association of more than one pathogens and hence similar symptoms are produced by more than one agents / pathogens. Hence it is essential to distinguish among different pathogens or between pathogenic and non-pathogenic strains. Bateman *et al.* (1992) detected the pathogenic strains of *Gaeumannomyces graminis* var. *tritici* (Ggt) and *G. graminis* var. *avenae* (Gga) causal agents of take all of wheat and barley and oat and turf grasses respectively. This probe was considered to be very useful, because the identification required only two days compared to other conventional method of oat leaf agar assay which required 4 to 5 weeks. Further, information on host combined with molecular data enables authentic identification of pathogens.

Similar probes have been developed for other soil borne pathogens such as *Phytophthora parasitica* (Goodwin *et al.*, 1989, 1990a), *Phytophthora citrophthora* (Goodwind *et al.*, 1990b), *Pythium* sp. (Martin 1991a, 1991b), *Verticillium alboartum* and *V. dahliae* Nazar *et al.*, 1991), *Leptosphaeria korrae* (Tisserat *et al.*, 1992), *Agrobacterium tumefaciens* (Dong *et al.*, 1992) and *Pseudomonas solanacearum* (Cook and Sequeira 1991). The RFLP analysis is still more rapid than many cultural tests, and is becoming more popular in the study of soil borne pathogens. It has been useful in detecting *Phytophthora citrophthora* in infected roots (Goodwin *et al.*, 1990b) and in assessing the population structure of *P. infestans* in potato fields in northern and central Mexico (Goodwin *et al.*, 1992).

5.2 Sources and Isolation of Nucleic Acid

For all the nucleic acid detection methods DNA or RNA must be accessible for analysis. The organismal nucleic acids used to identify the organismal can be obtained in a number of ways. Microbes can be isolated from solid or diseased plant tissues or alternate hosts by first culturing the organism and then purifying the DNA from the cultured cells. Many protocols for such nucleic acid purification are available.

Suspected pathogenic soil borne bacteria or fungi may be directly identified by a dot blot approach without prior laboratory culture of the organism. DNA can be prepared directly from a soil sample and probed, to determine whether the target organism is present (Holven *et al.*, (1988, Jonson *et al.*, 1989)

These are several innovations in hybridization technology developed which enhance the sensitivity of detection of soil borne organisms in soil or from plant samples. The most significant and remarkable of these is PCR (Polymerase chain reaction) technique, which has the potential of increasing the detection limits of particular sequences of DNA by several orders of magnitude. The PCR techniques have been employed to detect or distinguish a number of fungal pathogens such as *Pythium* species (Chen *et al.*, 1992), *Verticillium* (Nazar *et al.*, 1991) *Phytophthora* (Goodwin *et al.*, 1990, 1992) and *Gaeumannomyces* (Schesser *et al.*, 1991). *Agrobacterium tumefaciens* strains have been successfully detected with PCR techniques (Dong *et al.*, 1992).

7. CONCLUSION

Many of the advances in the molecular biology of plants and microbes have occurred only in the past 10 to 15 years and the usefulness or the products of recombinant DNA research are only just beginning to be extended for managing soil borne diseases. Most important areas that can be used for a disease management particularly include pathogenesis and ecology of soil borne pathogens and mechanism of biological control. Such studies would be the base for plant and microbial transformations.

Recombinant DNA techniques could serve as basic tool to understand interactions among soil borne microbes. Also equally important is understanding of molecular mechanism of pathogenesis. This is specifically useful in knowing the operation of promising bioagents and suitably manipulates them. Many a times the potential bioagents will be the disarmed version of the pathogen. The questions related to soil ecology, plant or pathogen or antagonist interaction are quite complex. These molecular approaches are likely to improve detection, enumeration, monitoring and manipulating the components of these systems, which were otherwise not possible with the conventional soil microbiological techniques.

Management of host genes for disease resistance has undergone a radical change. The conventional approaches may continue to dominate wherever natural genetic variability is available within the species but the novel approaches would occupy the central place in foreseeable future. Use of *Agrobacterium tumefaciens* for the dicots and partical bombardment technique for transforming the living cells/tissue irrespective of their family of even kingdom have opened few horizon to create a transgenic germplasm of the plants having genes from plants, animal, bacteria, fungi and viruses.

ACKNOWLEDGMENT

The authors are grateful to Dr. C.V. Patil, Director of Instruction (Agri.), Agriculture College Raichur for encouragement.

REFERENCES

Bateman, G.L., Ward, E. & Antonie, J.F. 1992. Identification of *Gaeumannomyces graminis* var. *tritici* and *G.graminis* var. avenae using a DNA probe and non-molecular methods. Mycol. Res. 96:737-742.

Boucher, C.A., Gough, C.L & Arlat, M. 1992. Molecular genetics of pathogenicity determinants of *Pseudomonas solanacearum*, with special emphasis on hrp genes. *Annu. Rev. Phytopathol.* 30:443-461.

Broglie, K., Chet,1., Holliday, M., Cressman, R., Biddle, P., Knowlton, S., Mauvais, C.J. & Broglie, R, 1991. Transgenic plants with enhanced resistance to the fungal pathogen *Rhizoctonia solani. Science.* 254:1194-1197.

Chen, W., Hoy, J.W. & Schneider, R.W. 1992. Species-specific polymorphisms in transcribed ribosomal DNA of five *Pythium* species. *Experimental Mycology..* 16:22-34.

Cook, D. & Sequira, L. 1991. The use of subtractive hybridization to obtain a DNA probe specific for *Pseudomonas solanacearum* race 3. *Mol. Gen. Genet..* 227:401-410.

Dickman, M.B. & Kolattukudy, P.E. 1987. Transofmation of *Fusarium solani* f.sp. *pisi* using the cutinase promoter (Abstr). *Phytopathology.* 77:1740.

Dong, L.C., Sun, C. W., Thesis, K.L., Luthe, D.S. & Graves, C.H., Jr. 1992. Use of polymerase chain reaction to detect pathogenic strains of *Agrobacterium. Phytopathology.* 82:434-439.

Ellis. J.G. Kerr, A., van Montagu, M. & Schell, J. 1979. *Agrobacterium.* Genetic Studies on agrocin 84 production and the biological control of crown gall. *Physicol. Plant Pathol.* 15:311-319.

Ford, Significantly & Olson, B.H. 1988. Methods for detecting genetically engineered microoranisms in the environment. Pages 45-79 in Advances in Mocrobial Ecology, Vol. 10, K.C. Marshall, ed. Plenum Press New York.

Fuchs, R.I., McPherson, S.A. & Drahos, D.J. 1986. Cloning of *Serratia marcescens* gene encoding chitinase. *Appl. Environ. Microbiol.*. 52:504-509.

Goldman, G.H., Van Montagu, M. & Herrera-Estrella, A. 1990. Transformation of *Trichoderma harzianum* by high-voltage electric pulse. *Curr. Genet.*. 17:169-174.

Goodwin, P.H., English, J.T., Neher, D.A., Duniway, J.M. & Kirkpatrick, B.S. 1990a. Detection of *Phytophthora parasitica* from soil and host tissue with a species-specific DNA probe. *Phytopathol* 80:277-281.

Goodwin, P.H., Kirkpatrick, B.C. & Duniway, J.M. 1989. Cloned DNA probes for the identification of *Phytophthora parasitica. Phytopathol.* 79:716-721.

Goodwin, S.B. Spielman, L.J., Matuszak, J.M., Bergeron, S.N. & Fry, W.E. 1992. Clonal diversity and gentic differentiation of *Phytophthora infestanS* population in northern and central Mexico. Phytopathol. 82:955-961.

Henson, J.M., Blanke, N.K. & Pilgeram. A. 1988. Transformation of *Gaeumannomyces graminis* to benomyl resistance. Curr. Genet.. 14:113-117.

Herrera-Estrella, A., Goldman, G.H. & van Montagu. M. 1990. High–efficiency transformation system for the biocontrol agents *Trichoderma* Spp. Mol. Microbiol.. 4:839-844.

Holben. W.E., Jansson, J.K., Chelm, B.K. and tiedje, J.M. 1988. DNA probe method for the detection of specific microorganisms in the soil bacterial community. Appl. Environ. Microbiol. 54:703-711.

Jansson, J.K., Holben, W.E.& Tiedje, J.M. 1989. Detection in soil of a deletion in a engineered DNA sequenced by using DNA, probes. Appl. Enviorn. Microbiol. 55:3022-3025.

Jaynes, J.M., Xanthopoulos, K.G., Destefan-Beltran, I., & Dodds, J.H. 1987. Increasing bacterial disease resistance in plants utilizing antibacterial genes from insects. Bioassays.. 6:263-270.

Jones, D.A. & Kerr, A. 1989. *Agrobacterium radiobacter* K 1026, a genetically engineered derivative of strain K 84, for biological control of crown gall. Plant Dis. 73:15-18.

Jones D.A. Ryder, M.H. Clare, B.G., Farrand, S.K. & Kerr, A. 1988. Construction of a Tra-deletion mutant of pAgK 84 to safeguard the biological control of crown gall. Mol. Gen. Genet. 212:207-214.

Jones, J.D.G., Grady, K.L., Suslow, T.V. & Bedbrook, J.R. 1986. Isolation and characterization of genes encoding two chitinase enzymes from *Serratia marcescens.* EMBO J. 5:467-473.

Kado, C.I. 1991. Molecular mechanisms of crown gall tumorigenesis. Crit. Rev. Plant Sci. 10:1-32.

Kafatos, F.C., Jones, C.W. & Efstratiadis, A. 1979. Determination of nucleic acid sequences homologies and relative concentration by adot hybridization procedure. Nuclear. Acids Res. 7:1541-1552.

Kotoujansky, A. 1987. Molecular genetics of pathogenesis by soft-rot Erwinias. Annu. Rev. Phytopathol. 25:405-430.

Leong. S.A. & Holden, D.W. 1989. Molecular genetic approaches to the study of fungal pathogenesis. Annu Rev. Phytopathol. 27:463-481.

Martin, F.N. 1991a. Characterization of circular mitochondrial plasmids in three *Pythium* speices. Curr. Genet. 20:91-98.

Martin, F.N. 1991b. Selection of DNA probes useful for isolation identification of two *Pythium* spp. Phytopathol. 81:742-746.

Matthews, D.E & Van Etten, H.D. 1983. Detoxification of the phytoalexin pisatin by a fungal cytochrome P450. Arch. Viochem. Viophys. 224:494-505.

Miller, S.A. & Martin, R.R. 1988. Molecular diagnosis of plant disease. Annu. Rev. Phytopathol... 26:409-432.

Montanelli, C& Nascari, G. 1991, Introduction of an antibacterial gene in potato (*Solanum tuberosum* L.) using a binary vector in *Agrobacterium rhizogenes*. J. Genet & Breed. 45:307-316.

Moore, L.W. & Warren, G. 1979. *Agrobacterium radiobacter* strain 84 and biological control of crown gall. Annu Rev. Phytopathol. 17: 163-179.

Nazar, R.N., Hu, X., Schmidt, J., Culham, D. & Robb, J. 1991. Potential use of PCR-amplified ribosomal intergenic sequences in the detection & differentiation of *Verticillium* wilt pathogens. Physiol. Molec. Plant Pathol. 39:1-12.

Naik, M.K. & reddy, M.V. 1996. Economic threshould level of inoculum in *Fusarium udum* In: D Rajagopal (ed.) soil biology. UAS, GK VK, Bangalore.

Naik, M.K. 1990. Wilt of watermelon caused by *Fusarium oxysporum* Schl. Ph. D. Thesis, Indian Agricultrual Research Institute, New Delhi.

Naik, M.K., Reddy, M.V. & D. Mc-Donald. 1997a. Wilt incidence in sole and intercropped pigeonpea at different inoculum densitites of *Fusarium udum*. Indian Phytotpathology. 50:337-341.

Naik, M.K. 1993. Ecology and Integrated management of *Fursarium wilt* of pigeonpea post doctoral report, ICRISAT Asia Center Patancheru.

Naik, M.K., Reddy, M.V., Raju, T.N. & D. Mc-Donald. 1997. Dynamics of *Fusarium udum* population in relation for resistance in pigeonpea. Proceedings of International conference on integrated plant disease management for sustainable Agriculture, IARI, New Delhi.

Naik, M. K., Reddy, M.V., Sinha Poonam & D. Mc-donald 1997b. Differential solarization-a decision making tool for management of soil borne plant pathogens. Symposium on Epidemiology and disease management held by the Indian Phytopathol. Society, Bangalore.

Ordentilich, A. Elad, Y. & Chet, I. 1988. The role of chitinase of *Serratia marscecens* in biocontrol of *Sclerotium rolfisii*. Phytopathol. 78:84-88.

Ossana, N. & Mischke, S. 1990. Genetic transformation of the biocontrol fungus *Glioladium virens* benomyl resistance. Appl. Environ. Microbial. 56:3052-3056.

Papavizas, G. C. 1985. *Trichoderma* and *Gliocladium*: Biology, ecology and potential for biocontrol. Annu. Rev. Phytopathol. 23:23-54.

Pe'er, Significantly & Chet. I. 1990. *Trichoderma* and protoplast fusion: application tool for improving biocontrol agents. Can. J. Microbiaol. 36:6-9.

Pentilla, M., Nevalamen, H., Ratto. M., Saliminen & Knowels, J. 1987. A versatile transformation system for the celluloytic filamentous fungus. *Trichoderma reesei*. Gene. 61:155-164.

Ryder, M. H. & Jones. D. A. 1990. Biological control of crown gall. Pages 45-63 in: Biological control of Soil-Borne Plant pathogens, D. Horn by, Ed., Cab international. Wallingford, UK..

Ryder, M. H. & Slota, J. E., Scarim, A. & Farrand, S. K. 1987. Genetic analysis of agrocin 84 production and immunity in *Agrobacterium* spp. J. Bacteriol. 169:4148-4189.

Schesser, K., Luder, A. & Henson, J. M., 1991. Use of polymerase chain reaction to detect the take- all fungus *Gaeumannomyces graminis* in infected wheat plants. Appl Environ. Microbiol. 57:553-556.

Shapira, R., Ordentilich, A., Chet, I. & Oppenheim, A. B. 1989. Control of plant diseases by chitinase expressed from cloned DNA in *Eschrichia coli* Phytopathol. 79:1246-1249.

Stasz, T. E., 1990, Genetic improvement of fungi by protoplast fusion for biological control of plant pathogens. Can. J. Plant pathol. 12 :322-327.

Stasz, T. E., Harman, G. E. & Weeden, N. F., 1988. Protoplast preparation and fusion in two biocontrol strains of *Trichoderma harzianum*. Mycologia. 80:141-150.

Sundheim. L., Poplawsky, R. A. & Ellingboe, A. H., 1988. Molecular coloning of two chitinase genes from *Serratia marcescens* and their expression in *Pseudomonas* species. Physiol. Mol. Plant Pathol. 33:483-491.

THOMAS, M. D. & Kennerly, C. M., 1989. Transformation of the *Letosphaeria korrae* by cloned DNA probes. Phytopathol. 81:917-921.

Toyoda, H., Hashimoto, H.,m Utsumi, R., Kobyashi, H,.& Ouchi, S. 1988. Detoxification of fusaric acid by a fusaric acid – resistat mutant of *Pseudomonas solonacearum* and its application to biological control of *Fusarium* wilt of tomato, Phytopathol. 78:1307-1311.

Trevosr, J. T., Van Elsas, J. D., Van, Overbeek. I. S. & Starodub, M. E. 1990, Transport of a genetically engineered *Pseudomonas fluorescens* strain through a soil microcosm. Appl. Environ. Microbiol. 56:401-408.

Turgeon, B. G., Garber, R. C. & Yoder, O. C., 1987. Development of a fungal transformation system basedon selection of sequence with promoter activity. Mol. Cell. Bniol. 17:3297-3305.

Utumi, R., Hadama, T., Noda, M., Toyoda, H., Hashimoto, H. & Ohuchi, S., 1988, Cloning of fusaric acid-detoxifying gene from *Cladosporium weneckii*: A new strategy for the prevention of plant disease, J. Biotechnol.8:311-316.

Van Elsas, J.D., Van Overbreek, I. S. & Fouchier, R. 1991, A specific marker pat for studying the fate introduced bacteria and their DNA in soil using a combination of detection techniques. Plant Soil. 138:49-60.

VAN Etten, H. D., Matthews, D. E, & Matthews, P. S. 1989, Phytoalexin detoxification: Importance for pathogencity and practical implication. Annu. Rev. Phytopathol. 27:143-164.

Van Ettern, H. D. Matthews, P. S., Tegtmeier, K. J., Dieteri, M. F. stein, J. I. 1980. The association of pisatin tolerance and demethylation with virulence on pea in *Nectria haematococca*, Physiol. Plant Patho. 16:257-268.

Weltring, K.M., Turgeon, B. G., Yoder, O. C., & Van Ettern H. D. 1988. Isolation of phytoalexin-detoxification gene from the plant pathogenic fungus *Nectria haematococca* by detecting its expression in *Aspergillus nidulans*. Gene. 38:335-344.

CHAPTER - 13

KERATIN COLONIZING FUNGAL FLORA

R.K.S. Kushwaha

Department of Botany, Christ Church College, Kanpur - 208 001

1. INTRODUCTION

Capability of soil inhabiting keratinophilic fungi and related dermatophytes to perform colonization of natural keratin such as human hair, skin, nails, hooves, horns and bird feathers is little known. Some fungi are capable to parasitize human being and animals and cause diseases [Gentles, 1962; Frey, 1965]. The prevalence of these fungi and their ecological back ground in Indian soils is not yet well documented. Keratinophilic and dermatophytic flora are also found rich in those habitats where bird feathers and hairs of animals are dropped off. Reports of the isolation of these fungi of the identical nature were composited [Table 1] from Andhra Pradesh, Assam, Bihar, Delhi, Goa, Gujrat, Himachal Pradesh, Jammu and Kashmir, Kerala, Karnatak, Madhya Pradesh, Maharashtra, Mysore, Orissa, Punjab, Rajasthan, Tamilnadu, Uttar Pradesh and West Bengal [Randhawa and Sandhu, 1965; Garg, 1966; Kushwaha and Agrawal, 1976; Jain *et al.*, 1985; Shrivastava *et al.*, 1987; Nigam and Kushwaha, 1986, 1989, 1990]. Some other reports on isolation of keratinophilic fungi were also made from Indian soils [Gugnani and Randhawa, 1973; Randhawa and Baruah, 1982; Dixit and Kushwaha, 1989; Awasthi and Gotewal, 1990].

The competition for the keratinous substrates in the soil is one of the most significant factors in the study of ecology of these keratinophilic fungi and dermatophytes. Many workers reported that frequency of occurrence of geophilic dermatophytes and related keratinophiles in a specified locality depends on the frequency of the animals visited [Randhawa and Sandhu, 1965; Garg, 1966 Sur and Ghosh, 1980].

The geophilic dermatophytes and other opportunistic fungi may actively survive out side of man and animals mainly in places where they can utilize various type of keratin. Keratin colonizing fungi have been commonly isolated from soil by hair baiting method. Griffin [1960], English

Table 1. Occurrence of Keratinophilic fungi in different states of India

Place	Samples	Keratinophilic fungi (Per cent occurrence)
Assam	277	M.gyp., M.aud.[0.9], E.flo.[30.9], T.rub.[31.9], T.menta.[19.6], T.vio.[5.6], T.ros[0.9], T.ver.[10.3]
Dibrugarh		
Bihar	40	C.ind.[8.5], Ch.sp.[5.7, K.aje[2.8], T.ton.[5.7].
Chhotanagpur		
Andhra Pradesh	10	T. menta.[1.0], T.rub.[41.0], T.vio.[6.0], T.var.[0.5]
Hazvrusahi		
Guntur		E.flo.[20.5]
Vellor		T.rub[33.9], T.men.[12.5], T.vio. [32.4], T.ton.[10.1], T.sim. [0.8], E.flo[2.7]
Warrangal		T.rub.[62.7], T.men.[16.1], T.vio.[4.2], T.ton[3.6], E.flo[12.0]
Vishakhapatnam	125	E.flo.[10.8], M.gyp. [1.6], T.men. [1.8], T.rub. [70.0], T.vio.[5.0]
Delhi	258	C.ser, K.ter., N.gyp., T.evol.
Goa	73	C. ind., C. trop., Ch. sp.I, II, C.ser., M.gyp. T.men.
Anjuna, Colva		
Colangate, Vagator		
Beaches		
Gujarat	277	C. ser[3.61], C.ker. [0.72], C. trop. [2.16], T.rub [84.5], T.men. [1.5], T. vio. [8.51], T. ton. [2.9], M. gyp.[0.41], E. flo. [2]
Ahmedabad		
Baroda		
Himachal Pradesh	100	T. evol., M. gyp., K. ter., T.ment.
Solan		
Shimla		
Jammu & Kashmir	84	M. gyp., [22.5], M.can. [9.0], K. aje [18.0] A. qud. [1.8], C.ker. [1.8]. C. ser. [9.1], M.coo. [0.9], T.men. [0.9], Unidentified sp. [2.7]
Srinagar		
Kerala	12	E.flo.[3.3], T.ment.[10.0], T.rub. [68.3], T.ver[1.6]
Cochin		
Trichur		
Trivendram		
Karnataka	26	K.ter., M.gyp., E.flo.[20.4], T.ment.[8.6], T.rub.[19.3], T.vio.[10.7]
Bangalore		
Maharashtra	423	C.ind.[25.4], C.trop.[22.5], C.evol.[2.9], C.ser.[1.9], A.tub.[3.9], M.gyp.[19.6]
Bombay		
Poona		
Dadar		
Mahim		
Burla		
Madhya Pradesh		C. Crass, C.trop., B.ker., G.agar,Micro.sp., M.gyp., T.men., T.rub., C.ele., K.agl., K.ter., T.ter., C.luck., K.aje., A.ter., M.ful., T.equi.
Sagar		
Patharia		
Chattarpur		
Bizapur		
Tikamgarh		
Gwalior		
Betul		
Bhopal		
Damoh		
Jabalpur		
Mandla		
Narsinghpur		

Table 1. contd...

Table 1 contd...

Place	Samples	Keratinophilic fungi (Percent occurrence)
Mysore	8	K.ter., M.gyp.
Gudadur		
Orissa	211	M.gyp., C.ker., C. trop., Ch.sp.I, II, A.fla., P.cit., P.hya., P.ros., P.nod., C.evol., A.ter., A.tub.
Jankia		
Baligadia		
Rameshwar		
Bhubneswar		
Cuttak		
Punjab	39	T.rub.[56.0], M.can.[1.7], T.men.[3.4], T.ver.[1.7], E.flo.[1.7], K.ter., M.gyp., C.trop., T.evol., M.van.
Chandigarh		
Amritsar		
Kasauli		
Ambala		
Jallandhar		
Rajasthan	112	K.ter., C.ind.[34.5], C.trop.[2.13], M.gyp.[11.3], M.coo.
Jaipur		
Mount Abu		
Tamilnadu	50	M.gyp.[0.7], T.vio.[33.7], T.sho.[0.37], T.ton[4.8], M.can.[0.7], E.flo.[6.6], T.rub.[33.6], T.men. [13.0], M.fer.[2.6]
Madras		
Uttar Pradesh	551	T.rub. [37.4], T.men.[23.3], M.aud.[1.1], M.gyp.[1.2], T.vio.[2.7], T.ton. [2.22], M.can.[1.2], C.ker.,C.ser., Ma.nan. K.aje, T.evol., T.sch,T.ton, K.tr, C.geo, A. ger., A.ter., C.tub., C.evol., C.far., C.luck., C.mer., C.pann., C.qun. Myceol ana, A.tub., Mycel.ana. C.novo.
Lucknow		
Dehradun		
Meerut		
Agra		
Bareilly		
Kanpur		
West Bengal	236	T.ros.[65.0], E.flo. [13.0], T.rub. [6.3], T.vio. [1.5], T.men. [2.4], C. ser., M.coo., K.aje, M.gyp.
Calcutta		
Darjeeling		
Fuleshwar		

A. flav.=Arachniotus flavoutues, A. qud.=Arthroderma quadrifidum., A. tub.=Arthroderma tuberculatum, C.ser = Ctenomyces serratus, Ch.sp= Chrysosporium species, C.ker. = keratinophilum, C.mer. = C. merdarium, C. crass. = C. crassitunicatum, C.luck = C. lucknowense, C.asp. = C. asperatum, C.tub.= C. tuberculatum, C.far.=C. farinicola, C.pann. = C. pannicola, C.qun. = C. queenslandicum, A.ter = Aphanoascus terreus, A.coj. = Auxarothron conjugatum, A. flav. = Archaneotus flavus, A.bru = Auxarothron bruneum, C.ele. = Cunninghemella elegans, B.ker = Botryotrichum keratinophilum, G.agr. = Gliocladium agrawalli, E.flo.=Epidermophyton floccosum, K.ter. = Keratinophyton terreum, K. aje. = Keratinomyces ajeoli, M. ful. = Microsporum fulvum, Mycel. ana. A. tub. Myceliophthora anamorph of C. novoguineensis, Myx. = Myxotrichum, Ony. = Onygena, M. gyp. = Microsporum gypsum, M.coo. = M. cookei, M.can. = M. canis, M.nan. = M. nanum, M.aud. = M. audounil, M.van. = M. vanbreuseghmeiim, M.sp. = Microsporum species, M.pl. = Malbranchea pulchella, T.gyp.= Trichophyton gypseum, T.ros= T. soaceum, T.evol. = T. evolceanui, T.sho.= T. shoelensis, T.sim. = T.simii, T. flavescens, T.ver. T. verrucosum, T.vio. = T. violaceum, Tter. = Tterrestre, T.ton= T. tonsurans, T.rub. = T. rubrum, T. men. = T. mentagrophytes, T.sp. Trichophyton species, P. hyl. = Pseudoarachneotus hylinosporus, P.ros. = P. roseus, P.nod = P. nodulses.

[1965] and Nigam and Kushwaha [1989] have marked their role in saprophytic colonization of keratinous substrates. There is frequent occurrence of non-keratinophilic fungi during isolation of keratinophilic fungi and related dermatophytes from the Indian soils. Majority of non - keratinophilic fungi are responsible for causing human diseases and reports of fungal infection by these fungi are suggestive of the potential danger posed by them.

The present study includes the prevalence of keratin colonizing fungal flora in soil collected from various habitats with a review of isolation of 51 fungal species from 19 states of India.

2. OCCURRENCE OF KERATINOPHILIC FUNGI FROM DIFFERENT HABITATS

A total of 120 soil samples were studied from 10 different habitats i.e. crop field, garden, hospital waste, soil with petrol residue, piggery house, poultry farm, sewage, tannery field and forest. The 76.6% samples were found positive for keratinophilic fungi. Five genera namely *Chrysosporium, Malbranchea, Microsporum, Myceliophthora* and *Trichophyton* representing 17 species were isolated by using nine types of Keratinous substrates from ten habitats. A detailed list of these fungi is given in Table 2-4. The fungi listed in Table 2 were deposited in Germ Plasm Center for Keratinophilic Fungi [GPCK], Department of Botany, Christ Church College, Kanpur.

Chrysoporium evolceauni was isolated from soil with petrol residue and from poultry farm exhibiting 20 and 2.5% distribution respectively. *C indicum* was isolated from all the habitats studied except sewage and tannery field. The distribution of this fungus was maximum [30.0%] in hospital waste and minimum [6.6%] in piggery house. The maximum distribution of *C. keratinophilum* was in soil with petrol residue and minimum in poultry farm. *C. lobatum* was isolated from forest soil and poultry farm soil while *C. lucknowense* was isolated from forest and piggery house. *C. pannorum* was found in crop field soil and poultry farm soil. The maximum distribution of this fungus was in crop field and minimum in crop field soil. *C. tropicum* was most prevalent and frequently trapped from all these habitats except soil with petrol residue, tannery field and forest. Maximum distribution of this fungus was marked in hospital waste and minimum in poultry from soil. *C. queenslandicum* was isolated from crop field and hospital waste and it was 20.0% in distribution in each habitat. *Chrysosporium* spp. were isolated from most of the habitats except crop field and soil with petrol residue. This species was maximum in garden, sewage, tannery field soil and minimum in poultry farm in its distribution.

Two species of *Malbranchea* were isolated from crop field and forest soils. Two species of *Microsporum* were also trapped from crop field, garden, hospital water, poultry farm and forest. *Microsporum fulvum* was isolated from garden and poultry farm. *Microsporum gypsum* occurred in hospital waste showing 10.0% distribution. *Microsporum* spp. were isolated from crop field, garden, and forest soil. Two *Myceliophthora* species were recorded from hospital waste and poultry farm. Maximum distribution of these fungi was found in hospital waste while minimum in poultry waste. Five species of *Trichophyton* were obtained during the isolation of keratinophilic fungi and related dermatophytes. *T. mentagrophytes* was isolated from hospital waste only. It was 20.0% in distribution. *T. rubrum* was frequently isolated from poultry farm and it was 2.5% in distribution. Unidentified species of *Trichophyton* were also isolated from crop field and poultry farm. Sterile mycelium was isolated from crop field, garden and soil with petrol residue. Maximum distribution of sterile mycelium was 40% in soil with petrol residue and 10.0% was in both crop field and garden soil.

The relative density of genus *Chrysoporium* was above 60% in all the habitats while that of *Malbranchea* ranged from 7-12.5%. *Myceliophthora* was 7.1% in hospital waste and *Trichophyton* was 11.7% in poultry farm. The highest relative density

Table 2. Keratinophylic fungi and related dermatophytes isolated from different habitats from various substrates.

Fungus	Accession Number*	Habitat	Substrate
Chrysosporium evolceanui	328	CF	HH
Chrysosporium evolceanui	363	HW	HH
Chrysosporium evolceanui	433	PF	BH
Chrysosporium indicum	427	SPRR	HH
Chrysosporium indicum	341	CF	HH
Chrysosporium indicum	357	G	CH
Chrysosporium indicum	407	PF	BH
Chrysosporium indicum	411	F	FB
Chrysosporium indicum	358	HW	BHA
Chrysosporium indicum	360	PH	CH
Chrysosporium indicum	428	SH	BH
Chrysosporium indicum	434	HW	GHA
Chrysosporium keratinophilum	342	PF	HH
Chrysosporium keratinophilum	430	F	BH
Chrysosporium keratinophilum	405	SPRR	CH
Chrysosporium lobatum	343	PF	BH
Chrysosporium lobatum	416	PF	CHA
Chrysosporium lobatum	406	F	HH
Chrysosporium lucknowense	437	G	HN
Chrysosporium lucknowense	436	SPRR	BHA
Chrysosporium lucknowense	432	F	HH
Chrysosporium lucknowense	426	F	FB
Chrysosporium lucknowense	431	PH	HH
Chrysosporium pannorum	417	CF	CH
Chrysosporium pannorum	421	PF	HH
Chrysosporium tropicum	420	CF	HH
Chrysosporium tropicum	346	G	CH
Chrysosporium tropicum	413	HW	HN
Chrysosporium tropicum	344	HW	BH
Chrysosporium tropicum	345	PH	CH
Chrysosporium tropicum	412	PF	CH,BH,BHA
Chrysosporium tropicum	439	SE	CH
Chrysosporium tropicum	392	SH	HH,FB,CHA
Chrysosporium tropicum	404	F	BH,HN,HH
Chrysosporium tropicum	312	G	HH
Chrysosporium tropicum	315	G	AH
Chrysosporium tropicum	316	Z	AH
Chrysosporium tropicum	318	CF	HH
Chrysosporium tropicum	319	CF	BF
Chrysosporium tropicum	321	F	HH
Chrysosporium tropicum	323	G	AH
Chrysosporium tropicum	326	F	HH
Chrysosporium tropicum	327	Z	BF
Chrysosporium queenslandicum	438	CF	CH
Chrysosporium queenslandicum	423	CF	HH
Chrysosporium queenslandicum	376	HW	CH, BHA
Chrysosporium queenslandicum	320	CF	AH
Chrysosporium queenslandicum	321	CF	AH
Chrysosporium sp.	349	TF	HH, HN

Table 2 contd...

Table 2 contd...

Fungus	Accession Number*	Habitat	Substrate
Chrysosporium sp.	347	G*	GH
Chrysosporium sp.	348	HW	BH,CH,FB,HN
Chrysosporium sp.	361	PH	BHA,CHA,HH
Chrysosporium sp.	374	PF	HH,HN,CH,CHA, BHA
Chrysosporium sp.	355	SE	CH,FB
Chrysosporium sp.	368	SH	HH,CHA
Chrysosporium sp.	399	F	FB,HH,GH
Chrysosporium sp.	304	G	AH
Chrysosporium sp.	307	G	AH
Chrysosporium sp.	309	Z	HH
Chrysosporium sp.	311	F	BF
Chrysosporium sp.	322	CF	BF
Chrysosporium anamorph of Arthroderma tuberculatum	310	G	BF
Chrysosporium anamorph of Gymnoascus demonbreunii	254	G	BF
Malbranchea sp.	351	CF	HH
Malbranchea sp.	370	F	CHA
Malbranchea sp.	213	CF	HH
Microsporum fulvum	410	G	CH
Microsporum fulvum	414	PF	BHA,CH
Microsporum gypseum	422	HW	CHA
Microsporum gypseum	385	PF	HH,BH
Microsporum gypseum	306	F	AH
Microsporum gypseum	331	F	HH
Microsporum gypseum	336	G	HH
Microsporum sp.	408	G	HH
Microsporum sp.	393	F	FB
Microsporum sp.	386	CF	HH
Myceliophthora vellerea	300	Z	HH
Myceliophthora sp.	379	PF	HH
Myceliophthora sp.	373	HW	CH
Trichophyton mentagrophytes	354	HW	BH
Trichophyton mentagrophytes	398	HW	CH
Trichophyton rubrum	400	F	BH
Trichophyton sp.	401	CF	HH
Trichophyton sp.	384	PF	CHA
Acremonium sp.		F	HH
Acremonium sp.		F	HH
Acremonium sp.		F	AH
Acremonium sp.		G	HH
Acremonium sp.		G	HH
Sterile mycelium	388	G	HH
Sterile mycelium	396	CF	CHA
Sterile mycelium	397	SPR	HH

CF: Crop field, G: garden, HW: hospital waste, SPR: soil with petrol residue, PH: piggery house, PF : poultry farm, SE: sewage, SH: slaughter house, TF : tannery field, F: forest, Z: zoo, HH : human hair, CH: cow horn, CHA: cow hair, BH: buffalo horn, BHA: buffaloo hair, GH: goat horn, GHA: goat hair, FB : bird feather, Germplasm center for Keratinophilic Fungi (GPCK) number.

Table 3. Per cent distribution of Keratin colonizing fungi in different habitats.

Habitats	A	B	C	D	E	F	G	H	I	J	K
Examined soil samples	10	10	10	5	15	40	5	10	5	10	120
Positive samples	8	9	9	2	10	35	2	7	3	7	92
Per cent occurrence	80	90	90	40	66.6	87.5	40	70	60	76.6	
Isolated fungi											
Chrysosporium evolceanui	0	0	0	20	0	2.5	0	0	0	0	1.7
C. indicum	10	10	30	20	6.6	5	0	10	0	10	9.2
C. keratinophilum	0	0	0	20	0	2.5	0	0	0	10	2.5
C. lobatum	0	0	0	0	0	7.5	0	0	0	10	3.3
C. lucknowense	0	10	0	20	6.6	0	0	0	0	20	4.2
C. pannorum	10	0	0	0	0	2.5	0	0	0	0	1.7
C. tropicum	20	20	40	0	6.6	5	20	10	0	0	10.8
C. queenslandicum	20	0	20	0	0	0	0	0	0	0	3.3
Chrysosporium sp.	0	20	10	0	6.6	5	20	10	10	10	8.3
Malbranchea sp.	10	–	0	0	0	0	0	0	0	10	1.7
Microsporum fulvum	0	10	0	0	0	2.5	0	0	0	0	1.7
M. gypseum	0	0	10	0	0	2.5	0	0	0	0	1.7
Microsporum sp.	10	10	0	0	0	0	0	0	0	10	2.5
Myceliophthora sp.	0	0	10	0	0	2.5	0	0	0	0	1.7
Trichophyton mentagrophytes	0	0	20	0	0	0	0	0	0	0	1.66
T. rubrum	0	0	0	0	0	2.5	0	0	0	0	0.8
Trichophyton sp.	110	0	0	0	0	2.5	0	0	0	0	1.7
Sterile mycelium	0	10	10	40	0	0	0	0	0	0	3.3
Total isolates	10	9	14	6	4	17	2	3	1	8	74
Average fungus/habitat	1	0.9	1.4	1.2	0.3	0.4	0.4	0.3	0.2	0.8	0.6

A: crop field, B: garden, C: hospital, D: soil withpetrol residue, E: piggery house, F: poultry farm, G: sewage, H: slaughter house, I: tannery field, J: Forest, K: Total

of sterile mycelium was recorded in soil with petrol residue.

The perusal of data showed that some species of *Chrysosporium, Microsporum* and *Trichophyton* were frequent in their distribution. *C. tropicum* and *C. indicum* were found to be common keratinophilic fungi in different habitats. The per cent distribution of *C. tropicum* was maximum as compared with *C. indicum, C. tropicum* was trapped from crop field, garden, hospital waste, piggery house, poultry farm, sewage and slaughter house soil. Randhawa and Sandhu (1965) isolated *C. tropicum* from cattle field of Punjab. Garg [1966 a,b] noticed its isolation from garden, cultivated field and river sand. Cattle field soils have been earlier described as a suitable habitats for the occurrence of *C. tropicum* [Deshmukh and Agrawal, 1983a, Singh and Agrawal, 1983]. Other reports have confirmed its isolation from poultry farm. Crop field and forest soil [Kushwaha and Agrawal, 1976; Jain, 1983; Dixit 1991]. Nigam and Kushwaha [1990] isolated *C. tropicum* from house dust nutritionally very poor habitat. These reports indicated the wide distribution of *C. tropicum* in Indian soils and its ability to survive in different climatic conditions. Reports on the isolation of this keratin colonizing fungus has also been made from other parts of the world such as Europe, New Guinea, Solomon Islands, United States, Belgium and Italy (Carmichael, 1962). Isolation of *C. tropicum* from lava bed soil of Galapogos Islands and organic rich matter soil of Easter Islands revealed its cosmopolitan distribution (Ajello and Alpert, 1972).

C. indicum was very common and isolated from most of the habitats except sewage and tannery field. *C. indicum* was

Table 4. Per cent distribution of Keratin colonizing fungi on keratinous substrates.

Fungus	Keratinous substrate								
	A	B	C	D	E	F	G	H	I
Chrysosporium evolceanui	1.35	0	0	0	1.35	0	0	0	0
C. indicum	2.7	0	2.7	0	2.7	1.35	0	1.35	1.35
C. keratinophilum	1.35	0	0	0	1.35	0	1.35	0	0
C. lobatum	1.35	0	0	0	1.35	0	0	1.35	0
C. lucknowense	2.7	1.35	0	0	0	1.35	0	0	1.35
C. pannorum	1.35	0	1.35	0	0	0	0	0	0
C. tropicum	5.4	2.7	5.4	1.35	4.05	1.35	1.35	0	1.35
C. queenslandicum	1.35	1.35	2.7	0	0	1.35	0	0	0
Chrysosporium sp.	6.75	4.05	4.05	4.05	1.35	2.7	4.05	0	5.4
Malbranchea sp.	1.35	0	0	1.35	0	0	0	0	0
Microsporum fulvum	0	0	2.7	0	0	1.35	0	0	0
M. gypseum	1.35	0	0	1.35	0	0	0	0	0
Microsporum sp.	2.7	0	1.35	0	0	0	0	0	1.35
Myceliophthora sp.	1.35	0	1.35	0	0	0	0	0	0
Trichophyton mentagrophytes	0	1.35	0	1.35	0	0	0	0	0
T. rubrum	0	0	0	0	1.35	0	0	0	0
Trichophyton sp.	1.35	0	0	1.35	0	0	0	0	0
Sterile mycelium	2.7	0	0	1.35	0	0	0	0	0
Average per cent per keratinous substrate	3.9	1.05	2.55	1.2	1.8	1.05	0.75	0.3	1.16

A: human hair, B: human nail; C: cow horn, D: cow hair, E: buffalo horn, F: buffalo hair, G: goat horn, H: goat hair, I: bird feather

earlier noticed from plains and hills of India. However, soil of plains exhibited the greater preponderance of this fungus in usual habitats than on the hills (Garg, 1966a). It is observed that the man and animals dwelling areas and soil rich in organic and keratinous substances favoured the distribution of this fungus. Isolation of this fungus is in accordance with previous reports from India (Padhye *et al.*, 1967; Sur and Ghosh, 1980; Verma *et al.*, 1982; Deshmukh and Agrawal, 1983 a, b). *C. indicum* was dominant in Egypt and Chilean Andes (Abdel Fattah *et al.* 1982) and has been reported to occur in soils of Galapogos Islands (Ajello and Padhye, 1974) and Mozambique (van Oorschot, 1980). *C. lucknowense* was isolated from garden, soil with petrol residue, piggery house and forest. Maximum distribution of this species was reported from forest followed by three more habitats. These findings correlate its survival in usual habitats with findings of earlier workers (Garg 1966; Jain 1983; Jain *et al.* 1985).

C. lobatum was isolated from poultry farm soil. Its occurrence has been reported from several other countries as USSR, Italy, Canada, Rumania and Czechoslovakia (Hubalek, 1972). *C. queenslandicum*, was, however, confined in distribution. This fungus was isolated from crop field and hospital waste. Nigam and Kushwaha (1986) trapped this fungus from house dust confirming its survivability in nutritionally poor soil. *C. keratinophilum* was isolated from forest, poultry farm and soil with petrol residue. This was isolated from India by Garg (1966a), Jain (1983), Deshmukh and Agrawal (1985) and Jain *et al.* (1985) and from abroad by Carmichael (1962), Ajello and Alpert (1972), Ajello and Padhye (1974) and Hubalek and Hornick (1977). *C. queenslandicum* was also isolated from house dust of Kanpur (Nigam and Kushwaha, 1986). *C. evolceauni* was restricted to poultry farm and soil with petrol residue only. Its isolation from soil of garden, farm, water course bank, and hills of India has already been reported (Garg,

1966a). Other reports showed the existence of this fungus in soils of various parts of India such as Punjab, Himachal Pradesh, Madhya Pradesh and Delhi (Randhawa and Sandhu, 1965; Kushwaha and Agrawal, 1976; Deshmukh and Agrawal, 1983a). This has been isolated from house dust (Nigam and Kushwaha, 1986). Its isolation from USA (Carmichael, 1962), Germany (Bohme and Koach, 1965; Bohme, 1966), Easter Islands (Ajello and Alpert, 1972), Israel (Fuerman *et al.* 1975) indicated wide spread geographical distribution. *C. pannorum* was found in soil of crop field and poultry farm. It was also isolated by Randhawa and Sandhu (1963). *Malbranchea* sp. was restricted to crop field and forest in its occurrence. This appears to be common in its distribution in India (Kushwaha and Agrawal, 1977; Jain and Agrawal, 1979; Jain 1983; Dixit, 1991) and in abroad (Orr and Kuehn, 1972; Ajello and Padhye, 1974; Sigler and Carmichael, 1976).

M. fulvum was isolated from soils of garden and poultry farm. This report agrees with earlier study of its isolation from cattle farm, poultry farm, crop field and forest soil (Kushwaha, 1976; Kushwaha and Agrawal 1976; Jain, 1983, Dixit, 1991). *M. gypsum* was isolated from hospital waste and poultry farm and was similar in distribution in both the habitats. This dermatophyte was first reported by Dey and Kakoti (1955) from animal house in Assam. Turner (1965) reported *M. gypsum* from cultivated and uncultivated soil. Garg (1966a) isolated this fungus from soils of garden, road side, water course bank and places contaminated with feathers of bird and animal horns. Randhawa and Sandhu (1965) also noticed this dermatophyte as second most common species in Indian soil. Desai *et al.*, (1959) reported its distribution in swimming pool, while Padhye (1961) isolated from farm yards, chicken house, dairy farm where soil presumably intermixed with keratin substrate. Jain (1983) also reported the colonization of peacock feathers buried in soil of poultry farm and cattle field. However, house dust of India also found to be positive for this dermatophyte (Nigam and Kushwaha, 1986). Dixit (1991) also isolated this dermatophyte from all the habitat, he studied *M. gypsum* have also been encountered in one or more countries of each continent (Frey, 1965; Rogers, 1971; Muhammad and Lalji, 1978; Mercantini *et al.*, 1980; Ulfig and Korecz, 1983; Calvo *et al.*, 1984).

High distribution of *Microsporum* spp. in crop field, garden and forest has been found by Kushwaha and Agrawal (1976). *Myceliophthora* spp. were reported from hospital waste and poultry farm in this study, Although this fungus was earlier reported by Nigam and Kushwaha (1990) and Dixit (1991). *T. mentagrophytes* was restricted to only hospital waste and showed ability to degrade human hair, bird feathers and animal horns. This dermatophyte was earlier reported from various parts of India. Occurrence of this fungus in soil of cattle field and garden have been reported by Gard (1966 a) and Deshmukh and Agrawal (1983). Other reports on occurrence of this fungus in soil are elsewhere (Desai *et al.*, 1959; Puri, 1961; Padhye *et al.*, 1966; Padhye and Carmichael, 1968; Kushwaha, 1984 a, b; Dixit, 1991). *T. mentagrophytes* is zoophilic and anthropophilic species, most common agent of tinea pedis and is often associated with other farms of tinea as well Emmon *et al.* (1977). Grin and Ozegovic (1963) discussed that it has a limited potential to survive in soil. Never the less many workers reported its geophilic occurrence (Puri, 1961; Padhye and Carmichael, 1968; Mercantini *et al.*, 1983). Isolation of *T.*

rubrum was extremely restricted to only poultry farm soil. The occurrence of this dermatophyte was supported by several workers (Ajello 1968; Deshmukh and Agrawal, 1983). Other report of its isolation from India were made (Garg 1966a, Deshmukh and Agrawal, 1983; Puri 1961; Desai *et al.*, 1959, Randhawa and Sandhu, 1962).

3. RECOVERY OF KERATINOPHILIC FUNGI FROM VARIOUS KERATINOUS SUBSTRATES

Colonization of various keratinic substrates i.e. human hair, human nail, cow horn, buffalo hair, buffalo horn, goat hair, goat horn, bird feathers by keratinophilic fungi and related dermatophytes was documented in Table 5. Five genera belonging to 14 species were isolated from nine keratinous substrates. The results presented here revealed that human hair favoured maximum growth and colonization of these keratinophilic fungi. Colonizing ability of *Chrysosporium* spp. was maximum on all these keratinous substrates. *C. tropicum* showed maximum colonization of most of the keratinous substrates except goat hair while minimum by *C. evolceauni* and *C. pannorum* on human hair, cow horn and buffalo hair. Unidentified species of *Myceliophthora* was found on human hair and cow horn, while *Malbranchea* sp. on human and cow hair but the per cent occurrence of both was relatively less as compared with rest of the isolated fungi. *T. mentagrophytes*, *T. rubrum* and an unidentified species of *Trichophyton* were frequently isolated from human hair, cow horn, buffalo horn, and cow hair. *T. rubrum* was capable of colonizing buffalo hair only. Two species of *Microsporum* were isolated from human hair, cow hair, buffalo horn and buffalo hair. *M. fulvum* and *M. gypseum* were similar in their per cent occurrence on keratinous substrates and unidentified species of *Microsporum* was found to colonize certain keratinous substrates.

Kushwaha and Agrawal (1976) proved the evidence of capability of *C. tropicum* to colonize various hard keratinous substrates such as hooves, horns and nails. Colonization of snake skin and feathers by several keratinophilic fungi and related dermatophytes was also reported earlier (Jain and Agrawal, 1977). Maximum colonization of human hair was shown by the species of *Chrysoporium*, *C. tropicum* colonized all the keratinous substrates used here such as human hair, human nails, cow horns, cow hair, buffalo horn, buffalo hair, goat horns and bird feathers. Kushwaha and Agrawal (1976) also isolated *C tropicum* from decomposing nails and confirmed its keratin colonization capacity. Padhye *et al.*, (1967) reported its isolation from feathers of partriges, qoails and crow peasent in marine areas of India. *C. indicum* and *C. lucknowense.* were commonly isolated from bird feathers. Jain (1983) reported *C. lucknowense* from peacock feathers and other keratin baits. However, Apinis and Rees (1976) reported some *Chrysosporium* from feathers of domestic fowl. *Malbranchea* sp. could colonize only human hair and cow horn used as keratinous substrates. Kushwaha and Agrawal (1976), Singh and Agrawal (1983) confirmed its keratinolytic nature. Dixit (1991) also isolated *M. pulchella* from cultivated field, forest and road side. *M. fulvum* was trapped from cow horn and buffalo hair used as keratinous substrates while *M. gypseum* was restricted to human, cow and buffalo hairs. Colonization of hairs, nails, horns, hooves, feather and wool by *M. gypseum* and *M. fulvum* have also been noticed by Kushwaha and Agrawal (1976), Jain (1983) and Dixit (1991). *Myceliophthora* species were isolated from human hair and cow horn. Nigam and Kushwaha (1990) isolated this fungus from horn used as keratinous substrate. *T. mentagrophytes* was restricted

Table 5. Relative density (%) of Keratin colonizing fungal genera isolated from different habitats.

Habitat	Keratin colonizing fungal genera					
	Chrysosporium	*Malbranchea*	*Microsporum*	*Myceliophthora*	*Trichophyton*	Sterile mycelium
Crop field	60	10	10	0	10	10
Garden	66.6	0	11.1	0	0	11.1
Hospital waste	71.42	0	7	7.1	0	0
Soil with petrol residue	66.6	0	0	0	0	33.3
Piggery house	100	0	0	0	0	0
Poultry farm	70.5	0	11	5.8	11.7	0
Sewage	100	0	0	0	0	0
Slaughter house	100	0	0	0	0	0
Tannery field	100	0	0	0	0	0
Forest	75	12.5	12.5	0	0	0
Relative density (%)	58.69	2.17	7.6	2.17	5.43	4.3

to cow and buffalo horns used as keratinous substrates. Kushwaha and Agrawal (1976), Jain (1983) and Dixit (1991) reported its capacity to colonize hair, human nail, bird feathers, horn and wool. *T. rubrum* was isolated from buffalo horn only. Deshmukh and Agrawal (1983) isolated this fungus from human hair used as keratinous substrates.

The capacity of keratinophilic fungi to colonize and degrade hair has been well known (Daniels, 1953; Galacozy and Novak, 1966). Davidson and Gregory (1934) reported that hair filament exposed to *T. mentagrophytes* were radially penetrated by organised group of hyphae that from wedge shaped perforation. Singh and Agrawal (1983) reported that colonization of keratinous substrates by keratinophilic fungi may be due to their preferential nutritional requirements. Hair keratin rich in oil and fats secreted by sebaceous glands of the animal skin hinders the colonization. Feathers contains some easily assimilated compounds other than keratin (Bolliger and Gross, 1952a, b). Hubalek (1976) and Kushwaha and Agrawal (1976) also reported that substrate specificity could be result of metabolic capacities of a fungus attributable to specific enzymatic system. On the part of dermatophytes the keratin is a vital dependence limited by specific nutritional demands (Otcenasek, 1978).

4. OCCURRENCE OF NON - KERATINOPHILIC FUNGI IN DIFFERENT HABITATS AND THEIR RECOVERY FROM KERATINOUS SUBSTRATES

Occurrence of non-keratinophilic fungi during the isolation of keratinophilic fungi and dermatophytes was frequently recorded from different habitats while using keratinous substrates (Table 6). Ten genera were isolated from all the habitats. *Chaetomium* was trapped directly from bird feathers. Crop field, garden and forest soil yielded the maximum number of non-keratinophilic fungi in comparison to other habitats. Species of *Acremonium, Alternaria, Fusarium and Aspergillus* were frequent in most of the habitats. Five species of *Aspergillus* isolated in current study are as follows : *A. flavus, A fumigatus, A. nidulans, A. ochraceus, A. niger*. The later fungus was found to be dominant species in soil with petrol residue. Unidentified species of *Acremonium, Alternaria, Botryotrichum, Chaetomium, Curvularia*

Table 6. Occurrence of non-keratinophilic fungi during the isolation of Keratin colonizing fungi from various habitats and keratinous substrates.

Fungus	Habitat	Keratinous substrate
Acremonium sp.	CF, HW, F,PF	HH, CH, FB
Alternaria alternata	CF, F, G, PF	BH, HH,HN
Aspergillus flavus	CF, F, SE, SH	BH, GH, FB
A. fumigatus	CF, F, G, PF	HN, CH, FB, HH
A. nidulans	CF, G, F, PH	CH, BHA, HH
A. niger	CF, SPR, G, SE, HW	CHA, HN, BH
A. ochraceous	CF, G, F, HW	BHA, HH, CH, HN
Aspergillus sp.	CF, SE, F, HH	HN, HH, FB
Botryotricum keratinophilum	CF, G, F	HN, CH
Cladosporium sp.	F, CF	HH, BH, HN
Chaetomium globosum	F, G	FB
Curvularia lunata	CF, G, F	HH, BHA
Fusarium sp.	G, HW, CF	HN, HH, BH
Helminthosporium sp.	CF, G	HH, BH
Penicillium sp.	F, CF, SH	HN, HH

and *Fusarium* were more or less common in their isolation and per cent distribution. *Cladosporium* and *Gliomastrix* were rare in their occurrence. The minimum occurrence of *Helminthosporium* spp. were noticed. Geophilic dermatophytes and keratinophilic fungi actively proliferates saprophytically out side man and animals under the conditions corresponding to the requirements of their metabolism, particularly in location where they find various types of keratinous substrates. Along with keratinophilic fungi some other fungi present in soil also colonizes keratinous substrates. While observing the colonization of keratinous substrates by saprophytic fungi, Griffin (1960) found that fungi with high saprophytic ability were followed by less competetive saprophytes. English (1965) found 31 saprophytic fungi capable of colonizing keratin. She observed that many of these fungi produced mycelium while colonizing human hair similar to those produced by dermatophytes. The current study showed that non- keratinophilic fungi were able to colonize keratin in different habitats. Non-Keratinophilic genera isolated from keratinous substrates can be listed as *Acremonium, Alternaria, Aspergillus, Botryotrichum, Cladosporium, Chaetomium, Curvularia, Fusarium, Helminthosporium and Penicillium.* Eight species of *Fusarium* were found to colonize different keratinous substrates. Kushwaha and Agrawal (1976) isolated *Fusarium* species from nail, horn and hair.

The colonization of keratinous substrates in house dust in India by *Aspergillus* and *Penicillium* may be a serious happening as Kishimato and Baker (1969) recognised their pathogenic behaviour. Kozak *et al.* (1980) during their house mold survey found *Penicillium* in 91.2% homes and *Aspergillus* in 48.5% homes in California. Some mucorales were recognised as pathogenic by Kishomoto and Baker (1969) and found involved in ear, nail and cornea infection in man. The infection potential of saprophytic *Absidia, Mucor, Rhizopus* and other *Mucorales* have also been discussed (Emmons *et al.*, 1977; Rippon, 1974). The kerationous substrates which are generally added to the environment as human and animal remain were differently colonized by non-keratinophilic fungi. Otcenasek (1978) showed that presence of keratinous material was not always necessary for the existence of Keratin colonizing fungi.

The comparative account of keratin colonizing ability of these fungi in different habitats exhibited that these habitats provided shelter to more fungi and some of them were with very high percent occurrence. Nigam and Kushwaha (1986, 1989) have also found that house dust of different localities was very rich in keratinophilic fungi and related dermatophytes as compared to the out door soil. Dixit (1991) also isolated some non-keratinophilic fungi during the isolation of keratinophilic and related dermatophytes from different habitats on different keratinous substrates. The capability of long term existence of these potentially pathogenic fungi in the soil of out door environment is basically important. It is, therefore, detailed survey for keratin colonizing non-keratinophilic fungi of out door environment of various localities seems to provide a better picture of their existence and their possible role in initial keratin degradation.

REFERENCES

Ajello, L., 1968. A taxonomic review of the dermatophytes and related species. Sabouraudaia 6 : 147-159.

Ajello, L. 1976. The zygomycete *Saksenia vasiformis* as a pathogen of humans with a critical review of etiology of Zygomycosis. Mycologia 68 : 52-62.

Ajello, L. & Alpert, E.M. 1972. Survey of easter Islands for keratinophilic fungi. Mycologia 64 : 161-166.

Ajello, L. & Padhye, A.A. 1974. Keratinophilic fungi of Galapogos Islands. Mykosen 17 : 239-243.

Apinis, A.E. & Rees, R.G. 1976. An undescribed keratinophilic fungus from Southern Queensland. Trans. Brit. Mycol. Soc. 67 : 522-524.

Awasthi, P.B. & Gotewal, S. 1990. Prevalence of keratinophilic fungi in certain soils of Rohikhand division in relation to Edaphic factors condition. Proc. Nat. Acad. Sci. 76 : 37. Abstract.

Bohme, H. & Koch, H.A. 1965. Nachweis von *Chrysosporium keratinophilum* [Frey] Carmichael, *Trichophyton indicum* Randhawa and Sandhu in Bodin Probenares, Deutschland. Mykosen 8 : 53-61.

Bohme, H. 1966. Verbreitung Keratinophiler Bodenpilze in derlan of licham Umbebung Berlines. Dermatol. Wochensehr. 152 : 266-277.

Bolliger, A. & Gross, R. 1952a. Quantitative studies on some water soluble organic constituents associated with vertebrate keratin. Austr. J. Exp. Biol. And. Med. Sci. 30 : 181-189.

Bolliger, A. & Gross, R. 1952b. Nitrogenous compounds in the aqueous extract of keratinous structures. Astr. J. Exp. Biol. And Med. Sci. 30 : 399-408.

Calvo, A. Vidal, M. & Guarro, J. 1984. Keratinophilic fungi from urban soil of Barcelona, Spain, Mycopathologia 85 : 154-157.

Carmichael, J.W. 1962. *Chrysosporium* and some other aleuriosporic hyphomycetes. Can. J. Bot. 40 : 1137-1173.

Daniels, G. 1953. The digestion of human hair keratin by *Microsporum canis*. J. Gen. Microbiol. 8 : 289-294.

Davidson, A.M. & Gregory, P.H. 1934. In situ culture of dermatophytes. Can. J. Res. 10 : 373-393.

Desai, S., Marquis, L. & Bhatt, M.L.A. 1959. Mycologic study on common superficial mycoses. Bull. Cal. Sch. Trop. Med. 7 : 37.

Deshmukh, S.K. & Agrawal, S.C. 1983a. Prevalence of dermatophytes and other keratinophilic fungi in soil of Madhya Pradesh (India). Mycoses 26 : 574-577.

Deshmukh, S.K. & Agrawal S.C. 1983b. Isolation of keratinophilic fungi from costal habitats of Goa (India). Kevaka, 11 : 53-54.

Dey, N.C. & Kakoti, L.M. 1955. *Microsporum gypseum* in India. J. Indian Med. Asso. 25 : 160-164.

Dixit, A.K. 1991. Occurrence and antagonistic studies of keratinophilic fungi with special reference to *Microsporum* species. Ph.D. Thesis Kanpur University, Kanpur.

Dixit, A.K. & Kushwaha, R.K.S. 1989. Prevalence of keratinophilic fungi in soil of Kanpur. Proc. Ind. Sci. Cong. 76 : 38.

Dixit, A.K. & Kushwaha, R.K.S. 1991. Occurrence of keratinophilic fungi on Indian birds. Folia Microbiologica (Czecslovakia) 36 : 383-386.

Dwivedi, R.S. & Garrett, S.D. 1968. Fungal competition on agar plate colonization from soil inocula. Trans. Brit. Mycol. Soc. 51 : 999-1000.

Emmons, C.W., Binford, C.H. & Utz. J.P. & Knowchung, K.J. 1977. Medical Mycology. Ed. Lea and Febiger. Phidelphia.

English, M.P. 1965. The saprophytic growth of non-keratinophilic fungi on keratinous substrates and a comparison with keratinophilic fungi. Trans. Brit. Mycol. Soc. 48 : 219-235.

Feurman, E.I., Alters, E. Hoing, & Lehrer, N. 1975. The isolation of keratinophilic fungi from soil of Israel. A preliminary report. Mycopathologia 56 : 41-46.

Frey, D. 1965. Isolation of Keratinophilic and other fungi from soil collected in Australia, New Guinea. Mycologia 57 : 202-215.

Galaoczy, J. & Novak, E.K. 1966. Investigation of hair digestion by dermatophytes. Mycopathol. Et. Mycol. Appl. 28 : 71-76.

Garg, A.K. 1966a. Isolation of dermatophytes and other keratinophilic fungi from soil in India. Sabouraudia 4 : 259-264.

Garg, A.K. 1966b. Occurrence of keratinophilic *Chrysosporium* Corda species in Indian soils. Mycopath. et Mycol. Appl. 29 : 189-192.

Gentles, J.C. 1962. Keratinophilic fungi and diseases caused by them. Fungi and Fungus Diseases. C. Thomas Spring Field.

Griffin, G. 1960. Fungal colonization of sterile hair in contact with soil. Trans. Brit. Mycol. Soc. 43 : 583-595.

Grin, E.T. & Ozegobic, L. 1963. Influence of soil on certain dermatophytes and their evolutionary trends. Mycopathol. et. Mycol. Appl. 21 : 23-28.

Gugnani, H.C. & Randhawa, H.S. 1973. An eizootic of dermatophytes caused by *Trichophyton simii* in poultry. Sabouraudia 1 : 1-3.

Gupta, P.K., Singh, R.P. & Singh, I.P. 1968. Dermatophyte from man, dogs and pigs with special reference to *Trichophyton simii* and *Microsporum nanum*. Ind. J. An. Health. 7 : 247-253.

Hubalek, Z. 1972. Keratinophilic Pilze an freilependen vogelin. Mykosen. 15 : 207-211.

Hubalek, Z. 1976. Seasonal distribution of fungi on house sparrows. Trans. Brit. Mycol. Soc. 66 : 509-516.

Hubalek, Z. & Hornik, M. 1977. Experimental infection of white mouse with *Chrysosporium* and *Paecilomyces*. Mycopathologia 62 : 173-178.

Jain, P.C. & Agrawal, S.C. 1977. Keratinophilic fungi from the soils of Mount Abu (India). Geobios 4 : 136-138.

Jain, P.C. & Agrawal, S.C. 1979. Some additions to Indian Malbranchea. Kavak 7 : 69-72.

Jain, P.C. 1983. Prevalence of keratinophilic fungi from soils of Madhya Pradesh and Uttar Pradesh, India. Geobios New Reports 2 : 10-13.

Jain, Madhu, Shukla, P.K. & Srivastava, O.P. 1985. Keratinophilic fungi and dermatophytes in Lucknow soils and their global distribution. Mykosen 28 : 98-101.

Kishimoto, R.A. & Baker, G.E. 1969. Pathogenic and potentially pathogenic fungi isolated from beach sands and selected soil of Ohio, Hawai. Mycologia 61 : 537-548.

Krempt-Lamprecht, L. 1965. Uederdas vorkommen von pilzen ausder gattung *Chrysosporium* auf der Hant Kran Kheiten and Tier. Spinger Verlag. Berlin.

Kozak, P.P., Gallup, J., Commins, L.H. & Gilman, S.A. 1980. Currently available methods for home mold surveys II Examples of problem of home surveys. Ann. Allergy. 45 : 167-176.

Kushwaha, R.K.S. 1976. Studies on Keratinophilic fungi from soils. Ph.D. Thesis, Saugar University, Sagar.

Kushwaha, R.K.S. & Agrawal, S.C. 1976. Some keratinophilic fungi and related dermatophytes from soil. Proc. Indian Natl. Sci. Acad. 42 : 108-110.

Kushwaha, R.K.S. & Agrawal, S.C. 1977. Isolation of *Malbranchea pulchella* from soil in India. Mykosen 20 : 163-164.

Kushwaha, R.K.S. 1984a. Keratinophilic fungi and related dermatophytes from Kanpur. Proc. Ind. Bot. Soc. 11 : 13.

Kushwaha, R.K.S. 1984b. Morphological studies of some dermatophytes and keratinophilic fungi. Bull. Bot. Soc. 30 & 31 : 48 - 51.

Mercantini, R., Marsella, R., Caprilli, F. & Dovgiallo, G. 1980. Isolation of dermatophytes and corelated species from the soil of public gardens and parks in Rome. Sabouraudia 18 : 123-128.

Mercantini, R., Marsella, R., Lambiase, L. & Fulvi, F. 1983. Isolation of keratinophilic fungi from floors in Roman primary schools. Mycopathologia 82 : 115-120.

Muhammed, S.I. & Lalji, N. 1978. The distribution of geophilic dermatophytes in Kenyan soils. Mycopathologia 63 : 95-97.

Nigam, Neeta & Kushwaha R.K.S. 1986. Termotolerance of keratinized spores of *Chrysosporium tropicum* and *C. keratinophilum.* Ind. J. Plant Sci. 4 : 5-8.

Nigam, Neeta & Kushwaha R.K.S. 1989. Occurrence of keratinophilic fungi with special reference to *Chrysosporium* species in India. Sydowia 42 : 200-208.

Orr, G.F. & Kuehn, H.H. 1972. Notes on Gymnoascaceae II Some Gymnoascaceae and keratinophilic fungi from Utah. Mycologia 64 :55-72.

Otcenasek, M. 1964. Isolation of *C. keratinohilum* (Frey) Carmichael and similar fungi from Czechoslovakian soil. Mycopathol. et. Mycol. Appl. 23 : 121-124.

Otcenasek, M. 1978. Ecology of dermatophytes. Mycopathologia 65 : 67-72.

Padhye, A. A. 1961. *Microspum gypseum* from soil in Poona. Prof. Agharkar Commemoration Volume 117-119.

Padhye, A.A., Mishra, S.P. & Thirumalachar, M.J. 1967. Occurrence of soil inhabiting dermatophytes and other keratinophilic fungi from soil in Poona. Hindustan Antibit. Bull. 9 : 90-93.

Padhye, A.A., Pawar, V.H., Sukapure, R.S. & Thirumalachar, M.J. 1967. Keratinophilic fungi from marine soil of Bombay, India. Part I Hindustan Antibiot. Bull. 10 : 130-141.

Padhye, A.A., Mishra, S.P. & Thirumalachar, M.J. 1967. Isolation of *Trichophyton simii* and *Cryptococcocus neoformans* from soil in India. Hindustan Antibiot. Bull. 10 : 200-202.

Padhye, A.A. & Carmichael, J.W. 1968. Mating reactions of *Trichophyton simii* and *T. mentagrophytes* strains from poultry farm soil in India. Sabouraudia 6 : 238-240.

Padhye, A.A. & Carmichael, J.W. 1971. The genus Berkeley. Can. J. Bot. 49 : 1525-1540.

Padhye, A.A., Sekhon, A.S. & Carmichael, J.W. 1973. Ascocarp production by *Nannizzia* and *Arthroderma* on keratinous and non keratinous media. Sabouraudia 11 : 109-114.

Park, D. 1961. Isolation of *Fusarium oxysporum* from soils. Trans. Brit. Mycol. Soc. 44 : 191-222.

Puri, H.R. 1961. Isolation of dermatophytes from Indian soil. Bull. Calcutta Sch. Trop. Med. 8 : 121-122.

Randhawa, H.S. & Sandhu, R.S. 1962. On the occurrence of two new species of *Trichophyton* in Indian soils. Mycopath. Mycol. Appl. 20 : 225-238.

Randhawa, H.S. & Sandhu, R.S. 1963. On the occurrence of two new species of Trichophyton in Indian soils. Mycopath. Mycol. Appl. 20 : 225-238.

Randhawa, H.S. & Sandhu R.S. 1965. Survey of soil inhabiting dermatophytes and related keratinophilic fungi of India. Sabouraudia 4 : 71-79.

Rippon, J.W. 1974. Medical Mycology. The pathogenic fungi and pathogenic actinomycetes. W.B. Saunders and Co. Philadelphia 587.

Rogers, A.L. 1971. Isolation of keratinophilic fungi from soil in viccinity of Bogota, Columbia. Mycopath. et. Mycol. Appl. 44 : 261-264.

Saxena, R.P. and Baruda, P. 1982. Keratinophilic fungi in soil. J. Indian Bot. Soc. 61 : 80 - 84.

Srivastava, O.P. 1984. Ecology of human pathogenic fungi. Progress in Microbial Ecology. Prof. J.N. Rai Festschrift 375- 390.

Shrivastava, J.N, Dixit, A.K. & Kushwaha, R.K.S. 1987. Keratinophilic fungi from U.P. Soil. Indian Bot. Soc. 10 :17.

Sigler, L. & Carmichael, J.W. 1976. Taxonomy of *Malbranchea* and some other hyphomycetes with orthroconidi. Mycotaxon 4 : 349-488.

Singh, K.V. & Agrawal, S.C. 1983. Keratinophilic fungi from some substrates on central India. Geobios 10 : 165-171.

Stockdale, P.M. 1964. Personal observations on the production of sexual forms of dermatophytes. Ann. Soc. Belge. Med. Trop. 44 : 112-123.

Sur. B. & Ghosh, G.R. 1980a. Keratinophilic fungi from Orissa, India : Isolations from soils. Sabouraudia 18 : 269-274.

Sur. B. & Ghosh G.R. 1980b. Keratinophilic fungi from birds and domestic fowls. Sabouraudia 18 : 275-280.

Turner, P.D. 1965. The occurrence of *Microsporum gypseum* in cultivated and uncultivated soil in Hong Kong. Trans. Brit. Mycol. Soc. 48 : 549-551.

Ulfig, K. & Krocz, M. 1983. Isolaton of keratinophilic fungi from sewage sludge. Sabouraudia 21 : 247-250.

Vanbreuseghem, R. 1952. Technique biologigue pautr I. isolement des dermatophytes dusoil. Ann. Soc. Belge. Trop. 32 : 173-178.

Van Oorschot, C.A.N. 1980. A revision of *Chrysosporium* and allied genera. Studies in Mycology 20 : 1-80.

Verma, T.N., Sinha, B.K. Das, U.L. 1982. Isolation of keratinophilic fungi from soil in Bihar (India). Mykosen 25 : 449-452.

CHAPTER - 14

ALKALIPHILES — MICROBES IN ALKALINE ENVIRONMENTS

S. Borkar and S. Bhosle***

* Dept. of Microbiology, P.E.S. College, Farmagudi, Ponda, Goa.
** Dept. of Microbiology, Goa University, Taleigao Plateau, Goa.

1. EXTREME ENVIRONMENTS AND THEIR INHABITANTS

Microorganisms are ubiquitous in nature occupying not only different moderate habitats on earth but are also found to adapt and colonize certain harsh environments, once assumed to be sterile by man. These amazing creatures, thriving in extreme environments are called Extremophiles (Krahe *et al.*, 1996; Madigen *et al.*, 1997) and are found in econiches with extremes of temperature (Thermophiles and Psychrophiles), pressure (barophiles), alkalinity (Alkali-philes), acidity (Acidophiles) and salinity (Halophiles). Some of them can also survive in niches/ media containing high levels of organic solvents (Organic solvent tolerant organisms), heavy metals radiation's or extra dry conditions (Table 1).

Research in this area was stimulated by the description of *Thermus aquaticus* by Brock in 1969 (Brock, 1986). Since then a surprising number and variety of isolates have been made from these environments. Table 2 (Prasad and Sarkar, 1997) depicts the examples of few microorganisms in different extreme environments. These organisms thus live in habitats that will support no other form of life and are perhaps oldest on the earth with a unique genome believed to have been adapted initially to suit the extreme environmental challenges that the earliest life forms faced on this planet.

Search for the extremophiles has intensified recently as scientists have recognized their potential use in an array of industrial applications. The enzymes from such organism, which function at unusual parameters, are referred to as "Extremozymes" (Pennisi, 1997). Such enzymes have already been commercialized example. Taq Polymerase, Cellulase

(Krulwich *et al.*, 1988) etc. (Table 3) (Aguilar, 1996; Cook *et al.*, 1996) and (Table 4) (Persidis, 1998).

Table 1. Extremophiles found in specific environments (Borkar and Bhosle, 2000)

EXTREMOPHILES	ECOSYSTEMS
Thermophiles	Hydrothermal vents
Psycrophiles	Sub-zero climates
Barophiles	Ocean floors
Alkaliphiles	Alkaline soda lakes
Acidophiles	Acid bogs
Halophiles	Hypersaline water
Organic solvent tolerant organismss	Presence of organic solvents
Heavy metal tolerant organisms	Presence of heavy metals
Radio-resistant forms	Presence of ionizing radiations
Xerophiles	Extra dry conditions

Most natural environments on the earth are essentially neutral pH values between 5.0 and 9.0 harboring neutrophiles. However, there are specified ecosystems that show unnaturally high pH ranging between 9.0 and 12.0. Interestingly, during the past 20 years, a number of studies have shown that microorganisms can indeed survive and grow at high pH of above 10.0 (Horikoshi and Akiba, 1982; Horikoshi, 1991) as seen in figure 1, termed as Alkaliphiles.

This article reviews the world of alkaliphilic microorganisms, their diversity, the fundamental bioenergetic puzzles of alkaliphily and their industrial applications.

2. DEFINITION AND CLASSIFICATION OF ALKALIPHILES

Alkaliphiles are a diverse group of organisms that thrive in highly alkaline environments with pH optima for growth in excess of pH 8.0 usually between 9.0 and 10.0. Based on pH preference, such alkaliphiles are grouped into two broad categories; alkaline tolerant organisms that show optimal growth in the pH range of 7.0-9.0 but cannot grow above pH 9.5 and alkaliphilic organisms that show optimal

Table 2. Microorganisms under extreme environments (Prasad and Sarkar, 1997).

Extreme Condition	Microorganisms	Habitat	Extreme Growth Conditions	Metabolic Characteristics
High temperature	*Pyrococcus furiosus*	Geothermal marine sediments	100°C	Anaerobic heterotroph
Cold temperature	*Bacillus* TA41	Antartic seawater	4°C	Aerobic heterotroph
High pressure	*Methanococcus janaschii*	Deep sea hydro-thermal vent	250 atm, 85°C	Growth and methane production stimulated by pressure
High pH	*Clostridium paradoxum*	Sewage sludge	pH 10.1, 56°C	Anaerobic heterotroph
Low pH	*Metallosphaera sedula*	Acid mine drainage	pH 2.0, 75°C	Facultative chemolithotroph
High salt	*Halobacterium halobium*	Hypersaline waters	4-5 M, NaCl	Aerobic heterotroph
Presence of organic solvents	*Pseudomonas putida* IH 2000	Mud samples of Kyushu island, Japan	Grows in culture media containing more than 50% toluene	Degrades polyaromatic hydrocarbons.
Presence of heavy metals	*Alcaligenes eutrophus*	Environment severely contaminated with heavy metals	Cd^{++}, Co^{++}, Zn^{++}, Ni^{++}, Cu^{++}	Plasmid borne resistance
Ionizing radiations	*Deinococcus radiodurans*	Soil	Ionizing radiation	Tolerates massive DNA damage.

Table 3. Industrial applications of Extremophiles (Aguilar, 1996; Cook *et al.*, 1996)

EXTREMOPHILIC GROUP	ENZYMES, ENDOGENOUS COMPOUNDS	APPLICATIONS, PRODUCTS
Thermophiles 50-110°C	Amylases	Glucose, fructose for sweetners
	Xylanases	Paper bleaching
	Proteases	Amino acid production from keratins food processing, baking, brewing, detergents, biosensors, bioremediation.
	Dna polymerases/Taq polymerase	Genetic enginnering, PCR, sequencing
Halophiles 3-20% salt	Antibiotics	Pharmaceuticals
	Carotene	Food coloring
	Glycerol	Pharmaceuticals
	Compatible solutes	Pharmaceuticals
	Membranes	Surfactants for Pharmaceuticals
	Enzymes	Waste treatment, peptide synthesis, enhanced oil recovery from wells
Barophiles Optimal growth >1 atm	Proteases	Proteolysis in leather treatment Food processing Oil-well stimulation
Psychrophiles 5-20°C	Neutral proteases	Cheese maturation, dairy production
	Proteases Amylases Lipases	Detergent for cold water washes
	Polyunsaturated fatty acids	Pharmaceuticals
	Ice-protein, w-3 fatty acids	Artificial snow, Production of Dietary supplements
Acidophiles pH <2.0	Sulfur oxidation	Desulfurisation of coal
Alkaliphiles pH >9.0	Proteases Amylases	
	Lipases	Detergents
	Cellulase	
	Cyclodextrins	Stabilization of volatile substances
	Antibiotics	Pharmaceuticals.

Table 4. Selected companies with extremophile organism or molecular Program (Persidis, 1998)

COMPANY	PROGRAM
Altus Biologics (Cambridge, MA)	Artificial extremozymes; cross-linked enzyme crystals.
Amersham Pharmacia Biotech (Cleveland, OH)	Thermostable alkaline phosphatase
Archaenzyme (Jerusalem, Israel)	Extremozymes from *Archaea*
Diversa (San Diego, CA)	Extremophile expression libaries
DuPont Merck (Nutley, NJ)	Protease stabilization in oxidizing, alkaline and hot environments.
Genencor International (Rochester, NY)	Protease stabilization in oxidizing, alkaline and hot environments
Hoffman La Roche (Basel, Switzerland)	Thermostable Taq polymerase for PCR
Maxygen (Santa Clara, CA)	DNA shuffing of extremozyme genes
Novo Nordisk Biotech (Davis, CA)	Extremozyme cloning and expression
Pfizer Central Research (Groton, CT)	Extremophile screening

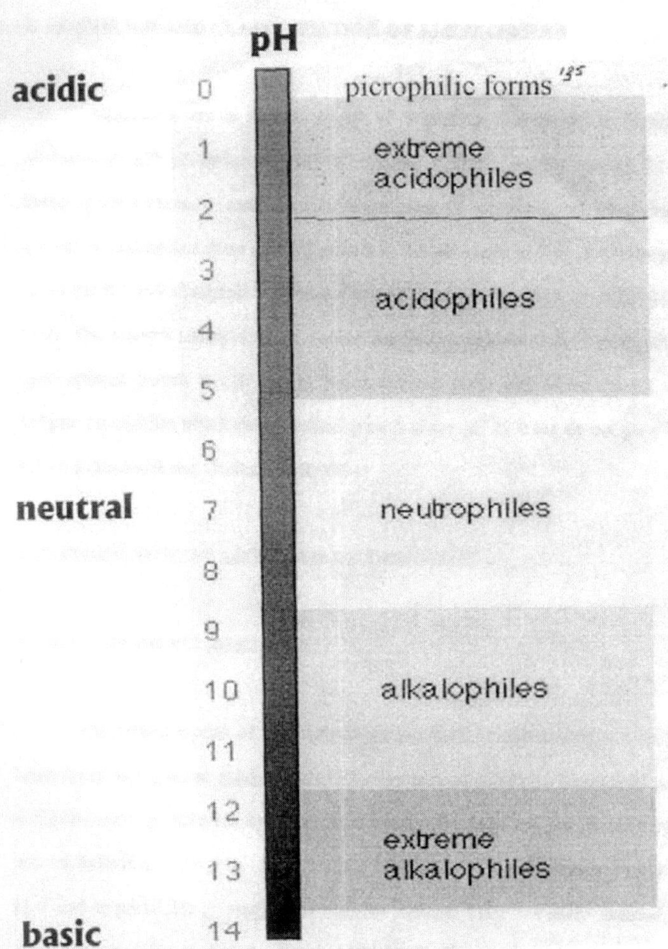

Fig. 1. Extremophilic groups at various pH ranges

growth between pH 10.0 and 12.0. The extreme alkaliphiles are further sub divided into facultative alkaliphiles, which show optimal growth at pH 10.0 or above but can grow well in neutral pH range and obligate alkaliphiles which show optimal growth above pH 10.0 but do not grow below pH 8.5-9.0 (Krulwich and Guffanti, 1989).

3. HISTORY, BIODIVERSITY AND DISTRIBUTION

3.1 History and Biodiversity

The earliest reports of microorganisms in alkaline environments appeared in 1922 when Meek and Lipman isolated alkaline tolerant nitrifying bacteria *Nitrosomonas* and *Nitrobacter* sp. followed by isolation of alkaliphilic *Streptococcus faecalis* by Downic and Cruikshank in 1928. Later a strain of *Bacillus pasteurii*, which grew well at pH 11.0 and required NH_3, was discovered by Gibson. Vedder isolated *Bacillus alkalophilus* growing at pH 8.6 - 11.0 in 1934. Twenty-five years later, Kushner and Lisson (1959) reported the discovery of many alkali tolerant strains, such as *Bacillus cereus* and *Bacillus circulans*. The use of such alkaliphilic microbes has a long history in Japan, since from ancient times, indigo has been naturally reduced under alkaline conditons in presence of sodium carbonate in a traditional process called "Indigo Fermentation". Takahara and co-workers in 1960 isolated the indigo reducing bacterium *Bacillus* sp. No. S-8 from an indigo ball and improved the indigo fermentation by adding this strain during fermentation under alkaline conditions.

Koki Horikoshi, "The explorer of extremophiles", conducted further microbiological studies was isolated a large number of alkaliphiles since 1969 and purified many alkaline enzymes (Horikoshi and Akiba, 1982; Horikoshi, 1991). Souza *et al.* (1974) studied the growth and reproduction of microorganisms under extreme

Table 5. Diversity of alkalophilic bacteria (Krulwich & Guffanti, 1989).

Group	pH Range for Growth	Natural Habitat	Comments
EUBACTERIA			
Gram-negative photoautotrophs			
Spiruluna sp.	8.0-11.0	Alkaline soda lakes, Rift valley, pH 10.5	Halotolerant cyanobacterium
Synechococcus sp.	6.5-10.0, optimum pH 8.0	Yellowstone, 55°C, pH 5.5	Alkalitolerant, Halotolerant *cyanobacterium*
Ectothiorhodospira sp.	8.0-10.0	Mud from alkaline salt lakes, pH 11.0	Anoxygenic purple sulfur bacterium
Gram-negative nonendosporeformers			
Aeromonas sp.	7.0-12.0	Soil	Facultative anaerobe
Flavobacterium sp.	7.0-11.2	Soil	Facultative anaerobic
Vibrio alginolyticus	6.0-9.2	Seawater	Halotolerant, alkalitolerant Facultative anaerobe
Gram-positive nonendosporeformers			
Exigubacterium sp.	7.0-11.5, optimum 8.5-9.5	Potato processing effluent	Facultative anaerobe
Actinomyces sp.	8.0-11.5, optimum 9.0-9.5	Soil	Heterotrophic facultative
Streptococcus sp.	5.0-11.0, optimum 8.0-9.0	Alkaline potato processing effluent	Fermentative facultative anaerobe
Coryneform bacteria	6.6-9.5	Seawater	Halotolerant facultative anaerobe
Gram-positive endosporeformers			
Clostridium paradoxum	8.0-11.3, optimum 9.5	Alkaline springs	Strict anaerobe, alkali thermophile
Bacillus alcalophilus	8.5-11.5, optimum 10.6	Soil (acid-alkaline)	Aerobic heterotroph
Bacillus strain A007	9.0-11.0	Soil	Aerobic heterotroph
Bacillus strain WN13	8.0-11.5, optimum 9.0-9.5	Soil	Aerobic, heterotroph halophile
Archaebacteria Aerobee			
Natronobacterium gregoryi *N. magadi* *N. pharaonis* *N. vacuolatum* *Natronococcus occultus.*	Optimum 9.0-10.0	Alkaline saline, lakes-Wadi Natrum, Egypt and Lake Magadi, Kenya	Halophilic (3M NaCl)
Anaerobes			
Methanobacterium thermoalcaliphilum	6.5-10.0	Biogas plant, optimum pH 7.5-8.5	Thermophilic autotroph

alkaline conditions. The alkali-philic microorganisms are now known as a diverse group comprising a heterogeneous collection of eubacteria (Aono and Horikoshi, 1983; Aono, 1989; Bahm *et al.*, 1998; Grant *et al.*, 1979; Hansen *et al.*, 1997; Khmelenina *et al.*, 1997; Kobayashi and Horikoshi, 1980;Li *et al.*, 1994; Shimao *et al.*, 1989; Sorokin *et al.*, 1996; Sorokin *et al.*, 1998; Sorokin *et al.*, 1998; Souza and Deal, 1997; Tsujibo *et al.*, 1988; Zhilina *et al.*, 1996; Zhilina and Zavarzin, 1994; Zhilina *et al.*, 1997) including *Cyanobacteria* (Buck and Smith, 1995; Gerasimenko *et al.*, 1996; Prasad and Kashyap, 1991; Singh, 1995; Xu *et al.*, 1994) to archaebacteria such as *Natronobacterium* and *Natronococcus* (Desmarais *et al.*, 1997; Keller *et al.*, 1995; Kobayashi *et al.*, 1994; McGenity and Grant, 1993; Studdert *et al.*, 1997; Xu *et al.*, 1999) Diversity of alkaliphilic bacteria is depicted in Table 5 (Krulwich and Guffanti, 1989). In addition to these prokaryotes, there are reports on eukaryotes growing at high pH (Goto *et al.*, 1981; Johnson, 1923; Souza *et al.*, 1974).

3.2 Distribution

Alkaliphilic microorganisms, particularly prokaryotes are widely distributed and can be found in non-alkaline such as neutral and acidic soils and alkaline environments, although counts of alkaliphiles are higher in alkaline environments. Transient alkaline conditions can be generated due to animal excreta and biological activities such as proteolysis, ammonification, sulfate reduction or photosynthesis (Horikoshi, 1991).

The most stable naturally occurring alkaline environments are caused by a combination of geological, geographical and climatic conditions such as eutrophic soda (Na_2CO_3) lakes, soda deserts and alkaline groundwaters (Grant *et al.*, 1990). Lake Magadi in Kenya, lake Wadi Natrum in Egypt of the East African rift valley and the western soda lakes of United States are probably the most stable and highly

Table 6. Specific Media for Alkaliphilic and Haloalkaliphic Microorganisms

INGREDIENTS	ALKALIPHILES			HALOALKALI-PHILES
	HORIKOSHI-I (g/l)[68]	HORIKOSHI-II (g/l)68	PPYG40 (g/l)	
GLUCOSE	10	–	5	–
SOLUBLE STARCH	–	10	–	–
POLYPEPTONE	5	5	5	5
TRYPTONE	–	–	–	1
YEAST EXTRACT	5	5	1.5	—
KH_2PO_4	1	1	–	13.7
K_2HPO_4	–	–	–	0.05
$MgSO_4$ $7H_2O$	0.2	0.2	5.03	–
Na_2CO_3	10	10	15	–
AGAR	20	20	15	–
Na_2HPO_4	–	–	1.5	–
NaCl	–	–	0.1	25
$MgCl_26H_2O$	–	–	0.1	–
SODIUM CITRATE	–	–	–	0.3

*Solutions of glucose, starch and Na_2CO_3 are sterilized separately by autoclaving as 10% solutions and then added to the cold basal medium to avoid the precipitation of salts.

alkaline environments on the earth with a consistent pH of 10.5 to 12.0. These environments are characterized by the presence of large amounts of sodium carbonate and sodium chloride, formed by the evaporative concentration, giving rise to environments that are both alkaline and saline.

The genesis of alkaline lakes is due to vulcanism, contributing to a source of sodium carbonate to the surface environment or biological activity and weathering, which produce CO_2, charged surface waters, thus forming a bicarbonate-carbonate solution that leaches surrounding minerals. Concentration of ions by evaporation leads to a shift in the carbon dioxide/ bicarbonate/carbonate equilibrium. In most environments, such ground waters rapidly become saturated with Ca_2 and Mg_2 resulting in the precipitation of calcite ($CaCO_3$), magnesite ($MgCO_3$) and dolomite ($MgCa(CO_3)_2$). Carbonate is removed from the solution and the genesis of alkaline brine is inhibited, as it is the case for most ground waters. However, in situations where the CO_3-2 concentration exceeds that of Ca^{+2} and Mg^{+2}, alkalinity develops, usually with Na^+ as the dominant cation. Figure 2, illustrates the schematic representation of the proposed mechanism involved in the formation of lake Magadi, an alkaline and saline lake, 200 Km south of Nairobi at the lowest point in the East African rift valley (Grant *et al.*, 1990).

Bacterial sulfate reduction in surrounding swamps has also been proposed to contribute towards the alkalinity in certain soda lakes such as those in Wadi Natrum depression and certain Hungarian soda lakes (Grant *et al.*, 1990).

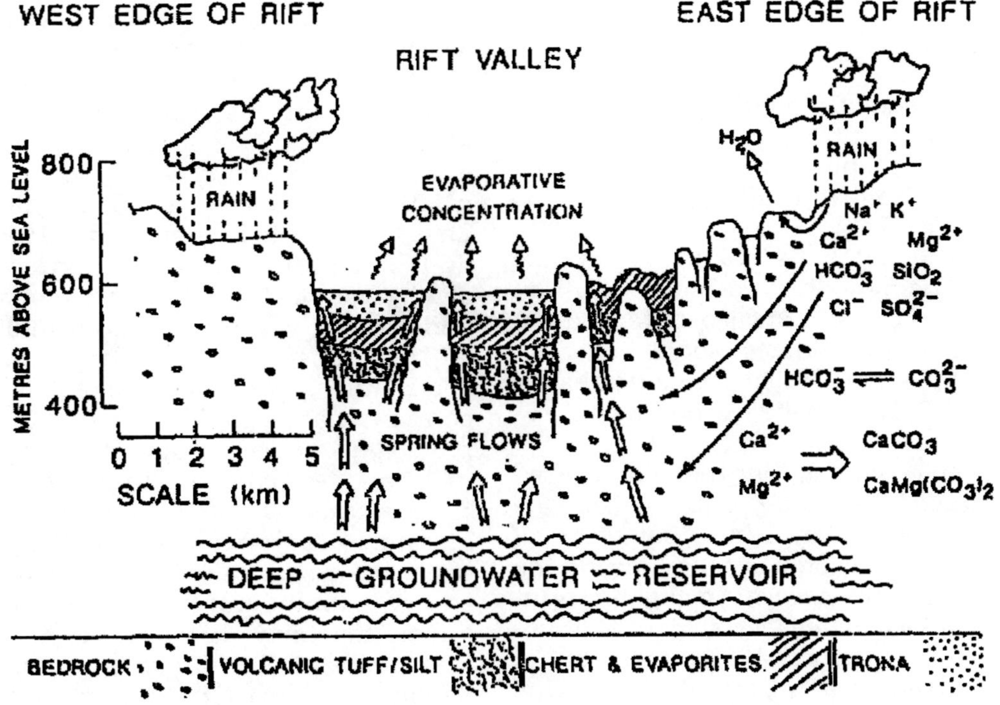

Fig. 2. Schematic representation of the possible mechanisms involved in the function of alkaline saline lakes (L. Magadi) (Grant *et al.*, 1990)

Most of the truly alkaliphilic microorganisms have been isolated from manmade alkaline environments such as indigo dye balls (Takahara and Tanabe, 1960) and alkaline effluents of various industries for example : (1) Potato or food processing plants (Gee *et al.*, 1980), (2) Paper processing factories, (3) Textile mills and rayon waste (Ikura and Horikoshi, 1997), (4) Detergent factories, (5) Battery cell manufacturing units, (6) Cement manufactures and casting, (7) Electroplating, (8) Lye treatment of animal hides, (9) Pharmaceuticals and (10) Agrochemical factories (Nagarsekar, 1996). These processes generate alkaline conditions and support the growth of alkaliphiles.

4. GROUPS OF ALKALIPHILES

Alkaliphiles consist of two main physiological groups; (A) *Alkaliphiles* and (B) *Haloalkaliphiles.*

(A) Alkaliphiles

These organisms require an alkaline pH of 9.0 or more for their growth and have an optimal growth pH of around 10.0. They can be isolated on alkaline media as indicated in Table 6 (Gee *et al.*, 1980; Horikoshi, 1991). Alkalinity can be generated using sodium carbonate, sodium bicarbonate, potassium carbonate, sodium hydroxide or sometimes sodium borate. These salts are added in concentration of about 0.5-2.0% depending on the micro-organisms used giving a pH of 8.5 to 11.0. It has been observed, that the diluent used to prepared the serial dilutions of the samples used for the isolation of alkaliphiles has a considerable effect on the count of alkaliphiles; with a higher count being obtained when serial dilutions are prepared in carbonate-bicarbonate buffer (pH 10.0) than when physiological saline (pH 7.0) is used as a diluent. Further, the counts of alkaliphiles obtained is higher on PPYG agar than Horikoshi I and II media (Table 7 and 8) (Borkar and Bhosle, 2000, Nagarsekar, 1996).

Alkaliphiles are found to coexist with neutrophilic microorganisms as well as occupy specific extreme environments in nature (Table 7 and 8) (Borkar and Bhosle, 2000; Nagarsekar, 1996). Many different kinds of aerobic alkaliphiles, both Gram positive and Gram negative including bacteria belonging to genera *Bacillus, Micrococcus, Corynebacterium, Pseudomonas, Flavobacterium, Actinomycetes, Streptomycetes, Nocardia* and Eukaryotes such as yeast and filamentous fungi have been isolated from neutral and alkaline environments (Horikoshi, 1991). *Exiguobacterium aurantiacum* (Clejan *et al.*, 1986), isolated from alkaline potato processing waste is one of the few isolates that have been subjected to a detailed taxonomic analysis.

Although most of the alkaliphiles are aerobic or facultatively anaerobic, there are reports of few alkaline tolerant strictly anaerobic strains e.g. *Clostridium* and *Methanobacterium* (Karlsson *et al.*, 1988; Li *et al.*, 1994; Lowe *et al.*, 1993; Nimura *et al.*, 1987; Nomoto *et al.*, 1984; Zhilina and Zavarzin, 1994; Zhilina *et al.*, 1997). The first facultative anaerobic xylan using alkalophile lacking cytochrome, quinone and catalase was reported by Nimura *et al.*, 1987. Subsequently, many anaerobic sporulating alkaliphiles were isolated by conventional methods. *Clostridium proteolyticum* was isolated from a chicken manure digestor and *Clostridium collagenovorans* from a sewage sludge digestor. *C. bifermentans* and *C. sporgenes* have been isolated from biodeteriorated casein containing building materials having maximal pH tolerance of 12.2 and 11.7 respectively (Karlsson *et al.*, 1988). Kevbrin *et al.*, (1998) isolated alkalophilic obligately anaerobic *Tindallia magadi* from deposits of lake Magadi Kenya.

Table 7. Total viable cell counts of Alkaliphiles from different samples (Borker and Bhosle, 2000)

Samples Analysed	pH	c.f.u./ml PPYG	Horikoshi-I	Horikoshi-II
A. NATURAL ALKALINE ENVIRONMENTS				
1. Saltpans from Ribandar area	8.00	2.57 x 10^6	1.55 x 10^4	1.20 x 10^4
2. Mangrove sediments	8.00	4.50 x 10^3	3.25 x 10^2	2.10 x 10^2
3. Mangrove water	7.80	1.00 x 10^2	2.50 x 10^1	1.00 x 10^1
4. Compost	8.70	2.50 x 10^{10}	5.20 x 10^8	3.00 x 10^{11}
5. Cow dung	8.50	4.56 x 10^{11}	2.40 x 10^8	2.80 x 10^8
6. Dropping of birds (Poultry)	8.70	1.55 x 10^8	1.80 x 10^8	2.00 x 10^8
7. Piggery waste	8.00	4.20 x 10^5	1.80 x 10^5	1.30 x 10^5
8. Coconut rhizosphere	7.80	4.20 x 10^5	1.50 x 10^8	4.00 x 10^4
9. Alkaline soil	8.50	2.50 x 10^4	1.20 x 10^4	—
B. MAN MADE ALKALINE ENVIRONMENTS				
10. Zuari effluent	10.0	5.30 x 10^9	1.50 x 10^5	1.30 x 104
11. Mining effluent	8.00	1.60 x 10^8	1.80 x 10^5	2.00 x 104
12. Dairy effluent	8.50	6.20 x 10^8	8.00 x 10^8	5.60 x 103
13. Biogas digester sample	8.67	3.60 x 10^8	2.40 x 10^4	4.00 x 105
14. Paper pulp	8.50	2.00 x 10^{10}	1.60 x 10^8	4.80 x 107
15. Polyester dyeing sample	8.01	2.00 x 10^8	1.60 x 10^3	1.20 x 103
16. Raw sewage	8.00	2.80 x 10^7	5.30 x 10^7	1.74 x 10^8
17. Treated sample	7.50	3.70 x 10^3	4.50 x 10^3	8.00 x 10^4
C. NON ALKALINE ENVIRONMENTS				
18. Marine waters	7.35	1.20 x 10^7	4.00 x 10^8	2.00 x 10^7
19. Acidic soils				
a. White	5.80	3.30 x 10^1	2.00 x 10^1	—
b. Green	6.00	2.06 x 10^3	1.20 x 10^3	1.50 x 10^1
c. Red	5.90	2.90 x 10^3	1.50 x 10^1	—
d. Yellow	5.80	2.00 x 10^3	1.70 x 10^1	—
20. Garden soil	7.00	7.00 x 10^8	6.20 x 10^8	6.50 x 10^8
21. Droppings of goat	7.50	7.50 x 10^8	5.00 x 10^8	5.50 x 10^8
22. Glutan water	3.19	2.20 x 10^4	1.00 x 10^2	4.00 x 10^2
23. Termite soil	7.00	2.00 x 10^4	1.10 x 10^4	1.56 x 10^5

* Serial dilutions were prepared in carbonate-bicarbonate buffer (pH 10.0)

Table 8. Total viable counts of Alkaliphiles and Haloalkaliphiles in samples collected from Sambhar salt lake, Jaipur - India (Borker and Bhosle, 2000)

Sample Discription	pH	Alkaliphiles cfu/ml			Haloalkaliphiles
		PPYG	Horikoshi-I	Horikoshi-II	PPYG + 25% NaCl
Sand	8.5	3.0 x 10^4	4.2 x 10^4	3.1 x 10^4	—
Black sediment	9.0	9.3 x 10^4	6.8 x 10^4	1.1 x 10^5	1.0 x 10^2
Red salt + Black sediment	9.0	7.8 x 10^4	6.5 x 10^4	4.0 x 10^4	6.0 x 10^2
Red salt	9.0	8.6 x 10^3	9.0 x 10^3	7.0 x 10^3	1.5 x 10^3
Red water	10.0	4.0 x 10^3	5.5 x 10^3	2.0 x 10^3	1.0 x 10^3

There are alkaliphiles reported to grow at high temperatures under aerobic (Kitada *et al.*, 1987) and anaerobic conditions (Engle *et al.*, 1995). Enzymes from these thermophilic alkaliphiles are both alkalostable and thermostable and are used for protein engineering and production of thermostable enzymes from aerobes, e.g. Thermostable xylanase (Dimitrov *et al.*, 1997; Okazaki *et al.*, 1985), Protease (Fujiwara *et al.*, 1993; Takami *et al.*, 1989; Tsuchiya *et al.*, 1991), Lipase (Wang *et al.*, 1995) and Chitinase (Bhushan and Hoondal, 1998).

Anaerobic alkalithermophiles have been isolated from alkaline hot spring in Yellowstone National Park in North America (Li *et al.*, 1994; Podkovyrov and Zeikus, 1992; Wiegel, 1998), shallow marine hydrothermal system at Vulcano island, Italy etc. Stetter's group isolated *Thermococcus alcaliphilus* sp. Nov. a new hyperthermophilic archaeon growing on polysulfide at alkaline pH and at temperatures between 56 and 90°C (Keller *et al.*, 1995).

Interestingly, Kimura and Horikoshi have isolated several strains of bacteria which can grow at high pH and temperatures as low as 0°C (Kimura and Horokoshi, 1988, 1989). *Micrococcus* strain 207, which is an aerobic coccus, 0.8 - 1.2 micron in diameter can grow at a temperature of -5 to -39°C at pH 8.5.

The existence of active microbial population in deep subsurface and deep sea has been recognised for several decades (Kato *et al.*, 1996; Myers and Anderson, 1992). Alkaliphiles and barophilic bacteria have also been isolated from such deep sea sediments colected from depth up to 10,898 m of Mariana trench (Georganta *et al.*, 1993; Takami *et al.*, 1997).

(B) Haloalkaliphiles

These organisms, isolated from alkaline and highly saline environments such as alkaline soda lakes like Wadi Natrum in Egypt, lake Magadi in Kenya (Tindall *et al.*, 1980) and Sambhar lake, India (Borkar and Bhosle, 2000) also require high salinity which is achieved by adding 20% NaCl to the isolation medium (Table 6). *Natronobacterium* and *Natronococcum* are two well recognized genera of haloalkaliphilic bacteria growing only at pH values above 8.0-9.0 and extremely high salt concentration (Table 5) (Krulwich and Guffanti, 1989). These red pigmented microbes reach number of 10^2 to 10^5/ml in soda lake brine. Table 8 depicts the counts of haloalkaliphiles from Sambhar lake-India. One of the isolates has been identified to belong to genus *Natronococcus* from rRNA analysis. A novel haloalkalophilic archaeon isolated from Lake Magadi, contains large gas vaculoes in stationary phase of growth forming bright pink colonies and is designated as *Natronococcus occultus* (Studdert *et al.*, 1997). Xu *et al.*, 1999 isolated two haloalkalophilic archea from soda lake in Tibet belonging to a new genus *Natronorubrum*. A number of anaerobic halophilic eubacteria have been isolated from hypersaline alkaline environments such as Big Sod Lake. Aqua de Ney spring etc. (Nimura *et al.*, 1987). A novel osmolyte. 2 sulfotrehalose was discovered in several *Natronobacterium* species of haloalkalophilic archaea (Desmarais *et al.*, 1997). The ecology, physiology and taxonomy of haloalkaliphiles has been recently reviewed by Jones *et al.*, 1998 and Duckworth 1996.

5. MOLECULAR AND BIOCHEMICAL ADAPTATIONS IN ALKALIPHILES

Organisms that grow at extreme pH values are faced with a central problem of pH homeostasis. The mechanisms by which the organisms solve and develop survival strategies have been extensively studied and investigated during the past decades. Yet, the understanding of any of the specific adaptation in all extreme alkaliphiles is incomplete. A number of adaptation and regulatory mechanisms have been elucidated in aerobic, alkaliphilic, *Bacillus* sp. which can be characterized as passive and active mechanisms.

(A) Passive mechanisms

(i) External pH change during cell growth

Alkaliphilic bacteria can change external pH values to a pH suitable for growth and create their own environment e.g. *Bacillus* sp. No 221 grows slowly at neutral pH changing the pH of culture broth to 9.0, further showing rapid growth and producing a large amount of alkaline protease (Horikoshi, 1991). Ueyma and Horkoshi (1981) isolated alkalophilic *Arthrobacter* sp. that was also capable of changing the broth pH to an optimum.

It is not only alkalophilic bacteria, but neutrophilic microorganisms can also change their environment pH values to optimum for growth (Johnson, 1923; Walksman and Jaffe, 1922). This phenomena is very interesting from the ecological point of view and may be the reason why alkaliphiles can create a microcosm and live in non-alkaline and acidic soils.

(ii) Alkaliphily and Buffering capacity

Studies done on buffering capacity of Bacilli that grow at different pH range (Krulwich *et al.*, 1985) reveal that alkalophilic bacteria have dramatically high values of cytoplasmic buffering capacity (Bi) at high external pH values than any of the non-alkalophiles tested. The Bi of *Bacillus firmus* RAB at pH 9.5 which is 1100 nmol of H^+ / pH unit/mgm protein was the highest value recorded in the study. Further, it has been observed that both the external and internal buffering capacities of extremely alkalophilic bacilli are high under alkaline conditions. It is postulated that this property is conferred by high levels of basic proteins or polyamines in alkaliphilic cells. These basic molecules may further be present in specific locations for the protection of a particular molecule, (e.g. Nucleic acids) or processes (e.g. Protein synthesis) (Horikoshi, 1991). The high polyamine content of alkaliphilic cell envelop has been confirmed by Yonezawa and Horikoshi (1978).

Hamana *et al.* (1989) further detected the presence of 3 Polyamines - putrescine, spermidine and spermine. Spermidine was the major polyamine in alkaliphilic *Bacilli* and no spermine was detected. Thus, the physiological function may be, that the cell membrane of alkaliphiles is stabilised by spermidine, which is more basic compound than spermine. The elevation of cytoplasmic buffering capacity at highly alkaline pH therefore has an important role in alkaliphily. A similar mechanism of pH regulation has also been observed in acidophiles (Guffanti and E. Hou, 1987; Zuchlinsky and Matin, 1983).

(iii) Cellular components

(a) Cell wall components : Since the protoplasts of alkaliphilic *Bacillus* strains lose their stability in alkaline environments, it has been suggested that the cell wall components may play a key role in protecting the cell from alkaline environments (Lewis *et al.*, 1982). Components of the cell walls of several alkaliphilic *Bacillus* sp. have been investigated in comparison with those of neutrophilic *Bacillus subtilis* (Aono and Horikoshi, 1983; 1984; Aono, 1985; 1989; Aono *et al.*, 1995).

All alkaliphilic *Bacillus* sp. examined so far have A-1 gamma peptidoglycan which is found in majority of neutrophilic *Bacillus* strains. However, the cell walls of *Bacillus lentus* C125 grown at pH 10.0 are thicker than those of the organisms grown at pH

Table 9. Composition of Cell Walls of Alkalophilic *Bacillus* strains (Horikoshi, 1991)

Gp.	Strain	Culture pH	Asp	Glu	L-Glu	Gly	Ala	DAP	Mur	GlcN	GalN	P	Glyc	Neutral sugars	Glc	Gal	Uronic acid
1	A-40	10	0.43				0.67	0.42	0.36	0.67	0.23	0.04	0.43	0.08			0.90
	2B-2	10	0.46				0.58	0.35	0.34	0.65	0.27	0.17	0.46	0.18	0.05		0.81
2	C-11	10		2.32	1.05		0.52	0.33	0.22	0.26	0.04	0.07		0.08			0.60
	C-125	7		1.75	1.08		0.88	0.52	0.48	0.46	0.04	0.03		0.18			0.46
		10		2.14	1.80		0.64	0.38	0.28	0.30	0.03	0.05		0.05			0.90
	Y-25	7		1.55	0.59		0.90	0.47	0.38	0.44	0.04	0.03		0.04			0.36
		10		1.97	0.86		0.54	0.34	0.29	0.32	0.03	0.05		0.06			0.79
	A-59	7	0.55	1.71	0.52		0.93	0.56	0.39	0.42	0.05	0.05		0.04			0.44
		10	0.71	1.82	0.70		0.61	0.38	0.27	0.30	0.04	0.06		0.07			0.62
	C-3	7	0.40	1.20	0.33	0.32	0.94	0.53	0.51	0.51	0.04	0.03		0.10	0.04		0.27
		10	0.72	1.50	0.54	0.52	0.80	0.39	0.37	0.37	0.06	0.06		0.22	0.08		0.49
3	C-59-2	10	0.21	0.26			0.35	0.24	0.18	0.45	0.05	1.65	0.56	0.96	0.45	0.49	0.11
	M-29	7		0.31			0.57	0.30	0.21	0.80	0.05	1.76	0.12	1.85	1.20	0.14	0.13
		10		0.33			0.59	0.31	0.30	1.01	0.05	1.70	0.13	1.73	1.21	0.11	0.12
	57-1	7		0.29			0.44	0.29	0.19	0.22	0.03	1.53	0.9	0.78		1.05	0.08
		10	0.01	0.30			0.45	0.30	0.21	0.25	0.05	1.82	1.23	0.67		1.08	0.08
	B. subtilis GSY 1026	7	0.01	0.54			0.80	0.49	0.48	0.50	0.19	1.77	0.83	1.34	1.02		0.11

7.2 (Aono *et al.*, 1995). Further, their composition was characterized by an excess of hexosamines and aminoacids in the cell walls with glucosamine, muramic acid, D- and L- alanine, D-glutamic acid, meso-diamino pimelic acid and acetic acid obtained in the hydrolysates. In addition to the peptidoglycan, alkaliphilic *Bacillus* sp. grown at pH 10.0 contain certain acidic polymers such as galacturonic acid, glucuronic acid, glutamic acid, aspartic acid and phosphoric acid which are more negatively charged than in cells grown at pH 8.0 (Table 9) (Horikoshi, 1991). The negative charges on the acidic non-peptidoglycan components may give the cell surface its ability to adsorb sodium and hydronium ions and repulse hydroxide ions and as a conseq- uence may assist cells to grow in alkaline conditions/environments/ econiches.

Aono *et al.*, have demonstrated the presence of a teichuronopeptide composed of glucuronic acid, galacturonic acid and fucosamine in the cell walls of alkalophilic *Bacillus* strain C-125 in a ratio of 1:1:1 and concluded that the acidic components in the outermost layer play an important role in supporting growth at alkaline pH 7.8.

(b) Cell membrane components : The surface of the plasma membrane in alkaliphiles must presumably be kept below pH 9.0 because the plasma membrane is very unstable at alkaline pH values. It may be the features in lipid and protein components, which may contribute to the alteration of the membrane properties of alkaliphilic bacteria.

Lipid

The cytoplasmic membrane, which is the actual pH barrier to keep the intracellular pH lower than the extra cellular pH in all extreme alkalophiles, shows differences in lipids. The first study on lipids in cell membranes of alkaliphilic bacteria was conducted by Koga *et al.*, 1982 followed by Nishihara *et al.*, 1982. The analytical studies on the lipids of alkaliphilic and neutrophilic bacteria bring out the following differences :

(a) Glycolipid and Phosphoglycolipids which commonly occur in Gram positive bacteria do not occur in alkaliphiles. Instead, BMP or bis (monoacylglycero) phosphate, a novel lipid which does not occur in neutrophiles is found in alkaliphiles (Nishihara *et al.*, 1982).

(b) Alkaliphiles have higher concentration of membrane lipid : protein ratio with a high cardiolipin content and fatty acid composition consistent with a very fluid membrane (Krulwich and Guffanti 1989). It is possible, that phase changes of membrane lipids and presence of polar isoprenoid lipids play a role in domain formation related to energy coupling and their properties help to maximize diffusion-based collsions.

(c) Phosphatidyl ethanol amine and Phosphatidyl glycerol are major phospholipids and have squalene and squalene derivatives C40 and C50 isoprenoids among neutral lipids. Studies (Clejan *et al.*, 1986; Dunkley *et al.*, 1991) indicate that fatty acid composition of obligate alkaliphile membrane is a factor precluding growth at low pH values. Specifically, the combination of high concentration of branched chain fatty acids and substantial fraction of unsaturated fatty acids may make the membrane leaky at near neutral pH, although it functions effectively at high pH. Facultative alkaliphiles lack desaturase activity and loose their ability to grow at near neutral pH when supplemented with an unsaturated fatty acid (Dunkley *et al.*, 1991).

Proteins

Alkalophilic bacteria also acquire alkaliphily by changing the protein composition of the membrane. They have more acidic amino acid composition of proteins and reduced basic amino acid content that are secreted into the external environment. The flagellin from *Bacillus firmus* RAB is acidic relative to its neutrophilic bacterial counterpart (Krulwich and Guffanti, 1989) and a cytochrome C that may have an exposed localisation is an acidic protein. Results of a study by Koyama *et al.* (1976b) indicaete that there is difference in patterns of membrane proteins as visualised after gel electrophroesis in facultative alkaliphile grown at neutral Vs alkaline pH. This was confirmed by Krulwich and Guffanti (1986) when studies of facultatively alkaliphilic *Bacillus* strains showed a marked difference in the polypeptide pattern in molecular weight of range 95,000-115,000 and 24,000-35,000. Studies on variants of *B. firmus* RAB, which grows well at high pH at lower Na^+ concentration than parent strain, shows an amplification of a 95 KD membrane protein. Similarly, the variants showed a marked enhancement in a 90,000 and 45,000 mol. wt membrane polypeptide. The findings suggest a genetic mechanism that allows the magnification of a certain gene or a group of genes that are related to extreme alkalophily, in view of existence of variants with different stabilities (Krulwich *et al.*, 1986).

Cytochrome content

One of the salient features of alkaliphiles is their high concentration of respiratory chain components (Krulwich *et al.*, 1988) which is believed to be crucial to life at high pH. This high concentration of respiratory chain components may provide the mechanism by which the alkaliphiles maximise productive proton tranferring collisions between respiratory chain components and the ATPase. Lewis *et al.* (1982) reported that the membranes of alkaliphilic *Bacilli* contained high cytochrome heme content while those of non alkalophilic derivates contained much less heme. The cytochrome centres of membrane residue from facultatively alkalophilic *Bacillus* strain OF4 was significantly higher at pH 10.5 than when cells grown at pH 7.5 (Guffanti *et al.*, 1986). Differences were specifically pronounced for cytochrome a and c; confirmed by spectral data. Studies with variants of *Bacillus alcalophilus* indicate that growth at pH 8.5 increases membrane cytochrome contents top 2 nmol from 1.27 nmole/mg of membrane protein at neutral pH. Thus, imposed changes in external pH can have an impact on cytochrome content (Lewis, *et al.*, 1982). It is suggested that a logical candidate for growth limitation at pH 10.5 may be a siderophore precursor (GasCoyne *et al.,* 1991). Cultures with high cytochrome contents compete better in batch culture molar growth yield studies at pH 10.5 whereas difference is less pronounced at pH 9.5 (Guffanti and Hicks 1991). Cytochrome oxidase from alkalophilic *Bacillus firmus* RAB has been isolated and characterised by Kitada and Krulwich (1984) whereas Davidson *et al.*, 1988 have purified and characterized two soluble cytochromes from the same organism. Gilmour *et al.*, 1996 have purified co-reactive b-type cytochromes from *Bacillus firmus* OF4 (Gilmour and Krulwich, 1996).

(B) Active mechanisms

(i) Na^+/H^+ antiporter system and solute transport

A typical alkaliphile which grows at an alkaline pH range of 9.0 to 11.0 maintains an internal pH 1 to 2 units lower than the external pH (Horikoshi, 1991; Krulwich *et al.*, 1997). As a result of this, the pH gradient is reversed (acid in) with respect to a chemiosmotic driving force. This mechanisms is found to be present in obligate as well as facultative alkaliphiles but is absent in alkali tolerant bacteria which cannot grow at pH values above 9.0. *Bacillus firmus* OF4 and other alkaliphilic *Bacillus* sp. have been shown to maintain a pH gradient of more than 2 full pH units during optimal logarithmic growth at alkaline pH values (Guffanti *et al.*, 1980). Neutrophilic prolaryotes do not generate a pH gradient of this magnitude during the growth or when challenged by sudden alkaline pH shift in the medium (Horikoshi, 1999).

Microbial cytoplasmic pH can be measured by assaying the distribution of weak bases across the cell membrane (Horikoshi, 1991, Sturr *et al.*, 1994). This method is based on the ability of the weak base such as methylamine to penetrate the membrane in its neutral form and pH is calculated by assuming that non charged amine concentration is equal on both sides due to permeability of this form. When the intracellular pH is expected to be lower than the external pH; the amine/base would be more concentrated in cells. Guffanti *et al.* (1978) used C^{14} methylamine to determine intracellular pH values in *B. alcalophilus* by flow dialysis and found that internal pH remained at 9.0-9.5 over a range of external pH values from 9.0 to 11.5. Table 10 shows the internal pH values in various microorganisms including acidophiles, neutrophiles and alkaliphiles at different external pH values (Horikoshi, 1991, 1999). Cytoplasmic pH has also been measured using the pH sensitive fluorescent probe 2, 7 bis (2 carboxyethyl) (5 and 6) carboxy fluorescein (BCECF) (Aono *et al.*, 1997; Breeuwer *et al.*, 1996; Bunthof *et al.*, 1999; Chitarra *et al.*, 2000; Collins *et al.*, 1983). Other indirect methods of pH measurements include 1. pH measurement of broke cell lysate or fluid from cells, 2. Use of visible indicator such as bromothymol blue, 3. High-resolution 31P nuclear magnetic resonance measurements (Horikoshi, 1991).

Internal cytoplasmic pH can also be

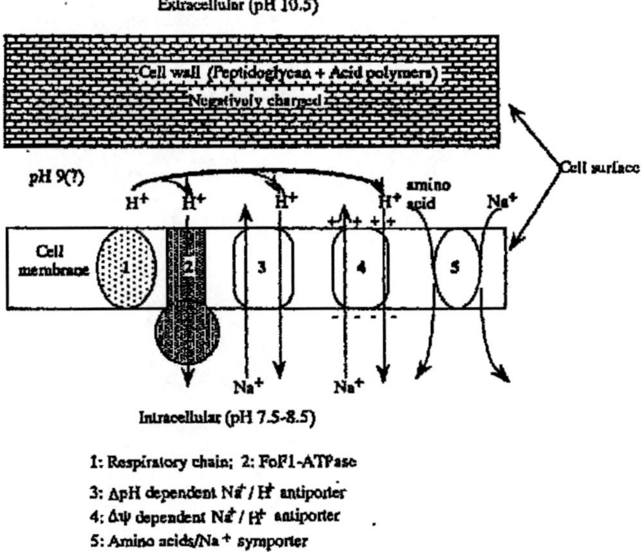

Fig. 3. Schematic representation of cytoplasmic pH regulation.

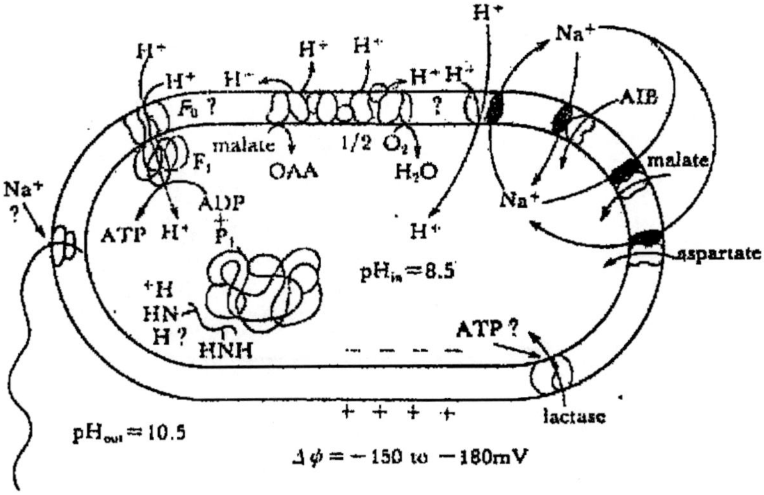

Fig. 4. Schematic representation of an alkalophile cell. Features of the model include : typical cytoplasmic pH and values during growth on malate at pH 10.5 : a proton - pumping respiratory chain; a Na^+/H^+ antiporter, catalyzing an electrogenic exchange of $H+_{out}$ (perhaps utilizing some protons that are transported along a localized pathway ?) for Na+in; Na+ solute symporters that complete the Na+ cycle; an F_1 F_0 = ATPase involved in ATP synthesis and, perhaps, utilizing some localized proton pathway; some basic molecules shown as protecting the chromosome, representative of unidentified, basic buffering components of the cytoplasm; an involvement of Na^+ in flagellar rotation; and an ATP-dependent lactose transport system. The Na^+ translocating porters are shown with a defined Na^+ site; that site is probably not, according to the most recent data; a distinct subunit. Some regulatory or common synthetic step may link the Na^+ translocating porters, or a pleiotropic nonalkalophilic phenotype may be produced by mutational generation of a Na^+ leak. (Horikoshi, 1991)

Table 10. Intracellular pH values in various microorganisms at different external pH values (Horikoshi, 1991, 1999).

Microorganisms	Optimum pH for growth	Internal pH	External pH	Method
A. ACIDOPHILIC MICROORGANISMS				
Thermoplasma acidophila	2 (59)°C	6.4-6.9	2,4,6	Cell lysis
		6.5	2,4,6	DMO
Bacillus acidocladarious	4	5.5 (55°C)	3	Flurescein diacetate
		6.5 (15°C)	3	DMO
Sulfolobus acidocladarious	3(70°C)	6.3		Cell lysis
Bacillus acidocladarious	3-4	6.15	2.0	
		5.85	3.0	Aspirin method
		5.89	4.0	
		6.31	4.5	
B. NEUTROPHILIC MICROORGANISMS				
Streptococcus faecalis		7.2	6.0	
		7.7	7.0	DMO
		8.0-8.2	7.5	
Escherichia coli MRE 600		7.55	6.0-8.0	^{31}P nuclear magnetic resonance
Bacillus megaterium		8.03	7.4	DMO
Escherichia coli ML 308-225		7.8,	6.3,	
		7.9,	6.8,	DMO
		8.4,	7.8,	
		8.8	8.7	
Escherichia coli K12		8.0,	6.0,	
		7.7,	7.0,	DMO
		7.7,	8.0,	
		8.5	9.0	
C. ALKALOPHILIC MICROORGANISMS				
Bacillus alcalophilus	10.5	8.0,	8.0,	
		7.6,	9.0,	Methylamine
		8.6,	10.0,	
		9.2	11.0	
Bacillus firrmus	10.5	7.7,	7.0,	
		8.0,	9.0,	
		8.3,	10.8,	Methylamine
		8.9,	11.2,	
		9.6	11.4	
Bacillus strain YN-2000	9.5	8.5,	7.5,	
		7.9,	8.5,	Methylamine
		8.1,	9.5,	
		8.4	10.2	
Bacillus halodurans C-125 infact cells	10.5	7.3,	7.0,	
		7.4,	7.5,	
		7.6,	8.0,	
		7.8,	8.5,	Methylamine
		7.9,	9.0,	
		8.1,	9.5,	
		8.2,	10.0,	
		8.4	10.5	
Protoplasts		7.5,	7.0,	
		7.9,	8.0,	
		8.2,	8.5,	Methylamine
		8.4,	9.0,	
		8.6	9.3	

estimated from the optimal pH of intracellular enzymes. For example alpha-galactosidase from an alkaliphilic *Micrococcus* sp. Strain 31-2 had its optimal catalytic pH at 7.5 suggesting that the internal pH is around neutral. Furthermore, the cell free protein synthesis systems from alkaliphiles optimally incorporate amino acids into proteins at pH 8.2 to 8.5, only 0.5 pH unit higher than that of the neutrophilic *Bacillus subtilis* (Horikoshi, 1999).

In accordance with chemiosmotic hypothesis of Mitchell, cells of alkaliphiles like other aerobic bacteria and mitochondria, extrude protons during respiration which is known to acidify a narrow region near the membrane. In alkaliphiles, however, primary proton extrusion is followed by a Na^+ dependent proton accumulation resulting in a proton gradient (pH) H^+ in >H^+ out and a Na^+ gradient Na^+ in >Na^+ out. This function is performed by Na^+/H^+ antiporter which exchanges internal Na^+ for external H^+, both the ions transported against their concentration gradients at high pH (Garcia *et al.*, 1983; Grant and Horikoshi, 1989; Hamamoto *et al.*, 1994, Ito *et al.*, 1997; Kitada and Horikoshi, 1977; Kitada *et al.*, 1982; Kitada *et al.*, 1989; Kitada *et al.*, 1994; Kitada *et al.*, 1997; Sturr *et al.*, 1994). This can occur because the antiporter is electrogenic and can admit more than one proton per Na^+ extruded. Exchange is therefore driven by the electrical potential across the membrane (Delta psi) rather than the proton gradient (Delta mu) H^+. The electrogenic antiporter thereby allows the bacteria to simultaneously acidify the cytosol and build sodium gradient (Delta mu) Na^+ to power Na^+ coupled transport processes (Grant and Horikoshi, 1989; Speelmans *et al.*, 1993). Non alkaliphilic mutants devoid of Na^+/H^+ antiporter loose the capacity to translocate solutes in the presence of Na^+ (Guffanti *et al.*, 1978), [Figures 3 (Horikoshi, 1999) and 4 (Horikoshi, 1991)]. Most alkaliphiles bacteria therefore require the presence of Na^+ in the growth medium and suspension of such a species in the absence of Na^+ results in the immediate loss of pH homeostasis. The first finding of Na^+ requirement was described by Horikoshi and Akiba (1982). The crucial involvement of Na^+/H^+ antiporter is supported by isolation of non-alkalophilic mutant strains that no longer grow well above pH 9.0 and have lost Na^+/H^+ antiporter activity. Na^+/H^+ antiporter has also been detected in alkaliphilic *Cyanobacterium synechocystis* (Buck and Smith, 1995). Properties of Na^+/H^+ antiporter include; (a) use of Na^+ or Li^+, (b) Linear dependance on trans- membrane electrical gradient and (c) Inhibition by a internal proton concentra- tion (Krulwich and Guffanti, 1989). The assays used have measured Na^+ dependent proton movement, pH induced Na^+ movements and Na^+ flux induced either by initiation of respiration or by imposition of Valinomycin mediated K^+ diffusion potential in cells, vesicles and proteolipo- somes (Guffanti, 1986; Garcia *et al.*, 1983).

Evidence for Na^+ coupled solute symport as a physiologically important route for Na^+ recentry for alkaliphilic pH homeostasis has been proved from the finding, that presence of alpha - amino isobutyric acid enhances the ability of cells to maintain a constantly acidified pH in (Guffanti *et al.*, 1978).

(ii) ATP synthesis in alkaliphiles

The pH gradient of alkaliphiles is reversed with respect to chemiosmotic driving force and is partially counter balanced by a substantial transmembrane electrical gradient positive outside but the overall proton motive force is only about -50 mV at pH 10.5. This represents a serious bioenergetic challenge if a proton coupled

ATPase is used for ATP synthesis. Alkalitolerant marine *Vibrio* sp. appear to harness a Na^+ motive force at high pH, which functionally is analogus to a H^+ motive force in neutrophiles. Skulachev and colleagues have shown that some of these organisms have a primary Na^+ pump coupled to respiration and probably a Na^+ linked ATPase (Grant and Horikoshi, 1989). The evidence for Na^+/ATPase is indirect and based on inhibition studies. Cells grown in the presence of inhibitors such as carboxy cyanide M- chlorophenyl Hydrazone (CCCP) which inhibits (Delta mu) H^+ still make ATP provided [Na^+] out [Na^+] in. Furthermore, ATP synthesis is abolished by an artificially produced reverse Delta pNa (Na^+) out (Na^+)in] under conditions where, Delta psi generation is unchanged. However, Krulwich and co-workers have repeatedly failed to find any evidence for Na^+ motive force in a range of alkaliphilic *Bacillus* sp. Instead, they have demonstrated that extreme, non-marine alkaliphiles possess a respiratory chain that pumps only protons (Krulwich and Guffanti, 1989; Krulwich *et al.*, 1997; Krulwich *et al.*, 1998) and an F_o-F_t- ATPase that is localised in cytoplasmic membrane using protons exclusively (Hicks and Krulwich, 1986). Compounding bioenergetic problems are early findings, that molar growth yields on malate for two obligate alkaliphiles are not usually low (Guffanti and Hicks, 1991) and that phosphorylation potential (Delta Gp) are also in conventional range in experiments in which starved cells or appropriate membrane preparations were energized in buffer (Guffanti *et al.*, 1980).

The molar growth yields of obligate and facultative alkaphiles on molate at pH 10.5 are comparable with those of neutrophiles (22-35 mg dry weight/nmol malate). *Bacillus firmus* OF4 gives the same molar growth yields on malate at pH 7.5 as at pH 10.5, a yield comparable to *B. subtilis* at pH 7.0. In competition experiments, the strict alkaliphiles *Bacillus alcaliphilus* and *Bacillus firmus* OF4 grow better than their facultative relatives at very alkaline pH (Dunkley *et al.*, 1991).

Results of molar growth yields and bioenergetic parameters indicate, that alkaliphiles are not simply oxidizing large amount of substrate in order to grow at very high pH values and low proton motive force, but are using substrates as efficiently as comaprable to aerobes at conventional pH values. Delta GP : ATP, ADP and Pi concentrations are not unusual.

Inspite of low electrochemical proton gradient at pH 10.5, which is a result of large reversed Delta pH (pH gradient) which detracts from the substantial transmembrane electrical gradient; the energy production from malate via respiratory chain through ATAase is at least as effective as at neutral pH. At near neutral pH, protons from bulk external environment might pass through the ATPase and energise ATP synthesis in a conventional chemiosmotic manner whereas at alkaline pH. Protons may be prevented from moving through F_o to and from the bulk. Protons from respiratory chain may be directly transferred to intramembrane residues in the ATPase (Guffanti and Hicks, 1991).

(iii) Requirement of an electrochemical sodium gradient for motility

The motility of extreme alkalophiles and moderately alkaline tolerant bacteria depends upon the presence of Na^+ and is energized by an electrochemical gradient of Na^+. The motility of alkalophilic bacilli and the swimming speed increases in a linear fashion with logarithmic increase in Na^+ concentration up to 100 mM Na^+. Amiloride, which is an inhibitor of various Na^+ fluxes and of electroneutral N^+/H^+ antiporter of eukaryotes, inhibits the motility of alkalophilic *Bacillus* strain (Horikoshi, 1991).

Flagellar movement of neutrophilic bacterias caused by H^+ driven motor and not by ATP. Hirota *et al.* (1981) however, reported on flagellar motors of alkalophilic *Bacillus*, powered by an electrochemical potential gradient of sodium ions, because the proton motive force is too low to driven flagellar motors in an alkaline environment. Hirota and Imae (1983) concluded that sodium motive force is an energy source for the flagellar motors.

(iv) Requirement of an electrochemical Na^+ gradient for sporulation and germination

Numerous environmental factors affecting the sporulation of *Bacillus* sp. have been reported by many investigators. Kudo and Horikoshi (1979) showed that alkaliphilic *Bacillus* sp. No. 2b-2, isolated from soil yielded spores in alkaline Schaffer medium. Chitarra *et al.* (2000) examined the effect of alkaline conditions and the development of the vegetative form from the spore and reported that the spore germination and outgrowth took much more rapidly under alkaline conditions (pH 9.5 to 10.3) in *Bacillus cereus* and *Bacillus circulans* than at pH 7.4. Kosono *et al.*, have also elucidated the role of Na^+/H^+ antiporter, Sha A for inhibitator of sporulation in *Bacillus subtilis*. (Kosono *et al.*, 2000).

6. INDUSTRIAL APPLICATIONS AND FUTURE PROSPECTS

Alkaliphilic microorganisms have made a large impact with their application and manufature of mass-market consumer products in industries. The so-called "Biological Detergents" contain enzymes obtained from alkaliphilic and alkali tolerant bacteria. The traditional craft industry of leather tanning uses a series of highly alkaline and unpleasant processes where the application of alkaline enzymes has brought significant process improvements. Alkaline enzymes have significantly been used in the food and waste treatment industries. Alkaliphilic bacterial sp. have also made a great impact in paper and textile processing for enzymatic pulping process under alkaline conditions, which help the plant tissues to swell and facilitate the degradation of pectic substances. Indigo fermentation is the first industrial application of alkaliphilic bacteria in the world where reduction of indigo from indigo leaves is broght about by alkaliphilic *Bacillus* sp.

6.1 Alkaline enzymes

Studies of alkaliphiles have led to the discovery of enzymes stable at high pH with numerous application. Table 11 (Horikoshi, 1991) depicts the examples of enzymes produced by alkaliphiles. Some of these enzymes have been produced on an industrial scale.

(i) Protease

These are the most commonly used enzymes in detergent formulation and have the largest market segment obtained from alkaliphilic microorganisms (Horikoshi, 1971; Nomoto *et al.*, 1984; Takami *et al.*, 1989). Serine proteinase produced from alkaliphilic *Bacillus* spp. are endopeptidase with a reactive serine molecule at the active site. Commercial products include Alcalse^R and Esperase^R (Novo industries) and Maxatase^R Maxacal^R (International Biosynthetics - IBIS). Maxatase^R is an alkaline protease active between pH 7.0 and 11.0 with an optimum activity at pH 9.5 to 10.0, the pH of many detergents. In their encapsulated form, they are added as 0.4-0.6% to detergents. Their use is to hydrolyze proteins, and remove proteinaceous strains as blood, egg, grass and those from body secretions. Besides lifting proteinaceous soil, proteinase ensure that coagulated protein is not redeposited on the fabric during the wash, which gives a grey unclear appearance to whites (Horikoshi, 1991; Horikoshi, 1996).

Table 11. Enzymes Produced by Alkaliphiles (Horikoshi, 1991)

Enzymes	Optimum pH	Stability pH	MW x 10^4	Enzymes	Optimum pH	Stability pH	MW x 10^4
Alkaline protease				B-1,3-glucanase			
No. 221	11.5-12.0	4.0-11.0	3	No. K-12.5	5.5-9.0	6.0-8.0	4
No. 8-1	10.5-11.0	6.0-9.0	3	No. 221	8.5	5.0-9.0	3.6
No. D-6	10.5-11.0	4.0-12.0	2-3	Xylanase			
Alkaline amylase				No. C-59-2	5.5-9.0	5.0-9.0	3.58
No. A-40-2	10.5	7.0-9.0	7	No. 212	5.7	4.0-9.0	2.3
No. III67	10-11	6.0-11.0	6		6.8		3.7
			7.3		7.8		1.45
			8		6.0-7.0	5.0-11.0	1.6
				No. C-125	6.0-10.0	4.0-12.0	4.3
No. 17-1	4.5-10.0	6.0-10.0	5-6				
No. 38-2	4.5	6.0-10.0	8.8	A-Galactosidase			
	7.0	6.0-9.0	8.5	No. 31-2	7.7	7.5-8.0	—
	8.0-9.0	—	8.5	B-Galactosidase			
No. 313	8.0	6.0-8.0	6.5	No. 31-2	6.5	5.5-9.0	18.5
Alkaline pectinase				Pencillinase			
No. P-4-N	10.0	5.0-9.0	6.-7	No. 170	6.0-7.0	7.0-10.0	2-3
Alkaline pullulanase				Maltose dehydrogenase			
No. 202-1	9.0	6.0-10.0	9.2	No. 93-1	10.2	6.0-10.0	3.9
					9.8	7.0-8.0	4.8
Alkaline cellulase							
No. N-4	6.0-11.0	5.0-11.0	4-8	Gluose dehydrogenase			
No. 212	6.0-8.0	5.0-11.0	5	No. 93-1	9.8	6.0-8.0	5.1
No. 1139	9.0	5.0-11.0	9.2	Uricase			
Alkaline alginase				No. H-3	9.0	10.0	10.0
No. M-2	9.0	8.0-10.0	4	Polyamine oxidase			
Alkaline catalase							
No. Ku-1	10.0	7.0-9.0	12.58	No. PO-1	4.0	3.0-6.0	6.4 x 2
					6.0		
Alkaline RNase				B-Mannanase			
No. 243	9.0	6.0-10.0	1.2	No. AM001	9.0	8.0-9.0	5.85
					9.0	8.0-9.0	5.95
Alkaline DNase							
No. M-29	9.0	6.0-10.0	4.0		8.5	8.0-9.0	4.02
Restriction enzyme				B-Mannosidase			
No. 170	7.5	—	—	No. AM001	6.0	6.0-5.0	9.4

Alkaline proteases have also been used in the hide dehairing process, where dehairing is carried out at pH values between 8.0 and 10.0 (Horikoshi, 1999). In the traditional process, the hides were placed in a bath containing calcium hydroxide and sodium sulfide at pH around 12.0. This process had the disadvantages of being detrimental to the hairs (which sometimes have commercial value), causing swelling of the skins resulting in difficulties in processing and causing unpleasant effluent problems. The application of bacterial enzymes in the manufacture of leather has now resulted in a quicker and more reliable process avoiding the use of sulfide and giving a higher quality product. Salt cured hides are soaked in alkaline liquor to swell the skins. The uptake of water and cleaning of the hides is improved by the addition of detergent proteinases such as Alcalase^R (Novo) and Milezyme^R (Miles Laboratories) and Reverdase^R (IBIS) (Grant and

Horikoshi, 1989). The dehairing process relies on dissolving the proteinaceous material binding the hair (keratin) in the hair follicle without damaging the fibres (collagen) of the skin and in some cases without damaging the hair (wool), which is also a valuable product.

The bating process, which is usually performed at pH 7.0-9.5, uses proteases such as Batinese[R] (IBIS) used to modify the matrix proteins, elastin and keratin making the leather supple or rigid depending on the desired character of the finished product.

An interesting application of alkaline protease was developed by Fujiwara and Coworkers (Fujiwara *et al.*, 1991) who reported the use of an alkaline protease to decompose the gelatinous coating of X-ray films from which silver recovered. Protease used has a higher optimum pH of 13.0 and temperature of 85°C. Thermostable alkaline proteases have also been purified by Fujiwara *et al.*, (1993, 1991) and Tsuchiya *et al.*, (1991).

Table 12. Industrial Applications of Cyclodextrin (Horikoshi, 1991)

FUNCTIONS	GUESTS	END PRODUCTS
FOODS		
1. Emulsification	Oils and Fats	Margarine, Cake, Whipping cream, French dressing
2. Stabilization	Flavors, spices, Colors and Pigments	Horse radish paste, Mustard paste, Cakes and cookies, Pickled vegetables, sDried vegetables
3. Masking of taste and odor		Juices, Soy milk, Bone powder, Boiled rice
4. Improvement of quality		Hard candy, Cheese, Soy sauce, Canned citrus fruits and juices
5. Reduce volatility	Ethanol	Food preservatives
6. Others		Breath mints
COSMETICS AND TOILETRIES		
1. Emulsification	Oils and Fats	Face creams, Face lotions, Tooth pastes
2. Stabilization	Flavors and Fragrances	Bath refresher crystals
AGROCHEMICALS		
1. Stabilization	Pyrolnitrin Pyrethroids	Fungicide, Insecticide
2. Reduce volatility	Organic phosphates (DDVP)	Insecticide
	Thiocarbamic acid	Herbicide
3. Reduce toxicity	2-Amino 4-methyl-phosphynobutyric acid	Fungicide
PHARMACEUTICALS		
1. Improve solubility	Prostagladins, Steroids, Cardiac glycosides, Non-steroidal anti inflammatory agents, Barbiturates, Phenytoin, Sulfonamides, Sulfonylureas, Benzodiazepines.	
2. Chemical stabilization		
(a) Hydrolysis	Prostacyclin, Cardiac glycosides, Aspirin, Atropine, Procaine	
(b) Oxidation	Aldehydes, Epinephrine, Phenothiazines	
(c) Photolysis	Phenothiazines, Ubiquinones, Vitamins	
(d) Dehydration	Prostaglandn E_1, ONO-802	
3. Improve bioavailability	Aspirin, Phenytoin, Digoxine, Acetohexamide, Barbiturates, Non- steroidal anti-inflammatories	
4. Powdering	ONO-802, Clofiberate, Benzaldehyde, Nitroglycerine, Vitamin K_1, K_2, methylsalicylate	
5. Reduce Volatility	Iodine, Naphthalene, *d*-Camphor, *l*-Menthol, Methylcinnamate	
6. Improve taste, smell	Prostaglandins, Alkylparabens	
7. Reduce irritation to stomach	Non-steroidal anti-inflammatory agents	
8. Reduce hemolysis	Phenothiazines, Flufenamic acid, Benzylalcohol, Antibiotics	

(ii) Amylases

These enzymes have industrial and pharmaceutical uses, particularly as detergent additives at high pH (Igarashi *et al.*, 1998; Kobayashi *et al.*, 1994; McTigue *et al.,* 1994). Amylases remove starch-based stains often working synergistically with proteinases to dissolve protein-starch combinations in food stains. Products include Termanyl[R] (Novo) and Maxamyl[R] (IBIS), which are effective up to 100°C and exhibit a good performance up to pH 10.0. Rapidase[R] is used in desizing and softening of denims (Horikoshi, 1999).

The first alkaline amylase was produced in Horikoshi II medium by cultivating alkaliphilic *Bacillus* sp. A-40-2 (Horikoshi, 1971). Boyer and Ingle reported an alkaline amylase in strain NRRL B-3881 that was the second report of an alkaline amylase (Boyer *et al.*, 1973).

Interestingly, there are reports of debranching enzymes from alkaliphiles such as Pullulanase discovered by Nakamura *et al.*, (1975) from an alkalophilic bacteria *Bacillus* sp. No. 202-1 and Alpha- Glucosidase as reported by Kelly *et al.*, (1983).

(iii) Cycloaltodextrin gluconotransferases (CGTase)

These enzymes catalyse the degradation of starch to form cyclodextrins, composed of six to eight glucose units linked by an alpha-1, 4 bond. The corresponding cyclodextrins formed are termed alpha, beta and gamma cyclodextrins. From the industrial point of view, alpha cyclodextrin is of particular interest because it is rare and can encapsulate large compound molecules. The centre of this doughnut is highly hydrophobic and is used to trap a variety of compounds, thus used to make "molecular capsules" that can wrap up fragrances or drugs packaging them for slow release.

Although cyclodextrins have been known for many years, commercial production was not economical until Horikoshi and Akiba (1982) discovered a new alkaliphilic *Bacillus* sp. No. : 38-2 producing a CGT with enhanced temperature and pH stability and high productivity compared with other microbial CGTases. Nakamura and Horikoshi further discovered many alkaliphilic *Bacillus* strains producing CGTases (Nakamura and Horikoshi, 1976d). Georganta *et al.* 1993 isolated CGTase producing psychrophilic alakliphilic bacteria from samples of deep sea bottom mud. Salva *et al.*, isolated 68 CGTase producing alkaliphilic bacteria from among 400 soil bacteria (Salva *et al.*, 1997). These enzymes have been commercially prepared and used in large quantities in foodstuffs, chemicals and pharmaceuticals (Table 12) (Horikoshi, 1991).

(iv) Cellulases

Alkaline cellulases with an optimum pH of 10.0 for activity were first reported by Horikoshi and coworkers who found two *Bacillus* sp. strains N4 and 1139 producing extra cellular alkaline CMcases (Horikoshi *et al.*, 1984). Bacillus sp. strain N-4 (ATCC 21833) was found to produce multiple CMCases active over a broad pH range of 5.0 to 10.0. This discovery led to an industrial application of cellulases as laundry detergent additives. Cellulase (Krulwich *et al.*, 1988), introduced by Genecor International (Rochester, NV) and launched as "Attack R" detergent with added cellulases, isolated from alkaliphilic bacterium is the first large scale commercial application of an extremomolecule. Ito *et al.*, 1989, 1997 and Shikata *et al.* 1990 have studied the use of alkaline cellulases in detergents.

Cellulases exhibit fabric softening and color brightening properties, besides removing soil (Horikoshi, 1991) by removing or opening up the microfibrils that appear on the surface of the cotton fabrics due to wear and repeated washing thereby restoring the original apperance and smooth fibre structure. Cellulases can also be used in waste treatment, since the undigested material is mainly cellulose.

(v) Lipases

These enzymes degrade fats into more hydrophilic fatty acids and find great applications as detergent additives as hydrophobic strains caused by cosmetics, body fats and oil based foods are more difficult to remove. Introduction of alkaline lipases as detergent additives is only recent, although their market value is as good as proteases. Extensive screening for alkaline lipase producing microorganisms from soil and water samples has been conducted by Watanabe *et al.* 1977. Wang and Shen reported purification and characterization of thermophilic and alkaphilic tributyrin esterase from *Bacillus* A-30-1 isolated from a hot spring area of yellow stone National Park (Wang *et al.*, 1995). Bhushan, 1994 found a lipase produced from an alkaliphilic *Candida* sp. by solid state fermentation. A Japanese company has produced a laundry detergent, containing a non-alkaliphilic fungus lipase, produced by recombinant DNA technology. Many alkaline tested with detergent are inhibited in the presence of either alkylbenzene sulfate or dodecyl benzene sulfonate.

(vi) Xylanases

The term xylanases denote the hydrolytic enzymes which attach the (Beta-1,4 linkages of xylan polysaccharide found in many plants, particularly in agricultural residues such as rice straw, barley straw and corn cob.

Alkaline xylanases would be interesting to study, because the enzyme can readily hydrolyze xylan, which is soluble in alkaline solution (Blanco *et al.*, 1995, Honda *et al.*, 1985; Nakamura *et al.*, 1993). The first paper of xylanases of alkaphilic bacteria was published in 1973 by Horikoshi and Atsukawa 1973b. The search for thermostable alkaline xylanases has been extensive for the biobleaching of the alkali treated wood pulp, instead of using environmentally damaging chemical processes, which involve the use of chlorine.

Many thermostable alkaline xylanases have been produced from various alkalophiles, isolated from geothermal areas (Dimitrov *et al.*, 1997). Use of these enzymes for debleaching in pulp milling industries would be a potential application in industrial scale plants.

(vii) Pectinases

These enzymes, which degrade pectic polysaccharides, are widely distributed in microorganisms and higher plants and find applications in the fruit and vegetable processing industry.

The first paper on alkaline endopoly galacturonase, produced by alkalophilic *Bacillus* sp. strain P-4-N was published in 1972 (Horikoshi, 1972) with an optimum pH of 10.0 for pectic acid. The first application of alkaline pectinase producing bacterial in the retting of Mitsumata bast was reported by Yoshihara and Kobayashi (Yoshihara and Kobayashi, 1982) who conducted that bacterial retting under alkaline condition is a potentially useful process for the production of pulp of excellent quality from non-woody pectocellulosic fibers. The pectic lyase (pH optimum 9.5) produced by the alkaliphilic *Bacillus* sp. strain GIR277 has been used in the production of a high quality, non-woody type of Japanese paper that is stronger than the paper produced by the conventional methods. Cao *et al.*, have used alkaline pectinases in degumming of ramie (Cao *et al.*, 1992).

Pectinases from alkaliphilic microorganisms have also been used for the treatment of pectic waste water from the citrus processing industry containing pectinaceous materials which are only slightly decomposed by microbes during activated sludge treatment. (Tanabe *et al.*, 1987, 1988).

(viii) Lignin degrading enzymes of alkaliphiles

Lignin is a complex polymer of phenylpropanoid subunits and is an essential component of woody tissue to which it imparts structural rigidity. Lignin is remarkably resistant to degradation to most microbes except for a few species of white rot fungi, which are able to catalyze, its oxidation to carbon dioxide, Its biodegradation is of great ecological significance, because next to cellulose, lignin is the most abundant, renewable polymer on earth. As lignin and lignin model compounds dissolve in water under alkaline conditions, Kavakami and Shimizu (1986) isolated alkalophilic microorganisms which degraded various lignin and model compounds such as vanillin, dehydrodivanillin. The organisms were identified to belong to genus *Corynebacterium* and *Acinetobacter* and possess potential application for biological pulping.

(ix) Phosphatase

The alkaline phosphatase of *E. coli* has been studied extensively and is found to be located in the periplasmic spase as a soluble enzyme. In Gram positive bacteria such as *B. subtilis* and *B. lichiniformis*, the alkaline phosphatase appears to be membrane bound. Nomoto *et al.* (1988) have isolated a strain of alkaliphilic *Bacillus* sp. OK-1 which excreted alkaline phosphatase into culture broth.

(x) Catalase

Certain strain of bacteria and yeast are known to produce catalase but are all intracellular enzymes located in the cells. From the industrial point of view, extracellular catalase, produced by microorganisms are of significance. Kurono and Horikoshi (1973) reported alkalophilic *Bacillus* strain No. KU-1 isolated from soil, which produced alkaline catalase in the alkaline medium. The great advantage of this alkaline culture method is that it eliminates the process of enzyme extraction from cells.

(xi) Chitinases

Tsujibo *et al.*, 1992 isolated chitinases from alkalophilic *Nocardiopsis albus* sub species *Prasina* OPC-131. Recently Bhushan and Hoondal isolated alkalophilic chitinase producing *Bacillus* sp. Strain BG-11 (Bhushan and Hoondal, 1998). The purified chitinase exhibited broad pH and temperature optima of 7.5 to 9.5 and 45 to 55°C respectively.

(xii) Beta-lactamase

Beta-lactamase, which cleaves the amide bond of Beta-lactam ring of penicillin or cephalosporin is used in the development of new semisynthetic antibiotics that are resistant to cleavage by the enzyme. The Beta-lactamase from alkalophilic *Bacillus* strain 670 was cloned into *E. coli* (Krulwich and Guffanti, 1989) and the sequence showed partial homology with *Bacillus cereus* Beta lactamase II.

6.2 Metabolites produced by alkaliphiles

(i) 2-phenylamine : Hamasaki *et al.*, 1993 found, that a large amount of 2-phenylamine was synthesised by the cells

of the alkalophilic *Bacillus* sp. Strain YN - 2000, secreted in the medium during cell growth.

(ii) Carotenoids of alkalophilic Bacillus strains : Aono and Horikoshi reported that alkalophilic *Bacillus* sp. Strain A-40-2; 2B-2, 8-1 and 57-1 produce yellow pigments in the cells and these are triterpenoid carotenoids (Aono and Horikoshi, 1991).

(iii) Siderophores : Gascoyne *et al.*, isolated siderophore producing alkalophilic bacterium that accumulated gallium and aluminum (Gascoyne *et al.*, 1991).

(iv) Cholic acid derivatives: Kimura *et al.*, isolated an alkalophilic bacterium *Bacillus* strain from soil that grew well in media containing cholic acid (Kimura *et al.*, 1994).

(v) Organic acids : During cultivation of alkaliphiles, the pH values of the culture media often decrease sharply due to the production of organic acids. Paavilainen *et al.*, reported the prodution of acetic acid by four alkalophilic *Bacillus* sp. (Paavilainen *et al.*, 1994).

(vi) Antibiotics and enzymes inhibitors : Since the rediscovery of alkaliphiles, many Japanese pharmaceutical companies have tried using alkaline media to isolate new microorganisms producing new antibiotics. Sato *et al.*, 1980 first reported the isolation of alkalophilic *Paecilomyces lilacinus* which produced antibacterial and antifungal compounds in alkaline medium (pH 10.5). Subsequently, *Nocardiopsis dassonvillei* tat produced phenazine antibiotics under alkaline culture conditions was isolated (Tsujibo *et al.*, 1988). Bahm *et al.* 1998 isolated a novel aldose reductase inhibitor from alkalophilic *Corynebacterium* sp. Strain YUA 25-1. Besides these antibiotics, pharmaceutical companies have screened many compounds. However, a weak point in the direct production of antibiotics by alkalophiles is their unstability under alkaline conditions. But, if we consider alkalophilic microorganisms as DNA sources for antibiotic production, it may be possible to use alkaliphiles for the production of biologically active substances.

6.3 Application of alkaliphiles in degradation of aromatic compounds

Aromatic compounds occur naturally as breakdown products of plant lignin, petroleum and man-made xenobiotics such as herbicides and pesticides. Due to the hazardous effects of aromatic hydrocarbons, it is necessary to keep a track on concentration of these substances in the environment. The prime control measure followed in recent times is the degradation of these substances by microorganisms.

Although aromatic compound degradation has been extensively studied, relatively little data is available on alkaliphilic, aromatic hydrocarbon degraders. One of the few available reports is on degradation of 4-Chlorobenzoate by a facultatively alkaliphilic *Arthrobacter* sp. strain SB8 9137). Further, since increase in pH can increase the bioavailability of chlorinated aromatics, with the resultant decrease in the toxicity, alkaliphiles appear as useful target for bioremediation of sites contaminated with chlorinated compounds. A combination of degradative traits and alkaliphilicity in the single cell would be ideal for biodegradation of aromatics under alkaline conditions.

7. FUTURE PROSPECTS

Alkaliphiles, by virtue of several unique characteristics, hold immense potential for the production of the enzymes that can be used as cleansing agents, in preparation of food and beverages, as pharmaceuticals and in production of fine chemicals. Since alkaliphilic bacterial strains are more tolerant to toxic compounds and have better bioavailability, this can improve the efficiency of biodegradation in control and removal of pollutants and in waste treatment systems (Teri, 2000). These organisms remain untapped resources and their potential should not be overlooked. Simultaneous research programs on intensive screening for recognition of novel enzymes as well as development of ideal bioprocesses and bioreactors need to be launched. The potential of gene cloning, PCR techniques and protoplast fusion will allow improved enzyme yield. The five major extreme groups of microorganisms i.e. Thermophiles, Halophiles, Acidophiles, Barophiles and Alkaliphiles have been termed by Horikoshi in 1991, as super bugs on account of their potentiality.

It has been estimated that 99% of microbes in most habitats are unidentified, unculturable and their enzymes are therefore unknown. The exploration of poorly characterized environments can be especially rewarding and extreme environments, in particular offer the robust enzymes for Protein engineering and directed evolution. Thus, it is absolutely obligatory that basic research, directed towards identification of extremophiles with unique extremozymes and bioactive products need to be continued, strengthened, and pursued with renewed scientific vigor so that the potential benefits are realized with time. Knowledge about "Life in Extreme Environments" will provide the basis to understand, not only how life originated and evolved on earth, but also how life may thrive on other planets.

REFERENCES

Aguilar, A. 1996. Extremophile research in the European union : from fundamental aspects to industrial expectations. FEMS. Microbiol. Rev. 18. 89-92.

Aono, R. and K. Horikoshi 1983. Chemical composition of cell walls of alkalophilic strains of *Bacillus* J. Gen. Microbiol. 129, 1083-1087.

Aono, R. and K. Horikoshi, 1984. Composition of peptidoglycan of Alkaliphilc *Bacillus* sp. J. Bacteriol. 157, 688-689.

Aono, R. 1985. Isolation and partial characterization of structural components of the walls of alkalophilic *Bacillus* strain, C-125. J. Gen. Microb. 131. 105-111.

Aono, R. 1989. Characterization of cell wall components of the Alkalophilic *Bacillus* strain C-125 Identification of a polymer composed of polyglutamate and polyglucoronate. J. Gen. Microb. 135. 265-271.

Aono, R. and K. Horikoshi, 1991. Carotenes produced by the alkalophilic yellow pigmented strains of *Bacillus*. Agric. Biol. Chem. 55, 2643-2645.

Aono, R., M. Ito and K. Horikoshi, 1993. Occurrence of teichuronopeptide in cell walls of group 2alkalophilic *Bacillus* sp. J. Gen. Microbiol. 139. 2739-2744.

Aono, R., M. Ito, K.N. Joblin and K. Horikoshi 1995. A high cell wall negative change is necessary for the growth of the alkalophilic *Bacillus lentus*. C-125 at elevated pH. Microbiology. 141. 2955-2964.

Aono, R., M. Ito and K. Horikoshi 1997. Measurement of cytoplasmic pH of the alkalophilic *Bacillus lentus* C-125 with a fluorescent pH probe. Microbiology. 143, 2531-2536.

Bahm Y.S., J.M. Park, D.H. Bai, S. Takase and J.H. Yu. 1998. YUA001, a novel aldose reductase inhibitor isolated from alkaliphilic *Corynebacterium* sp. YUA 25. Taxonomy, fermentation, isolation and characterization. J. Antibiotics. 51. 902-907.

Bhushan, B., N.S. Dosanjh, K. Kumar and G.S. Hoondal. 1994. Lipase production from alkalophilic yeasts sp. by solid state fermentation. Biotechnol. Lett. 16, 841-842.

Bhushan, B., and G.S. Hoondal, 1998. Isolation, purification and properties of a thermostable chitinase from an alkalophilic *Bacillus* sp. Strain BG-11. Biotechnol Lett 20, 157-159.

Blanco A., T. Vidal, J.F. Colom and F.I.J. Pastor. 1995. Purification and properties of xylanase A. from alkali tolerant *Bacillus* sp. Strain BP-23. Appl. Environ. Microbiol. 61, 4468-4470.

Borkar, S. and S. Bhosle 2000. Biodiversity of alkaliphiles in natural and man made alkaline environments of Goa and Studies on their enzyme activity. Abstracts and souvenir MICROBIOTECH 2000 A1st. Annual conference of Association of Microbiologists of India held at Birla Institute of Scientific research - Jaipur, India.

Borkar, S. and S. Bhosle, 2000. Biodiversity of alkaliphiles in natural and man made alkaline environments of Goa. Report of Research project submitted to UGC, University Grants Commission, Pune - India.

Boyer, E.W., M.B. Ingle and G.D. Mercer, 1973. *Bacillus alkalophilus* sub sp. *Halodurans* sub sp. nov. : an alkaline amylase-producing alkalophilic organisms.Int. J. Syst. Bacteriol. 23, 238-242.

Breeuwer, P., J.C. Drocourt, F.M. Rombouts and T. Abee. 1996. A novel method for continuous determination of the intracellular pH in bacteria with the internally conjugated fluorescent probe 5 (and 6)- carboxyfluoresceing Succinimidyl ester. Appl. Env. Microb. 62, 178-183.

Brock, T.D., 1986. Thermophiles : General Molecular and applied Microbiology. John Wiley and sons. New York.

Bunthof J.; S. Braak; P. Breeuwer, F.M. Romobouts and T. Abee 1999. Rapid fluorescence assessment of the viability of stressed *Lactococcus lactis* Appl. Env. Microb. 65 : 3681-3689.

Buck, D.P. and G.D. Smith. 1995. Evidence for Na^+ H^+ electrogenic antiporter in an alkalophilic *Cyanobacterium synechocystis.* FEMS Microb. Lett. 128 : 315-320.

Cao, J., L. Zheng and S. Chen. 1992. Screening of pectinase producer from alkalophilic bacteria and study on its potential application in degumming of ramie. Enzyme Micro. Technol. 14 : 1013-1016.

Chitara, L.G., P. Breeuwer, R.W. Vanden Bulk and T. Abee 2000. Rapid fluorescence assessment of intracellular pH as a viability indicator of *Clavibacter michiganensis* subsp. Michiganensis. J. App. Microbiol. 88 : 809-816.

Chislet M.E. and D.J. Kushner 1961b. Germination under alkaline conditions and transmission of alkali resistance by endospores of certain strains of *Bacillus cereus* and *Bacillus circulans.* J. Gen. Microbiol. 25, 151-156.

Clejan S., T.A. Krulwich, K.R. Mondrus and Seto D. 1986. Membrane lipid composition of obligately and facultatively alkaliphilic strains of *Bacillus* sp. J. Bacteriol. 168. 334-340.

Collins M.D., B.M. Lund, J.A.E. Farrow and K.H. Schliefer. 1983. Chemotaxonomic study of an alkaliphilic bacterium *Exiguobacterium aurantiacum* gen. nov. sp. nov. J. Gen Microbiol. 129, 2037-2042.

Cook M.G., J.B. Russel, A. Reichert and J. Wiegel. 1996. The intracellular pH of *Clostridium paradoxium,* an anaerobic alkalophilic and thermophilic bacterium. App. Env. Microbiol. 62 : 4576-4579.

Davis, C.M. 1998. Making a living at the extremes. TIBTECH. 16 : 102-104.

Davidson, M.W., K.A. Grayt, D.B. Knaff and T.A. Krulwich, 1988. Purification and characterization of two soluble cytochromes from the alkalophilic *Bacillus firmus* RAB. Biochem. Biophy. Acta. 933, 470-477.

Desmarais, D.P., E. Jabalonski, N.S. Fedarko and M.F. Roberts. 1997. 2-sulfatrehalose a novel osmolyte in haloalkaliphilic archea. J. Bacteriol. 179, 3146-3153.

Dimitrov, P.L., M.S. Kambourova, R.D. Mandeva, and E.I. Emanuilova. 1997. Isolation and characterization of xylan- degrading alkali-tolerant thermophiles. FEMS Microbiol. Lett. 157. 27-30.

Downie, A.W. and J. Cruickshank 1928. The resistance of *Streptococcus faecalis* to acid and alkali media. Brit. J. Exp. Pathol. 9, 171-173.

Duckworth, A.W., W.D. Grant, B.E. Jones and R. Vansteenbergen. 1996. Phytogenetic diversity of soda lake alkaliphiles. FEMS Microbiol. Ecol. 19 : 181-191.

Dunkley, E.A., A.A. Guffanti, S. Clejan and T.A. Krulwich 1991. Facultative alkaliphiles lack fatty acid desaturase activity and loose the ability to grow at near neutral pH when supplemented with unsaturated fatty acid. J. Bacteriol. 173 (3), 1331-1334.

Engle, M., Y.H. Li, C. Woese, and J. Wiegel. 1995. Isolation and characterization of a novel alkali tolerant thermophile. *Anaerobraca Horikoshii* gen. nov., sp. nov. Int. J. Syst. Bacteriol. 45 : 454-461.

Fujiwara, N., A. Masui, and T. Imanaka 1993. Purification and properties of the highly thermostable alkaline protease from an alkaliphilic and thermophilic *Bacillus* sp. J. Biotechnol. 30 : 245-256.

Fujiwara, N., K. Yamamoto, and A. Masui, 1991. Utilization of thermostable alkaline protease from an alkalophilic thermophile for the recovery of silver from used X-ray film. J. Ferment. Bioeng. 72 : 306-308.

Gas Coyne, D.J., J.A. Connor, and A.T. Bull. 1991. Capacity of siderophore producing alkalophilic bacteria to accumulate iron, gallium and aluminium. Appl. Microbiol. Biotechnol. 36 : 136-141.

Gas Coyne, D.J., J.A. Connor, and A.T. Bull. 1991. Isolation of bacteria producing siderophores under alkaline conditions. Appl. Microbiol. Biotechnol. 36 : 130-135.

Garcia M.L., A.A. Guffanti and T.A. Krulwich 1983. Characterisation of Na^+/H^+ antiporter of alkalophilic *Bacilli* in vivo; Delta psi dependent 22 Na^+ efflux from whole cells. J. Bact. 156. 1151-1157.

Gee, J.M., B.M. Lund, G. Metealf and J.L. Peel. 1980. Properties of a new group of alkalophilic bacteria. J. Gen. Microb. 117 : 9- 17.

Georganta, G., T. Kaneko, N. Nakamura, T. Kudo and K. Horikoshi 1993. Isolation and partial properties of cyclomaltodextrin glucanotransferase producing alkalophilic *Bacillus* sp. from a deep sea mud sample starch. 45 : 95-99.

Gerasimenko, L.M., A.V. Dubinin, and G.A. Zavarzin. 1996. Alkaliphilic *Cyanobacteria* from soda lakes of Tuva and their ecophysiology. Microbiology. 65 : 736-740.

Gibson, T. 1934. An investigation of *Bacillus pasteurii* gp. II, Special physiology of organisms. J. Bacteriol. 28, 313- 322.

Gilmour, R. and T.A. Krulwich 1996. Purification and characterization of the succinate dehydrogenase complex and co- reactive b-type cytochromes from the facultative alkalophile *Bacillus firmus* OF4. Biochem. Biophy. Acta, 1276 : 57-63.

Goto, S., R. Aono and K. Horikoshi 1981. *Exophiala alcaliphtla*, a new black yeast like hypomycete with an accompanying *Phaeococcomyces alkaliphilus*, morphology and its physiological characteristics. Trans. Mycol. Soc. Japan. 22, 429-439.

Grant, D.W., A.A. Mills and A.K. Schofield 1979. An alkalophilic sp. of *Ectothio rhodospira* from a Kenyan soda lake. J. Gen. Microb. 110, 137-142.

Grant, D.W. and K. Horikoshi, 1989. Alkaliphiles in microbiology of Extreme environments and its potential for biotechnology. Elseiver Science publishers. p. 346-366.

Grant, W.D., W.E. Mwatha and B.E. Jones 1990. Alkaliphiles : Ecology diversity and applications FEMS. Microbio. Rev. 75 : 255- 270.

Guffanti, A.A., P. Susman, R. Blanco and T.A. Krulwich 1978. The proton motive force and Alpha-amino iso-butyric acid transport in an obligately alkaliphilic bacterium. J. Biol. Chem. 253, 708-715.

Guffanti, A.A., R. Blanco; R. A. Benenson and T.A. Krulwich 1980. Bioenergetic properties of alkaline tolerant and Alkalophilic strains of *Bacillus firmus*. J. Gen. Microbiol. 119 : 79- 86.

Guffanti, A.A., O. Finkelthal, D.B. Hicks., L. Falk., A. Sidhu., A. Garro and T.A. Krulwich 1986. Isolation and characterization of New facultatively alkalophilic strains of *Bacillus* sp. J. Bact. 167, 766-733.

Guffanti, A.A. and E. Hou 1987. The effect of pH on the passive proton conductance of *Bacillus acidocaldarius*. FEMS Microb. Lett. 41 : 275-278.

Guffanti A.A. & D.B. Hicks. 1991. Molar Growth yields and bioenergetic parameters of extremely alkalophilic *Bacillus* sps in batch cultures & growth in a chemostat at pH 10.5. J. Gen. Microb. 137 ; 2375-2379.

Hamana, K., T. Akiba, F. Uchino and S. Matsuzaki. 1989. Distribution of spermine in Bacilli and lactic acid bacteria. Can. J. Microbiol. 35, 450-455.

Hamamoto, T., M. Hashimoto, M. Hino, M. Kitaba. Y. seto, T. Kudo and K. Horikoshi 1994. Characterization of a gene responsible for the Na^+/H^+ antiporter system of alkalophilic *Bacillus* sp strain C-125. Mol. Microbiol 14 : 939-946.

Hamasaki, N., S. Shirai, M. Niitsu, K. Kakinuma and T. Oshima 1993. An alkalophilic *Bacillus* sp. Produces 2- phenylethylamine. Appl. Environ. Microbiol. 59 : 2720-2722.

Honda, H., T. Kudo and K. Horikoshi 1985. Purification and partial characterization of alkaline xylanase from *E. coli* carrying pCX311. Agric. Biol. Chem. 49 : 3165-3169.

Hansen, S. Jacob and B.K. Ahring 1997. Anaerobic microbiology of an alkaline Icelandic hot spring. FEMS microbiol. Ecol. 23 : 31- 38.

Hicks D.B. and T.A. Kurlwich 1986. The membrane ATPase of alkaliphilic *Bacillus firmus* RAB is an F_1-type ATPase. J. Biol. Chem. 261, 12896-12902.

Hirota, N., M. Kitada and Y. Imae. 1981. Flagellar motors of alkaliphilic *Bacillus* are powered by an electrochemical potential gradient of Na^+. FEBS Lett. 132, 278-280.

Hirota N. and Y. Imae. 1983. Na^+ driven flagellar motors of an alkalophilic *Bacillus* strain YN - 1. J. Biol. Chem. 258. 10577-10581.

Horikoshi, K. 1971. Production of alkaline enzyme by alkaliphilic microorganisms II. Alkaline amylase produced by alkaline *Bacillus* No. A-40-2. Agric. Bio. Chem. 35, 1783-1791.

Horikoshi, K. 1971. Production of alkaline enzymes by alkaline microorganisms part I. Alkaline protease produced by Bacillus No. 221. Agric. Bio. Chem. 36, 1407-1414.

Horikoshi, K. 1972. Production of alkaline enzymes by alkaline microorganisms part III. Alkaline pectinase of *Bacillus* No. P-4-N. Agric. Bio. Chem. 36, 285-293.

Horikoshi, K. and Y. Atsukawa. 1973b. Xylanase produced by alkaliphilic *Bacillus* No. C-59-2. Agric. Bio. Chem. 37, 2097-2103.

Horikoshi, K. and T. Akiba, 1982. Alkaliphilic microorganisms : A new microbial world. Springer - Verlag KG; Heidelberg; Germany.

Horikoshi, K. M. Nakao, Y. Kurono, and n. Sashihara. 1984. Cellulases of an alkalophilic *Bacillus* strain isolated from soil. J. Microbiol. 30 : 774-779.

Horikoshi, K. 1991. Microorganisms in Alkaline environments. Kodansha, VCH, Tokyo, Japan.

Horikoshi K. 1996. Alkalophiles from an industrial point of view. FEMS. Microb. Rev. 18 : 259-270.

Horikoshi, K. 1999. Alkalophiles : Some applications of their products for Biotechnology, Microbiol. and Mol. Biol. 63 : 735- 750.

Igarashi, Y. Hatada, H. Hagihara, K. saeki, M. Takaiwa, T. Uemura, K. Ara, K. Ozaki, S. Kawai, T.Kobayashi & S. Ito. 1998. Enzymatic properties of novel liquefying alpha-amylase from an alkalophilic *Bacillus* isolate and entire nucleotide and amino acid sequences. Appl. Environ. Microbiol. 64 : 3284-3289.

Ikura Y. and K. Horikoshi 1997. Isolation and some properties of alkalophilic bacteria utilizing rayon waste. Agri. Biol. Chem. 41, 1373-1377.

Ito, S., S. Shikata, K. Ozaki, S. Kawai, K.I. Okamoto, A. takei, Y. Ohta, and T. Satoh 1989. Alkaline cellulase for laundry detergents : production by *Bacillus* sp. KSM-635 and enzymatic properties. Agri. Biol. Chem. 53 : 1275-1281.

Ito M., A.A. Guffanti, J. Zemsky., D. Mackivey and T.A. Krulwich 1997. Role of the nha. C encoded Na^{+}/H^{+} antiporter of the alkalophilic *Bacillus firmus* OF4. 179, 3851-3857.

Ito, S. 1997. Alkaline cellulases from alkalophilic *Bacillus* : Enzymatic properties, genetics, and application to detergents. Extremophiles. 1 : 61-66.

Johnson H.W. 1923. Relationships between hydrogen ion : hydroxyl ion and salt concentration and growth of seven soil molds. Iowa. Agri. Exp. Stn. Res. Bull. 76, 307-344.

Jones B.E., W.D. Grant, A.W. Duckworth and G.G. Owenson. 1998. Microbial diversity of Soda Lakes. Extremophiles 2, 191-200.

Karlsson; S.Z.G. Banhide and A.C. Alberssen 1988. Identification and characterization of alkali-tolerant *Clostridia* isolated from biodeteriorated biodeteriorated casein-containing building material. Appl. Microb. Biotech. 28 : 305-310.

Kato, C., A. Inoue and K. Horikoshi 1996. Isolating and characterizing deep-sea marine microorganisms. TIBTECH 14 : 6-12.

Kawakami H. and S. Shimizu. 1986. Degradation of lignin and lignin related aromatics by bacteria in alkaline media. Bitoechnology in pulp and paper industry. The third international conference in Stockholm. Proceedings. Pp 167-169.

Keller M., F.J. Braun, R. Dirmeir., D. Hafenbradl., S. Burggraf., R. Rachel and K.U. Stetter 1995. *Thermophilus alcaliphilus* sp. nov., a new hyperthermophilic archaeum growing in polysulfide at alkaline pH. Arch. Microbiol. 164, 390-395.

Kelly, C.T., O. Reilly and M.W. Fogarty. 1983. Extracellular alpha glusidase of an alkalophilic microorganism *Bacillus* sp. ATCC 21591. FEMS Microb. Lett. 20, 55-59.

Kevbrin, V.V., T.N. Zhillina, F.A. Rainey, and G.A. Zavarzin 1998. *Tindallia magadi* gen. nov., sp. novo, an alkalophilic anaerobic ammonifier from soda lake deposits. Curr. Microbiol. 37 : 94-100.

Khmelenina, V.N., M.G. Starostina, N.E. Suzina & Y.A. Trotsenko 1997. Isolation & Characterization of halotolerant alkalophilic methanotrophic bacteria from Tuva soda lakes. Curr. Microbiol. 35 : 257-261.

Kimura, T. and K. Horokoshi 1988. Isolation of bacteria which can grow at both high pH and low temperature. Appl. Env. Microb. 54, 1066-1067.

Kimura, T. & K. Horokoshi 1989. Carbohydrate use by alkalopsychrotrophic bacteria. Agri. Biol. Chem. 53 : 2019-2020.

Kimura, H., Okamura A and H. Kawaide 1994. Oxidation of 3, 7, and 12 hydroxyl groups of cholic acid by an alkaliphilic *Bacillus* sp. Bioc. Biotechnol. Biochem. 58, 1002-1006.

Kitada, M. and K. Horokoshi 1977. Na^+ion stimulated alpha (14c) amino isobutyric acid uptake in alkaliphilic *Bacillus* sp. J. Bacteriol. 131 (3) : 784-788.

Kitada M., A.A. Guffanti and T.A. Krulwich 1982. Bioenergetic properties and viability of alkalophilic *Bacillus firmus* RAB as a function of pH and Na^+ contents of the incubation medium. J. Bast. 152 : 1096-1104.

Kitada, M. and T.A. Krulwich. 1984. Purification and characterization of the cytochrome oxidase from alkaliphilic *Bacillus firmus* RAB. J. Bacteriol. 158, 963-966.

Kitada, M. l. Wijayanti and K. Horikoshi 1987. Biochemical properties of a thermophilic alkalophilic. Agri. Biol. Chem. 51, 2429-2435.

Kitada, M., K. Onda & K. Horikoshi 1989. The sodium/proton antiport system in a newly isolated alkalophilic *Bacillus* sp. J. Bact. 171. 1879-1884.

Kitada, M., M. Hashimoto, T. Kudo and K. Horikoshi 1994. Properties of two different Na^+/H^+ antiport systems in Alkalophilic *Bacillus* sp strain. C-125. J. Bact. 176 : 6464-6469.

Kitada, M., S. Morotomi, K. Horokoshi, and T. Kudo. 1997. K^+/H^+ antiporter in alkalophilic *Bacillus* sp. no. 66 (JCM 9763). Extremophiles 1: 135-141.

Kobayashi, Y. and K. Horikoshi, 1980. Identification and growth characterization of alkaliphilic *Corcynebacterium* sp. which produces NAD(P)- dependent maltose dehydrogenase and glucose dehydrogenase. Agri. Biol. Chem. 44 (1), 41-47.

Kobayashi, T., H. Kanal, R. Aono, K. Horikoshi, & T.Kuo 1994. Cloning expression and nucleotide sequence of the amylase gene from the haloalkalophilic archaeon *Natronococcus* sp. strain Ah - 36. J. Bacteriol. 176 : 5131-5134.

Koga, Y., M. Nishihara and H. Morii. 1982. Lipids of alkaliphilic bacteria. Identification, composition and metabolism. J. Univ. occu. and Env. Heal. 4, 227-240.

Kosono, S., Y. Ohashi, F.Kawamura, M. Kitada and T. Kudo 2000. Function of a principal Na^+/H^+ antiporter, Sha A is required for initiation of sporulation in *Bacillus subtilis*. J. Bact. 182, 898-904.

Koyama, N and Y. Nosoh. 1976b. Effect of pH of culture medium on the alkalophilicity of a species of *Bacillus*. Arch. Microbiol. 109. 105-108.

Krahe, M.G. Antranikian and H. Markl 1996. Fermentation of extremophilic microorganism. FEMS Microb. Rev. 18 : 271-285.

Krulwich, A.T., R. Agus, M. Schneier and A.A. Guffanti 1985. Buffering capacity of *Bacilli* that grow at different pH ranges. J. Bact. 162: 768-722.

Krulwich, T.A., A.A. Guffanti, M.Y. Fong, L. Falk and D.B. Hicks 1986. Alkalophilic *Bacillus firmus* RAB generates variants which can grow at lower Na^+ concentration that the parental strain J. Bact. 185 : 884-889.

Krulwich, T.A., D.B. Hicks, D. Seto Young and A.A. Guffanti. 1988. The bioenergetics of alkaliphilics bacteria. Crit. Rev. Microbiol. 16, 15-36.

Krulwich, T.A. and A.A. Guffanti, 1989. Alaklophilic Bacteria. Annu. rev. Microbiol. 43. 435-463.

Krulwich, T.A., M.Ito and R. Gilmour 1997. Mechanism of cytoplasmic pH regulation in alkalophilic strains of *Bacillus*. Extremophiles 1 : 163-169.

Krulwich, T.A., M. Ito, D.B. Hicks, R. Gilmour, and A.A. Guffanti. 1998. pH Homeostasis and ATP synthesis : Studies of two processes that necessitate inward proton translocation in extremely alkaliphilic *Bacillus* species. Extremophiles 2 : 217-222.

Kurono Y. and K. Horikoshi 1973. Alkaline catalase produced by *Bacillus* No. Ku-1, Agri. Biol. Chem. 37, 2565-2570.

Kushner D.J. and T.A. Lisson 1959. Alkali resistance in *Bacillus cereus*. J. Gen. Microbiol 21, 96-108.

Kudo, T., and K. Horikoshi 1979. The environmental factors affecting sporulation of an alkaliphilic *Bacillus* sp. Agri. Biol. Chem. 43, 2613-2614.

Lewis, R., J.E. Kaback and T.A. Krulwich 1982. Pleotrophic properties of mutation to non alkalophily in *Bacillus alcaliphtlus* J. Gen Microbiol 128, 427-430.

Li, Y.H., M. Engle, N. Weiss, L. Mandelco and J. Wiegel 1994. *Clostridium thermoalcaliphilum* sp. Nov., an anaerobic and thermotolerant facultative alkaliphile. Int. J. Syst. Bacteriol. 44 : 111-118.

Lowe *et al.*, 1993. Anaerobic bacteria adapted to environmental stress. Microb. Rev. 57 : 483-492.

Madigan, T. Michael and B.L. Marrs. 1997. Extremophiles. Scientific American, April 1997. 66-71.

Mc Genity, T.J. and W.D. Grant. 1993. The haloalkaliphilic archea *Natronococcus occultus* represents a distinct lineage within the halobacteria closely related to the other haloalkaliphilic lineage (*Natronobacterium*). Syst. Appl. Microbiol. 16, 239-243.

McTigue, M.A., C.T. Kelly, W.M. Fogarty, and E.M. Doyle, 1994. Production studies on the alkaline amylases of three alkalophilic *Bacillus* sp. Biotechnol. Lett 16 : 569-574.

Meek, C.S. and C.B. Lipman 1922. The relation of the reactions of the salt concentrations of the medium to nitrifying bacteria. J. Gen. Physiol. 5, 195-204.

Myers, F.S. and A. Anderson 1992. Microbes from 20,000 feet under the sea. Science, 255 : 28-29.

Nagarsekar, Y. studies on alkalophilic bacteria isolated from spent waters of an Agro Chemical factory. Msc Desertation 1996. Dept. of Microbiology, Goa Univ.

Nakamura, N., K. Watanabe and K. Horikoshi 1975. Purification and some properties of alkaline pullulanase from a strain of *Bacillus* No. 202-1- an alkaliphilic microorganism. Biochem. Biophy. Acta. 397, 188-193.

Nakamura, S.K. and K. Horikoshi 1976d. Purification and properties of neutral cyclodextrin glycosyl transferase of an alkaliphilic *Bacillus* sp. Agri. Biol. Chem. 40, 1785-1796.

Nakamura, S.K. Wakabayashi, R. Nakai, R. Aono and K. Horikoshi 1993. Purification and some properties of an alkaline xylanase from alkaliphilic *Bacillus* sps. Strain 41 M-1. App. Env. Microb. 59, 2311-2316.

Nimura, Y., F. Yanangida, T. Uchimura, N. Ohara, K. Suzuki and M. Kozaki. 1987. A new facultative anaerobic xylan using alkalophile lacking cytochrome, quinone and catalase. Agric. Biol. Chem. 51, 2271-2275.

Nishihara, M., H. Mori and Y. Koga 1982. Bis - (monoacylglycero) Phosphate in alkaliphilic bacteria. Agri. Biol Chem. 52, 1643- 1647.

Nomoto, M.T. - C. Lee, C.W. Liao. T.M. - Yen, and C.P. Yang 1984. Alkaline proteinases from alkalophilic bacteria of Taiwan. Agric. Biol. Chem. 48 : 1627-1628.

Nomoto, M., M. Oshawa, H.L. Wang, C.C. Chen and K.W. Yeh 1988. Purification and characterization of extracellular alkaline phosphatase from an alkaliphilic bacterium. Agri. Biol. Chem. 52, 2643-1647.

Okazaki, W., T. Akiba K. Horikoshi, and R. Akahoshi 1985. Purification and characterization of xylanases from alkalophilic thermophilic *Bacillus* sp. Agric. Biol. Chem. 49 : 2033- 2039.

Paavilainen, S., P. Helisto and T. Korpela 1994. Conversion of carbohydrates to organic acids by alkaliphilic *Bacilli* J. Fern. Bioeng. 78, 217-222.

Pennisi E. 1997. In Industry Extremophiles begin to make their mark, Science, 276 : 705-706.

Persidis, A. 1990. Extremophiles. Unusual and useful molecules are found in life on the edge of environmental tolerance. Nature Biotech. 16 : 593-594.

Podkovyrov, S. and J.G. Zeikus. 1992. Structure of the gene encoding cyclomaltodextrinase from *Clostridium thermohydrosulfuricum* - 39E and characterization of the enzyme purified from *E. coli* J. Bacteriol 174, 5400-5405.

Prasad, P., and A.K. Kashyap. 1991. Ammonium transport in the alkalophilic diazotropic Cyanobacterium *Nostoc calcicola* - influence of phosphate limitation and metabolic inhibitors. J. Plant Physiol. 138 : 244-247.

Prasad, R. and S. Sarkar 1997. Evolutionary extremophilic Archaeal domain of life. Curr. Sc. 73 : 842-853.

Salva, T.D. G., V.B. Delima and a. P. Pagana 1997. Screening of alkaliphilic bacteria for Cyclodextrin glycosyl transferase production. Rev. Microbiol. 28, 157-164.

Sato, M., T. Beepu and K. Arima 1980. Properties and structure of a novel peptide antibiotic No. 1970. Agri. Biol. Chem. 44, 3037- 3040.

Schleper, C., G. Puhler, I. Holz, A. Gambacorta, D. Janekoric, U. Santarium, H. P. Klerk and W. Zillig 1995. Picrophilus genus nov. fam. Nov., a novel aerobic heterotrophic thermoacidophilic genus and family comprising archaea capable of growth around pH O. J. Bact. 177, 7050-7059.

Shikata, S., K. Saeki, H. Okoshi, T. Yoshimatsu, K. Ozaki, S. Kawai, and S. Ito. 1990. Alkaline cellulase for laundry detergents : Production by alkalophilic strains of *Bacillus* and some properties of the crude enzymes. Agric. Biol. Chem. 54 : 91-96.

Shimao, M., S. Onishi, S. Mizumori, N. Kato and C. Sakazawa 1989. Degradation of 4-chlorobenzoate by facultatively alaklophilic *Arthobacter* sp. Strain SB-8. App. Env. Microb. 55, 478- 482.

Singh, S. 1995. Partial Purification & Some properties of urease from the alkaphilic Cyanobacterium *Nostoc calcicola.* Folia Microbial. Prague 40 : 529-533.

Sorokin, D.Y., A.M. Lysenko, and L.L. Mityushina. 1996. Isolation and Characterisation of alkalophilic chemoorganoheterotrophic bacteria oxidizing reduced inorganic sulfur compounds to tetrathionate. Microbiology 65 : 326-338.

Sorokin, D.Y., & L.L. Mityshina, 1998. Ultra structure of alkalophilic heterotrophic bacteria oxidizing sulfur compounds to tetrathionate. Microbiology 67 : 78-85.

Sorokin, D.Y., G. Muyzer, T. Brinkhoff, J.G. Kuenen an M.S.M. Jetten 1998. Isolation and Characterisation of a novel facultatively alkalophilic *Nitrobacter* sps. *N. alkalicus* sp. nov. Ach. Microbiol. 170. 345-352.

Souza, K.A. P.H. Deal, H.M. Mack and C.E. Turnbill 1974. Growth and reproduction of micro-organisms under extremely alkaline conditions. App. Microbiol. 28 : 1066-1068.

Souza, A.K. and P.H. Deal, 1997. Characterisation of a novel extremely alkalophilic bacterium. J. Gen. Microbiol. 101, 103- 109.

Speelmans G.B. Poolman and W.N. Konings 1993. Amino acid transport in the thermophilic anaerobe *Clostridium fervidus* is driven by an electrochemical sodium gradient. J. Bact. 175, 2060-2066.

Studdert, C.A., R.E. Decastro, K.H. Seitz, and J.J. Sanchez 1997. Detection and preliminary characterization of extracellular proteolytic activities of the haloalkalophilic archaeon *Natronococcus occultus*. Arch. Microbiol. 168, 532-535.

Sturr, M.G., A.A. Guffanti and T.A. Krulwich 1994. Growth and bioenergetics of alkalophilic *Bacillus firmus* OF4 in continuous culture at high pH. J. Bacteriol. 176, 3111-3116.

Takami, H., T. Akiba, and K. Horikoshi 1989. Characterization of an alkaline protease from *Bacillus* sp. no. AH-101. Appl. Microbl. Biotechnol. 33 : 519-523.

Takami, H., T. Akiba, and K. Horikoshi, 1989. Production of extremely thermostable alkaline protease from *Bacillus* sps. no AH-101. Appl. Microbl. Biotechnol. 30 : 120-124.

Takami, H., A. Inoue, F. Fuji, and K. Horikoshi 1997. Microbiol flora in the deepest sea mud of the Mariana Trench. FEMS Microbiol. Lett. 152 : 279-285.

Tanabe, H., K. Yoshihara, K. Tamura, Y. Kobayashi, I. Akamatsu, N. Niyomwan and P. Footrakul. 1987. Pretreatment of Pectic wastewater from orange canning process by an alkalophilic *Bacillus* sp. J. Ferment. Technol. 65 : 243-246.

Tanabe, H., Y. Kobayashi, and I. Akamatsu. 1988. Pretreatment of Pectic wastewater with pectate lyase from an alkalophilic *Bacillus* sp. Agric. Biol. Chem. 52 : 1855-1856.

Takahara, Y. and O. Tanabe 1960. Studies on the reduction of Indigo in industrial fermentation vat (VII). J. Ferm. Technol. 38, 329-331.

Teri-Biotechnological exploitation of extremophiles for industrial applications. 2000. Proceedings of the Brainstorming session sponsored by DBT, India.

Tindall, B.J., A. Allan, A. Mills & W.D. Grant 1980. An alkalophilic Red halophilic Bacterium with low magnesium requirement from a Kenyan soda lake. J. Gen. Microb. 116 : 257- 260.

Tsuchiya, K., H. Sakashita, Y. Nakamura and T. Kimura. 1991. Production of thermostable alkaline protease by alkalophilic *Thermoactinomyces* sp. HS682. Agric. Biol. Chem. 55 : 3125- 3127.

Tsujibo, H., T.Sato, M. Inui, h. Yamamoto, and Y. Inamori 1988. Intra-cellular accumulation of phenazine antibiotics production by an alkalophilic *Actinomycete*. Agric. Biol. Chem. 52 : 301-306.

Tsujibo, H., Yoshida, Y., Miyamato, K., Hasegawa T., and Y. Inamore 1992. Purification and properties of two types of chitinases produced by an alkaliphilic actinomycete. Biosci. Biotechnol. Biochem. 56 : 1304-1305.

Ueyma, H. and K. Horikoshi. 1981. JPN Kokai 56 - 186153.

Vedder, A. 1934. *Bacillus alcalophilus* n sp. benevens enkle ervaringen met sterk alcalische voedingsbodems. Anton. Van Leeu. I, 141-147.

Walksman, S. A. and Jaffe J.S. 1922. Microorganisms concerned in the soil II. *Thiobacillus thioxidans*, a new sulfur oxidizing organism isolated from soil. J. Bacteriol. 7, 239-256.

Wang, Y.X., K.C. Srivastava, G.J. Shen, and H.Y. Wang 1995. Thermostable alkaline lipase from a newly isolated thermophilic *Bacillus*, strain A30-1 (ATCC 53841). J. Ferment. Bioeng. 79 : 433-438.

Watanabe, N., Y. Ota, Y. Minoda, and K.Yamada. 1977. Isolation & identification of alkaline lipase-producing micro-organisms, culture conditions and some properties of crude enzymes. Agric. Biol. Chem. 41 : 1353-1358.

Wiegel, J. 1998. Anaerobic alkalithermophiles, a novel group of extremophiles. 2 : 257-267.

Xu, C.H., a Nejidat, S. Belkin and S. Boussiba. 1994. Isolation and characterization of the plasma membrane by two phase partitioning from the alkaliphilic *Spirulina platensis*. Plant Cell Physiol. 35, 737-741.

Xu, Y., P.J. Zhou and X.Y. Tian. 1999. Characterization of two novel haloalkalophilic archaea *Natronorubrum bangense* gen. Nov., sp. Nov., and *Natronorubrum tibetense* gen. Nov. Int. J. Syst. Bacteriol. 49 : 261-266.

Yonezawa, Y. and K. Horikoshi 1978. Polyamines in alkaliphilic *Bacillus* Y-25. Agri. Biol. Chem. 42, 1955-1956.

Yoshihara, K. and Y. Kobayashi, 1982. Retting of Mitsumata bast by alakophilic *Bacillus* in papermaking, Agri. Biol. Chem. 46, 109-117.

Zhilina, T.N., G.A. Zavarzin, F. Rainey, V.V. Kevbrin, N.A. Kostrikina, and A.M. Lysenko 1996. *Spirochaeta alkalica* sp. nov., *Spirochaeta africana* sp. Nov. and *Spirochaeta asiatica* sp. nov., alkalophilic anaerobes from the continental soda lakes in Central Asia and the East African Rift. Int. J. syst. Bacteriol. 46 : 305-312.

Zhilina, T.N., and G.A. Zavarzin, 1994. Alkalophilic anaerobic community at pH 10.0. Curr. Microbiol. 29 : 109-112.

Zhilina, T.N., G.A. Zavarzin, F. Rainey, E.N. Pikuta, G.A. Osipov, and N.A. Kostrikina 1997. *Desulfonatronovibrio hydyogenovorans* gen. nov., sp. nov., an alkalophilic, sulfate- reducing bacterium. Int. J. Syst. Bacteriol. 47 : 144-149.

Zychlinsky E. and A. Matin 1983. Cytoplasmic pH homeostasis in an acidophilic bacterium *Thiobacillus acidophilus* J. Bact. 156. 1352-1355.

CHAPTER – 15

A NOVEL INEXPENSIVE TECHNIQUE FOR ISOLATION OF ANAEROBES

Nandita Ashtaputre and R.H. Balasubramanya

Central Institute for Research on Cotton Technology, (ICAR), Adenwala Road, Matunga, Mumbai 400 019

1. INTRODUCTION

The development of various media led to the culturing of different groups of microorganisms followed by isolation, characterization and their maintenance in pure culture. An exhaustive study has been done on the culturing of aerobic microorganisms which do not involve specific gas phases since oxygen is used as the terminal electron acceptor in the synthesis of ATP via the electron transport chain. Contrary to this, anaerobic organisms cannot be cultured with the usual procedures since the presence of oxygen in the system limit their growth.

The roll tube technique described by Hungate (1969) set the ball rolling in the field of isolation of anaerobes. The method involves the use of highly purified gases like hydrogen, carbon dioxide and nitrogen. Various modifications were suggested as described by Miller and Wolin (1974), Edwards and Mc. Bride (1975), Balch and Wolfe (1976) and Hermann *et al.*, 1986). Aranki and Freter (1972) designed and developed an anaerobic glove box for culturing and handling anaerobic micro-organisms. All these techniques invariably require ultra pure gases passed over preheated copper turnings (35°C) to remove traces of oxygen to be flushed into the tubes, bottles or chambers to expel the air. Gassing manifold assembly and glassware like serum bottle-neck tubes, serum bottles, butyl rubber seals, syringes, needles, aluminium seals, crimping machines are needed to isolate anaerobes. Apart from having all this, lot of expertise is required. The entire process is expensive and takes a number of days to obtain pure cultures.

During the incubation period the bottles have to be checked with great care as there is every possibility of negative pressure in the sealed bottles due to utilisation of the filled gases or there may be a build up of pressure due to production of different gases on catabolism of acetate, formate, etc.

Alternatively, it is also suggested to plate out the organism on recommended media and incubate them in anaerobic jar and flush with oxygen free gases. Cold catalyst, palladized alumina removes the traces of oxygen by combining with hydrogen to form water. A sachet placed in the jar produces carbon dioxide and hydrogen on contact with water.

The methods described above are expensive and time consuming.

It is well known that a functioning biomethanation batch fed digester produces methane, carbon dioxide and hydrogen. The gas phase on the top of such batch fed digester which is 100% anaerobic can be utilised for incubating inoculated petriplates for culturing cellulose degrading, sulfate reducing and methanogenic organisms.

2. APPARATUS

A simple assembly from Borosil glass bottles was fabricated. A five litre bottle cut from the top served as a digester. A ten litre bottle cut from the bottom was placed on the digester. Three outlets have been provided on the top of the digester cover which are closed with butyl rubber and aluminium seals. This facilitates sampling of gases for analysis on GLC. The mouth of the bottle was closed with a one-holed rubber cork through which a rubber tube was taken and connected to an air trap. The entire assembly was kept in a plastic tub.

The digester was charged with alkali pretreated and aerobically digested willow-dust which produces biogas soon after closing, Balasubramanya *et al.*, (1981). A perforated stainless steel plate was kept on the top of the digester on which petridishes plated with inoculum from an actively working biogas digester on recommended media, namely, Smith and Hungate (1958) medium for methanogens, Hungate (1950) medium for cellulolytic organisms and Postgate (Lapage *et al.*, 1970) medium for sulfate reducers were incubated. Tap water was added in the plastic tub till the level reached the top of the digester bottle. This process removes most of the air inside the digester. Bubbling continued in the air trap indicating the onset of biogas production. Complete anaerobiosis was achieved in about 4-6 hours.

The cultures were identified upto genus level by consulting Prokaryotes (1992). The composition of the biogas was analysed by Gas Chromatography.

Table 1. Composition of Biogas in the Digester

Day	Nitrogen (%)	Methane (%)	Carbon-dioxide (%)
1	45	16	38
7	5	46	48
13	2	54	43
20	2	60	35
27	2	70	27
34	2	70	27
48	2	70	27

H_2 and H_2S = Traces

3. RESULTS AND DISCUSSION

The composition of the biogas in the chamber is given in Table 1. The methane percentage goes on increasing and reaches to a maximum of 70% whereas carbon dioxide percentage decreases and stabilizes at around 27%. The nitrogen gas comes down to 2% from 45%. H_2S and H_2 were in traces. Well developed colonies were observed after 7-10 days. They were found to be genuine cellulose degraders, sulphate reducers and methane producers. The cultural identity is given in Tables 2 & 3. The apparatus is simple, inexpensive and

Table 2. Cultural Characteristics of Cellulose Degraders and Sulphate Reducers

Culture	Gram nature	Motility	Colony Character	Spore
Clostridium cellobioparum	Negative	+	White	Terminal
Clostridium sp.	Negative	+	Hollow centre	Cylindrical
Clostridium sp.	Positive	+	Yellow	Terminal
Clostridium sp.	Positive	+	Yellow	Curved/straight rods with round spores
Desulfovibrio sp.	Negative	+	Black	No endospore

Table 3. Cultural Characteristics of Methanogens

Culture	Morphology	Gram nature	Motility	Colony Character
Methanomicrobium sp	Short Rods	Variable	Motile	Pinpoint, Colourless
Methanothrix sp.	Long Rods	Negative	Non-motile	Colourless

can be set up in any laboratory without depending on expensive gases with cumbersome procedure. Hydrogen dependent anaerobes fail to grow in this system since the percentage of hydrogen is in traces. Under high solid loadings, it is possible to increase the percentage of hydrogen in the gas phase at which time hydrogen dependent anaerobes can be recovered if need arises. During winter months when atmospheric temperature goes down below 20°C, it has been observed that the rate of gas production comes down and the growth of the organisms is also affected. In order to overcome this problem a unit made of acrylic similar to the one devised by Edward Balish *et.al.* (1977) may be set-up and used to incubate the entire assembly.

REFERENCES

Aranki, A. and Freter, R. 1972. Use of anaerobic glove boxes for cultivation of strictly anaerobic bacteria. Am. J. Clin Nutr., 25, 1329-1334.

Balish Edward, Brown, J.F. & Wilkins, T.D. 1977. Transparent plastic incubator for the anaerobic glove box. Appl. Env. Microbiol., 33, 525 –527.

Balasubramanya, R.H., Khandeparkar, V.G. and Betrabet, S.M. 1981. Production of biogas from willow-dust by a batch fermentation process. J. Text. Assn., 42, 145-48

Balch,W.E. and Wolfe, R.S. 1976. New Approach to the cultivation of methanogenic bacteria:2-mercaptoethane sulfonic acid (HS-COM)- dependent growth of methanobacterium ruminantium in a pressurised atmosphere. Appl. Environ. Microbiol., 32, 781-791.

Balows, A., Truper, H.G., Dworkin, M., Harder, W. and Schleifer, K.H. 1992. The Prokaryotes, 2nd Edn., Springer-Verlag, New York, Vol.1-4.

Edwards, T and Mc Bride, B.C. 1975. New method for the isolation and identification of methanogenic bacteria. Appl. Microbiol., 29, 540-545.

Hermann, M., Noll, K.M. and Wolfe, R.S. 1986. Improved agar bottle plate for isolation of methanogens or other anaerobes in a defined gas atmosphere. Appl. Environ. Microbiol., 51, 1124-1126.

Hungate, R.E. 1950. The anaerobic mesophilic cellulolytic bacteria. Bacteriol. Rev., 14, 1-48.

Hungate, R.E. 1969. A roll tube method for cultivation of strict anaerobes. p. 117-132. In Methods in Microbiology. Vol 3B. J.R. Norris and D.W. Ribbons(ed.). Academic Press Inc., New York.

Lapage, S.P., Shelton, J.E. and Mitchell, T.G. 1970. Media for maintenance and preservation of bacteria. P.120.In Method in Microbiology, Vol. 3A. J.R. Norris and D.W. Ribbons (ed.). Academic Press Inc., New York.

Miller, T.L. and Wolin, M.J. 1974. A serum bottle modification of the Hungate technique for cultivating obligate anaerobes. Appl. Microbiol., 27, 985-987.

Smith. P.H. and Hungate, R.E. 1958. Isolation and characterization of Methanobacterium ruminantiumn. Sp. J.Bacteriol.,75, 713-718.

CHAPTER – 16

DEVELOPMENT AND STRUCTURE OF NODULES IN LEGUMINOUS WEEDS

P.S. Jain

Department of Botany, University of Rajasthan, Jaipur – 302 004 (India)

1. INTRODUCTION

The importance of Legumes in the improvement of nitrogen status of both the plant and soil is due to the presence of nodule, a symbiotic structure, on the roots. The nodule contains bacteria which fix atmospheric nitrogen and thus it enriches both the plant and the soil with nitrogen.

Beijerinck (1888) isolated and named the causal organism of nodule in *Vicia* as *Bacillus radicicola*. Frank (1890) designated it, the casual organism of Pea nodule as *Rhizobium leguminosarum*.

Peirce (1902) observed the infection of root through the root hair in *Medicago denticulata*. Spratt (1919) investigated the internal structure of nodule of a large number of genera and classified it into four types.

McCoy (1929, 1932) in her histological study of the nodule of *Phaseolus vulgaris* stated that the root hairs were not the chief portals of entry. The entry into the root could be a case of wound infection. A cluster of small infection threads were seen in each root hair. The infection threads penetrated five to six cells deep and the nodule was exogenous. The increase in infected cells was brought by the divisions of the infected cells.

Thornton (1930) observed the infection of *Medicago sativa* through the root hairs. Two or more infected root hairs contributed infection threads to a single nodule. The infection thread often branched within the root hair, and it penetrated the cortex up to the endodermis. The cellulose sheath around the infection thread was vivid in the cortex but not in the root hair. The infected cells enlarged with their nuclei hypertrophied.

Biederdorf (1938) studied the Cytology and histology of root nodule in *Glycine, Vigna, sinensis, Melilotus, Medicago sativa, Vicia villosa* and *Arachis hypogaea*. In

Glycine the infection was through the broken epidermal cells besides the root hairs.

Allen and Allen (1940) showed that in *Arachis hypogaea* normal root hairs were extremely rare and the root hair infection could not be made out. The nodule originated in the pericycle and subsequently out of the host root like a lateral root.

Chen and Thornton (1940) studied the structure of ineffective and effective nodules in clover, pea, and soybean. The early course of development of both effective and ineffective nodules were the same. The process of disintegration of bacterial tissue in soybean could start from any point in the central tissue instead of the base.

Bond (1948) gave an exhaustive account of the root anatomy and development of nodules of *Pisum sativum*. The nodules were pericyclic in origin. The endodermal layer with casparian bands surrounded the vascular bundles and the outer layer of the nodule.

Nutman (1948, 1956, 1962) studied the physiology of nodule formation and observed that the rate of nodule formation and of lateral roots was determined by the same hereditary factors in the plant.

Lawrie and Ray (1981) studied the nodule development and infection process in some native legumes of Australia. They found that the root hairs showed marked deformation and branching within six days of contact with rhizobia.

2. OBSERVATIONS

In *Lathyrus aphaca* Linn. and *Lathyrus sativus* Linn. the first nodule is observed on the primary root. The nodules are spherical to begin with subsequently they become cylindrical and fan-shaped. The branched nodules are seen after 15-20 days of nodule initiation. The host root is infected through the root hair. The microsymbiont enters the curled root hair and forms the infection thread. The main infection thread is quite massive and dark on staining with haematoxylon and it is not enclosed by the cellulosic sheath. A funnel-shaped expansion is formed by the infection thread as it passes through the host cell wall. The same is indistinct in young infected cells. The surface of infection thread is slightly uneven.

The nodule meristem is organised in young nodule. In the early stage the nodule is surrounded by the root cortex, when the nodule grows, the root cortex sloughs off and is replaced by the nodule cortex formed by the nodule meristem.

The mature nodule consists of nodule meristem, nodule cortex, vascular system and bacteroid zone. The nodule meristem has one layer of small, thin walled cells each with a conspicuous nucleus embedded in dense cytoplasm. The nodule is spherical to start with become cylindrical and branches into fan shaped structure. In *Lathyrus sativus* one arm of the branched nodule is smaller than the other. The meristem mostly adds cells internally towards the bacteroid tissue and a few cells externally to the over arching protective cortex. The cells which are added internally mature, some of them are invaded by the infection threads developed from the infected cells of the bacteroid tissue. In *Lathyrus aphaca* generally the nodule meristem is opposite to root at the distal end. In both the members the nodule cortex is homogenous consisting of four to eight layers of thin - walled parenchymatous cells. There is an endodermal layer enclosing collectively the nodule meristem, vascular system and bacteroid zone and root stele. There is no cork cambium developed in the mature nodule.

The vascular supply of nodule is formed as a procambial strand by the radial divisions of cortical cells situated between the region of nodule initiation and root stele. The procambial strand gets connected with protoxylem of the root and subsequently extends distally and branches enclosing the bacteroid zone. The vascular supply of the nodule consists of two traces cleparting from root stele one on either side of the nodule base. They branch and extend distally enclosing the central bacteroid zone. They branch into seven or eight bundles in *Lathyrus aphaca* and six bundles in *L. sativus*. These vascular bundles are separated from the bacteroid tissue by one to three layers of parenchymatous cells. They are mostly inversely collateral in both the members.

The bacteroid tissue is heterogenous, it consists of infected and uninfected cells. The infected cells are of 85 to 86% in both the species. The increase in bacteroid tissue takes place by the bacterial invasion of newly formed cells. The infected cells increase their size two or three times the size of uninfected cells. Their nucleus may centric with vacuoles all around. These vacuoles merge into one central large vacuole pushing the host nucleus to one side. The bacteroid line along the wall. The nucleus of the infected host cell is hypertrophied, spherical in the beginning and amoeboidal and dark-stained subsequently. A thin layer of small starch grains occupies the place between the wall of infected host cell and its contents. The uninfected cell has a small eccentric nucleus with a large central vacuole. It is filled with abundant starch grains.

The process of degeneration commences from the base upwards. The nucleus becomes small and loses stain. The bacteria clump together followed by degeneration and disappearance of host nucleus and other contents leaving absorption spots or spaces.

The ineffective nodules are fewer than the effective nodules. The nodule meristem is indistinct. The central tissue of large cells contain ramifying infection threads. The nodule has a cortex of two to three cells thick. The vascular supply is ill-developed.

In *Lathyrus aphaca* nodules develop occasionally in close succession. A group of two nodules develop on either side of the root or a bunch of three nodules develop around the root. Instances of necrosis developing in the bacteroid zone of nodules are also seen. Some effective nodules show premature degeneration of the bacteroid zone. The bacteroid cells do not clump but degenerate as such by losing stain. Its nuclear membrane disappears followed by its nucleolus. The infection thread present in the cells of bacteroid zone become prominent by their dark staining.

2. (a) *Medicago polymorpha* Linn. (Syn. *M. denticulata*)

The first nodule is observed on the primary root in *Medicago polymorpha* on 8th day of the seedling in culture tubes. Subsequently more nodules develop both on primary and lateral roots.

The nodules are spherical to begin with and become cylindrical and then branched. The branched nodules are seen after 15 days of nodule formation and are not specific to roots of any particularl order.

The mature nodule is differentiated into nodule meristem, nodule cortex, vascular system and bacteroid region.

The nodule meristem consists of one layer of small, thin-walled cells each with a conspicuous nucleus embedded in dense cytoplasm. The nodule is spherical to start with becomes cylindrical and branches into lobed and digitate structure. The branching is caused by the cessation of meristematic activity of certain groups of cells in the nodule meristem and subsequent growth. Some of the cells produced by the nodule meristem towards the bacteroid zone are invaded by the infection threads and bacteria are released into them. In this way the nodule grows and the bacteroid zone also increases in volume.

The nodule cortex is homogeneous consisting of two to five layers of thin-walled parenchymatous cells. The cells are large. In the mature nodule cambium for the formation of periderm is formed in the basal layer of the cortex.

The vascular supply of the nodule consists of two traces one on either side of the nodule base. These branch further repeatedly as they extend to the distal end of the nodule. The number of these vascular bundles increases to as many as 18 or more. The vascular bundles are mainly inversely collateral and a few are biocollateral or amphicribal. Each bundle has a pericycle and an endodermis enclosing the conducting elements.

The bacteroid tissue is heterogeneous. It consists of infected and uninfected cells. The infected cells are of 78 to 79%. The increase in bacteroid tissue is caused by the bacterial invasion of new cells formed by the meristem. The infected cells are either similar to or twice or thrice the size of uninfected cells. In the early stage numerous vacuoles of different sizes appear surrounding the nucleus. These vacuoles disappear as the bacteria increase in quantity. The nucleus of the infected host cells also increases along with the nucleolus in size, stains densely and becomes amoeboidal in the old cells. The uninfected cells are either similar in size to or smaller than the infected cells. They have small nuclei dense cytoplasm lining the wall, a large central vacuole and abundant starch grains.

The process of degeneration starts from the base upwards. The bacteria loose the stain and clump together. The nucleus of the host cell also loses the stain, degenerates and disappears. The contents of the infected host cell disappear leaving the absorption spot or space.

(b) *Melilotus indica* (Linn) (Syn. *M. parviflora*)

In *Melilotus indica* the first nodule is observed on the primary root on 10th day of the seedling in culture tubes. Subsequently more nodules develop on lateral roots than on primary root.

The nodules are spherical to start with. They become cylindrical and then digitate. The fan shaped (digitate) nodule is seen after 15 to 20 days of nodule formation and it is not specific to roots of any order.

The mature nodule is differentiated into nodule meristem, nodule cortex, vascular system and bacteroid region. The nodule meristem consists of one layer of small, thin-walled cells each with a conspicuous nucleus embedded in dense cytoplasm. The spherical young nodule becomes cylindrical by the activity of nodule meristem and subsequently lobed or digitate by the formation and activity of many groups of meristematic cells. Some of the cells produced internally to the meristem are invaded by the infection threads and the bacteria are released into them forming part of bacteroid zone.

The nodule cortex is homogeneous consisting of four to five layers of thin-walled parenchymatous cells. The inner most layer becomes meristematic forming a continuous ring of cambium producing periderm.

The vascular bundles that surround the bacteroid tissue, are 7 or more. They are separated from the central bacteroid tissue by one or two layers of parenchymatous cells. The vascular bundles are chiefly inversely collateral. Each bundle is surrounded by a ring of pericycle and of an endodermis.

The bacteroid tissue is heterogeneous. It consists of infected and uninfected cells. The infected cells are 78%. The increase in bacteroid tissue is caused by the bacterial invasion of new cells produced by the nodule merkistem. The infected cells are either of the same size or twice or thrice the size of uninfected cells. In the early stage numerous vacuoles of different sizes appear around the nucleus of the infected cell. These vacuoles disappear as the bacteria increase in quantity. The nucleus of the infected host cell also increases in size and stain densely. It becomes amoeboidal with increase in age of the cell. The uninfected cells are polygonal, thin-walled, either similar to or smaller than the infected cells with thin cytoplasm lining the wall inclosing a large vacuole, small nucleus and abundant starch grains.

The process of degeneration is acropetal. The bacteria clump together. The nucleus of the host cell reduces in size and loses stain. The contents of the host cell disappear leaving an absorption spot.

(c) *Tephrosia apollinea* (Del.) Link.

In *Tephrosia apollinea* the first nodule is observed on the primary root on 10 day of the seedling in culture tubes. Subsequently more nodules develop on roots of other orders than on primary roots.

The nodules are spherical to start with. Later they become cylindrical in the first instance and then branched. The branched nodules are seen after 15 to 20 days of nodule initiation and are not specific to roots of any particular order. The nodules may be in pairs. They are of two types : (i) effective nodules which are slightly pink and (ii) ineffective nodules which are white or brown. The latter are smaller than the former and remain spherical.

The root hairs are the chief part of entry of *Rhizobium* in *Tephrosis apollinea.* The microsymbiont enters the curled root hair and forms the infection threads. The infection threads are more than one (two) in the root hair. Adjacent root hairs may also be infected to form a nodule. There is no definite cellulosic sheath seen surrounding the infection threads in the root hair or cortical cells. The surface of infection thread bears blister-like swellings. The infected root hair enlarges three or more times their original size and its radial and inner tangential walls become thickened. The hypodermal cortical cells do not enlarge. The infection thread penetrates upto the middle cortex of root, ramifies and initiates nodule formation. The bacteria may be released into the host cell by the bursting of blister-like swellings.

There is no nodule meristem organised in the young stage. When the nodule becomes 10 to 15 days old it organises the nodule meristem at the distal end. In the early stage the nodule is surrounded by two layers of root cortex. When it enlarges and differentiates, the root cortex sloughs off and is replaced by the nodule cortex formed by the activity of distinct apical nodule meristem.

The mature nodule is differentiated into nodule meristem, nodule cortex, vascular system and bacteroid region.

The nodule meristem consists of one layer of small thin-walled cells each with a conspicuous nucleus embedded in dense cytoplasm. The nodule is spherical to start with, becomes cylindrical and branches into a lobed or digitate structure. The branching is caused by the cessation of meristematic activity of certain groups of cells in the nodule meristem and subsequent growth. Some of these cells produced by the meristem are invaded by the infection thread and bacteria are released into them. In this way the infection threads move towards the apex of the nodule as new cells are formed by the apical meristem.

The nodule cortex is homogeneous consisting of six to eight layers of thin-walled parenchymatous cells. In the periphery of nodule cortex certain cells become meristematic forming a continuous ring of cambium producing periderm.

The vascular supply of the nodule is formed as a procambial strand by the radial divisions of the cortical cells situated between the region of nodule initiation and the root stele. The procambial strand gets connected to the protoxylem of the root stele. It subsequently branches and extends towards the distal end of nodule enclosing the bacteroid tissue. In *Tephrosia apollinea* the vascular supply consists of two traces one on either side of the nodule at the base. These branch further repeatedly as they extend to the distal end of the nodule. Their number increases to eight. These strands are separated from the central bacteroid tissue by three or four layers of parenchymatous cells. The vascular bundles are mainly bicollateral and a few are amphicribal. In each bundle the vascular elements are surrounded by a layer of pericycle and an endodermis.

The bacteroid tissue is heterogeneous. It consists of infected and uninfected cells. The infected cells are 78%. The increase in bacteroid tissue is caused by the bacterial invasion of new cells formed by the meristem. The infected cells are either of the same size or twice or thrice their original size. In the early stage numerous vacuoles of different sizes appear surrounding the host cell nucleus. These vacuoles disappear as the bacteria increase in quantity. The nucleus may become amoeboidal and stains densely with the increase in age of infected cell. The uninfected cells are either similar in size to or slightly smaller than the infected cell. Their nuclei are comparatively small with thin cytoplasm, large vacuole and abundant starch grains.

The bacteroid zone is surrounded by a few one- or two celled rarely three-celled structures. These are like secretory cells with large size, dense cytoplasm and a prominent nucleus with large nucleolus. The presence of such a cell element is unique in *Tephrosia apollinea*.

The process of degeneration starts from the base upwards. The bacteria clumps together. The nucleus of the host cell degenerates. The other contents of the infected cells gradually disappear leaving absorption spots or spaces.

The ineffective nodules are fewer than the effective nodules. The mode of infection is through the root hair the base of which enlarges many times its original size. There is no nodule meristem. One or two layers of root cortex enclose the central group of infected cells which are traversed by infection threadly. Which either contact or pass very close to the host cell nucleus. Either the infected cell nucleus nor the host cell enlarges. Generally the ineffective nodule is away from the effective nodule. In one instance the ineffective nodule is organised on the root close to the effective nodule. The vascular supply is organised, but less extensive. This may be in keeping with the small size of the nodule.

(d) *Trigonella corniculata* Linn. and *T. polycerata* Linn.

The first nodule is observed on the primary root in *Trigonella corniculata* on 8th day and in *T. polycerata* on 10th day of the seedling in culture tubes.

The nodules are spherical in the beginning. Later they become cylindrical in the first instance and branched or digitate subsequently. The branched nodules are seen after 15 to 20 days of nodule initiation. They are present both on primary and lateral roots. The ineffective nodule is scarce in *Trigonella corniculata* while it is present in *T. polycerata*.

The root hair is the chief portal of entry of *Rhizobium* in both these host members. The mature nodule is differentiated in to nodule meristem, nodule cortex, vascular system and bacteroid region. The nodule meristem is generally at the end opposite to root. In one instance it is lateral in position. The nodule meristem consists of one layer of small thin-walled cells each with a conspicuous nucleus surrounded by dense cytoplasm. The nodule is spherical to begin with, becomes cylindrical and branches into a lobed or digitate structure. The branching is caused by the cessation of meristematic activity of certain cells and subsequent growth in the nodule. The meristem adds cells more towards the bacteroid tissue than towards the nodule cortex. The cells added on the inner side (towards the bacteroid region) gradually enlarge. Some of them are penetrated by the infection threads which usually show affinity with the host cell nucleus. The infection threads release the bacteria into the cell which enlarges more than the uninfected cell. In this way the infection threads move towards the apex of nodule in both the members.

The nodule cortex is homogeneous consisting of six to eight layers in *T. corniculata* and four layers in *T. polycerata* of thin-walled parenchymatous cells as large as those of the root cortex. Certain cells in the fourth or fifth layer of the nodule cortex become meristematic organising phellogen to produce periderm.

The vascular supply of the nodule originates as a procambial strand by the radial divisions of the cortical cells situated between the region of nodule initiation and root stele. It establishes contact with the protoxylem of root stele and extends towards the nodule and branches enclosing the bacteroid zone. In both *Trigonella corniculata* and *T. polycerata* the vascular supply consists of two traces one on either side of nodule base. Both the traces divide as soon as they enter the nodule and ramify. The number of vascular bundles that surrounds the bacteroid tissue is variable from 8 to 12. The vascular bundles are separated from the central bacteroid tissue by three or four layers of parenchymatous cells rich in starch grains. They are chiefly inversely collateral and a few are biocollateral. The vascular elements are surrounded by a layer of pericyclic parenchyma and an endodermis.

The bacteroid tissue is heterogeneous. It consists of infected and uninfected cells. The infected cells are of 89% in *Trigonella corniculata* and 91% in *T. polycerata.* The increase in infected tissue is caused by the bacterial invasion of new cells formed by the meristem. The infected cells are two to eight times the size of uninfected cells. The cytoplasm of the infected host cell may enclose many vacuoles around the centric, hypertrophied nucleus with its large darkly stained nucleolus. The bacteroids line themselves along the wall. The host nucleus subsequently gets laterally displaced by the development of a large central vacuole. The uninfected cells are small with thin cytoplasm lining the wall, a large central vacuole, small acentric nucleus and abundant starch grains. The infected cells also contain starch grains between the host cell wall and the bacteroids.

The process of degeneration starts from the base upwards. To begin with, the host nucleus degenerates. The central vacuole disappears. The bacteria clump together in the host cell and ultimately disappear leaving absorption spots or space.

The ineffective nodule in *Trigonella polycerata* is not having the nodule meristem. The central bacteroid tissue contains large cells and spaces with no bacteroid cells. The vascular bundles of the nodule are ill developed. The nodule cortex is also thin.

(e) *Vicia hirsuta* Linn. and *V. sativa* Linn.

The first nodule is observed on the primay root in *Vicia hirsuta* on 10th day and in *V. sativa* on the 12th day of the seedling in culture tubes. Subsequently more nodules develop on the roots of other orders than on primary root.

The nodules are spherical to start with later they become cylindrical and brached. The nodules may developed juxtaposed or in a group of four in close proximity with one another. The nodules may be in pairs. The branched nodules are seen after 15 to 20 days of nodule initiation. They are not specific to roots of any order. There are both effective (pink) and ineffective (brown or white) nodules. The latter are small and spherical.

The mature nodule is differentiated in to nodule meristem, nodule cortex, vascular system and bacteroid region.

The nodule meristem consists of one layer of small, thin walled cells each with a conspicuous nucleus embedded in dense cytoplasm. The nodule is spherical to start with and becomes cylindrical and branched into digitate structure. The branching of the nodule is caused by the cessation of meristematic activity of certain cells in the nodule meristem.

The nodule meristem produces more cells proximally than distally. Of the cells produced proximally, a few are invaded by the infection threads developed from the infected cells. The others are uninfected. Thus, the infection threads move towards the distal end of the nodule to infect some of the new cells produced by the nodule meristem. The infection threads show funnel shaped expansion as it crosses the host cell wall and exhibit affinity to the host cell nucleus which usually becomes hypertrophied and densely stained.

The nodule cortex is homogenous consisting of four to five layers of thin-walled parenchymatous cells both in *V. hirsuta* and *V. sativa*. The inner most layer is the nodule endodermis enclosing the vascular bundles and the bacteroid zone. The phellogen is not seen in the nodule cortex of fairly mature nodules.

The vascular supply of the nodule develops as a procambial strand by the radial divisions of the cortical cells situated between the region of nodule initiation and the root stele and gets connected with the protoxylem of the root stele. It branches and extends towards the distal end of the nodule around the central bacteroid zone on the inner side of nodule endodermis. In *Vicia hirsuta* and *V. sativa* the vascular supply consists of two traces on either side of nodular base. These traces branch repeatedly during their journey to the distal end of nodule. There is no vascular bundle over arching the nodule meristem. Where there are two distinct bacteroid zones in the root the vascular supply between the two zones functions common to both the zones besides the distinct vascular branches on the free side. The number of vascular bundles that surrounds the bacteroid zone varies from 9 to 11 in *V. hirsuta* and as many as 17 in *V. sativa*. These vascular bundles are separated from the central bacteroid zone by the two or three layers of parenchymatous cells rich in starch grains. The vascular bundles are generally inversely collateral in both the members. Each bundle has the vascular elements surrounded by distinct layers of pericycle and endodermis.

The central bacteroid tissue is heterogeneous. It consists of infected and uninfected host cells. The infected cells are 83% in *Vicia hirsuta* and 92 to 93% in *V. sativa*. The bacteroid tissue increases in quantity as the nodule grows by the activity of nodule meristem which produces many cells on the inner side and a few cells on the outer side. Some of the cells produced internally are invaded by the infection threads developed from the bacteroid tissue. In *V. sativa* more than one infected thread invades the host cell and blister like swellings on the infection threads and funnel-shaped expansion of infection thread close to the wall are discernible. In this way the infection thread moves to the distal end of the nodule by infecting the newly produced cells. The infected host cell enlarge twice or thrice the size of uninfected cells. They possess a central large vacuole with accentric nucleus from the beginning. the bacteroids line all along the walls. The host nucleus is hypertrophied, spherical with large nucleolus and dense stain. It become amoeboidal when the host cell becomes old. The space between the host cell wall and the bacteroid layer is usually occupied by starch grains. This is an interesting feature. This situation does not change with increase in size of infected cells. The uninfected cells are small and thin walled with eccentric, small, weakly stained nucleus and abundant starch grains. These get depleted as the nodule becomes old and some of the starch grains are stored in the infected host cell itself.

The process of degeneration in nodules of both *V. hirsuta* and *V. sativa* commences from the base upwards. The host cell nucleus losses its size and stain and finally degenerates. The bacteria clump together. The entire content of the cell gradually disappears leaving absorption spot or space.

The ineffective nodules in both *V. hirsuta* and *V. sativa* are fewer than the effective nodules. The early development of ineffective nodule is very much similar to that of the effective nodule. There is nodule meristem but its activity is very much limited. The central tissue is ineffected but bacteroids are not formed. The infection thread passes through many cells of the central tissue. It shows affinity towards the host nucleus. However, the latter neigher gets hypertrophied nor takes dense stain. There is neither starch grain in this tissue nor the funnel-shaped swellings of the infection thread in the region of intercellular passage. The nodule cortex is four layered and homogenous. The vascular supply is well developed. The nodule meristem, vascular bundles, bacteroid zone and root stele are enveloped by the common endodermal layer.

3. DISCUSSION

The chief mode of entry of infection thread into the root in *Tephrosia apollinea, Lathyrus aphaca* and *L. sativus* is through the root hair.

The number of infection threads in the root hair is variable. Generally, a single thread is seen in the root hair. In the present studies two infection threads are seen in *Tephrosia apollinea.* Similarly a cluster of two or more infection threads are reported in the root hair of *Phaseolus vulgaris* (McCoy, 1929), *Glycine max* (Bieherdorf, 1938).

In *Tephrosia apollinea* the infected root hair enalrges three or more times and its wall becomes thickened. The penetration of cortex by the infection thread is inter cellular. The depth of penetration of infection thread in *Tephrosia apollinea, Lathyrus aphaca* is upto fourth or fifth layer of the root. However, in *Glycine max* (Bieherdorf, 1938), *Trigonella foenum-graecum* and *Canavalia gladiata* (Narayana and Gothwal, 1964).

The infection thread bears an uneven or blister like swellings along its surface in the root hair of *Tephrosia apollinea* and in the bacteroid zone of *Vicia sativa.* The blister like swellings were also reported in *Medicago sativa* (Thornton, 1930).

The spread of *Rhizobium* in host tissues is either by the invasion of newly produced cells by the distinct meristem present at the distal end of nodule in *Tephrosia apollinea, Lathyrus aphaca, L. sativus, Vicia hirsuta, V. sativa, Trigonella polycerata, Medicago polymorpha* and *Melilotus indica.* The nodule cortex is homogenous consisting of parenchymatous cells in all the weeds (Jain, 1981). The periderm may be formed in *Teprhosa apollinea* and *Melilotus indica* and *Trigonella polycerata.* The homogenous parenchymatous nodule cortex is reported in *Vigna sinensis* and *Dolichos biflorus* (Narayan, 1964).

The vascular supply of the nodule consist of one strand as in *Glycine max* (Bieberdorf, 1938), *Sesbania grandiflora* (Harris, *et al.*, 1949), Narayana and Gothwal, 1964) and *Erythrina suberosa* (Mahna, 1981), two strand in *Vicia villosa* (Bieberdorf, 1938). *Pisum sativum* (Bond 1948), *Cajanus cajan* (Arora, 1956b, Kapil and Kapil 1971). Two to five in *Crotalaria juncea* (Arora 1956a); three to five in *Cicer arietinum* (Arora, 1956c) and four in *Trigonella foenumgraecum* (Narayana and Gothwal, 1964) at different levels originating from more than one protoxylem point. In the leguminous weeds in the present investigation (Jain 1981) the vascular supply consist of two traces originating from more than one protoxylem point of the root stele at different levels. This contradicts the observation of single nodule trace originating from the root-stele in *Erythrina suberosa* (Mahna 1981). The number of vascular bundles that surround the bacteroid zone is variable in different plants, 6 in *Aeschynomene indica* (Arora, 1954), 7 to 8 in *Phaseolus mungo*, 10 to 11 in *Canavalia gladicala* (Narayana and Gothwal, 1964) and *Erythrina suberosa* (Mahna, 1981); 7 to 18 in *Dolichos biflorus* and 12 in *Vigna sinensis* (Narayana, 1964), 12 to 21 in *Prosopis juliflora* and 5 to 23 in *P. cineraria* (Mahna, 1981), 20 to 24 in *Trigonella foenum graecum* (Narayana and Gothwal 1964) and 126 in one year old nodule of *Sesbania grandiflora* (Harris *et al.*, 1949). In *Vicia hirsuta* number of the vascular bundles that surround the bacteroid zone is 9 to 11, as many as 17 in *V. sativa*, 8 to 12 in *Trigonella corniculata* and *T. polycerata* 7 or more in *Melilotus indica*, 7 to 8 in *Lathyrus aphaca* and 6 in *Lathyrus sativus*, 13 to 15 in *Erythrina suberosa*, 18 or more in *Medicago polymorpha* and 8 in *Tephrosea apollinea* (Jain, 1981).

These vascular bundles are reported to exhibit different orientations, inversely collateral in *Pisum sativum* (Bond, 1948), *Erythrina suberosa* (Mahna 1981), *Sesbania grandiflora* (Harris *et al.*, 1949) both collateral and amphicribal at the base and collateral in higher-up in *Phaseolus mungo* (Narayana and Gothwal 1964), in *Vicia hirsuta, V. sativa, Lathyrus aphaca, L. sativus* is inversely collateral. Inversely collateral and bicollateral in *Trigonella corniculata* and *T. polycerata*; inversely collateral, bicollateral and amphicribal in *Medicago polymorpha* and bicollateral and amphicribal in *Tephrosia apollinea* and amphicribal and collateral in *Erythrina suberosa* (Jain, 1981).

The amount of infected cells in bacteroid tissue is variable in different legumes about 100% in *Arachis hypogaea* (Allen and Allen, 1940) and *Aeschynomene indica* (Arora, 1954). 50 to 70% in *Sesbania grandiflora* (Harris *et al.*, 1949); 42% in *Erythrina suberosa*, 49% in *Prosopis juliflora* and 53% in *P. cineraria* (Mahna 1981) 20 to 60% in *Cajanus cajan, Cicer arietinum* (Arora, 1956b, and c; Kapil and Kapil 1971) and 25 to 55% in *Dolichos biflorus* and *Vigna sinensis* (Narayana 1964). 58% in *Erythrina suberosa*, 85 to 86% in *Lathyrus aphaca* and *L. sativus*, 78 to 79% in *Medicago polymorpha*, 78% in *Melilotus indica, Trigonella corniculata*, 91% in *T. polycerata*, 83% in *Vicia hirsuta* and 92 to 93% in *Vicia sativa* (Jain 1981).

REFERENCES

Allen, O.N. & Allen, E.K. 1940. Response of peanut plant to inoculation with rhizobia with special reference to morphological development of nodule. Bot. Gaz. 102 121-142.

Arora, N. 1954. Morphological development of root and stem nodules of *Aeschynomene indica* L. Phytomorphology G 211- 215.

Arora, N. 1956a. Morphological development of root nodules on *Crotolaria juncea*. Proc. 43rd Indian Sci. Congr. 244.

Arora, N. 1956b. Morphological study of the root nodules on *Cajanus indicus*. Proc. 43rd Indian Sci. Congr. 244-245.

Arora, N. 1956c. Histology of the root nodules on *Cicer arietinum* L. Phytomorphology 6 367-378.

Beijerink, M.W. 1888. Die Bacteriem der Papilionaceen knolleher. Bot Ztg. 46 726-735; 741-750; 757-771; 781-790; 797-804.

Biederdorf, F.W. 1938. The cytology and histology of the root nodules of some leguminaceae. J. Am. Soc. Agron. 30 : 375-389.

Bond, L. 1948. Origin and developmental morphology of root nodules of *Pisum sativum*. Bot. Gaz. 109 411-434.

Chen, H.K. & Thornton, H.G. 1940. The structure of ineffective nodule and its influence on nitrogen fixation. Proc. Roy. Soc. Lond. 129B 208-229.

Frank, B. 1890. Under die pilzsymbiose der Leguminosen. Landw. Jb 19 523-640.

Harris, J.O., Allen, E.K.; Allen, O.N. 1949. Morphological development of nodules on *Sesbania grandiflora* Poir. with reference to the origin of nodule rootlets. Ann. J. Bot. 36 : 651-661.

Jain, P.S. 1981. Studies on the casual organisms, development and structure of root nodules and effect of passage in a few summer and winter leguminous weeds. Ph.D. Thesis. Univ. Raj. Jaipur. India.

Kapil, R.N. & Kapil, N. 1971. Root nodules of *Cajanus cajan* : origin, structure and ontogeny. Phytomorphology. 21 191-202.

Mahna, V.K. 1981. Studies on the causal organism, origin, development and biology of root nodules of tree legumes. Ph.D. Thesis, Univ. Raj. Jaipur. India.

McCoy, E.F. 1929. A cytological and histological study of the root nodules of the bean, *Phaseolus vulgaris* L. Z. Bakt. Abt. 79 : 394-412.

McCoy, E.F. 1932. Infection by *Bact. radicicola* in relation to the microchemistry of the host's cell walls. Proc. Roy. Soc. Lond. 110 514-533.

Narayana, H.S. 1964. Morphology and anatomy of the root nodules in Agiosperms. Indian Bot. Soc. Memoir 4 : 208-218.

Narayana, H.S. & Gothwal, B.D. 1964. A contribution to the study of root nodules in some Legumes. Proc. Indian Acad. Sci. 59 : 350-359.

Nutman, P.S. 1948. Physiological studies on root nodule formation. I. The relation between nodulation and lateral root formation in red clover. Ann. Bot. 12 : 81-86.

Nutman, P.S. 1956. The influence of legume in root nodule symbiosis. A comparative study of host determinants and functions. Biol. Rev. 31 : 109-151.

Nutman, P.S. 1962. The relation between root hair infection by Rhizobium and nodulation in *Trifolium* and *Vicia*. Proc. Roy. Soc. Lond. 156 : 122-137.

Peirce, G.J. 1902. The root tubercles of bur clover (*Medicago denticulata*) and of some other leguminous plants. Proc. Calif. Acad. Sci. Ser. III. 132 : 245-328.

Purchase, H.F. 1953. Studies of Rhizosphere of leguminous plants and its relation to nodule formation. Ph.D. Thesis, Univ. London, London.

Thornton H.G. 1930. The early development of root nodules of Lucerne (*Medicago sativa* L. Ann. Bot. 44, 385-392.

CHAPTER – 17

EMERGING NEW APPROACHES FOR DEVELOPMENT OF VACCINES

G. Prasad, Minakshi and Yashpal Malik

Department of Veterinary Microbiology, CCS Haryana Agricultural University, Hisar – 125 004, India

1. INTRODUCTION

Discovery of the concept of vaccination by Edward Jenner more than 2 centuries ago laid the foundation for immunoprophylaxis against infectious diseases. Jenner's observation that milkmaids milking cows that had cowpox infection did not suffer from small pox led to the belief that cowpox and causative agent of smallpox were related. Because of relatedness, the milkmaids got immunised by the cowpox agent naturally when they came in contact with the infected animals, thereby developing cross-protective immunity against smallpox. Thus the science of vaccinology was born in 1798 with the discovery of concept of vaccination against smallpox by Jenner. The success of Jenner's concept in preventing smallpox by vaccination of human being by cowpox led to the realisation that related pathogens can induce cross protection. Almost a century latter, Louis Pasteur, one of the most influential microbiologists, coined the term 'vaccine' in honour of Jenner, to refer generically to immunising agent.

The last three decades of nineteenth century witnessed emergence of bacteriology as an important branch of science. One after another causative agents of infectious diseases were discovered during this period. Pasteur's ideas backed with the findings of laboratory experiments had revolutionary impact on bacteriology. Development of artificial media for the cultivation of different kinds of bacteria paved the way for development of vaccines. Pasteur was first to demonstrate that old cultures of *Pasteurella septica* (now *Pasteurella multocida*), causative agent of lethal fowl cholera, loose their virulence and did not produce disease when inoculated in chickens. Subsequently, he showed that the chickens inoculated with old cultures were protected against challenge with the fresh cultures of *Pasteurella septica*. This observation led to the conclusion by Pasteur that, old cultures under go certain changes that result in attenuation but no loss of antigenicity. Using this concept, Pasteur attenuated *Bacillus anthracis* and demonstrated that, keeping cultures at 42°C to 43°C for 2 weeks resulted in loss of virulence. Later Pasteur deliberately attenuated causative agent of rabies by serially passaging it in rabbit brain. Thus Pasteur pioneered the concept of attenuation as method for decreasing virulence in 1885 which is being used extensively through out the world even today. However, progress in commercialisation of vaccine technologies remained slow for 150 years since Jenner's discovery. During this period newer concepts in microbiology and immunology were established. The progress in the development of viral vaccines had been limited due to non-availability of methods for *in vitro* cultivation of viruses and production of immunizing agent. Propagation of viral vaccines was relied on infected animal organs and tissues as sources of antigens till 1950. The decade of 1950s brought Enders cell culture breakthrough that made possible a number of human and animal viral vaccines.

Subsequently, a number of live attenuated and inactivated/killed vaccines were developed against viral and bacterial diseases. The extensive use of traditional vaccines based on whole organisms resulted in prevention of a variety of human and animal diseases and averted disease epidemics. The classical vaccines have helped in global eradication of smallpox and regional elimination of other diseases including yellow fever, poliomyelitis, rinderpest, plague, tuberculosis, etc. Nevertheless, several viral, bacterial and parasitic diseases acquired immunodeficiency disease syndrome (AIDS), dengue, sheeppox, bluetongue and malaria still continue to pose serious threat and remain unresolved.

Protection from infectious diseases has conventionally depended on the use of either live attenuated or killed vaccines. Nevertheless, some of the currently available traditional vaccines are inadequate for reasons of safety, efficacy and cost effectiveness. Live attenuated vaccines may lead to clinical disease following vaccination due to incomplete attenuation or the vaccine agent may revert back to virulence. Further more, over attenuation of infectious agent may render it less immunogenic. The inactivated/killed vaccines are often unable to generate protective levels of immune response due to loss of important antigenic determinants during inactivation and poor antigen load in the vaccinates. Therefore, more efficacious and safer vaccines are required to effectively deal with infectious diseases.

Advances in molecular biology and unprecedented development of molecular techniques such as recombinant DNA (rDNA) technology have provided hope for developing better vaccines which would be devoid of the problems associated with currently available traditional vaccines. The applications of molecular biology and synthetic chemistry technologies for designing vaccines are also advancing our understanding of the complex mechanism of cellular and humoral immune responses, the nature of antigenic epitopes on the infectious agents and the presentation of such epitopes to the immune system. The immune system recognises those antigenic determinants which are revealed to it. A vaccine is collection of immunological determinants that are presented to the immune system as killed or live or subunit antigens which provoke protective immune responses. There are three compartments of immune system which participate in a classical immune response against vaccines. These are: antigen presenting cells such as macrophages, B cells that make antibodies and T cells which generate cell mediated immune response. Vaccines, to be highly efficacious, must induce immediate effect or responses in addition to long term memory B and T cells, capable of rapid anamnestic protective recall on exposure to the specific antigen.

The unprecedented advancement in our understanding of genetic and antigenic organisation of infectious agents during the past three decades have provided innovative approaches for development of new generation vaccines. A variety of approaches leading to development of new generation vaccine such as subunit synthetic or recombinant peptide immunogen vaccines, deletion mutants based live vaccines, vectored vaccines, nucleic acid vaccines and plant vector based edible vaccines, have been extensively studied during past two decades. It is almost impossible to do justice with the enormous research data that have been generated on new approaches to development of vaccines in the past over three decades in this chapter. Therefore, this chapter aims at briefly analysing the emerging trends in vaccine technologies using selected important diseases as models.

2. PEPTIDE VACCINES

(a) Synthetic peptide vaccines

In early 1960s, it was for the first time demonstrated that carboxyl terminus fragment of the capsid protein of tobacco mosaic virus produced neutralising immune response in rabbits (Anderer, 1963). Subsequently, Anderer and Schlumberger (1965) demonstrated that chemically synthesised peptides corresponding to this fragment would also produce the virus neutralising antibody. This observation led to an extensive research on the synthetic antigens representing fragments of immunogenic peptides of infectious agents. Langebeheim *et al.* (1976) demonstrated that a synthetic fragment of 20 amino acids corresponding to 89-108 of the coat protein of MS2, an RNA containing bacteriophage, elicited antibodies which reacted with the intact virus particle. These initial observation provided foundation for the subsequent exploration of peptide immunogens as vaccines. Most immunogenic determinants of proteins are thought to be conformational rather than sequential. They depend on the three dimensional structure of the molecules rather than on linear sequence of amino acids. The fact that alteration of conformation by denaturation or enzymatic digestion greatly changes the immunogenicity suggests that most antigenic determinants of proteins are conformational rather than sequential. However, studies carried out in the past two decades have revealed that small synthetic peptides having sequences of immunogenic epitopes on a protein will react with antibody to the intact native antigen and in some cases neutralise the biological activity.

The understanding that only a few epitopes on structural proteins of infectious agents play an important role in inducing protective immune responses led to the development of synthetic peptide vaccine concept. A peptide vaccine would have several advantages including long shelf life (even at the temperatures in tropical countries), well defined composition, no possibility of the adventitious presence of live infectious agents and no requirement for containment facilities necessary for handling large amounts of highly infectious agents.

Now application of computer in prediction of T and B cell epitopes on the surface proteins of infectious agents has greatly enhanced our ability to design synthetic peptide vaccines. In 1981, a computerised analysis of the amino acid sequence of the hepatitis B virus surface antigen was used to predict the location of an antigenic epitope in a limited portion of the molecule comprising residues 138-149 (Hopp, 1981). This sequence was synthesised and its ability to bind the hepatitis B virus specific antibodies was tested. The result indicated that the synthetic peptide has ability to bind with the specific antibodies (Hopp, 1981). Attempts to produce the native conformation of the antigenic epitopes by producing a cyclic peptide did not yield an antigen clearly superior to the linear peptides (Lerner *et al.*, 1981).

The identification of genes encoding immunogenic proteins and availability of nucleic acid sequences allowed derivation of amino acid sequences of the immunogenic peptides. Since the synthetic peptides are reasonably safe and free of the problems associated with the classical vaccines, a great deal of efforts are now being made to develop synthetic peptide vaccines (Arnon, 1986; Sela & Arnon, 1992; Sela, 1996). A number of different infectious diseases including protozoan have been targeted for synthetic vaccine design and extensive research has been done on this area. Synthetic peptide vaccines with built-in adjuvanticity have been reported to evoke high level and long-lasting immunity.

Foot-and-mouth disease (FMD) is the most dangerous disease of livestock. Recent outbreaks of the disease in United Kingdom played havoc economically as well as politically. Considering its immense global importance Foot-and-mouth disease virus (FMDV) has been extensively studied and still an ideal vaccine is not available against it. Early immunological studies with FMDV allowed recognition of two major antigenic sites on capsid protein VP1: site A, located within the G-H loop of VP1 position (140 to 160) and site C, located in the carboxy-terminal region of VP1 (Bittle *et al.*, 1982; Strombmaier *et al.*, 1982). A synthetic peptide vaccine based on site A or on constructs comprising sites A and C induced significant levels of anti-FMDV neutralising antibodies and protection in guinea pigs, swine and cattle (Bittle *et al.*, 1982; Di Marchi *et al.*, 1986; Morgan & Moore, 1990). The previous studies on the immunogenicity of FMDV peptides were carried out with limited number of animals and the peptides were designed to contain only VP1 regions characterised as including mainly B cell epitopes. However, the immune response against FMDV is T cell dependent (Collen, 1994), and therefore synthetic vaccines should incorporate T cell epitopes to adequately induce humoral and cellular immune responses (Collen *et al.*, 1991; Francis *et al.*, 1987).

Recently a number of FMDV T cell epitopes which have become potential candidates to be incorporated in to vaccine peptides have been identified and demonstrated to be recognised by a significant number of individual cattle and pigs (Collen *et al.*, 1991; Rodriguez *et al.*, 1994). Taboga and associates (Taboga *et al.*, 1997) reported a study carried out on the immune response and protection of cattle afforded by a number of synthetic peptides which included B and T cell epitopes corresponding to FMDV C3 Argentina 85 (Arg 85). In contrast with previous studies, this large scale vaccination experiment with an important FMDV host revealed considerable difficulties for vaccines based on synthetic peptides to achieve the required levels of efficacy. The emergence of escape mutants of lymphocytic choriomeningitis virus in mice vaccinated with peptide consisting of T cell epitopes has been reported (Weidt *et al.*, 1995), again indicating the limitations of simple, peptide vaccines to achieve protection against highly variable RNA viruses. In the case of FMDV, since deviant antigenic forms of antigenic site A are as immunodominant as the parenteral sequences (Borrego *et al.*, 1995), a possible increase in the efficacy of peptide vaccine may be attained by enlarging the repertoire of independent B cell and T cell epitopes together with variant forms of some of them. This may expand the over all response as well as the ability to protect against mutants that may potentially arise in the course of viral replication (Taboga *et al.*, 1997).

Malaria is one of the most important protozoan disease of human beings in tropical countries. Though chemotherapy is available, drug resistance has become a major problem due to indiscriminate use. Therefore, efforts are being made to develop an effective vaccine using molecular approaches. Patarroyo *et al.* (1988) reported that a synthetic peptide vaccine could protect human volunteers against challenge with blood stage *Plasmodium falciparum.* This study set in motion an intense research on synthetic peptide vaccines. Many field trials were conducted using Colombian synthetic peptide polymer (SPf 66) containing epitopes from at least four distinct proteins and includes sequences that have not been defined as being from *Plasmodium* origin. The mechanism of the vaccine is not well understood but may involve targeted interaction between the malaria parasite and the red blood cells (Millet *et al.* 1993). Recent vaccine trials with SPf 66 have shown at least 31% efficacy (News and Comments, Science, Vol. 267 (1995), p 3207). The success of the Colombian synthetic malaria vaccine paved the way for development of synthetic peptide vaccines against other protozoan parasites of man and animals.

Robinson *et al.* (1995) reported that 40-mer synthetic peptide vaccine against internal nematode parasite *Trichinella spiralis* produced high level of protection against subsequent challenge in female NIH mice. The 40-mer synthetic peptide was prepared from 40 kDa immunodominant glycopeptide secreted by *T. spiralis.* This was the first synthetic peptide vaccine against an intestinal cestode. Su *et al* (1999a) immunised mice with synthetic multiple antigen peptide on 2 known antigens of *S. japonicum.* Each mouse was inoculated with 200 mg peptides, 3 times at one week intervals. Two weeks after immunisation they were challenged with 40 2 cercariae of *S. japonicum.* The immunised mice showed 24.3 to 47.9 % worm reduction and 35.1 to 50.4 % egg reduction as compared with control. The study suggested that multiple antigen peptides of *Schistosoma* are quite effective in inducing immunity and need further studies.

Though significant protective immune responses have been achieved in experimental conditions, the synthetic vaccines are far from the commercialisation mainly due to poor immunogenicity, short term immunity and high cost of vaccination. In the past couple of years, the researchers have been focusing on improving the immunogenicity by incorporating B and T lymphocytes epitopes and better adjuvants in the synthetic peptides. The availability of cloning and expression techniques enable researchers to produce full length and chimeric immunogenic peptides by recombinant DNA technology. The recombinant peptides have shown better promise than synthetic peptides as vaccine.

(b) Recombinant peptide vaccines

The understanding that isolated proteins of viruses, bacteria and parasites can provoke protective immune response, led to application of genetic engineering in cloning and expression of critical genes encoding immunogenic proteins in prokaryotic and eukaryotic expression system. The concept of a subunit vaccine containing only those immunogenic components necessary to elicit a protective response and excluding others not necessary for immunity, has several attractive features. The first and the foremost is the element of safety. Subunit vaccines lack infectivity and hence no complication of infection arising out of vaccination in immunocompromised individuals. Extragenous antigens are lacking, thus reducing or eliminating pyrogenic, allogenic, immunosuppressive and other undesirable reactogenicity. Subunit vaccines also exhibit a great deal of stability.

(i) *Vaccines against viral diseases*

All most all pathogenic viruses have been targeted for peptide vaccine development but due to limitation of space only a few have been described here. Foot-and-mouth disease virus is a small RNA virus which consists of only 4 proteins. FMDV because of small genome size and its great economic importance to livestock industry, became handy model to test the feasibility of using recombinant capsid protein VP1 as a peptide vaccine as early as 1981 (Kleid *et al.*, 1981). This was probably the first demonstration of use of recombinant protein as vaccine in animals. Even though initial experiments indicated the potential of recombinant capsid protein VP1 of FMDV to protect swine, it soon became clear that VP1 was less immunogenic as compared to the equivalent amount of antigen administered in the form of virus particles (Kleid *et al.*, 1981). Subsequently, a number of other peptide vaccines against bacterial, viral and parasitic agents have been developed and evaluated in laboratory experiments and some of these have undergone field trials with great promise. The most successful recombinant peptide vaccine is against hepatitis B which is being used through out the world. The success of the first recombinant vaccine against hepatitis B for use in human has paved the way for development of other recombinant peptide vaccines.

Different genetically engineered expression systems (insect cells, yeast or mammalian cells) have been used to produce immunologically important proteins of a variety of human and animal pathogenic viruses, bacteria and parasites. For example, human immunodeficiency virus has been extensively researched with a view to develop suitable peptide vaccine. The viral envelope protein gp120 or gp160, produced by recombinant DNA technology, have been formulated with alum or novel adjuvants. The recombinant envelope proteins produced in insect and yeast cell system were found to be very poor in eliciting neutralising immune response in human volunteers. However, gp120 vaccine produced in mammalian cells could provoke neutralising antibodies in 100% human volunteers and these antibodies also cross neutralised the strains other than that used for vaccine development (Schultz, 1996). Efforts are being made to improve the immunogenicity of envelope protein vaccines. It has been observed that envelope protein containing the entire gp120 plus only part of gp41 forms oligomeric structures of these proteins and that would mimic the truly native conformation of viral spikes and therefore be more immunogenic with respect to inducing neutralising antibodies (Broder *et al.*, 1994). Several HIV recombinant peptide vaccines are at various stages of development (Table 1).

Table 1. Human immunodeficiency virus recombinant peptide vaccines under development

Vaccine	Manufacturer
gp120, CHO expressed	Genetech
gp120, CHO expressed	Biocine
gp120, yeast expressed	Biocine
gp160, baculovirus exp.	MicroGeneSys
gp 160, vaccinia exp.	Immuno-Ag
+ gp120, CHO expressed	+ Genetech
+ gp120, CHO expressed	+ Biocine
+ gp120, yeast expressed	+ Biocine
+ gp 160, vaccinia exp.	+Immuno-Ag
Octameric V3 peptide	UBI
Oral, V3 peptide cocktail	UBI

Table 1 shows how many different commercial organisation are engaged into development of recombinant peptide vaccine against one of the most feared diseases of human beings.

Bluetongue is an important viral disease of ruminants against which considerable efforts have been made to develop recombinant peptides vaccine. It is now well known that VP2 and VP5 of bluetongue virus (BTV) together generate neutralising immune response. This has raised the possibility of using recombinant immunogens as vaccine against BTV. Roy and co-workers (1990) reported that the VP2 obtained through recombinant baculovirus also elicited immune response that protected the immunised sheep against homologous virulent virus challenge. These workers further reported that combination of VP2 and VP5 elicited better neutralising immune response which conferred protection. It has been suggested to be due to synergistic effect of two proteins which probably offers better conformational presentation of the neutralising epitopes. The purified VP2 and VP5, either produced traditionally or through recombinant DNA technology, produce protective immune response which is short lived. Thus peptide vaccines have limited utility and are not economically attractive vaccine to manufacturers. Another possibility of making hybrid VP2 and VP5 representing the immunogenic epitopes of different serotypes also remains workable alternative.

Tsibezov *et al.* (2000) evaluated a recombinant peptide vaccine against swine fever virus. The authors tested immunogenicity of a recombinant major surface glycoprotein E2 from a virulent Shimen strain of classical swine fever virus (CSFV) and reported that immunisation of 3 months old piglets with 200 mg of recombinant glycoprotein E protected the animals against lethal challenge with virulent CSFV strain.

There are several other human and animal viral pathogens against which considerable research efforts have been made to develop recombinant peptide vaccines.

(ii) *Vaccines against bacterial diseases*

A number of bacterial diseases have been targeted to produce effective peptide vaccines. However, only few have been discussed here to demonstrate the potential of peptide vaccines against bacterial diseases. Enterotoxins of *Vibrio cholerae*, *E. coli* and *Staphyloccoci* have been evaluated as immunizing agents after attenuation. The fusion protein representing immunodominant antigens of ESAT-6 and 85B has been recently evaluated as a candidate vaccine against tuberculosis in mouse model. The immunisation of mice with the fusion protein along with an adjuvant induced high level of protective immunity comparable to *Mycobacterium bovis* BCG induced protective immunity. The fusion protein vaccine induced efficient immunological memory, which remained stable for 30 weeks post vaccination (Olsen *et al.*, 2001).

Recently, West and co-workers (2001) demonstrated that recombinant transferrin binding proteins (Tbps) of *Neisseria meningitidis* elicit humoral immune response in mouse model. The protection induced by TbpA was broad based and may provide protection against wider range of meningococcal strains. Martin *et al.* (2001) evaluated recombinant antigen enterotoxin A2/B chimeric mucosal immunogens for intranasal immunisation of BALB/c mice. The chimeric proteins produced significantly higher immune response characterised by elevated levels of IgA and IgG antibodies. The study suggested that chimeric peptide vaccines have better co-stimulatory effect on $CD4^+$ T cells. Recently, Stiles *et al.* (2001) evaluated recombinantly attenuated staphylococcal enterotoxin B

(SEB) as vaccine in murine model. This study showed that nasal or oral immunisation of mice with SEBv and immunoadjuvant cholera toxin induced IgG and IgA response in saliva and IgG, IgA and IgM response in serum. The mice immunised mucosally were protected against 50% lethal dose of wild type staphylococcal enterotoxin B. Thus demonstrating that nasal or oral administration of SEBv with cholera toxin elicits systemic and mucosal antibodies to SEB that protect against SEB induced lethal shock.

Conlon *et al.* (2000) used recombinant outer surface A (OspA) vaccine (Recombitek) of *Borrelia burgdorferi* to immunise dogs and reported that the vaccinated dogs withstood the virulent challenge. No clinical signs and negative isolation of spirochete on biopsy indicated that the recombinant vaccine blocked infection with *B. burgdorferi* in all (20/20) vaccinated dogs. The authors reported that the adjuvanted recombinant *B. burgdoreri* vaccine protected 100 per cent of vaccinates against infection after severe challenge with *B. burgorderferi.*

(iii) *Vaccines against parasitic diseases*

Control of parasitic diseases is one of the most difficult tasks. Though there are drugs but due to side effects and development of drug resistance, vaccines are an important international goal. A number of approaches are currently being used to realise this goal. Recombinant peptide vaccines have also been produced and evaluated in laboratory trials against a number of human and animal parasites. A few salient examples are briefly described here. *Theileria* is one of the important protozoan pathogen of cattle in tropical and semitropical countries. Molecular and immunological studies on *Theileria parva* have led to the identification of a 67 kDa polymorphic immunodominant antigen (PIM) of sporozoites and a 85 kDa complex antigen of *T. annulata.* The p67 gene has been expressed in pMGI, where it is fused with first 85 amino acids of non-structural protein (NS1) of influenza virus A. The recombinant antigen (p67-NS1) has been reported to induce high titre neutralising antibodies in cattle and protected 6 out of 9 animals against homologous challenge and 6 out of 11 against heterologous challenge (Musoke *et al.* 1992). Suitable modification of recombinant fusion protein antigen and optimisation of the vaccination regime might provide protection in larger proportion of the vaccinated cattle. This could eventually form the basis for a successful commercial vaccine against *T. parva.* Wright and his colleagues have been pursuing the goal of developing a defined antigen vaccine against *Babesia bovis* over past 10 years. Some 80 vaccination trials have been conducted in cattle using fractionated *Babesia* antigens. This led to the isolation of several recombinant host protective antigens (Wright *et al.* 1992). Two antigens, 42 and 58 kDa of *B. bovis* and *B. bigemina,* respectively, were found to induce protective immune response. These merozoite antigens have been expressed in various expression systems and the recombinant antigens were tested for immunogenicity in cattle (Musoke 1991).

Taenia ovis, an important pathogenic parasite of sheep, has been targeted for development of peptide vaccine. Originally, Johnson *et al.* (1989) described use of 45 W recombinant molecule as vaccine against *T. ovis* in sheep. The expression of GST-45W antigen using plasmid vector similar to pGEX-1 produced a soluble glutathion-S-transferase fusion protein (GST-45W) which provided 94% protection in vaccinated sheep. Subsequently, Richard *et al.* (1995) examined a number of issues related to commercialisation of GST-45B/X recombinant immunogen and reported that (i) the immunogen retains its potency at 2-8°C for at least 18 months, (ii) saponin adjuvant was effective and does not leave residual lesions after subcutaneous inoculation, (iii) two doses of GST-45B/X given 4 wks apart provide solid protection for 3-6 months and partial protection for 6-12 months (iv) transfer of maternal antibody was quite efficient as evidenced by high level protection up to 6 wks of age in lambs. (v) GST-45B/X does not cross protect against other species of *Taenia.* (vi) Doses of vaccine from 10-100 mg are equally effective. The field trials using GST-45 B/X vaccine antigen have been conducted by Harrison *et al.* (1993). The result of the vaccine trial suggested that the recombinant vaccine could produce upto 98% protection. Heath and co-workers (1993) have cloned two additional antigens of *T. ovis* in *E. coli* having molecular weight of 16 kDa and 18 kDa. Fusion protein GST-16 and GST-18 produced 73-100% and 96-100% protection, respectively in sheep in vaccination trials. GST-18 was often found to be better than GST-45W (B/X). There is no homology in the sequences of these three proteins. Further studies are required to be carried out on GST-16 and GST-18 antigens to enable their commercial production. Similar recombinant products could be developed against *T. saginata* and *T. solium.*

A recombinant Ts 39 fusion protein (FP) derived from *Trichinella spiralis* induced protective immune response in BALB/c mice. The mice were immunised with PF emulsified with Freund's complete adjuvant (FCA) three times at weekly intervals. The immunised mice were challenged with a lethal dose of 1200 larvae of *T. spiralis.* The findings suggested that the recombinant Ts 39FP could be a potential immunogen for vaccine development (Sun *et al.* 1994).

Liu *et al.* (1995) reported that pigs vaccinated with 26 kDa-GST antigen (Sj 26-GST), a recombinant peptide of *Schistosoma japonicum,* produced anti-fecundity response. Vaccinated animals had reduced worms load after challenge infection. The most striking observation of the vaccine trial was the significant reduction in the number of eggs in livers of the vaccinated pigs. Boulanger *et al.* (1994) reported that injection of recombinant rSb28GST of *S. bovis* in FCA resulted in good antibody response in goats. Further studies are required to explore the potential of this recombinant antigen as vaccine. Baras *et al.* (2000) evaluated antigens entrapped microparticles produced by spray drying technique and various polyesters polymers. The recombinant 28 kDa glutathione-s-transferase of *S. mansoni* (rSM28GST) characterised by major epitopes involved in the active site of this enzyme was selected as model antigen. A single intra-peritoneal injection of rSm28GST entrapped in microparticles resulted in specific antibody response. This study indicated that microencapsulation of antigens by spray-drying preserved their crucial characteristics required for generating effective immune response after single dose administration.

GST antigens of *Fasciola hepatica* have also been selected as target antigens for development of a recombinant vaccine against liver fluke because recombinant GSTs of schistosomes have been successfully used to induce significant protection against infection with *S. japonicum* and *S. mansoni* in animal models. Vaccination of sheep with affinity purified GST from adult *F. hepatica,* induced 57% reduction in the number of flukes developing following a challenge infection with metacercariae (Sexton *et al.* 1990).

Vaccination against haemonchosis offers an attractive alternative to anthelmintics. Vaccines mainly are of two categories, the "hidden" antigens (usually derived from gastrointestinal tract of the adult *Haemonchus contortus*), or the "natural" antigens (those exposed to the host immune system during infection, usually derived from the larval stage). Promising results have been obtained using the hidden gut antigens H11 or H110D and more recently with the other hidden antigen, H-gal-GP (Smith *et al.* 1994, Newton *et al.*, 1995). Chances of development of a commercial vaccine using H11 are quite good because it provides quite high level of protection against haemonchosis.

Seven common species of *Emeria* cause coccidiosis in poultry. The conventional methods for development of vaccines against coccidiosis have not been successful in developing efficacious vaccine. Therefore, recombinant peptide vaccine approach is being pursued. Jenkins *et al.* (1988) have characterised the immunodominant surface antigens of E*meria acervulina* sporozoites and merozoites. The 35 kDa polypeptide derived from p250 merozoite antigen was shown to induce specific T cell stimulation *in vitro*. This 35 kDa peptide was able to induce partial protection in chickens against challenge with sporulated oocysts of *E. acevulina*. Oocyst production in vaccinated birds was decreased 30% compared with controls. Danforth *et al.* (1989) achieved significant protection by vaccinating birds with B-gal fusion peptide containing 31 kDa protein of *E. tenella*. Miller *et al.* (1989) working with A.H. Robbins Co. and Genex Corporation also achieved partial protection against *E. tenella* using recombi- nant antigen. Subsequently, Jenkins *et al.* (1991) used expression vector and the host bacterium capable of controlling *in vivo* expression of the antigen and achieved significant protection against challenge. All the recombinant vaccine trials have achieved partial protection so far, hence it is difficult to predict future for recombinant vaccines against coccidiosis.

(iv) *Vaccines against arthropods*

The development of acariside resistance and increased environmental concern about residues in the food chain calls for the exploration of alternative environment friendly control strategies for vector-borne diseases. The most attractive alternative is the anti-insect vaccination approach which has advanced from the possibility to reality in the past 10 years. The anti-insect vaccines have the advantages of target species specificity, ease of administration, lack of human health risk, environmental safety and cost effectiveness of immunisation. Considerable progress has been made in identifying the protein components of ticks which might serve as immunogens to augment host defences. *Boophilus microplus* (cattle tick) transmits babesiosis and anaplasmosis in livestock. Bm86, an 89 kDa glycoprotein with extracelullar location on digestive cells of tick gut (Gough and Kemp 1993), has been expressed in a variety of expression systems and the recombinant Bm86 has been evaluated as vaccine.

Recombinant Bm86 antigen expressed in *E. coli, Aspergillus nudulans* or *A. niger* (Turnbull *et al.* 1990) or *Pichia pastoris* (Rodriguez *et al.*, 1994) produced protective immune response. A vaccine based on recombinant Bm86 (expressed in *E. coli*) named "tickgard" was commercialised in Australia in 1994. Subsequently, another recombinant vaccine, GavacIM/Heber Biotech S.A.-Cuban has been commercialised in Cuba. (Rodridguez *et al.*, 1995). It contains sterile protein of high purity formulated in oil adjuvant and stable at 2-8°C for at least 3 years. Recommended dose is 2ml which contains 50 mg of antigen. Sixteen field trials involving over 3500 cattle were conducted before commercial registration of the recombinant vaccine. Both these vaccines are based on recombinant Bm86. The extensive field trials have revealed that both the vaccines effectively control *B. microplus* tick population in grazing cattle.

Though the recombinant subunit vaccines are quite safe, there are still considerable problems like poor immunogenicity, high cost of production, short-term immunity, etc. Until these problems are tackled effectively, it is unlikely that the recombinant subunit vaccines will hit the market in near future.

3. VIRUS LIKE PARTICLES (VLPS) AND CORE LIKE PARTICLES (CLPS) VACCINES

To avoid chances of reverting back to virulence or possibility of integration of genes of infectious agents into host cell genome, efforts have been made to produce non-infectious VLPs and CLPs devoid of genetic material using yeast, mammalian, insect and plant expression systems by genetic engineering. In the past 10 years remarkable success has been achieved in this area. A few models are briefly describes here.

Because of close association of immunogenicity and formation of VLPs of infectious hepatitis B virus, Wampler *et al.* (1985) expressed infectious hepatitis B virus gene encoding HBs Ag in yeast and reported that the yeast expressed recombinant HBsAg was similar to HBsAg particle found in infected human serum. HBsAg -VLPs produced in yeast could be successfully converted into efficacious commercial vaccine for human use.

Another important viral pathogen, Norwalk virus has also been cloned in baculovirus expression vector and has been demonstrated to produce VLPs similar to authentic Norwalk virus (NV). Jiang *et al.* (1992) expressed capsid protein of NV in insect cells produced 58 kDa peptide that self assembled into noninfectious VLPs lacking viral genome. The VLPs synthesised in insects cells strongly reacted with sera from humans infected with NV. The electron microscopy revealed that VLPs were similar to the authentic virus particle and quite stable at 4°C after lyophilization and resistant to pH 3.0 (Jiang, 1992). Bansal *et al.* (1992) developed VLPs against parvovirus B19 by expressing VP1 and VP2 in baculovirus system which resulted into synthesis of VLPs. The VLPs thus obtained have been shown to be immunogenic and are under clinical trials. Perhaps this is the only way to produce vaccine against this virus which normally requires differentiating erythroid cells to replicate and hence very difficult to grow *in vitro.*

VLPs and CLPs have also been produced using critical genes of BTV. The genes corresponding to outer capsid proteins (VP2 and VP5) and major core proteins (VP3 and VP7) have been cloned in two separate baculovirus expression vectors. The simultaneous co-infection of the *Spodoptera frugiperda* cells with both the recombinant baculoviruses produced the VLPs not morphologically distinguishable from the conventionally cell culture produced authentic BTV particles. The small scale studies conducted in sheep in South Africa and Australia using the BTV-VLPs as candidate vaccine, have clearly suggested that the VPLs when used in combination with appropriate adjuvant elicit protective immune response and the vaccinated animals resist the homologous virulent virus challenge (Johnson *et al.*, 1996). Recent preliminary studies carried out in Merino sheep in Australia using recombinant CLPs along with adjuvant have suggested that the immunised sheep partially resist the virulent virus challenge (Hasseini *et al.*, 1996). Unlike VLPs the partial protection induced by CLPs is not serotype restricted. This makes CLPs very attractive vaccine candidate because the protection is broad based possibly due to cell mediated immune response. The CLPs have been proposed as potential broad spectrum BTV vaccine. It has been suggested that in the face of outbreak BTV infection, where serotype specific vaccines are not available or the serotype is unknown, CLPs could be used to reduce the impact of the infection (Hasseini *et al.*, 1996). Since these particles are non-infectious and devoid of genetic material, these do not have the risks associated with live attenuated vaccines. More over, CLPs can be engineered to distinguish between vaccinated and BTV infected sheep by developing appropriate ELISA.

Similarly VLPs of rotavirus have also been produced by co-infecting insect cells with baculovirus recombinants expressing rotavirus structural proteins VP2, VP4, VP6 and VP7 (Crawford *et al.*, 1994). These VLPs have demonstrated to produce protective immune response in mice when administered parenterally (Crawford *et al.*, 1994; Fernandez *et al*, 1996). Recently, the rotavirus VLPs have been shown to produce strong immune response when administered mucosally using adult mouse model (Fernandez *et al*, 1996). The authors opined that VLPs offer a promising, safe and non replicating vaccine for rotavirus.

The VLPs lack the genetic material hence these can not cause infection to the vaccinated animals and can not pass on to the foetus. Since these are non-infectious, there is no risk of reverting back to virulent form and transmission by the vector. Another advantage of VLPs is that a combination of different serotypes can be made in the form of cocktail which protects against a number of serotypes prevalent in a particular geographical region. One can visualise potential of the technology to produce vaccine chimeras representing different serotypes.

4. GENETICALLY MANIPULATED LIVE VACCINES

Naturally found avirulent mutants or deliberately attenuated infectious agents have been successfully used as vaccines against several bacterial and viral pathogens. Most of the attenuated vaccines are very immunogenic. Now with the advances in molecular biology and biotechnology, it is possible to knockout virulence character(s) by deleting or altering certain gene(s). The deletion of certain gene(s) may render the pathogen avirulent but capable of replication in host. Thus genetic engineering has made it possible to develop live vaccines quickly with precisely defined modifications. Using this approach, Daniel *et al.* (1992) attenuated simian immunodeficiency virus (SIV) by deleting *nef* gene. The *nef* gene deleted SIV vaccine induced strong immune response in macaques which withstood highly virulent SIV challenge 2 years after vaccination. This observation showed the potential of deletion mutants as live vaccines. Subsequently, Mettenlieter (1995) applied the same concept to develop a live vaccine against pseudo rabies virus of pigs. The thymidine kinase (TK) gene deleted pseudo rabies virus strain has been licensed for use by United States Department of Agriculture. In addition, with a view to develop live vaccines which could be differentiated from the natural infection, the vaccine strains have been genetically manipulated in such a way that they do not express one or more of the non-essential glycoproteins gE, gC, or gG. Thus, the animals immunised with these strains do not produce antibodies against these glycoproteins. Hence, vaccinated animals can be easily differentiated from the naturally infected ones. Similar approach has also been used to design a live vaccine against bovine herpes virus 1 (BHV 1) which lacks gE (Strube *et al.*, 1995).

Deletion of poly C tract at 5' portion of encephalomyocarditis virus (a picornavirus), renders it avirulent but retains immunogenicity. The truncated virus has been found to induces strong protective immune response and stable in host. Similar approach has been used in FMDV. Several such vaccines including improved polio (Nomoto, 1995) and dengue vaccines (Lai *et al.*, 1992) are likely to be used. Perkin *et al.* (1997) demonstrated the feasibility of using genetically modified live vaccine against influenza A.

Recently, Walsh *et al.* (2000) have reported use of a genetically modified rinderpest vaccine expressing green fluorescent protein (GFP). This vaccine was evaluated as a recombinant marker vaccine due to its potential to express GFP in the infected cells. All the vaccinated cattle and goats were completely protected when challenged with the virulent rinderpest virus. The marker vaccine has potential applications in global eradication of rinderpest.

Similarly bacterial strains can also be genetically manipulated by creating defined mutations in genes. For example, *Bacillus anthracis* strain which do not produce capsule, are avirulent, although they are capable of replication in host. However, the *B. anthracis* strains that are defective in aromatic amino acid biosynthesis are less virulent than non-encapsulated strains and yet produce protective immunity in experimental disease models (Ivins, 1986, 1990). Therefore, in theory it is possible to attenuate organism to varying degree depending upon the mutations that are introduced. This technology can be applied to virtually any species of bacteria although it is easier in organism that are genetically characterised. Ideally atleast 2 deletion mutations should be introduced in the organism in order to ensure safety. For example Munson (1989) developed *cya/crp* double mutants of *Salmonella typhimurium* that when used as vaccine resulted in protection against both *Salmonella* and *E. coli* infection in poultry.

Different issues related to the use of genetically manipulated human pathogens as vaccines have been debated for the past few years and the consensus has emerged that the precisely manipulated clones may not pose any serious problem. However, it is likely that this approach could be used more effectively for development of veterinary vaccines as the perceived risk-benefit considerations for human vaccines are very different including residual virulence and reversions. The clinical trials of genetically manipulated vaccines have not yet been conducted in humans.

5. RECOMBINANT VECTORED VACCINES

This approach involves expression of heterologous gene(s) encoding immunogenic protein of the target pathogen in an avirulent viral/bacterial organism which are harmless to vaccinated animals and elicits protective cellular and humoral immune responses against the foreign proteins. This approach also enables expression of more than one foreign genes encoding immunogenic proteins, thus paving the way for development of multivalent recombinant vaccines against human and animal viral infectious diseases. In the past over two decade, a wide range of viruses and bacteria have been evaluated for expression of relevant foreign genes. Vaccinia virus, because of its use as vaccine against smallpox and greater understanding of its biology, has been extensively used as live vector for experimental vectored vaccines against several diseases including, rabies, HIV, measles, influenza, tuberculosis, malaria, etc. Subsequently, other pox viruses such as capripox (Romero *et al.*, 1993), fowl pox and cannarypox viruses (Taylor *et al.*, 1992, Taylor *et al.*, 1995) have also been used for development of animal and poultry vectored vaccines. Other viruses which are being used as vectors include adenovirus, herpesvirus, papova virus. Similarly, several bacterial vectors including *Salmonella*, BCG,

Shigella, Vibrio cholerae are also being extensively studied with a view to develop bacterial vectored vaccines.

Recombinant attenuated vaccinia virus encoding immunogenic protein of rabies virus (Taylor, *et al.*, 1992) and rinderpest virus (Taylor, *et al.*, 1995) have been developed and evaluated in some clinical trials. The genes of several other viruses have been also expressed in a variety of live vectors and their vaccine potential has been studied extensively. Fowl pox virus has been used for expression of poultry pathogens. The fowl pox vectors containing gene(s) encoding immunogenic proteins of Newcastle disease, turkey rhinotracheitis, avian influenza, Marek's disease, reticulo-endotheliosis and infectious bursal disease viruses. Similarly turkey herpes virus (HVT) vectors have been employed to express genes of Marek's disease virus as well as Newcastle disease virus. A double recombinant of HVT expressing genes of both Marek's and Newcastle disease viruses has also been reported. These vector recombinants have been shown to produce immunity against the target viral diseases of poultry.

Lagrandderie *et al.* (1997) evaluated *Mycobacterium bovis* BCG-SIV *nef* recombinant and showed that it induces local and systemic cytotoxic T lymphocyte (CTL) responses. These findings provided an evidence that orally administered rBCG SIV *nef* may contribute to local defences against the viral infections. Recently, Stephensen and co-workers (1997) have evaluated canine distemper virus ferret model for testing vectored vaccines against morbilliviruses. An and Whitton (1997) have successfully expressed a number of epitopes from several pathogens in the form of minigene using vaccinia virus as vector. The immunological results of minigene vectored vaccine have suggested the feasibility of developing a multivalent minigene vaccine. Tang and co-workers (1997) have showed an innovative way of immunising animals by applying the adenovirus vector carrying gene encoding human carcinoembryonic antigen (Ad-hcea) onto the skin of mice (strain C57BL/6). The serum of the mice immunised this way contained antibodies which reacted with human carcinoembryonic antigen in immunoblots. The authors suggested that this method of immunisation of animals by genetic vectors is simple, painless and economical. This approach has potential for developing vaccines that could be administered by individuals without specific medical training or equipment. Townsend and coworkers (1997) have recently demonstrated that recombinant retroviral vectors expressing HBcAg and eAg antigens genes of hepatitis B virus induce long lived CTL responses in the immunised animals. However, concerns have been raised about the use of retroviral vectors in human beings because of the ability of retroviruses to get integrated in host cell genome.

The vaccinia virus has been proposed for development of human vectored vaccines. However, it suffers from several drawbacks such as complex *in vivo* replication demand, less than perfect virulence characteristics, undesirable host range traits. Further research is required to define and characterise these limitations. For example, the persistence of adenovirus in lymphatic tissues and herpes viruses in neuronal tissues and progressive course of vaccinia virus infection in immunocompromised individuals must be studied further (Minor, 1996). To overcome some of these problems replication defective and restricted host range vectors have been developed in the past couple of years. A modified vaccinia virus Ankara (MVA) has been generated by over 500 passages of the parental strain in chicken embryo fibroblast, during which it became severely host restricted and unable to propagate efficiently in mammalian cells. In this respect, MVA resembles avipox viruses which are also being developed as safe vaccines (Romero *et al.*, 1993). The block in replication of MVA in human and other mammalian cells occurs at a step in virion assembly, allowing unimpaired expression of early and late viral and recombinant genes (Sutter & Moss, 1992). Thus, MVA is an efficient as well as safe vector. Moss *et al.* (1996) reviewed the use of MVA as vector for development of vaccines and suggested that replication defective recombinant MVA can be used as host restricted vector for vaccines development. Intra-muscular inoculation of MVA-IN-$F_{ha/np}$ induced cellular and humoral immune responses in mice and protected against lethal challenge with influenza virus. Similarly intra-muscular immunisation of rhesus macaque with MVA-SIVenv/gag/pol greatly reduced severity of the disease caused by SIV. These observation suggested that replication defective host restricted vaccinia virus can be used as safe vaccine vector (Moss *et al.*, 1996).

Recently, Monteil *et al.* (2000) evaluated a recombinant replication defective adenovirus harbouring gD, gC and gB genes under control of cytomegalovirus promoter as vaccine against pseudo rabies virus (Aujeszky virus). The piglets vaccinated i/m with recombinant adenovirus, developed comparable neutralising antibodies response against pseudo rabies virus. This study suggested that one shot vaccination of piglets at birth could provide life long protection. Like live viral vectors, several attenuated bacterial species have been evaluated as vectors. The most extensively studied bacterial vectors include *Salmonella*, *Shigella*, BCG and *Vibrio cholerae*. The most common attenuated *S. typhimurium*, *S. dublin* and *S. typhi* are *aro* mutants and *cya*, *crp* mutants. Several human and animal protozoan genes have been cloned and expressed in attenuated *Salmonella* spp. The serine-rich *Enomeba histolytica* protein (SREHP) has been expressed in *Salmonella* (Zhang *et al.*, 1996). The gene encoding p67 protein of *T. para* has also been expressed in *Salmonella typhimurium*. Preliminary vaccination trials Salmonella-p67 recombinant protected 2 out of 3 cattle immunised under laboratory experimental conditions (Dr. Subhash Morzaria and Dr. Musoke, 1996, Personal communication). Attenuated BCG has also been used as live attenuated vector for expression of HIV-1 VP3 (Honda *et al.*, 1995), Measles virus nucleoprotein (Fennelly *et al.*, 1995), *S. pneumoniae* PsA antigen (Stover *et al.*, 1995) and *Lieshmania major* gp63(Connell *et al.*, 1993).

Though spectacular progress has been made in the field of vectored vaccines, these are a far from being licensed for use in human beings due to a number of regulatory, social, technical, psychological and ethical issues need to be resolved. The specific mutations in the vector viruses/ bacteria and their possible impact on virulence, cell tropism, immunogenicity and host range need to be clearly understood. The regulatory requirements for human and veterinary vectored vaccines are quite different. In the case of veterinary vaccines, the main consideration is on cost-benefit ratio while in case of human vaccines, the main focus is upon safety. Not only are adverse reactions objectionable in human vaccines but there can be the legal basis for liability claims that influence all other aspects of vaccine development and use (Minor, 1996). A limited amount of information is available based on which it is possible to anticipate what reasonable requirements might be necessary based on past precedents and special information. The regulatory concerns for licensable recombinant vector vaccines have been documented by Minor (1996).

6. NUCLEIC ACID VACCINES

Nucleic acid as a vaccine is a relatively recent approach. It is based on the observation that when plasmid containing suitable promoter and the gene(s) encoding immunogenic protein(s) is injected into animals, the gene may be expressed and generate humoral and cellular immune responses (Tang *et al.*, 1992; Wolff *et al.*, 1990, Ulmer *et al.*, 1993; Cox *et al.*, 1993). The introduction of naked nucleic acid *in vivo* as a plasmid for the purpose of generating immune response is called 'naked DNA vaccine'. A number of terms such as DNA vaccine, genetic vaccine and genetic immunisation have been used by different workers. To avoid confusion, the term 'nucleic acid vaccine' was coined and recommended by the scientists initially at the World Health Organisation meeting at Geneva in 1994 for any procedure involving DNA or RNA based immunisation. The ability of naked DNA to elicit immune response was accidentally discovered by Wolff *et al.* (1990) during an experiment that evaluated the potential of liposomal delivery of DNA as an immunogen. Observation in this study suggested that naked DNA was more efficient in provoking immune response as compared to the DNA packaged in liposomes. This concept has generated a great deal of commercial interest due to several advantages including duration of immunity generated and stability of the vaccines. Nucleic acid vaccine approach provides an exciting new vista for prevention and control of a variety of viral, bacterial and parasitic diseases of humans, animals and poultry. Potential of nucleic acid vaccines against infectious and parasitic organisms has been discussed with regard to method of administration, advantages and safety issues, cellular and humoral immune responses by Waine and McManus (1995).

There are two strategies of genetic immunisation which are currently under development through which introduction of DNA to somatic cells can influence the host. The first is generally referred to as gene therapy. Under this approach, DNA encoding gene sequences that are absent from the particular cell type or for some reasons dysfunctional in a particular individual, is inoculated to replace the defective gene. The use of this type of gene delivery is beyond the scope of the present review. In the second major strategy, usually called nucleic acid or DNA immunisation, protein products of the introduced gene(s) are recognised as foreign by the body's immune system to provoke immune response. This may not at first glance appear to provide any obvious benefit over administering the same product by conventional means as purified protein vaccine, but the route and method of administration can have great influence in the presentation of the immunogen to the immune system (Simmond *et al.*, 1997).

If a protein, encoded by foreign gene, is expressed in the host body cells, it enters major histocompatibility (MHC) class 1 pathway of that cell. The foreign proteins are cleaved into short peptides within cystolic proteasome and the fragments are transported to endoplasmic reticulum (ER) by transport associated proteins. In side the ER, peptide fragments bind to MHC class 1 molecules and the peptide-MHC class 1 complex is transported through the Golgi apparatus to the cell surface where these are recognised by T cells and stimulate MHC class 1 restricted CD^{8+} cytotoxic T cells. The relevance of CD^{8+} cytotoxic T cells stimulation is that these cells can invoke cell mediated immune response that is generally important for protection against intracellular pathogens like viruses, bacteria and certain parasites. In contrast, if a protein is inoculated from out side as is the case with non-replicating subunit vaccine immunogens, it is processed by macrophages through MHC class II pathway. In this pathway, the protein is fragmented to short peptides within endosomes. The protein fragments bind to MHC class II antigens and the complex is transported to the surface of the antigen presenting cells where it is recognised by CD^{4+} helper T cells. These cells primarily stimulate B lymphocytes resulting in humoral immune response (Simmond *et al.*, 1997). Thus, nucleic acid immunisation offers better prospect against infectious diseases where cell mediated immunity plays predominant role.

(i) *Composition*

A typical nucleic acid vaccine contains a plasmid harbouring gene encoding immunogenic protein. The gene of interest must by flanked by strong viral promoter and polyadenylation transcriptional termination sequence. Following inoculation of the recombinant plasmid, the viral promoter would initiate the transcription of the gene and termination codon would complete the transcription process. The transcribed mRNA would be translated. Conventionally, SV40 promoter has been used to prepare nucleic acid vaccines. The studies conducted by Davis *et al.* (1993) have demonstrated that the Rous sarcoma virus (RSV) promoter confers about 1000-fold increase in the expression of protein as compared to SV40 promoter. Recently, it has been shown that cytomegalovirus promoter is even more efficient than RSV promoter in nucleic acid vaccines (Lee *et al.* 1997). Based on the gene of interest, a number of other modifications can be introduced in the recombinant plasmid to be used as nucleic acid vaccine.

Delivery of nucleic acid vaccines requires expression of immunogenic proteins in tissues accessible to the immune system such as muscle or skin or mucous membranes. For effective transfection and expression of the gene within these tissues, it is imperative that the introduced DNA is super coiled and it includes strong tissue specific promoter and the transcribed mRNA has poly A tail to ensure its stability in the host cells. The most frequently used promoters are from viruses. The commonly used viral vectors include vaccinia, capripox, swinepox, fowl pox, feline herpes virus 1 and herpes virus of turveys. Depending upon their origin promoters differ in tissue specificity and efficiency in initiating mRNA synthesis (Xiang *et al.*, 1994). To date most DNA vaccines in mammalian system have relied on promoters derived from cytomegalovirus (CMV). The promoters have had good efficiency in both muscle and skin inoculations in a number of mammalian species but may not be as effective in chicken and fish. Another factor known to affect the immune response provoked by DNA immunisations is method of DNA delivery. Parenteral routes may yield low rate of transfection and produce considerably variable gene expression (Montgomery *et al.*, 1993). The high velocity inoculation of plasmid coupled to gold particles into epidermis by gene gun elicits significantly higher level of protective immune response (Fynan *et al.*, 1993). Presumably this could be due to greater efficiency of DNA transfection and more effective presentation of antigen by dendritic cells.

Almost all important viral, bacterial and parasitic diseases of humans and livestock have been targeted for development of efficacious nucleic acid vaccines. A few important ones of each class of pathogens are briefly described here (Table 2).

(ii) *Vaccines against viral diseases*

The simplicity of DNA vaccine technology, its ability to induce antibody

Table 2. Partial list of DNA vaccines against infectious diseases

Disease	Etiologic agent	Reference
Herpes genital disease	Herpes simplex virus 1	Stresser *et al.* (2000)
Hepatitis B	Hepatitis virus	Chow *et al.* (1997)
AIDS	HIV	Wang *et al.* (1993)
Influenza	Influenza virus	Fu *et al.* (1997)
Rabies	Rabies virus	Fodor *et al.* (2000)
Pseudo rabies	Pseudo rabies virus	Roou *et al.* (2000)
Tuberculosis	*Mycobacterium tuberculosis*	Feng *et al.* (2001)
Periodontitis	*Porphyromonas gingivalis*	Yonezawa et al. (2001)
Listeriosis	*Listeria monocytogenes*	Yamada *et al.* (2001)
Malaria	*Plasmodium* species	Doolan & Halffman (1997)
Chaga's disease	*Trypanosoma cruzi*	Costa *et al* . (1998)
Theileriosis	*Theileria annulata*	Onuma *et al.* (1998)
Schistosomiasis	*Schistosoma japonicum*	Yang et al. (1995)
Mycoplasmosis	*Mycoplasma* sp	Barry *et al.* (1995)

and CTL response has led to extensive studies on this concept using various virus-animal models (Ulmer *et al.*, 1993; Cox *et al* , 1993; Donnelly *et al.*, 1997; Fu *et al.*, 1997). The mice immunised with plasmid DNA encoding wild type nucleoprotein (NP) and plasmid DNA encoding a full length NP of influenza virus in which the anchor residues for the immunodominant NP147-155 epitope were altered, were protected from the lethal cross strain virus challenge, and the protection could be adoptively transferred by immune splenocytes. this study suggested that DNA immunisation is capable of inducing protective cellular immune response against both immunodominant and immunorecessive CTL epitopes (Fu *et al.*, 1997).

In the past few years DNA immunisation has been done generally by systemic route which fails to induce immune responses at mucosal surfaces (Ulmer *et al.*, 1996). Recently, Kuklin *et al.* (1997) studied the effect of intranasal (i.n.) and intra-muscular (i.m.) inoculation of plasmid DNA encoding herpes simplex virus (HSV) glycoprotein B (gB) and reported that i.m. immunisation produced both cellular as well as humoral immune response while i.n. immunisation produced mucosal antibodies. However, i.n. immunisation was inferior in providing protection against HSV lethal challenge via the vaginal route. In addition, both i.n. and i.m. plasmid immunisation failed to generate an immune barrier to viral invasion of the mucosa. Recently, Strasser *et al.* (2000) evaluated efficacy of DNA vaccine encoding glycoprotein D of HSV and reported that gD2 produced high level of antibodies and significantly reduced acute disease in mice and guinea pigs. The plasmids co-expressing hepatitis B surface antigen and interleukin-2 can greatly improve the efficacy of DNA vaccine. The study also suggested that co-expression of certain cytokine genes could substantially enhance the immune response against the DNA vaccine (Chow *et al.*, 1997) .

Similar to plasmid DNA, naked mRNA encoding immunogenic protein can also be directly inoculated into host to induce immune response against the protein product. The mRNA encoding nucleoprotein of influenza virus can efficiently elicit virus specific cell mediated immune response *in vivo* when administered encapsulated in simple liposomes (Martinon *et al.*, 1993). Self replicating Semliki forest virus (SFV) RNA has also been used as recombinant vaccine (Zhou *et al.*, 1994). The nucleoprotein of influenza virus expressed in SFV RNA replicon produced humoral immune response with high antibody titre when *in vitro* synthe- sised recombinant RNA was injected directly into the quadriceps muscle of mice. These studies suggest that self replicating recombinant SFV RNA may be quite useful in nucleic acid vaccines (Zhou *et al.*, 1994). Naked RNA as vaccine can be used as an alternative to DNA in certain situations. RNA vaccination appears attractive in the sense that it would remain in the host system transitory hence, the possibility of integration and undesirable mutagenic consequences can be avoided. In addition, SFV replicons can be engineered to express multiple subgenomic RNAs allowing coexpression of several protective antigens along with cytokines or other immunomodulators to enhance the generation of cellular and humoral protective responses (Rice, 1996).

Nucleic acid immunisation approach has been very extensively pursued in relation to livestock and poultry pathogens in the past couple of years leading to generation of enormous information. Fodor *et al.* (1999) immunised chickens with a plasmid encoding infectious bursal disease virus (IBDV) antigens. The authors cloned VP2 gene of GP40 strain of IBDV and VP2-VP4-VP3 encoding segment of strain D78 in an expression vector which consisted of human cytomegalovirus (HCMV) immediate early enhancer and promoter, adenovirus tripartite leader sequence and SV40 polyadenylation signal sequence. The chickens were immunised by i/m and i/p routes. Two booster inoculation were given with the recombinant plasmids. Low to undetectable level of antibodies and no protection to virulent virus challenge was observed in chickens vaccinated with plasmid encoding VP2 gene. However, plasmid encoding VP2-VP4-VP3 induced IBDV specific antibodies and protection in chicken. This demonstrated the feasibility of using VP2-VP4-VP3 segment of IBDV for genetic vaccination. Subsequently, Liu *et al.* (2000) used pcDNA3 vector consisting CMV immediate early promoter and enhancer and the bovine growth hormone poly A stop signal sequences to clone VP2/VP2-VP4-VP3 gene segments of IBDV. The chicken embryo fibroblast cells transfected with the plasmid containing VP2/VP2-VP4-VP3 expressed VP2 and VP2-VP4-VP3 proteins of IBDV which were detected by ELISA and immunoblotting. This suggested that multiple genes of the virus could be expressed in a single expression vector which could be used as potential DNA vaccine.

Pseudo rabies virus (PRV) or Aujeskey's disease virus, an important pathogen of pigs, has been recently targeted for development of DNA vaccine. Takada *et al.* (1999) have reported that intranasal immunisation of mice with the plasmid encoding PRV glycoprotein B induced both serum and secretary antibody responses. The immunised mice were found to be resistant to the lethal virulent challenge with PRV intranasally. However, the intranasally immunised mice could not resist the intraperitoneal challenge with virulent PRV. The study suggested that intranasal immunisation induces local secretary antibodies which protect animals from the pseudo rabies virus. In order to enhance the immune response against the target gene product in the immunised animals, Somasundaram *et al.* (1999) evaluated the effect of GM-CSF (granulocyte macrophage-colony stimulating factor) on DNA vaccination in swine against pseudo rabies virus. This study indicated that simultaneous immunisation of pigs with plasmid DNA encoding gB and gD glycoproteins of PRV and the plasmid encoding GM-CSF enhances immune response significantly. The study demonstrated potential of GM-CSF as adjuvant in DNA vaccinations. Roou *et al.* (2000) also immunised pigs intradermally with plasmid construct carrying gB or gC or gD gene of PRV. Six weeks after 3rd vaccination, vaccinated and control pigs were challenged intranasally with virulent NIAS-3 strain of PRV. Vaccination with plasmid encoding gB induced strongest cellular immune response including cytotoxic T cell response where as plasmid encoding gD induced strongest neutralising antibody responses. This is the first study to demonstrate cytotoxic T cell response in pigs following DNA vaccination.

Equine herpes virus 1 (EHV 1) is one the important pathogens of horses causing reproductive as well as respiratory infections leading to abortions and foal mortality. Recently, Ruttenberg *et al.* (1999) explored the possibility of DNA mediated vaccination against EHV 1 using murine model. These workers reported that immunisation of mice with the plasmid DNA encoding EHV 1 envelope glycoprotein D(gD) induced specific antibody response detectable by 2 weeks and maintained upto 23 weeks after i/m inoculation. Following intranasal challenge with EHV 1, mice immunised with 50 mg of EHV 1 gD DNA were able to clear the virus more rapidly as compared to control mice. In another recent study Romito *et al.* (1999) demonstrated that DNA encoding VP2 gene of African horse sickness virus induced mild humoral and strong cell mediate immune response in horses. This study showed that DNA mediated immunisation strategy could be useful in developing an effective vaccine against this economically important livestock disease.

Rabies is one of the most horrifying zoonotic diseases. The studies conducted recently by Perkin *et al.* (1999) using this approach have yielded quite encouraging results. The dogs vaccinated i/m with plasmid DNA encoding rabies virus (PV strain) glycoprotein induced neutralising antibody response against both wild type rabies virus and European bat lyssa viruses (EBL1 and EBL2). All the vaccinated dogs were protected against the lethal rabies virus challenge. This is one of the first studies to show that the dogs could be vaccinated with DNA vaccine. Fodor *et al.* (2000) immunised mice with plasmid DNA encoding full length glycoprotein G of rabies virus. The i/m immunised mice were challenged with virulent rabies virus. The study showed that 83 to 91% mice were protected against the challenge. The study concluded that DNA vaccination is simple and effective method against rabies.

The nucleic acid immunisation concept is not only limited to immunisation of livestock, poultry and humans but its application has been demonstrated in fish as well. Recently, Corbeil *et al.* (1999) immunised rainbow trout (*Oncorhynchus mykiss*) with plasmid carrying nucleoprotein (N), phosphoprotein (P), matrix protein (M), non-virion protein (NV) and glycoprotein (G) of rhabdovirus infectious haemopoietic necrosis virus (IHNV). The DNA vaccine was prepared by cloning each gene in pcDNA3.1 plasmid. Only the plasmid expressing glycoprotein gene was protective against the virulent challenge.

Uptake and expression of foreign gene has been a major problem associated with nucleic acid vaccines. To overcome this problem, electroporation technique has been utilised to enhance the uptake of DNA by the host cell. The authors demonstrated that *in vivo* electroporation of DNA vaccine enhances delivery of the DNA to cells substantially, resulting in increased expression and elevated immune responses. Potency of weakly immunogenic hepatitis B surface antigen DNA vaccine was increased in mice. In addition, the immunogenicity of a potent HIVgag DNA vaccine was increased significantly by *in vivo* electroporation technology (Widera *et al.*, 2000).

(iii) *Vaccines against bacterial diseases*

Porphyromonas gingivalis is a Gram-negative anaerobic bacterium which has been frequently isolated from subgingival lesions from patients with adult periodontitis. The bacterium is thought to be associated with this disease. Recently a DNA based vaccine against *P. gingivalis* using BALB/c mouse model has been evaluated. The mice were immunised with plasmid DNA encoding arginine-specific cysteine proteinase (rgpA). The antibody response against *P. gingivalis* was analysed by ELISA. The rgpA DNA vaccine induced high levels of serum antibody against *P. gingivalis.* The challenge studies in mouse model suggested that DNA vaccination could confer protection against *P. gingivalis* infection (Yonezawa *et al.*, 2001) .

Another DNA vaccine against *Mycobacterium tuberculosis* has been evaluated. A DNA vaccine (D) and a recombinant modified vaccinia virus Ankara (M) in which DNA encodes early secreted antigen target 6 and mycobacterial protein 63 synthesised, and each was found to generate specific gamma interferon (IFN- g) secreting CD^{4+} T cells. Enhanced CD^{4+} IFN-g T cell responses were produced by both D-M and M-D immunisation regimens. Significantly higher levels of IFN-g were seen with D-D-D-M immunisation regimen. Most of the immunisation regimens were assessed in challenge studies and found to produce protection equivalent to that produced by *Mycobacterium bovis* BCG (McShane *et al.*, 2001). Futher, sequential immunisation of mice with mycobacterial antigen Ag85B-expressing DNA and *Mycobacterium bovis* BCG was found to be more effective than BCG alone immunisation in protecting the animals against *M. tuberculosis* infection. The depletion of CD^{8+} T cells in mice, impaired protection suggesting the role of CD^{8+} T cells in DNA induced protection (Feng *et al.*, 2001). Tanghe *et al.* (2001) vaccinated C57BL/6 mice with a plasmid encoding Ag85 from *M. tuberculosis* and subsequently boosted with Ag85 protein adjuvant, while DNA immunisation produced strong Th1-type cytokine responses, protein adjuvant elicited very poor cytokine response which was 10 fold lower than those observed with DNA immunisation alone. Injection of Ag85 DNA primed mice with 30-100 mg of purified Ag85 adjuvant increased interleukin-2 and gamma interferon in spleen two to four fold. The study clearly indicated that the protein boosting increased protective efficacy of the DNA vaccine.

Yamada *et al.* (2001) evaluated a DNA vaccine against *Listeria monocytogenes* and reported that immunisation of mice with plasmid DNA encoding a single dominant T cell epitope is sufficient for induction of protective immunity against *L. monocytogenes.*

(iv) *Vaccines against parasitic diseases*

Despite the great progress made in the field of recombinant DNA technology, there are relatively fewer DNA vaccines development efforts are reported against parasitic disease. The DNA immunisation approaches has been predominantly tried in protozoan diseases. Several candidate DNA vaccines have been evaluated against human and animal *Plasmodium* species (Sedegah *et al.* 1994, Doolan *et al.*, 1996, Grifantini *et al.*, 1998). The studies on evaluation of *P. falsiparum* DNA vaccines and reported that recombinant plasmids encoding pfCSP, pfSSP2, HEP17 (pfEXP-1) and the liver-stage antigen (pfLSA-1) could generate protective immune responses in rodents and primates (Doolan *et al.*,1996). Subsequently, futher evaluation of combination of all these plasmids as multi-gene vaccine and suggested that immunisation with different combination of genes induces higher rate of protection (Doolan & Hoffman, 1997). The DNA vaccines generating humoral as well as cellular immune responses would be ideal candidate vaccines against malaria. The gene encoding surface antigen P30 of tachyzoits of *Toxoplasma gondi* has been evaluated as DNA vaccine. Under laboratory conditions immunisation of mice with plasmid harbouring P30 gene has yielded strong protective immune response (Ramaswamy, 1999).

Trypanosoma, causative agent of Chaga's disease, has also been targeted for development of DNA vaccine. Recently, Costa *et al.* (1998) constructed a plasmid containing a gene encoding catalytic domain of trans-sialidase (TS) enzyme. This enzyme is critical for survival of parasite in the host and may be associated with virulence. The immunisation of mice with this plasmid yielded antibody response which neutralised the enzyme *in vitro*. The challenge studies showed that TS DNA vaccine protected mice against *T. cruzi*. Leishmaniasis is another important human disease in many parts of the world. Based on the protective nature of gp63 glycoprotein of the parasite, a plasmid encoding the gene for gp63 was constructed and used for immunisation of experimental animals. The authors reported that the immunised animals exhibited much smaller lesions and delayed onset of the disease (Xu & Liew, 1995, Walker *et al.*, 1998). Using a different approach, Gurunathan *et al.* (1997) prepared a candidate DNA vaccine using immunodominant *Leishmania* homologue of receptors for activated C kinase (LACK) and demonstrated that it conferred significant protection.

The plasmid DNA encoding allelic forms of the major merozoite surface antigens of *T. annulata*, Tams1-1 and Tams1-2 conferred significant protection in calves (Onuma *et al.*, 1998). This study indicated potential of DNA vaccine for use in cattle against *T. annulata*. Jenkins *et al.* (1995) evaluated a DNA vaccine against *Cryptosporidium parvum* and reported that sheep immunised with genes encoding two surface antigens of sporozoit stage of parasite via i/m or intrammary route, substantial amounts of neutralising antibodies were produced.

This approach is now being used for multicellular parasites also. Naked DNA of *T. ovis* encoding the host protecting antigens 45W was used to immunise sheep with a protection rate of 94% (Rothel *et al.* 1997). Yang *et al.* (1995) reported that intra-muscular injection of plasmid DNA encoding fragments of *S. japonicum* paramyosin (Sj 97), a 97 kDa molecule, induces antibody response in mice. This raised the hope for development of nucleic acid vaccine against schistosomiasis. Shi *et al.* (1999) developed a DNA vaccine against *S. japonicum* and evaluated it in mice. The mice immunised with plasmid (pCD-Sj32) developed considerable immune response. Subsequently, Yi *et al.* (2000) used *S. japonicum* catthepsin BDNA vaccine in mice and reported an egg reduction rate of 59.3-65.0% in liver, 57.6-62.1% in intestine and 86.5% in faeces, in addition marked reduction in eggs in uteri. This study suggested that cathepsin B DNA vaccine has potential application as antifecundity immunising agent.

With a view to develop nucleic acid vaccine, Harrison and Bianco (2000) used DNA encoding *Oncocerca volvulus* antigen ov-TMY-1(tropomyosin) and ovB20 (a nematode specific gene product). Intramuscular and gene gun immunisation of mice with ov-tmy-1 induced expected Th1/Th2 associated IgG isotypes profiles. The study suggested that DNA immunisation has good potential for inducing humoral immune response against nematodes but this is not sufficient for protection.

DNA vaccination approach is still in infancy for many protozoan and helminth parasite of medical and veterinary significance. A number of parasites are currently being targeted for development of DNA vaccines.

(v) *Limitations*

Though the nucleic acid vaccine approach appears very attractive, there are a number of concerns, including the possibility of integration of the nucleic acid resulting in tumourigenesis, effect of nucleic acid persistence and immuno-tolerance, which need to be resolved. These concerns are reduced if the target tissue is skin and the DNA vaccine is introduced through gene gun. The microparticle transfected skin cells are lost in 10-14 days as part of the normal sloughing of keratinised skin tissue. However, in some instances dermal cells are transfected and the foreign gene and these may express the product for longer time. In worst case scenario, the chances of integration are at least three orders of magnitude below the spontaneous mutation frequency (Nichols *et al.*, 1995). Certain autoimmune diseases such as systemic lupus erythematosus are associated with the production of anti-DNA antibodies by the patients, causing some scientists to suggest that ever persistent non-integrated DNA might pose problem by inducing anti-DNA response. Safety issues concerning nucleic acid vaccines in humans and animals have not been effectively addressed so far. The lack of potential oncogenecity and toxicity need to be established before they can be readily recommended for clinical use. There are many unknown links in the mechanism of immune response against DNA immunisation. Further optimisation of plasmid construction, route of immunisation, doses of immunisation need to be clearly worked out.

These appear to be theoretical concerns regarding the DNA vaccines. It is too early to predict whether DNA vaccines will be a practical reality in near future. Nevertheless, potential advantages are considerable.

7. VACCINES PRODUCED IN PLANTS

Recent progress in plant molecular biology and refinement in gene transfer technology together have enabled the researchers to produce transgenic plants at will. These unprecedented developments have raised the possibility of commercial scale production of a wide range of products with potential medical and veterinary applications. One of the most exciting applications of transgenic plant technology is production of vaccines. During the past one decade, plants have been shown to reliably express genes from diverse sources including animals, insects, fungi, bacteria, viruses and other plants. In addition, the plants are also endowed with the ability to perform complex post-translational modifications such as glycosylation. The scientific reports accumulated in the past few years have suggested that level of expression of foreign proteins in plants could go in excess of 1% which is quite high. Intensive global research efforts are being made to manipulate foreign gene expression in plants through design of more efficient vectors. With the recent developments in the understanding of protein folding, assembly and processing of proteins in transgenic plants, it is hoped that level of protein expression in plants could be enhanced many folds to make the venture commercially attractive.

Tobacco plant was the first plant system used in 1983 for expression of foreign gene. Since then a wide range of plants have been utilised for transgenic manipulations successfully. There are mainly two approaches to transfer genes in plants. The first and the most versatile approach utilises *Agrobacterium tumefa-*

ciens, a soil bacterium, for transfer of genes. The plasmid of *A. tumefaciens* contains a DNA sequence which allows incorporation of foreign gene into plant genome. Second approach is biolistics system that accelerates DNA-coated gold or tungsten particles, which penetrate plant cell to deliver DNA that integrates into nuclear or organelle genome.

Mason *et al.* (1992) wanted to know whether transgenic plants would produce VLPs. The results of the study indicated that transgenic tobacco plant leaves synthesised recombinant hepatitis B virus antigen (rHBsAg) in abundance. The electron microscopic examination of the leaf extracts revealed presence of 22-nm particles analogous to authentic HBsAg particles. From this study, they concluded that transgenic tobacco plants allowed self assembly of peptide subunits into VLPs. Due to similarity of tobacco produced VLPs with the serum HBsAg particles, recombinant VLPs have tremendous potential as promising vaccine candidates. Similarly, Norwlk virus capsid protein (NVCP) has been successfully expressed in transgenic tobacco plants (Mason *et al.*, 1996). Electron microscopic studies on the purified leaf extracts clearly demonstrated presence of VLPs. The authors concluded that transgenic plants allow expression and self assembly of NVCP into VLPs. The immunological studies carried out on the antigens derived from the plant cells contained both T and B cell epitopes.

The recent developments in plant biotechnology have opened new vistas to our vision for development of edible plant vaccines against man and animal diseases. The results of recent studies have suggested that genetically engineered plants and plant viruses could be used as vaccines. Haq and co-workers (1995) have demonstrated that transgenic potato and tobacco plants expressing LT-B toxin of cholera organism, produced antibodies in mice which neutralised LT-B enterotoxin in cell protection assays. Similarly these workers expressed hepatitis B virus surface protein in transgenic tobacco and showed that when mice were inoculated with transgenic tobacco, they produced hepatitis B virus specific antibodies. Subsequently, Mason *et al* (1996) reported that CD1 mice immunised with partially purified rNVCP produced in transgenic tobacco plants elicited both serum and secretary immunoglobulins that reacted with NV as well as NVCP produced by insect cell-expression system. The authors concluded that rNVCP contains the epitopes necessary for oral immunisation with ability to stimulate mucosal immune system.

Cow pea mosaic virus (CPMV) has been identified as a potential vector for expression of foreign antigens and development of novel vaccines because of several important features of the virus. Firstly, the virus is strongly immunogenic and therefore, may be an effective vehicle for presentation of foreign antigens to immune system. Secondly, the virus is stable at temperatures up to 65°C therefore, an adequate shelf life without the need for cold storage can be anticipated. Using CPMV chimaeric virus particle (CVP) technology has been developed which is based on expression of multiple copies of an epitope on the surface of CPMV (Usha *et al*, 1993). The immunological studies conducted with CVPs expressing FMDV, HRV and HIV epitopes have suggested that CVPs produce specific antibody response. Recently Wigdorovitze *et al.* (1999) reported use of VP1 structural protein of FMDV expressed in transgenic tobacco plants for immunisation of mice in order to assess protective immune response. The mice were immunised with foliar extracts of leaves of tobacco plants expressing VP1 gene. All the immunised mice were protected against the virulent virus challenge. The study suggested that the plant expressed immunogens could be used as vaccines.

The stability of CPMV at temperature up to 65°C pH as low as 1, makes this plant virus highly suitable for CVP technology. Furthermore, the stability of CVP-HIV at acid pH in the presence of pepsin indicates the CVPs could retain their structural integrity during passage through the stomach, thus raising the possibility of edible vaccine based on CVP technology (Xu *et al.*, 1996). Edible vaccines have also provided protection against *Helicobacter pylori*, rabies and mink enteric virus in animal models (Langridge, 2000).

The findings of the research carried out in the past few years have amply demonstrated potential of plant based vaccines for medical and veterinary use. Nevertheless, several technical, regulatory, economic, social and psychological issues remain to be addressed. One of the important technical issues is the choice of the plant expression system. The plant expression system should produce stable immunogen, it should not be toxic to human and animals. The plant used for production of edible vaccines should be eaten raw. The test has to be worked for quantitation of antigen dosages. The yield of the recombinant immunogens has to be enhanced in order to make plant based antigen production commercially viable. Banana is grown quite extensively as staple fruit crop throughout the world and could be used for expression of immunogens. It may be possible to produce 1-10 mg recombinant protein in each fruit. However, lots of information is needed in order to fully exploit banana for production of edible plant vaccines for human use. Similarly, important fodder crops have to be selected for developing edible vaccines for veterinary use. Degradation of antigen in digestive tract is another major concern in development of oral vaccines. This problem could be eliminated by use of bioencapsulating agents. Development of tolerance to antigens in food is another problem that has to be resolved. Even if the technological feasibility of developing edible plant vaccines has been clearly demonstrated, acceptability of genetically modified (GM) plants by the people remain major issue. Recent violent protests throughout the world against GM foods have shaken the plant biotechnology companies venturing into area of transgenic edible plant vaccines.

8. FUTURE PERSPECTIVE

Advances in our understanding of molecular organisation of pathogenic organisms and molecular mechanisms of immune response against peptides as well as whole organisms have opened new vistas to our vision. Though majority of vaccines continue to be produced by conventional technologies, several novel approaches are currently under extensive research. Already some efficacious recombinant vaccines have been commercialised and some are under different stages of development. The major drawback with subunit vaccines is poor immunogenicity. The efforts are being made to tackle this problem using cytokines and superior quality new adjuvants. Vectored vaccines harbouring gene(s) encoding immunogenic proteins along with the immunostimulatory cytokines hold great potential. In the past few years newer approaches such as DNA vaccines, ribozyme and antisense RNA therapeutic vaccines have been evaluated with considerable success against infectious diseases.

Quite a few transgenic plants expressing immunogens have been developed which demonstrated feasibility of plant based vaccines. Preliminary studies conducted using cholera toxin gene and its expression in potato have been highly successful. With the availability of excellent plant virus expression vectors, it would not be difficult to express a number of relevant genes encoding immunogenic proteins in edible plants (fruits, roots, leaves and stem). Such recombinant plants carrying immunogens could easily immunise humans and animals against the targeted viral, bacterial and parasitic diseases. This approach offers an easy way to eliminate the need for parenteral administration of vaccines. However, induction of immunotolerance to antigens given via oral route presents a formidable but not insurmountable obstacle to the use of such a strategy. From the emerging trends in vaccine technologies, it is clear that no single approach can be used for development of future vaccines. The developments in the past 10 years have clearly pointed towards immense possibility of developing more efficacious, safer, thermostable and economically affordable vaccines using molecular approaches. Despite the great strides in vaccinology, a number of problems related to stability, safety, efficacy, environment and economic feasibility remain unresolved.

REFERENCES

An LL and Whitton J L. 1997. A multivalent minigene *Vaccine,* containing B-cell, cytotoxic T-lymphocyte, and Th epitopes from several microbes, induces appropriate responses *in vivo* and confers protection against more than one pathogen. *Journal of Virology* 71: 2292-2302.

Anderer FA & Schlumberger HD. 1965. Properties of different antigens immunologically related to tobacco mosaic virus. *Biochemica Biophysica Acta* 97: 503-509.

Anderer FA. 1963. Preparation and properties of an artificial antigen immunologically related to tobacco mosaic virus. *Biochemica Biophysica Acta* 71: 246-248.

Arnon R. 1986. Synthetic peptides as the basis for future Vaccines. TIBS 11: 521-524.

Bansal G P, Hartfield J A, Dunn F E, Warrener P, Young J F, Top F H Collet M S, Anderson S, Rosenfeld S, Kajigaya S and Young N S. 1992. Immunogenicity human recombinant parvovirus B19 protein. *Modern Approaches to New Vaccine Including prevention of AIDS,* Eds. Brown F, Chanock R M, Ginsberg H S, Lerner R A, Cold Spring Harbour Laboratory Press, PP 315- 320.

Baras B, Benott, MA, Poulan-Godefroy O, Schacht AM, Capron A, Gillard J & Riveau, G. 2000. Vaccine properties of antigens entrapped in microparticles produced by spray-drying technique and using various polyster polymers. *Vaccine* 18:1495-1505.

Barry MA, Lai WC & Johnston SA. 1995. Protection against mycoplasma infection using expression library immunization. *Nature* 377: 632-636.

Bittle JL, Houghten RA, Alexander H, Shinnick TM, Sutckiffe JG, Lerner RA Rowlands DJ & Brown F. 1982. Protection against foot-and-mouth disease by immunization with a chemically synthesized peptide predicted from the viral nucleotide sequence. *Nature* 298: 30-33.

Borrego B, Camrero JA, Mateu MG & Domingo E. 1995. A highly divergent antigenic site of foot-and-mouth disease virus retains its immunodominance. *Viral Immunology* 8: 11-18.

Boulanger D, Reid G D F, Storrock R F, Wolowczuk I, Balloul J M, Grezel D, Pierce R J, Otieno M F, Guerret S, Grimaud J A, Butterworth A E and Capron A. 1994. Immunization of mice and baboons with the recombinant Sm28GST affects both worm viability and fecundity after experimental infection with Schistosoma mansoni. *Molecular and Biochemical Parasitology* 13: 473-490.

Broder RC, Earl PL, Long D, *Et al.*1994. Antigenic implication of human immunodefficiency virus type 1 envelope quarternery structure: oligomer specific and sensitive monoclonal antibodies. *Proceedings of National Academy of Sciences* (USA) 91:11699-11703.

Chow Y H, Huang W L, Chi W K, Chu Y D and Tao M H. 1997. Improvement of hepatitis B virus DNA *Vaccines* by plasmids co-expressing hepatitis B surface antigen and interleukin-2. *Journal of Virology* 71: 169-178.

Collen T, Di Marchi R & Doel T. 1991. T cell epitope in VP1 of foot and mouth disease virus is immunodominant for vaccinated cattle. *Journal of Immunology* 146: 749-755.

Collen T. 1994. Foot-and-mouth disease virus (aphthovirus): viral T cell epitopes. In : Cell Mediated Immunity in Ruminants (Godeevis BML & Morrison I, eds) CRC Press, Boca Raton, Fla , p 173-197.

Conlon JAR, Mather TN, Tanner P, Gallo G & Jacobson RH. 2000. Efficacy of a nonadjuvanted, outer surface protein A, recombinant *Vaccine* in dogs after challenge by ticks naturally infected with *Borrelia burgdorferi. Veterinary Therapeutics* 1: 96-107.

Connel ND & Medina-Acosta E, McMaster WR, *et al.* 1993. Effective immunization against cutaneous leishmaniasis with recombinant bacille Calmette-Guerin expressing leishmania surface proteinasegp63. *Proceedings of National Academy of Sciences* (USA) 90: 11473-11477.

Corbeil S, Lapatra, SE, Anderson ED, Jones J, Vincent B, Hsu YL & Kurathi G. 1999. Evaluation of the protective immunogencity of the N, P, M, NV and G proteins of infectious haemopoietic necrosis virus in rainbow trout (*Oncorhynchus mykiss*) using DNA Vaccine. *Disease of aquatic organisms* 39: 29-36.

Costa F, Franchin G, Pereira-Chioccola VL, Riberao M, Schekman S & Roddrigues MM. 1998. Immunization with a plasmid DNA containing the gene of trans- sialidase reduce *Trypanosoma cruzi* infection mice. *Vaccine* 16: 768-774.

Cox G J M, Jamb T J and Babuik L A. 1993. Bovine herpes virus 1: immune response in mice and cattle injected with plasmid DNA. *Journal of Virology* 67: 5664- 5667.

Crawford SE, Labbe M, Cohen J, Burroughs MH, Zhou Y & Estes MK & Conner ME. 1994. Characterization of virus like particles produced by the expression of rotavirus capsid proteins in insect cells. *Journal of Virology* 68: 5945-5952.

Cremer KJ, Mackett M, Wohlenberg C, Notkins AL & Moss B. 1985. Vaccinia virus recombinant expressing herpes simplex virus in mice. *Science* 228: 737-740.

Danforth H D, Augustine P E, Ruff M D, McCandiss R, Strausburg R L and Likel M. 1989. Genetically engineered antigen confers partial protection against avian coccidial parasites. *Poultry Science* 68: 1643-1652.

Daniel MD, Kirkchhoff F, Czajak SC, Sehgal PK & Desroiers RC. 1992. Protective effect of a live attenuated SIV Vaccine with a deletion in the *nef* gene. *Science* 258 :1938-1941.

Davis HL, Michel ML & Whalen RG. 1993. DNA based immunization induces continuous secretion of hepatitis B surface antigen and high levels of circulating antibodies. *Human Molecular Genetics* 2 :1847-1851.

Di Marchi R, Brooke G, Gale C, Cracknell V, Doel T & Mowat N. 1986. Protection of cattle against foot and mouth disease by a synthetic peptide. *Science* 232 : 639-641.

Donnelly J J, Ulmer J B, Shiver J W and Liu M A. 1997. DNA Vaccines. Annual *Review of Immunology* 60: 163-172.

Doolan DL, Sedegah M, Hedstrom RC, Hobart P, Charoenvit Y & Hoffman SL. 1996. Circumventing genetic restriction of protection against malaria with multi-gene DNA imminization: CD 8 T cell, intereferon gamma and nitric oxide dependent immunity. *Journal of Experimental Medicine* 183: 1739-1746.

Doolan DL & Hoffman SL. 1997. Multi-gene vaccination against malaria: a multistage, multiimmune response approach. *Parasitology Today* 13: 171-178.

Feng CJ, Palendira U, Demangel C, Spratt JM, Malin AS & Britton WJ. 2001. Priming by DNA immunization augments protective efficacy of *Mycobacterium bovis* Basille Calmette-Guerin against tuberculosis. *Infection and Immunity* 69: 4174- 4176.

Fennelly GJ, Flynn JL, ter Meulen V, *et al.* 1995. Recombinant bacille Calmette-Guirin priming against measles. *Journal of Infectious Diseases.* 172: 698-705.

Fernandez FM, Conner MF, Parwani AV, Todhunter D, Smith KL, Crawford SE, Estes MK & Saif LJ. 1996. Isotype specific antibody responses to rotavirus and virus proteins in cows inoculated with with subunit vaccines composed of recombinant SA11 rotavirus core-like particles (CLP) or virus like particles (VLP) *Vaccine* 14: 1303-1312.

Fodor I, Horvath E, Fodor N, Nagy E, Rencendorsh A, Vakharia VN & Dube SK. 1999. Induction of protective immunity in chickens immunised with plasmid DNA encoding infectious bursal disease virus antigens. *Acta Veterinaria Hungarica* 47: 481-492.

Fodor I, Kuscera L, Fodor N, Palfi V & Grabko V. 2000. Gene immunization of mice with plasmid DNA expressing rabies virus glycoprotein. *Acta Veterinaria Hungaria* 48: 229-236.

Francis MJ, Hastings GZ, Syred AD, McGinn B, Brown F & Rowlands DJ. 1987. Non-responsiveness to a foot and mouth disease virus peptide overcome by addition of foreign helper T cell determinants. *Nature* 300: 168-170.

Fu TM, Friedman A, Ulmer JB, Liu MA & Donnelly JJ. 1997. Protective cellular immunity: Cytotoxic T-lymphocyte responses against dominant and recessive epitopes of influenza virus nucleoprotein induced by DNA immunization. *Journal of Virology* 71: 2715-2721.

Fynan EF, Webster RG, Fullar DH, Hyanes JR, Santoro JC & Robinson HL. 1993. DNA Vaccine: protective immunizations by parenteral, mucosal and gene gun inoculations. *Proceedings of National Academy of Sciences* (USA) 90: 11478-11482

Gough JM & Kemp DH. 1993. Localisation of a low abundance membrane protein (Bm86) on the gut cells of the cattle tick Boophilus microplus by immunogold lebelling. *Journal of Parasitology* 79: 900-907.

Grifantini R, Finco O, Barrolini E, Draghi M, Giudice GD, Kocken C, Thomas A, Abrignani S & Grnadi G. 1998. Multi-plasmid DNA vaccination avoids antigenic competetion and enhances immunogensity of a poorly immunogenic plasmid. *European Journal of Immunology* 28: 1225-1232.

Gurunathan S, Sack DL, Brown DR, Reiner SL, Charest H, Glaichenhaus N & Seder RA. 1997. Vaccination with DNA encoding immunodominant LACK parasite antigen confers protective immunity to mice infected with *Leishmania major. Journal of Experimental Medicine* 186: 1137-1347.

Haq TA, Mason HS, Clements JD & Arntzen CJ. 1995. Oral immunization with a recombinant bacterial antigen produced in transgenic plants. *Science* 268: 714- 716.

Harrison GBL, Dempster R P, Robinson CM, Heath DD, Lightowlers MW & Richards M D. 1993. Pilot scale field trial to test the efficacy of Taenia ovis recombinant Vaccine. *Proceedings of Vaccines*, Lorne, Australia, p 123-130.

Harrison RA & Bianco AE. 2000. DNA immunization with *Onchocerca volvulus* gene, Ov-tmy-1 and OvB20: serological and parasitological outcomes following intramuscular or GeneGun delivery in mouse model of onchocerciasis. *Parasite Immunology* 22: 249-257.

Heath DD, Lawrence SB, Ralston M J, Harrison GBL, Dempster RP, Robinson C, Lightowlers M W, Gaucci CWG, & Newton S. 1993. New protective cloned antigens of *Taenia ovis*. Proceedings of New Zealand Society for Parasitology, p. 211-216. *Journal of Infectious Diseases* 182: 1311-1320.

Honda M, Kazuhiro M, Nakazone T, *et al.* 1995. Protective immune responses induced by secretion of a chimeric soluble protein from recombinant *Mycobacterium bovis* bacille Calmette-Guerin vector candidate Vaccine for human immunodefficiency virus type 1 in small animals. *Proceedings of National Academy of Sciences* (USA) 92: 10693-10697.

Hopp TP. 1981. *Molecular Immunology* 18: 869-874.

Hosseini M, Dixion RJ, Kirkland PD & Roy P. 1996. Studies on vaccination of sheep with bluetongue virus core-like particles produced by recombinant baculovirus. In :*Proceedings of the First Southeast Asia and Pacific Regional Bluetongue Symposium* (St. George ST & Kagao P, eds) p 198-202.

Ivins BE, Welkos SL, Knudson GB, Little SF. 1990. Immunization against anthrax with aromatic compoun-dependent (Aro-) mutants of *Bacillus anthracis* and with recombinant strains of *Bacillus subtilis* that produce anthrax protective antigen. *Infection and Immunity* 58: 303-308.

Ivins BE, Ezzell JW Jr, Jemski J *et al.* 1986. Immunization studies with attenuated strains of *Bacillus anthracis. Infection and Immunity* 52: 454-458.

Jenkins M, Kerr D, Fayer R & Wall R. 1995. Serum and colostrum antibody responses induced by jet injection of sheep with DNA encoding a *Cryptosporidium parvum* antigen. *Vaccine* 13: 1658-1664.

Jenkins MC, Lillehoj H S and Dame J B. 1988. E*meria acervulina* DNA cloning and characterization of recombinant sporozoite and merozoite antigens. Experimental *Parasitology* 66: 96-107.

Jenkins M C, Castle M D and Danforth H D. 1991. Protective immunization against intestinal parasite *Emeria acervulina* with recombinant coccidial antigen. *Poultry Science* 70: 539-547.

Jiang X, Wang M, Graham DY & Estes MK. 1992. Expression, self assembly and antigenecity of Norwalk virus capsid protein. *Journal of Virology* 66: 6527- 6532.

Johnson SJ & Roy P. 1996. Bluetongue recombinant Vaccines. In : *Proceedings of the First Southeast Asia and Pacific Regional Bluetongue Symposium* (St. George ST & Kagao P, eds) p 192-197.

Jonson K S, Harrison G B L, Lightowlers M W, O'Hoy S B, Caucci W G, Dempster R P, Lawrence S B, Vinton J G, Heath D D and Richards M D. 1989. Vaccination against ovine cysticercosis using a defined recombinant antigen. Nature 338: 585-587.

Kleid DG, Yansura D, Small B, Dowbenko D, Moore DM, Grugman MJ, McKercher PD, Morgan DO, Robertson BH & Bachrach HI. 1981. Cloned viral protein Vaccine for foot-and-mouth disease: response in cattle and swine. *Science* 214: 1125-1129.

Kuklin N, Daheshia M, Karem K, Manickan E and Rouse B T. 1997. Induction of mucosal immunity against herpes simplex virus by plasmid DNA immunization. *Journal of Virology* 71: 3138-3145.

Lagranderie M, Balazuc A M Gicquel B and Gheorghiu M. 1997. Oral immunization with recombinant Mycobacterium bovis BCG simian immunodeficiency virus *nef* induces local and systemic cytotoxic T-lymphocyte response in mice. *Journal of Virology* 71: 2303-2309.

Lai C J, Ruhe M, pethel M and Bray M. 1992. infectious RNA transcribed from stably cloned dengue virus cDNA , construction of growth restricted dengue virus mutants. *Modern Approaches to New Vaccine Including prevention of AIDS, Vaccine*, Eds. Brown F, Chanock R M, Ginsberg H S, Lerner R A, Cold Spring Harbour Laboratory Press, PP 265-270.

Langebeheim H, Arnon R & Sela M. 1976. Antiviral effect on MS-2 coliphage obtained with a synthetic antigen. *Proceedings of National Academy of Sciences* (USA) 73: 4636-4540.

Langridge WHR. 2000. Ediable Vaccines. *Scientific Amarican* (September): 48-53.

Lee AH, Suh YS, Sung SH, Yang SH & Sung YC. 1997. Comparisonof various expression plasmids for induction of immune respone by DNA immunization. Molecular Cells. 7: 12-18.

Lerner RA, Green N, Alexander H, Liu FT, Sutcliffe GJ & Shinnick TM. 1981. Chemically synthesized peptide predicted from the nucleic acid sequence of hepatitis B virus genome elicit antibodies reactive with the native envelope protein of Dan particles. *Proceedings of National Academy of Sciences* (USA) 78: 3403-3407.

Li A, & Jiang J. 2000. Vaccination against coccidiosis with SO7 recombinant antigen of *Emeria tenella* BJ strian. *Chinese Journal of Veterinary Sciences* 20: 167- 170.

Lightowlers M W. 1994.Vaccination against animal parasites. *Veterinary Parasitology* 54: 177-204.

Liu SX, Song GC, Xu YX, Wangh W & McManus DP. 1995. Anti-fecundity immunity induced in pigs vaccinated with recombinant Schistosoma japonicum 26 kDa glutathione-S-transferase. *Parasite Immunity* 17: 335-340.

Liu Y, Jin Ningy, Gu Changquing, luo Kun, Fang Houhua, WangWei, Li Ping & Yin Zhen. 2000. Construction and expression of IBDV VP2/VP243 genes DNA Vaccine expressing plasmids. *Chinese Journal of Veterinary Science* 20: 321- 323.

Martin M, Hajishengallis G, Metzer DJ, Michalek SM, Connel TD & Russel MW. 2001. Rcomibinant antigen-enterotoxin A2/B chimeric mucosal immunogens differentially enhance antibofy responses and B7-dependent costimulation of CD4+ T cells. *Infection and Immunity* 69: 252-261.

Martinon F, Krishnana S, Lenzen G, Magne R, Gomard E, Guillet JG, Levy JP & Meulien P. 1993. Induction of virus-specific cytotoxic T lymphocytes *in vivo* by liposome-entrapped mRNA. *Eupean Journal of Immunology* 23: 1719- 1722.

Mason HS, Lam DMK & Arntzen CJ. 1992. Expression of hepatitis B surface antigen in transgenic plants. *Proccedings of National Academy of Sciences* (USA) 89: 11745-11749.

Mason HS, Ball J, & Shi J-J *et al.* 1996. Expression of Norwalk virus capsid protein in transgenic tobacco and potato plants and its oral immunigenecity. Proccedings of National Academy of *Sciences* (USA) 93: 5335-5340.

McShane H, Brooks R, Gilbert SC & Hill AVS. 2001. Enhanced immunogeneciry of CD4+ T-cell responses and protective efficacy of a DNA-modified vaccinia virus Ankara prime-boost vaccination regimen for murine tuberculosis. *Infection and Immunity* 69: 681-686.

Mettenlieter T C. 1995. New developments in construction of safer and more versatile pseudorabies virus *Vaccine*. Developments in Biological Standardization. Basel, S. Karger, 84: 83-88.

Michel ML, Davis HL, Schieef M, Mancini M, Tiollais P & Whalen RG. 1995. DNA mediated immunization to hepatitis B surface antigen in mice: aspects of the humoral immune response mimic hepatitis B virus infection in humans. *Proceedings of National Academy of Sciences* (USA) 92 : 5307-5311.

Miller G A, Bhogal B S, McCandiss R, Strausberg R L, Jessee E J, Anderson A C, Fuchs C K, Nagle J, Likel M H. Strauser J M and Strausberg S. 1989. Characterisation and *Vaccine* potential of a novel recombinant coccidial antigen. *Infection and Immunity* 57: 2014-2020.

Millet P, Cambell G H, Sulzer A J, Grady K K, Pohl J, Aikawa M and Collins W H. 1993. Immunogenicity of the Plasmodium falciparum a sexual blood stage synthetic peptide *Vaccine* SPf 66. *American Journal of Tropical Medicine and Hygiene* 48: 424-431.

Minor P D. 1996. Recent developments in the design of viral *Vaccines*. *Developments in Biological Standardization*. Basel, S. Karger, 87: 51-57.

Monteil M, Pottier MF Le, Ristov AA, Cariolet R, Hospitalier R, Klonikowsky B & Eloit, M. 2000. Single inoculation of replication-defective adenovirus-vectored Vaccines at birth in piglets with maternal antibodies induces high level of antibodies and protection against pseudorabies. *Vaccine* 18: 1738-1742.

Montgomery D L, Shiwer JW, Leander KR, Perry HC, Friedman A, Martinez D, Ulmer J B, Donnelly J J & Liu M A. 1993. Heterologous and homologous protection against influenza A by DNA vaccination: optimisation of vectors. *DNA Cell Biology* 12 :777-783.

Morgan DO & Moore DM. 1990. Protection of cattle and swine against foot-and- mouth disease, using biosynthetic peptide *Vaccine*. *American Journal of Veterinary Research* 51: 40-45.

Moss B, Carroll MW, Wyatt LS, Bennink JR, Hirsch VM, Goldstein S, Elkins WR, Fuerst TR, Lifson JD, Piatak M, Restifo NP, Overvijk, W, ChamberlainR, Rosenberg SA & Sutter G. 1996. Host range restricted, non-replicating vaccinia virus vectors as Vaccine candidates. In : *Novel Strtegies in Design and Production of Vaccines* (Cohen S & Shafferman A, eds) Plenum Press, New York, p 7-13.

Munson M, kelly SM, Curtiss R. 1989. Oral immunisation with a virulent *Salmonella typhimurium* to induce cross protective immunity to *Escherichia coli* air saccultitis and septicemia in chickens. *Conference of Research Workers in Diseases*, Chicago, p61.

Musoke A J. 1991. Novel *Vaccines* against tick-borne diseases. *Proceedings of a Joint OAU, FAO and ILRAD Workshop*, Kampala, p. 27.

Musoke AJ, Morzaria S, Nkonge C, Jones E & Nene V. 1992. A recombinant merozoite surface antigen of *Theileria para* induces protection in cattle. *Proceedings of National Academy of Sciences*, (USA) 89: 514-518.

Newton SE, Morrish LE, Martin PJ, Montague PE & Rolph T P. 1995. Protection against multiple drug resistant and geographically distant strains of *Haemonchus contortus* by vaccination with H11, a gut membrane derived protective antigen. International *Journal for Parasitology* 25: 511-521.

Nomoto A. 1995. Recombinant polioviruses as candidates strains of oral polio*Vaccines*. *Developments in Biological Standardization*. Basel, S. Karger, 84: 123-128.

Olsen WA, Laurens AH, van Pinxteren, Okkels LM, Rasmussen PB & Andersen P. 2001. Protection of mice with a tuberculosis subunit Vaccine based on a fusion protein of antigen 85B and ESAT-6. *Infection and Immunity* 69: 2773-2778.

Olson CA & Werner PA. 1999. Oral rabies Vaccine contact by roccoons and nontarget specieis in field trial in Florida. *Journal of Wildlife Diseases* 35: 687-695.

Onuma M, Kakuda T & Sugimoto C. 1998. *Theileria* parasite infection in east Asia and control of the disease. *Comparative Immunology, Microbiology and Infectious Diseases* 21: 165-177.

Perkin N T, Chiu P and Coelingh K. 1997. Genetically engineered live attenuated influenza A virus *Vaccine* candidate. *Journal of Virology* 71: 2772-2778.

Patarroyo M E, Amador R, Clanj P, Moreno A, Guzman F, Romero P, Tascon R, Franco A, Murrilo L A, Ponton G and Thujillo G. 1988. A synthetic peptide *Vaccine* protects human against challenge with sexual blood stages of *Plasmodium falciparum* malaria. *Nature* 332: 158-161.

Perrin P, Jocob Y, Agular-Setien A, Loza-Rubio E, Jallet C, Desmezieres E, Aubert M, Cliquet F & Torido N. 1999. Immunisation of dogs with a DNA *Vaccine* induces protection against rabies virus. *Vaccine* 18: 479-486.

Ramaswamy K. 1999. Progress towards development of naked DNA *Vaccine* against parasitic infections. *Journal of Veterinary Parasitology* 13: 1-12.

Rice CM. 1996. Alphavirus based expression systems. In : *Novel Strtegies in Design and Production of Vaccines* (Cohen S & Shafferman A, eds) Plenum Press, New York p 31-40.

Richard M D, Harrison G B L, Heath D D and Lightowlers M W. 1995. *Taenia ovis* recombinant Vaccine-'quo vadit'. *Parasitology* 110: 85-89.

Robinson K, Bellaby T, Chan W C and Wakelin D. 1995. High levels of protection induced by a 40-mer synthetic peptide *Vaccine* against the intestinal nematode *Trichinella spiralis*. *Immunology* 86: 495-498.

Rodriguez A, Saiz JC Nombela AR, Andreu D & Sobrini F. 1994. Antigenic specificity of porcine T cell respone against foot-and-mouth disease virus: identification of T helper epitopes in VP1. *Virology* 205 :24-33.

Rodriguez M, Rubiera R, Pinchet M, Montesinos R, Cremeta J, Falcon V, Sanchez GBringas R, Cordoves C, Valdes M, Leonart R, Herrera L & De La Fuente J. 1994. High level expression of the *Boophilus microplus* Bm86 antigen in yeast *Pichia pastoris* forming highly immunogenic particles for cattle. *Journal of Biotechnology* 33: 135-146.

Rodriguez M, Penichet M, Mouris A E, Labarta V, Luaces L L, Rubiera R, Cordoves C, Sinchez P A, Ramos E, Soto A, Canales M, Palenzuela D, Triguero A, Leonart R, Herrera L & De La Fuente J. 1995. Control of *Bophilus microplus* population in grazing cattle vaccinated with a recombinant Bm86 antigen preparation. *Veterinary Parasitology* 57:339-349.

Romero CH, Barrett T, Evans SA, Kitching RP, gershop PD, Bostock C & Black DN. 1993. Single capripox recombinant vaccine for the protection of cattle against rinderpest and lumpy skin disease. *Vaccine* 11: 737-742.

Romito M, Plessis DH Du & Viljoen GJ. 1999. Immune response in a horse inoculated with the VP2 gene of African horse sickness virus. *Onderstepoort Journal of Veterinary Research* 66: 139-144.

Roou EMA, van Hagmans BL, Glansbeek HL, Visser YE De, Brun MGM De , Boersma

W & Bianchi AT. 2000. A DNA Vaccine coding for glycoprotein B of pseudorabies virus induces cell mediated immunity in pigs and reduces virus excretion early after infection. *Veterinary Immunology and Immunopathology* 74: 121-136.

Rothel JS, Boyale DB, Both GW, Pye AD, Waterkeyn JG, Woods, PR & Lightowlers MW. 1997. Sequential nucleic acid and recombinant adenovirus vaccination induces host-protective immune response against *Taenia ovis* infection in sheep. *Parasite Immunology* 19: 221-227.

Roy P, Urakawa T, Van Dijk AA & Erasmus BJ. 1990. Recombinant virus *Vaccine* for bluetongue disease in sheep. Journal of Virology 64 :1198-2003.

Ruttenberg KM, Walker C, Wellington, JE, Love DN & Whalley JM. 1999. Potential of DNA-mediated vaccination for equine herpes virus 1. *Veterinary Microbiology* 68: 35-48.

Sedegah M, Hedstrom R, Hobert P and Hoffman P. 1994. Protection against malaria by immunization with plasmid DNA encoding circumsporozoite protein. Proceedings of National theAcademy of *Sciences*, USA 91 : 9866-9870.

Sela M & Arnon R. 1992. Synthetic approaches to Vaccines for infectious and autoimmune diseases. *Vaccine* 10: 991-999.

Sela M. 1996. Synthetic Vaccines for infectious and autoimmune diseases. Novel Strategies in Design and Production of *Vaccines* Eds. Cohen S and Shafferman A, Plenum Press, New York, PP 1-5.

Sexton J L, Milner A R, Panaccio M, Waddington J, Wijffels G, Chandler D, Thompson C, Wilson L, Spithill T W, Mitchell G F and Campbell N J. 1990. Glutathione-s-transferase: novel *Vaccine* against Fasciola hepatica infection in sheep. *Journal of Immunology* 145: 3905-3910.

Shi Y, Xie Laiping, Chuanming L & Jiajun H. 1999. DNA Vaccine pCD-Sj32 and its efficacy of protective immunity against infection of *Schistosoma japonicum. Chinese Medical Journal* 112: 677-680.

Simmonds RS, Shearer MH & Kennedy RC. 1997. Parasitology Today 13: 328-335.

Smith N C. 1992. Concepts and strategies for anti-parasite immunoprophylaxis and therapy. *International Journal for Parasitology* 22: 1047-1082.

Smith W D, Smith S K, Murray J M, Liddell S and Kent D P.1994. *Vaccine* against metazoan parasites. Patent No. WO 94/02169.Sun S, Xu W, He N, and Sugane K. 1994. An antigenic recombinant fusion protein from *Trichinella spiralis* induces a protective respone in BALB/c mice. *Journal of Helminthology* 6: 89-91.

Somasundaram C, Takamatsu H, Andreoni C, Audonnet JC, Fischer L, Lefevre F & Charley B. 1999. Enhanced protective response and immuno-adjuvant effects of porcine GM-CSF on DNA vaccination of pigs against Aujeszky's disease virus. *Veterinary Immunology and Immunopathology* 70: 277-287.

Stephensen C B, Welter j, Thaker S R, Taylor J Tartaglia J and Paoletti E. 1997. Canine distemper virus (CDV) infection in ferrets as a model for testing Morbillivirus Vaccine strategies: NYVAC and ALVAC-based CDV recombinants protect against symptomatic infection. *Journal of Virology* 71: 1500-1513.

Stiles BG, Garza AR, Urich RG & Boles JW. 2001. Mucosal vaccination with recombinantly attenuated Staphylococcal enterotoxin B and protection in a murine model. *Infection and Immunity* 69: 2031-2036.

Stover CK, Bansal GP, Hanson MS, *et al.* 1995. Protective immunity elicited by recombinant bacelle Calmette-Guerin (BCG) expressing outer surface protein A (Osp

A) lipoprotein: A candidate Lyme disease *Vaccine. Journal of Experimental Medicine* 178: 197-209.

Strasser JE, Arnold RL, Pachuk C, Higgins TJ & Bernstein DI. 2000. Herpse simplex virus DNA *Vaccine* efficiency: effect of glycoprotein D plasmid construct. *Journal of Infectious Diseases* 182: 1304-1310.

Strombmaier K, Franze R & Adam KH. 1982. Location and characterization of the antigenic portion of the FMDV immunising protein. *Journal of General Virology* 59: 259-306.

Strube W, Abar B, Bergle R D, Block W, Heinsen E, Kretzdon D, Roden B C and Schmeer N. 1995. Safety aspects in the development of an infectious bovine rhinitrcheitis marker Vaccine. *Developments in Biological Standardization.* Basel, S. Karger, 84: 75-82.

Su C, Jiang Y, Xiong G, Shi Y, Ning C, He C, Cao S, Cai M & Qin Z. 1999. The effect of protective immunity of multiple antigen peptide of *Schistosoma* and change of blood cells in mice. *Chinese Journal of Parasitic Disease Control* 12: 297-299.

Sun S, Xu W, He N & Sugane K. 1994. An antigenic recombinant fusion protein from *Trichnella spiralis* induces a protective response in BALB/c mice. *Journal of Helmenthology.* 6: 89-91.

Sutter G & Moss B. 1992. Non-replicating vaccinia virus efficiently expresses recombinant genes. *Proceedings of National Acdemy of Sciences* (USA) 89: 10847-10851.

Taboga O, Tami C, Carrillo E, Nunez JI, Rodriguez A, Saiz JC, Blanco E, Valero M- L, Roig X, Camarero JA, Andrue D, Mateo MG, Giralt E, Domingo E, Sobrino, F & Palma EL. 1997. A large scale evaluation of peptide Vaccines against foot-and-mouth disease: lack of solid protection in cattle and isolation of escape mutants. *Journal of Virology* 71: 2606-2614.

Takada A, Okazaki K & Kida H. 1999. Protective effects of intranasal vaccination with plasmid encoding pseudorabies virus glycoprotein B in mice. *Japanese Journal of Veterinary Research* 47: 25-33.

Tang D, DeVit M & Johnston SA. 1992. Genetic immunization is a simple method for eliciting an immune response. *Nature* 356 :152-154.

Tang D, Shi Z & Curiel DT. 1997, *Nature* 388 (1997) 729.

Tanghe A, D'Souza S, Russels V, Denis O, Ottenhoff THM, Dalemans W, Wheeler C & Huygen K. 2001. Improved immunogenicity and protective efficacy of a tuberculosis DNA *Vaccine* encoding Ag85 by protein boosting. *Infection and Immunity* 69: 3041-3047.

Taylor J, Weinberg R, Tartiglia J, Richardson C, Alkhatib G, Breidis D, Appel M, Norton E & Paoletti, E. 1992. Non replicating viral vectors as potential Vaccines: recombinant canarypox virus expressing measles virus fusion (F) and haemagglutin (HA) glycoproteins. *Virology* 187: 321-328.

Taylor J, Meigner B, Tartiglia, Languet B, VanderHoeven J, Franchini G, Trimarchi C& Paoletti, E. 1995. Biological and immunogenic properties of canarypox- rabies recombinant ALVAC-RG (vCP65) in non avian species. *Vaccine* 13: 539-549.

Townsend K, Sallberg M, O'Dea J, Banks T, Driver D, Sauter S, Chang SM, Jolly DJ, Mento SJ, Milich DR & Lee WTL 1997. *Journal of Virology* 71:3365.

Tsibezov VV, Bogdnova, VS, Zaberrezhnyi AG, Grebennikova TV, Krivonos AV, Dunnikov LA, Kalnov SL, Aliper TI, & Nepoklonov YA. 2000. Use of recombinant swine fever virus protein E2 for immunization of animals. *Vaprosy Virusologii* 45: 3641.

Turnbull I F, Smith D R, Sharp P J, Cobon G S and Hynes M J. 1990. Expression and secretion in *Aspergillus nidulans* and *Aspergillus niger* of a cell surface glycoprotein from the cattle tick, Boophilus microplus, by using fungal amdS promoter system. *Applied and Environmental Microbiology* 56: 2847-2851.

Ulmer J B, Deck R R, Yawman A, Friedman A, Dewitt C M, Martinez D, Montgomery D L, Donnelly J J and Liu M A. 1996. DNA Vaccines for bacteria and viruses. *Novel Strategies in Design and Production of Vaccines* Eds. Cohen S and Shafferman A, Plenum Press, New York, PP 49-53.

Ulmer J B, Donnelly J J, Parker S E, Rhodes G H, Feligner P I, Dwarki V J, Gromkwaski S J, Deck R R, Dewitt C M, Friedman A, Hawe I A, Leander K R, Martinez D, Perry H C, Shiver J W, Montgomery D L and Liu M A. 1993. Heterologous protection against influenza by injection of DNA encoding a viral protein. *Science* 259:1745-1749.

Usha R, Rohil J B, Spall V E, Shanks M, Maule A J Johnson J E and Lomonossoff G P. 1993. Expression of an animal virus antigenic site on the surface of a plant virus particle. *Virology* 197: 366-374.

Waine G J and McManus D P. 1995. Nucleic acid Vaccines of the future. *Parasitology Today* 11: 113-116.

Walker PS, Scharton-Kersten T, Rowton ED, Hengge U, Bouloc A, Udey MC & Vogel JC. 1998. Genetic immunization with glycoprotein 63 cDNA results in a helper T cell type 1 immune response and protection in murine model of leishmaniasis. *Human Gene Therapy* 9: 1899-1907.

Walsh EP Baron MD, Anderson J & Barret T. 2000. Development of a genetically marked recombinant rinderpest Vaccine expressing green fluorescent protein. *Journal of General Virology* 81: 709-718.

Wampler DE, Lehman ED, Boger J, *et al.* 1985. Multiple chemical forms of hepatitis B surface antigen produced in yeast. *Proccedings of National Academy of Sciences* (USA) 82: 6830-6834.

Wang B, Ugen KE, Srikantan V, Agadjanyan MG, Dang KRefacli Y, Sato AI, Boyer J, Williams WV. 1993. Gene inoculation generates immune response against human immunodefficiency virus type 1. *Proceedings of National Academy of Sciences* 90: 4156-4160.

Weidt G, Deppert W, Utermoblen O, Heukeshoven J & LehmanGrube F. 1995. Emergence of virus escape mutants after immunization with epitope Vaccine. *Journal of Virology* 69: 7147-7151.

West D, Reddin K, Mathesonson M, Heath R, Funnel S, Hudson M, Robinson A & Gorringe A. 2001. Recombinant *Neisseria meningitidis* transferrin binding protein A protects against experimental menigococcal infection. *Infection and Immunity* 69: 1561-1567.

Widera G, Austin M, Rabussay D, Goldbeck C, Barnett SW, Chen M, Leung L, Otten GR, Thudium K, Selby MJ & Ulmer JB. 2000. Increased DNA Vaccine delivery and immunogenicity by electroporation *in vivo*. *Journal of Immunology* 164: 4635-4640.

Wigdorovitz A, Perez- Filgueira DM, Robertson N, Crrillo C, Sadir AM, Morris TJ & Borca MV. 1999. Protection of mice against challenge with foot and mouth disease virus (FMDV) by immunization with foliar extracts from plants infected with recombinant tobacco mosaic virus expressing FMDV structural protein VP1. *Virology* 264: 85-91.

Wolff JA Malone RW, Williams P Chong W, Acsadi G, Jani A & Felgner PL. 1990. Direct gene transfer into mouse muscle *in vivo*. *Science* 247: 1465-1468.

Wright I G. 1991. Towards a synthetic Babesia Vaccine. *International Journal for Parasitology* 21: 155-159.

Wright I G, Casu R, Commins M A, Dalrymple, B P, Gale K R, Goodgar B V, Riddles P W , Waltishuhl D J, Abetz I, Berrie D A, Bowles, Dimmock C. Hayes T, Kalnins H, Leatch G, McCrae R, Montogue P E, Nisbet I T, Parrodi F, Peters J M, Scheiwe P C, Smith W, Rode-Bramnis K and White M A. 1992. The development of recombinant Babesia vaccine. *Veterinary Parasitology* 44: 3-13.

Xiang Z, Spitanlik QS, Tran M, Wunner WH, Cheng J & Ertl HCJ. 1994. Vaccination with a plasmid vector carrying the rabies virus glycoprotein gene induces protective immunity against rabies virus. *Virology* 199:132-140.

Xu D and Liew F Y. 1995. Protection against leishmaniasis by injection of DNA encoding a major surface glycoprotein, gp63, of L- major- rapid communication. *Immunology* 84: 173-176.

Xu F, Jones TD and Rodgers PB. 1996. Potential of chimaeric plants virus plant virus particls as novel, stable Vaccines. *Developments in Biological Standardization.* Basel, S. Karger, 87: 201-205.

Yamada T, Uchiyama H, Nagat T, Uchijima M, Suda T, Chida K, Nakamura H & Koide Y. 2001. Protective cytotoxic T lymphocyte responses induced by DNA immunization against immunodominant and subdominant epitopes of *Listeria monocytogenes* are noncompetitive. *Infection and Immunity* 69: 3427-3430.

Yang W, Waine G J and McManus D P. 1995. Antibodies to Schistosoma japonicum (Asian blood fluke) paramyosin induced by nucleic acid vaccination. *Biochemical and Biophysical Research Communications* 212: 1029-1039.

Yi X, Zeng, X, Zeng X, Wang S, Shu X & McReynolds L. 2000. Studies on antifecundity induced by *Schistosoma japonicum* cathepsin B DNA Vaccine. Chinese Journal of Zoonosis. 16: 45-47.

Yonezawa H, Ishihara K & Okuda K. 2001. Arg-gingipain A DNA Vaccine induces protective immunity against infection by *Porphyromonas gingivalis* in murine model. *Infection and Immunity* 69: 2858-2864.

Zhang T & Stanely SL Jr. 1996. Oral immunization with an attenuated Vaccine strain of *Salmonella typhimurium* expressing the serine-rich *Entamoeba histolityca* protein induces an anti-amoebic immune response and protects gerbils from amoebic liver abscess. *Infection and Immunity* 64: 1526-1531.

Zhou X, Berglund P, Rhodes G, Parker SE, Jonad M & Liljestron. 1994. Self replicating Semliki Forest virus RNA as recombinant Vaccine. *Vaccine* 12: 1510-1514.

CHAPTER - 18

PHOSPHATE SOLUBILISING MICROBIAL SYSTEMS

Geeta Singh

Senior Scientist, Division of Microbiology, Indian Agricultural Research Institute, New Delhi 110 012

1. INTRODUCTION

Phosphorus is one of the important macro nutrients required for plant growth, metabolism and the synthesis of various biomolecules of all living organisms. Phosphorus is in short supply in many agricultural soils and may limit the production of arable crops. Soil contain both organic (inositol phosphates) and inorganic forms of 'P' in the alkaline and acidic soils, they occur as insoluble Ca, Fe, and Al phosphates (Marschner, 1995). Soluble phosphates have a low mobility in soil and their uptake creates a depletion zone around root (Hoffland *et al.*, 1989). Much of the inorganic P applied to soil as fertilizer is rapidly converted to unavailable forms with low solubility (Stevenson, 1986; Sanyal and De Datta, 1991). As more than 95% of soils in India are deficient in P they require a large amount of phosphatic fertilizer of which only 10-15% is available to plants. According to Finck (1992) the utilization rate of 'P' in mineral fertilizer is 10-25% during the first year. Rest is available at the rate of 1-2% per year during the following decade. Precipitation is an important mechanism for converting soluble forms of 'P' into insoluble forms. Initially the precipitate are amorphous and 'P' is moderately available to plants (Rajan, 1976). The precipitates are slowly converted to less soluble forms such as calcium orthophosphates [$Ca_3(PO_4)_2$] , hydroxyapatite [$Ca_5(PO_4)_3$ (OH)] and fluoroapatite [$Ca_{10}(PO_4)_6F_2$] in alkaline soils or crystalline ferric phosphate [$FePO_4$ $2H_2O$ (Cr) or strengite)] and crystalline aluminium phosphate [$AlPO_4$ $2H_2O$ (Cr) or Variscite] in acidic soils (Lindsay *et al.*, 1962; Rajan, 1976; Sample *et al.*, 1980).

The soluble 'P' is released from insoluble phosphates mainly by three mechanisms. First acid production (of plant or microbial origin) may cause a lowering of rhizosphere pH (Kapoor *et al.*, 1989, Kuely *et al.*, 1989). Second, organic acids capable of chelating metal ions resulting in competition with 'P' for adsorption sites in the soil. Third, organic acids may form soluble complexes with metal ions associated with insoluble 'P' (Ca^{2+}, Al^{3+} and Fe^{3+}) thus releasing 'P' (Kepert *et al.*,1979). Soluble phosphates have low mobility and their uptake creates a depletion zone around the roots (Hoffland *et al.*, 1989).

Mechanisms of 'P' solubilisation Bacteria, yeasts, actinomycetes, mycorrhizal fungi and free living fungi have been reported to cause increase in plant available 'P' in the soil (Table 1). These microbial biofertilizers are cost effective ecofriendly, renewable and generally capable of supplementing chemical fertilizers (Gold stein *et al.*, 1993; Isher- wood, 1998).

The phosphate solubilizing soil and rhizosphere microorganisms have been distinguished by the relative abilities to dissolve calcium phosphate and apatites in pure culture (Gaur *et al.*, 1973; Kucey, 1987), and in association with (Gerretsen, 1948) plants roots. Microbial solubilization of phosphate minerals has been achieved by removing phosphate contaminants from iron ore (Park *et al.*, 1990) and as an alternative process for producing phosphates from fluorapatite (Nahas *et al.*, 1990).

This activity was attributed to organic acid and chelating metabolites (Gaur, 1980; Molla and Chowdhury, 1984; Tinker, 1984) or direct acidulation or through proton

Table 1. List of some microorganisms and the acid produced during phosphate solubilisation

Microorganism	Phosphate solubilization (mg P litre^{-1} culture)	Organic acid produced	References
Aspergillus terreus	1000	Oxalate and H_2S production	Rose, 1957
Aspergillus niger	460	Citrate, glycollate, succinate, gluconate	Sperber, 1958
Penicillium sp.	599	ND	Subba Rao and Bajpai, 1965
Penicillium lilacinum	78	ND	Chhonkar and Subba Rao, 1967
Aspergillus candidus	20	Oxalate, tartarate	Banik and Dey, 1982
Aspergillus niger	666	Lactate, glycollate, citrate and succinate	Venkateshwarlu *et al.*, 1984
Aspergillus sp.	148	ND	Thomas *et al.*, 1985
Penicillium sp.	41	Oxalate	Parks *et al.*, 1990
Penicillium spp.	65	Gluconate, lactate, citrate	Illmer & Schinner (1992)
Aspergillus awamori	190	Oxalate	Gupta *et al.*, 1994
Aspergillus japonicus	180	Oxalate	Singal *et al.*, 1994
Aspergillus aculeates	470	ND	Narsian *et al.*, 1995
Acetobacter diazotrophicus		Gluconic	Galar and Boiardi, 1995
Arthrobacter		Oxalic and Malic	Banik & Dey, 1982
Bacillus firmus		2-keto gluconic, succinic	Banik & Dey, 1982
Bacillus megaterium		Lactic, malic	Taha *et al.*, 1969
Escherichia fruendii		Lactic	Sperber, 1958
Pseudomonas striata		Oxalic, tartaric	Krishamurti, 1989
P. striata		Tartaric, citric	Gaur, 1990

extension (Illumer & Schinner, 1992, 1995a, Gupta *et al.*, 1994; Gyaneshwar *et al.*, 1998. Whitelaw *et al.*, 1991). The organic acids produced (Table 1) and the mechanism by which organic acids solubilize phosphates in liquid media are reported to be due to either lowering of pH or the chelation of cations bound to P. chelation of cations has been shown to be an important mechanism of 'P' solubilization when organic acid structure is favourable. Many investigators have reported that organic acids were more efficient solubilizing agents than inorganic acids at the same pH with the difference presumed to be primarily due to chelation (Cunningham & Kuiack 1992; Kim *et al.*, 1997; Whitelaw *et al.*, 1999).

The extent to which an organic acid is capable of chelating metal cations is governed by the hydroxyl and carboxyl groups present in it (Johnston, 1956; Johnston & Miller, 1959). The decreasing order of cation chelating capacity is tricarboxylic (citric acid), dicarboxylic (malic & tartaric acids) followed by monobasic acids such as lactic acid, which posses a hydroxyl groups adjacent to carboxylic groups. (Bolan *et al.*, 1994) found the ability of organic acids to prevent 'P' adsorption in soil decreased in the following order tricarboxylic acid dicarboxylic acid monocarboxylic acid. A direct correlation of the fall in pH and increase in available 'P' although has been observed in certain cases (Sperber, 1957; Agnihotri, 1970; Liu *et al.*, 1992), largely the correlation does not exist (Asea, 1988). Other postulated mechanism involved reduction of Fe in anaerobic conditions and flooded soils (Alexander, 1977), liberation of H_2S which reacts with ferric phosphate to yield ferrous sulphide and liberate phosphate (Sperber, 1957). Table 1 list the organic acid secreted by mineral phosphate solubilising microorganism.

2. MYCORRHIZAE AS 'PHOSPHOLYTIC MICROORGANISM

Use of mycorrhiza in agroforesty has immense potential to increase crop yields both in low input sustainable agricultural systems and on low fertility soil where 'P' is growth limiting, a common situation in tropical regions (Gianinazzi and Schuepp, 1994). Mycorrhizal fungi forms symbiotic association with roots of most terrestrial plants, including many agricultural crops. (Mahadevan *et al.*, 1988). The fungi colonize the cortex of roots to form the multicellular structure called the ectomycorrhizal mantle and forms a new exchange inter face called Hartig net, which obtains carbon from their host plant while assisting the plant with the uptake of phosphate and other minerals from the soil. Several studies have clearly revealed that mycorrhizal fungi play an important role in mineral cycling particularly in mobilizing and translocating P and N to host plants (Lanfranco *et al.*, 1999). The interpretation of this phenomenon is that the mycelia bridge the depletion zone and have a high affinity for soluble phosphates. They have high 'P' solubilizing and mobilizing potential. Mycorrhizal fungi release phosphorus from rock phosphate by reducing the pH through excreting organic acid eg., Oxalic acid (Malajezuk and Cromack, 1982 Lapyrie *et al.*, 1991). Oxalate production has an important role in biomineralization (Snetselaar and Whitney, 1990) and phosphate solubilization (Cannon *et al.*, 1995). High concentration of organic acids were detected in the rhizosphere soils than in bulk soils and in soils containing mats of ectomycorrhizal fungi than in non mat soils (Caldwell *et al.*, 1991; Grierson, 1992). Concentration of organic acid were high in the microorganism colonizing around fine roots or fungal hypae (Ochs *et al.*, 1993; Drever and Vance, 1994). Production of organic acids especially oxalate by ectomycorrhizal fungi is well documented (Lapeyrie *et al.*, 1991). Griffiths *et al.* (1994) found that mycelial mats formed by *Gautieria monticola* (high production of oxalic acid) weathered soil minerals more efficiency than mycelial mats formed by *Hysterangium setchellii* (low production of oxalic acid). To confirm the secretion of organic acids by *L. fraterna* HPTLC analysis showed secretion of oxalic acids when grown on mineral phosphate (Lapeyrie *et al.*, 1991). Watteau and Berthelin (1994) studied the weathering of biotic and the effect of the ectomycorrhizal fungus *Suillus granulatus* growing in a nutrient medium free of iron. Iron was released from the mineral in response to release of malic and citric acids by the fungus. *Suillus variegatus* reduced soil pH more than non-mycorrhizal *Pinus sylvestris* seedlings or seedlings colonized by *P. involutus* (Wallander *et al.*, 1997). Oxalic acid was detected in the water droplets added to the surface of mycelium of *S. bovines* (Sun *et al.*, 1999). As compared to other organic acids, oxalic acid is most efficient in mineral weathering because of its strong complexing property. By forming complex with metal ions, such as Ca, Al and Fe, oxalate would release phosphate from insoluble phosphate (Cormack *et al.*, 1979).

Paxillus involutus 1 and 2, *L. laccata, C. geophilum, H. clyindrosporum, H. crustuliniforme* 1 and 2, *L. bicolor* and *P. tincotorius* 1 and 2 isolated from carpophore were classified into 8 groups on the basis of their phosphate solubilizing activity (Lapeyrie *et al.*, 1991). Phosphates were applied in the form of synthetic mineral phosphates such as $AlPO_4$. Calcium phytat, $CaHPO_4$, $Ca_3(PO_4)^2$ and natural crystalline phosphate such as orthoclase feldspar, plagioclase, apatite, fluorapatite and iron phosphate. Nitrogen sources such as ammonium sulphate or nitrate nitrogen were supplied along with 'P'. The fungi were ranked into 8 groups such as no visible solubilization (marked 0) to high solubilization range (marked 200) times the diameter of the colony. Ammonium nitrogen or nitrate nitrogen significantly influenced phosphate solubilization by fungi except *P. involutus* 1. Calcium phytate and $CaHPO_4$ were easily solubilized especially in the presence of ammonium nitrogen, by most of the strains, while $AlPO_4$ was toxic to *H. crustuliniforme* 1 and *L. laccata*. Ammonium nitrogen stimulated proton extrusion and solubilization was due to acidification (Salsac *et al.*, 1982). After 15 d at pH 6 with nitrate nitrogen, *P. involutus* I secreted oxalic acid, citric acid, tartaric acid and fumaric acid, while *L. laccata* and *H. clyindrosporum* produced oxalic acid. The reasons for this phenomenon are that they have a high P solubilzing and mobilizing potential and mineral phosphates could be solubilized by excreted protons or organic acids and organic phosphates are mobilised by phosphatase enzyme (Lapeyrie, 1988).

3. GENETICS OF MINERAL PHOSPHATE SOLUBILIZATION

Genes play an important role in bacterial phosphate starvation metabolism. MPS (mineral phosphate solubilization) phenotypes was induced /repressed by low/high levels of exogenous Pi (Goldstein, 1986) The gene (s) responsible for this trait were termed as 'MPS'. Shotgun cloning experiments using *Erwinia herbicola* DNA, enabled Goldstein and Liu (1987) to clone gene(s) to convert, MPS *E. coli* to MPS^+ strain. On screening a cosmid pHC 76 library obtained from *E. herbicola*, it was found that a 55 kb insert was able to transform *E. coli* HB 101 to MPS^+ phenotype. On the basis of sequence comparison and mini cell analysis, Liu *et al.*, 1992 deciphered that the 'gene' codes for a co-enzyme pyrroloquinoline quinone (PQQ) for the enzyme glucose dehydrogenase that enabled the transgenic *E. coli* HB 101 to produce gluconic acid. This gluconic acid was responsible for the MPS phenotype.

Mineral phosphate solubilizing genes so far isolated from bacteria such as *Acinetobacter calcoaceticus* by Goosen *et al.* (1987, 1989), *Erwinia herbicola* and *P. cepacia* (Liu *et al.*, 1992 and Khan *et al.*, 1995) cloned on to *E. coli* K-12 and HB 101. The ability of some gram negative bacteria to dissolve poorly soluble calcium phosphates (Mps$^+$ phenotypes) is the result of periplasmic oxidation of glucose to gluconic acid via quinoprotein glucose dehydrogenase (GDH), components of the direct oxidation pathway. *E. coli* K12 synthesis Apo GDH but not the co-factor pyrolo-quinoline-quinone (PQQ) essential for the formation of the holoenzyme. Therefore, in the absence of exogenous PQQ, these strains do not produce gluconic acid and are designated MPS$^-$. As a result of gluconic acid production, *E. coli* HB 101 (PMCG 898) showed enhanced mineral phosphate solubilizing phenotypes due to acid dissolution of hydroxy apatite substrate. The pH of the medium declined from 5.2 to 3.5-4 at the end of 24 h in minimal medium amended with tricalcium phosphate. In *Erwinia herbicola* and in clone PMCG 898 most of the mineral phosphate solubilization occurred on third day which was evidenced by the detection of gluconic acid by HPLC (Liu *et al.*, 1992).

In *Pseudomonas cepacia*, an insert of 2.4 kb designated as clone pSLY 4 in *E. coli* JM 109) (K-12 derivatives) was sufficient to induce the MPS$^+$ phenotype and the production of gluconic Acid (Khan *et al.*, 1995). A larger deletion of 2.4 kg insert to 0.692 kb resulted in a sub clone designated as PGABY possessing Mps$^+$ phenotypes and gluconic acid production. Disruption of the ORF (open reading frame) at the unique Xho I site within the ORF changed the phenotypes to MPS$^-$ and eliminated gluconic acid production. In ectomycorrhizae *Laccaria fraterna* genomic library was constructed in *E. coli* DH5 using cosmid vector PANUMV2 at the Hind III site. This vector contained HY^R (hygromycin resistance) as selectable marker. Using restriction enzyme Hind III genomic library of *Laccaria fraterna* was constructed from which two clones designated as MLA 22 and MLA 46 that utilized rock phosphate as sole 'P' source were isolated. The recombinant plasmid was isolated from the clones and digested with Hind III to release the insert. The 5 kb insert was further restricted using Bam HI enzyme. The sub clone containing 3 kb insert showed 80% identity with mineral phosphate solubilization sequence present in *Erwinia herbicola* mps genes responsible for biosynthesis of the co-enzyme PQQ. In case of *Pseudomonas putida* and *Burkholderia cepacia* 78.8% identity in 33 nucleotide overlap and 58.7% identity in 121 nucleotide overlap was observed between mps sequences of *L. fraternae* and mps genes involved for the biosynthesis of the co-enzyme PQQ. Using HPTLC oxalic acid was identified in both wild *L. fraterna* and 3 kb containing subclone MAP 50. The wild subclone secreted 1490 and 2950% g/ml of oxalic acid. The homology similarity of mps sequences of *L. fraternal* with mps sequence of other prokaryotes suggest that such sequence arose form a common ancestors by divergent evolution (Marger and Seair, 1993). Ecotmycorrhizal fungus are versatile mineral phosphate solubilizing agents. Complete sequencing of the fragment involved in mineral phosphate solubilization can help to design specific probes. These probes can form the basis of diagnostic kit for large scale screening and isolation of phospholytic ectomycorrhizal fungi.

4. CONCLUSION

Phosphorus is a major nutrient for crop plants and is involved in various essential processes such as cell division, development, photosynthesis breakdown of sugar; energy transfer nutrient transport within plant and expression and maintenance of the genetic material (Tandon, 1987). Major fraction of the phosphatic fertilizers applied to agricultural fields get fixed into unavailable forms in soil. Thus soils are rich in 'P' which is to be harnessed. Microbial biofertilizers are an important means of converting the unavailable forms into available forms. One way of improving the mineral phosphate solubilizing microorganism is through selection of isolates with improved MPS activity. A great potential, thus, exists for metabolic engineering of mineral phosphate solubilizing phenotypes to accomplish better transformation of unavailable 'P' forms.

REFERENCES

Agnihotri, V.P. 1970. Solubilization of insoluble phophates by some soil fungi isolated from nursery seed beds. Can. J. Microbiol., 16 : 877-880

Alexander, M. 1977. Introduction to Soil Microbioloy. Wiley Eastern Limited, New Delhi.

Asea, P.E.A., Kucey, R.M.N. and Stewart, J.W.B. 1988. Inorganic phosphate solubilization by two Penicillum species in solution culture and soil. Soil Biol Biochem, 20 (4) : 459-464.

Banik, S. and Dey, B.K. 1982. Available phosphate content of an alluvial soil as influenced by inoculation of some isolated phosphate solubilizing microorganisms. Plant Soil 69 : 353-364.

Bieleski, R.L. 1973. Phosphate pools, phosphate transport, and phosphate availability. Ann. Rev. Plant Physiol., 24 : 225-252.

Bolan, N.S,. Naidu, R., Mahimairaja, S., and Baskaran, S. 1994. Influence of low-molecular weight organic acids on the solubilization of phosphates, Biol Fertil. Soil, 18 : 311-319.

Caldwell, B.A., R.P. Griffiths, K. Cromack. Jr, and M.A. Castellano. 1991. Enzyme activities soil present in ectomycorrhizal mat communities. In : Abs. of Soil Ecology Society Meeting, Corvallis, Oregon. pp. 23.

Cannon, J.P., E.B. Allen, M.F. Allen, L.M. Dudley and J.J. Jurinak. 1995. The effect of oxalates produced by Salsola tragus on the phosphorus nutrition of Stipa pulchra. Oecologia, 102 : 265-275.

Chhonkar, P.K. and Subba Rao, N.S. 1967. Phosphate solubilization by fungi associated with legume roots nodules. Can. J. Microbiol., 13 : 749-753.

Cromack, K., Jr., P. Sollins, W.C. Graustein, K. Speidel, A.W. Todd, G. Spycher, C.Y. Li and R.L. Todd. 1979. Calcium oxalate accumulation and soil weathering in mats of the hypogrous fungus Hysterangium crassum. Soil Biol Biochem., 11 : 463-468.

Cunningham, J.E. and C. Kuiack 1992. Production of citric and oxalic acid and solubilization of calcium phosphate by Penicillium bilaii. Appl. Environ. Microbiol., 58 : 1451-1458.

Dalal, R.C. 1977. Soil organic phosphorus. Adv. Agron., 29 : 83-117.

Drever, J.I. and G.F. Vance 1994. Role of organic acid in mineral weathering processes.

In : Organic acids in geological processes. (eds. Pittman E.D. and Lewman E.D.). Springer, Berlin, pp. 138-161.

Finck, A. 1992. World Fertilizer Use Manual, IFA, Paris.

Gaur, A.C. 1990. Phosphate solubilizing micro organisms as biofertilizer. Omega scientific publishers, New Delhi, pp. 176.

Gaur, A.C., M. Madan and K.P. Ostwal 1973. Solubilization of phosphatic compounds by native microflora of rock phosphates. Indian J. Exp. Biol., 11 : 427-429.

Gay, G. and J.C. Debaud 1987. Genetic study on indole-3-acetic acid production by Hebeloma species : inter and intraspecific variability in homo and di-karyotic mycelium. Appl. Microbiol. Biotechnol., 26 : 141-146.

Gerresten, F.C. 1948. The influence of microorganism on the phosphate intake by the plant. Plant Soil. 1 : 51-81.

Gianinazzi, S. and H. Schuepp. 1994. Impact of Arbuscular Mycorrhizas on sustainable Agriculture and Natural Ecosystems. Birkauser Verlag, Basel.

Grierson, P.F. 1992. Organic acids in the rhizosphere of Banksia integrifolia L.F. Plant Soil, 144 : 259-265.

Goldstein, A.H., R.D. Rogers and G. Mead 1993. Separating phosphate from ores via., bioprocessing. Biot/Technology, 11 : 1250-1254.

Goldstein, A.H. 1986. Bacterial solubilization of mineral phosphate : Historical prespectives and future prospects. Am. J. Act. Agric. 1 (2) : 51-57.

Goosen, N., H.P.A. Horsman, R.G.M. Huinen and P.V. Putte 1989. Acinetobacter calcoaceticus genes involved in biosynthesis of the coenzyme pyrrolo-quinoline-quinone : nucleotide sequence and expression in Escherichia coli K-12. J. Bacteriol., 171 : 447-455.

Goosen, N., D.A.M. Vermaas and P.V. Putte 1987. Cloning of the genes involved in synthesis of coenzyme pyrorlo-quinoline-quinone from Acinetobacter calcoaceticus. J. Bacteriol., 169 : 303-307.

Gupta, R., Rekha, S., Aparna, S., and Kuhad, R.C. 1994. A modified plate assay for screening phosphate solubilizing microorganisms. J. Gen. Appl. Microbiol., 40 : 255-260.

Gyaneshwar, P., G.N. Kumar and L.J. Prakesh 1998. Cloning of mineral phosphate solubilizing genes from Synechocystis PCC 6803. Curr. Sci, 74 : 1097-1099.

Hajra, J.N. and S.C. Kole 1998. Occurrence and activity of tricalcium phosphate and rock phosphate solubilzing microorganisms in mechanical plant compost of Calcutta and an alluvial soil of West Bengal. Environ Ecol., 16 : 344-349.

Harrison, M.J. and M.L. van Buuren 1995. A phosphate transporter from the mycorrhizal fungus Glomus versiforme. Nature, 378 : 626-629.

HO, 1987. Comparison of eight Pisolithus tinctorius isolates for growth rate, enzyme activity and phytohormone production. Can. J. For. Res., 17 : 31-35.

Hoffland, E. 1992. Quantitative evaluation of the role of organic acid exudation in the mobilization of rock phosphate by rape. Plant Soil, 140 : 279-289.

Hoffland, E., G.R. Findenegg and J.A. Nelemans 1989. Solubilization of rock phosphate by rape. I Evaluation of the role of the nutrient uptake pattern. Plant Soil, 113 : 155-160.

IFA, 1998. The fertilizer industry, world food supplies and the environment. IFA, Paris pp. 57.

Isherwood, K.F. 1998. Fertilizer use and the environment. IFA, Paris. pp. 26.

Illmer, P. and Schinner, F. 1992. Solubilization of inorganic phosphates by microorganisms isolated from forest soils. Soil Biol. Biochem., 24 (4): 389-395.

Illmer, P. and Schinner, P. 1995. Solubilization of inorganic calcium phosphate-solubilization mechanims. Soil Biol Biochem., 27 (3) : 257-263.

Illmer, P. Barbato, A. and Schinner, F. 1995. Solubilization of hardly soluble AlPO4 with P-solubilizing microorganisms. Soil Biol Biochem., 27 (3) : 265-270.

Johnston, H.W. 1959. The solubilization of 'insoluble phosphate' V. The action of some organic acids on iron and aluminium phosphates. N.Z.J. Sci., 2 : 215-218.

Johnston, H.W. and Miller, R.B. 1959. The solubilization of 'insoluble' phosphate . IV. the reaction between organic acids and tricalcium phosphate. N.Z.J. Sci., 2 : 109-120.

Junghans, D.T., E.A. Gomes, W.V. Guimaraes, E.G. Barros and E.F. Araujo 1998. Genetic diversity of the ecotmycorrhizal fungus Pisolithus tinctorium based on RAPD-PCR analyis. Mycorrhiza 7 : 243-248.

Jurinak, J.J., L.M. Dudely, M.F. Allen and W.G. Knight 1986. The role of calcium oxalate in the availability of phosphorus in soils of semiarid region : a thermodynamic study. Soil Sci., 142 : 255-261.

Kapoor, K.K., M.M. Mishra and K.Kukreja, 1989. Phosphate solubilization by soil microorganisms. A. Review. Indian J. Microbiol., 29 : 119-127.

Khan, S.B., T.C. Yeo, W.L. Martin, M.R. Duron, R.D. Rogers and A.H. Goldstein 1995. Cloning of a mineral phosphate solubilizing gene from Pseudomonas cepacia. Appl. Environ Microbiol., 61 : 972-978.

Krishnamurty, H.A. 1989. Effect of pesticides on phosphate solubilizing microorganisms. M.Sc. Thesis UAS Dharwad, India.

Kepert, D. G. Robson, A.D, and Posner, A.M. 1979. The effect of organic root product on the availability of phosphorus to plants. In "The Soil Root Interface (J.L. Harley an dR. Scott-Russell Eds.). pp. 115-124. Academic press, London.

Kim, K.Y., Mc Donald, G.A. and Jordan, D. 1997. Solubilization of hydroxypatite by Enterobacter agglomerans and cloned Escherichia coli in culture medium Biol. Fertil. Soils, 24 : 347-352.

Kucey, R.M.N. 1987. Increased phosphorus uptake by wheat and field beans inoculated with a phosphorus-solubilizing Penicillium bilaji and vesicular -arbuscular mycorrhizal fungi. Appl. Environ. Microbiol., 53 : 2699-2703.

Kucey, R.M.N., H.H Janzen and M.E. Leggett 1989. Microbially mediated increase in plant available phosphorus. Adv. Agron., 42 : 199-228.

Lanfranco, L., L. Gancro and P. Bonfante 1999. Chitin synthase genes in the arbuscular mycorrhizal fungus Glomus versiforme : full sequence of a gene encoding a class IV chitin synthase. FEMS Microbiol. Lett., 170 : 59-69.

Lapeyrie, F 1988. Oxalate synthesis from soil bicarbonate by the mycorrhizal fungus Paxillus involutus. Plant Soil, 110 : 3-8.

Lapeyrie, F.G, G.A. Chilvers and C.A,.Behm 1987. Oxalic acid synthesis by the mycorrhizal fungus Paxillus involutus. New Phytol., 106 : 139-146.

Laperyie, F., J. Ranger and D. Vairelles, 1991. Phosphate-solubilizing activity of ectomycorrhizal fungi in vitro. Can. J. Bot., 69 : 342-346.

Liu, S.T,. L.Y. Lee, C.Y. Tai, C.H. Hung, Y.S. Change, J.H. Wolfram, R.D. Rogers, and

A.H. Goldstein 1992. Cloning of an *Erwinia herbicola* gene necessary for gluconic acid production and enhanced mineral phosphate solubilization in *Escherichia coli* HB 101 : nucleotide sequencing and probable involvement in biosynthesis of the coenzyme pyrroloquinoline quinone. J. Bacteriol., 174 : 5814-5819.

Mahadevan, A., N. Raman and K. Natarajan 1988. Mycorrhizae for green Asia. University of Madras, Madras, India.

Malajczuk, N. and K. Cromack. Jr. 1982. Accumulation of calcium oxalate in the mantel of ectomycorrhizal roots of Pinus radiata and Eucalyptus marginata. New Phytol., 92 : 527-531.

Marger, M.D. and M.H. Saier, Jr. 1993. A major super family of transmembrane facilitators that catalyse uniport, symport and antiport. TIBS 18 : 13-20.

Marschner, H. 1995. Mineral nutrition of higher plant. Academic Press, London.

Molla, M.A.Z. and A.A. Chowdhary 1984. Microbial mineralization of organic phosphate in soil. Plant Soil, 78 : 393-399.

Moog, R.S., M.A. Mc Guirl, C.E. Cote and D.M. Dooley 1986. Evidence for methoxatin (pyrroloquinolinequinone) as the cofactor in bovine plasma amine oxidase from resonance. Raman Spectroscopy. Proc. Natl. Acad. Sci., 83 : 8435-8439.

Nahas, E. 1996. Factors determining rock phosphate solubilization by microorganisms isolated from soil. World J. Microbiol. Biotechnol., 12 : 567-572.

Nahas, E., D.A. Banzatto and L.C. Assis. 1990. Fluorapatite solubilization by Aspergillus niger in vanasse medium. Soil Biol. Biochem., 22 : 1097-1101.

Narsian, V., Thakkar, J. and Patel, H.H. 1993. Solubilization of natural rock phosphates and pure insoluble inorganic phosphates by Aspergillus awamori. Indian J. Exp. Biol., 31 : 747-749.

Ochs, M., I. Brunner, W. Stumm and B. Cosovic 1993. Effects of root exudates and humic substance on weathering kinetics. Water Air Soil Pollut., 68 : 213-229.

Park, E.J., G.J. Olson, F.E. Brinckman and F. Baldi 1990. Characterization by high performance liquid chromatography (HPLC) of the solubilization of phosphorus in iron ore by a fungus. J. Indian Microbiol., 5 : 183-190.

Pikovskaya, R. I. 1948. Mobilization of phosphorus in soil in connection with the vital activity of some microbials species. Mikrobiologia, 17 : 362.

Pritsch, K., J.C. Munch and F. Buscot. 2000. Identification and differentiation of mycorrhizal isolates of black alder by sequence analysis of the ITS region. Mycorrhiza, 10 : 87-93.

Rose, R.E. 1957. Techniques for determining the effect of microorganisms on insoluble phosphates. N. Z.J. Sci. Technol., 38 : 773-780.

Raman N. and A. Mahadevan 1996. Mycorrhizal research-A priority in agriculture. In : Concepts in Mycorrhizal Research (eds. K.G. Mukerji), Kluwer, Dordrecht, Netherlands . pp. 41-75,

Sample, E.C., Soper, R.J. and Racz, G.J. 1980. Reactions of phosphate fertilizers in soils In : "The Role of Phosphorus in Agriculture (F.E. Khasawnch, E.C. sample and E.J. Damprath, Eds.), pp. 263-310 ASA-CSSA-SSSA, Madison, WI.

Sanyal, S.K. and S.K. De Datta 1991. Chemistry of phosphorus transformations in soil Adv. Soil Science, 16 : 1-20.

Signal, R, Gupta, R, and Saxena, R.K. 1994. Rock phosphate solubilization under alkaline conditions by *Aspergillus japonicus* and *A. foetidus*. Folia Microbiol. 39 : 33-36.

Sneteslarr, K.M. and K.D. Whitney 1990. Fungal calcium oxalate in mycorrhizae of *Monotropa uniforma.* Can. J. Bot., 68: 533-543.

Sperber, J.I. 1957. Solution of mineral phosphate by soil bacteria. Nature, 180: 994-995.

Sperber, J.I. 1958. Solution of apatite by soil microorganisms producing organic acids. Aust. J. Agril. Res., 9: 778-781.

Stevenson, F.J. 1986. Cycles of Soil Carbon, Nitrogen, Phosphorus, Sulfur, Micro-nutrients. Wiley, New York.

Subba Rao, N.S. and Bajpai, P.D. 1965. Fungi on the surface of legume root nodules and phosphate solubilization. Experientia. 21: 386-387.

Sun, Y.P., T. Unestam, S.D. Lucas, J. H. Johanson, K.J. Kenne and R. Finlay 1999. Exudation-reabosrption in a mycorrhizal fungus, the dynamic interface for interaction with soil and other miroorganism. Mycorrhiza, 9: 137-144.

Taha, S.M. Mahmoud, S.A.Z. Halini EL-Damaty, A. Abd EL-Hofez, A.M. 1969. activity of phosphate dissolving bacteria in Egyptian soils. Plant Soil., 31 : 149-160.

Tandon, H.L.S. 1987. Phosphorus research and agricultural production in India. Fertilizer development and consultation organisation, New Delhi pp. 160 + xii.

Thomas, G.V., Shantaram, M.V. and Saraswathy, N 1985. Occurrence and activity of phosphate solubilizing fungi from coconut plantation soils. Plant soil, 87:357-364.

Tinker, P.B. 1984. The role of microorganism in mediating and facilitating the uptake of plant nutrients from soil. Plant Soil, 76 : 77-91.

UNEP, 1998. Mineral fertilizer production and environment. Part I. the Fertilizer Industry's Manufacturing Processes and Environment Issue. pp. 47-52.

Venkateswaralu, B. Rao, A.V. and Raina, P. 1984. Evaluation of phosphorus solubilization by micro-organisms isolated from aridsoils. J. Indian Soc. Soil Sci., 32 : 273-277.

Vora, M.S. and H.N. Shelat 1998. Impact of addition of different carbon and nitrogen sources on solubilization of rock phosphate by phosphate solubilizing micro-organisms. Indian J. Agric. Sci., 68 : 292-294.

Watteau, F. and J. Berthelin 1994. Microbial dissolution of iron and aluminium from soil minerals: efficiency and specificity of hydroxamate siderophores compared to aliphatic acids. Eur. J. Soil Biol., 30 : 1-9.

Wallander, H., T. Wickman and G. Jacks 1997. Apatite as a source in mycorrhizal and non-mycorrhizal Pinus sylvestris seedlings. Plant soil., 196 :123-131.

Whitelaw, M.A., Harden, T.J. and Helyar, K.R. 1999. Phosphate solubilization in solution culture of soil fungus Penicillum radicum. Soil Biol. Biochem., 31 : 655-665.

CHAPTER - 19

BIOTECHNOLOGICAL APPLICATION OF MICROBES FOR IMPROVED PLANT PRODUCTIVITY

S.K. Dwivedi and Padmanabh Dwivedi***

*Department of Environmental Science, B.B. Ambedkar University, Rae Bareli Road, Lucknow-226 025, India

**Department of Botany, Arunachal University, Rono Hills, Itanagar-791 111, India

1. INTRODUCTION

The history of biotechnology dates back to the time when microorganisms were used to produce yoghurt, vinegar, bear, cheese etc. The biotechnological expertise, with the growing awareness, spread and employed for the manufacture of products like ethanol, butanol, glycerol etc. The discovery of penicillin and other antibiotics during 1930 and 1940's augmented the prospects of biotechnological approaches, and this was later culminated to the implication in the fields of amino acids, enzymes and detergents. The concept of new biotechnology came into existence in 1970's; the European Federation of Biotechnology (FEB) defined biotechnology as "The integrated use of biochemistry, microbiology and engineering sciences in order to achieve technological application of the capabilities of microorganism, cell cultures and their components". The role of modern biotechnology includes, in the area of pharmaceutical industry, the production of antibiotics, enzymes, vaccines, hormones, etc; in the field of biomass and bioenergy, it includes biogas, fuel, alcohol, production of hydrogen and hydrocarbon, etc; in the field of food industry, it includes mass production of yeasts, mushroom and single cell proteins, etc. Enhanced competition in the world markets and rising costs of inputs have underlined the cause for efficiency in plant agriculture. Excessive use of chemical fertilizers, herbicides, pesticides, and increased soil salinity not only threatens the soil chemistry and its fertility but also leads to ecological destruction. Due to the environmental constraints like environmental stress conditions, the plant productivity is prone to be more susceptible. All these factors demand greater exercise over the characteristics of the plant production system. There is a need of proper management system to increase the efficiency of plant productivity. The exploitation of new tools of genetic engineering and biotechnology will aid in the tailoring of plant production. Here, keeping in view the length of the article, we focus mainly on the role of biotechnological application of microbes in the field of agro-forestry aimed at sustainable plant productivity.

2. BIOLOGICAL NITROGEN FIXATION

In recent years the use of chemicals as fertilizers and pesticides is being practised to increase the agricultural production. But these chemicals have adverse effects on the environment and the whole biota including ground water and food-stuffs too. This led to the attention of scientists to explore other methods of improving plants' growth and yield. The biological nitrogen fixation offers an eco-friendly approach for improving plant growth. A group of bacteria is capable of fixing gaseous nitrogen into ammonia and then assimilates it into organic compounds which plants can absorb; this conversion of atmospheric nitrogen into inorganic or organic forms through living organism is known as biological nitrogen fixation. It provides a non - toxic way of fertilization of agricultural fields. Biologically fixed nitrogen requires about 30% less energy compared to the chemical process. Table I lists some of the important genera of microorganisms capable of fixing atmospheric dinitrogen. Both symbiotic and asymbiotic form of nitrogen fixation involves reduction of molecular N_2 to ammonium by the nitrogenase. This enzyme complexes with molecular nitrogen and help convert it into ammonia. The symbiotic genes encoding for nodulation are called *nod* genes, and the genes responsible for synthesis of nitrogenase, or nitrogen fixation are called *nif* genes. These *nif* genes are linked, and this attribute makes it possible to nick the desired piece of DNA of the bacterium and insert it into another organism. From the study of Rhizobial genome, it appears that nodule specific plant genes code for the synthesis of flavonoids that trigger the expression of nodulation (nod) genes in the bacteria. There are two sets of nod genes: common *nod* genes (such as A, B, C, I, J and D) and the host-specific *nod* genes (like *nod* E,F). These nodulation genes along with other symbiotic genes are located on a large plasmid termed symbiosis or sym plasmid. The nod genes are regulated by the *nod* D-gene. The *nif* genes are located on a Sym plasmid adjacent to *nod* genes. Studies on *nif* genes were made for the first time in *Klebsiella pneumoniae*, and later *nif* genes were discovered in *Rhizobium*, cyanobacteria and other N_2-fixers. Biotechnologists are serious about the prospects of (a) transferring *nif* genes into higher plants, particularly monocots so that the transformed plants are able to fix their own nitrogen directly from air, (b) transforming *nif* genes from N_2-fixers into other microorganisms that inhabit the roots of important plants but fail to fix atmospheric nitrogen. Nif genes from *K. pneumoniae* have been transformed to *Escherichia coli* which was then able to fix molecular nitrogen. Since *nif* genes are prokaryotic in origin, therefore, their cloning in non-leguminous plants is a tedious job. However, various cloning vectors like *Agrobacterium* plasmids (Ti and Ri-plasmids) and bacteriophage vectors can be used for such purposes. Application of tissue culture technique could also solve the problem of transforming *nif* genes into higher plants; infecting callus cells, produced from root explants of soybean on a particular medium, with *Rhizobium japonicum* formed infection threads. Also, presence of nitrogenase in soybean callus-*Rhizobium* system growing on solid medium has been observed. Despite much advancement, the expression of *nif* genes in newer hosts is yet to be accomplished. The interaction of associative symbionts, which establish in rhizosphere and rhizoplane with plant system, is yet to be investigated at the molecular level.

Table I. Examples of Nitrogen fixing genera

(1)	Free living bacteria:
(a)	Aerobic forms: *Azotobacter chroococcum, Derxia, Beijerinkia*
(b)	Anaerobic forms: *Bacillus polymyxa, Pseudomonas spp., Klebsiella pneumoniae, Enterobacter*
(c)	Photosynthetic forms: *Rhodospirillum, Chlorobium, Rhodocyclus*
(d)	Non-photosynthetic forms: *Methylococcus, Methanobacterium*
(e)	Chemosynthetic forms: *Thiobacillus, Desulphovibrio*
(2)	Cyanobacteria: *Synechococcus, Gloeotheca, Lyngbya, Nostoc, Scytonema, Anabaena*
(3)	Fungi: *Yeasts, Pullularia*
(4)	Symbiotic association of actinomycetes with angiosperms: *Frankia sp., Alnus* *Frankia sp., Myrica (Sweet gale)*
(5)	Symbiotic association of bacteria with legumes: *Rhizobium meliloti* - alfalfa (*Medicago sativa*), *R. leguminosarum* - Cowpea (*Vigna sp.*) etc

3. BIOFERTILIZERS

The growing population problem in India and other parts of the world has severely affected agricultural productivity. The excessive use of chemical fertilizer and industrialization have made cultivatable lands less productive, and badly damaged the population dynamics of soil-microflora. Increased health hazards and soil pollution have led scientists to realize for a non-chemical way of supplying fertilizers to the field in order to enhance crop productivity. Biofertilizers, as a result, were recommended. Biofertilizers represent the nutrient inputs of biological origin destined for plant growth (Subba Rao, 1982). This has led to the mass production of *Rhizobium, Azotobacter,* Cyanobacteria, *Azospirillum,* phosphate solubilizing bacteria and vesicular-arbuscular mycorrhiza (VAM). Biofertilizers have opened new vistas for sustainable management of available resources and have provided impetus to crop production.

(a) Bacterial biofertilizers

Two methods, namely bacterization and green manuring, are used to increase amount of nitrogen in the soil, as a result of enhanced activity of several free-living and symbiotic bacteria in the soil. Bacterization is a technique whereby seeds are treated with bacterial suspension of *Rhizobium, Azotobacter, Bacillus* etc. These bacteria are active in rhizosphere of plants and help improve the growth of host plants by secreting growth promoting substances and secondary metabolites (eg. antibiotic) (Subba Rao, 1982; Dwivedi *et al.*, 1989). 'Azotobacterin' and 'Phosphobacterin' are the two most important bacterial biofertilizers containing cells of *Azotobacter chroococcum* and *Bacillus megaterium* var. *phosphaticum,* respectively. Their application in the field enhances the crop yield by about 20%. Rhizobial biofertilizers contain the rhizobial cells. The best studied nitrogen fixing symbiotic association is between *Rhizobium spp.* and various leguminous plants. 'Rhizotecka', 'Nitrogerm' and 'Nodin' are some important biofertilizers prepared from rhizobial inocula. Rhizobial biofertilizer fixes about 50-150 Kg N/ha/year. Inoculation of seeds with *Azotobacter brasliense* not only increases the grain yield of rice and sorghum but also saves nitrogen to about 30 kg/ha (Tilak, 1991).

Green manuring is a common practice since very long time by farmers to cultivate leguminous plants in their crop fields which fix atmospheric nitrogen in soil in association with *Rhizobium* in their root system. When non-leguminous crops are grown in such crop fields, these plants derive benefits of such soil enriched with already fixed nitrogen along with decomposed parts of plant body. Tropical legumes can provide about 80 kg N/ha when grown as green manures, according to a report published by International Rice Research Institute. *Sesbania rostrata, Cyanosis pamas, Trifolium alexandrinum* are some of the important leguminous plants used as green manures. Green manures not only contribute nitrogen to the soil but they are also the source of potassium, phosphorus and organic matter (Ghai and Thomas, 1989).

Azolla, which is an aquatic fern, contains an endophytic cyanobacterium, *Anabaena azollae*, in its leaf cavity. A total of six species of *Azolla* are known so far viz., *A. caroliniane, A. filiculoides, A. maxicana, A. microphylla, A. nilotica, A. pinnata* and *A. rubra*. The method for the mass cultivation of *Azolla* is as follows: Microplots (about 20 m^2) are prepared in nurseries which is then irrigated up to 5 - 10 cm. About 4 - 20 kg P_2O_5 per hectare is also added to these plots. The temperature (15 - 30 $C) and pH (8.0) are maintained. These microplots are then inoculated with fresh *Azolla* (0.5 kg / m^2). After about three weeks growth mat formed by *Azolla* is harvested, which is then dried to be used as green manure. Application of *Azolla* has been successfully tried in rice fields; these can be applied in rice fields in two ways (i) application of *Azolla* in soil before rice plantation, and (ii) transplantation of rice followed by water draining and incorporation of *Azolla* (Singh, 1979). Application of *Azolla* in rice field before rice transplantation increases the yield equivalent to that obtained from 30 kg N/hectare as urea or ammonium phosphate (A Report from International Rice Research Institute, Philippines). Also, *Azolla* shows tolerance against heavy metals.

(b) Cyanobacterial biofertilizers

The concept of using cyanobacterial cultures as biofertilizers came into existence with their application in paddy fields, through a process called as 'algalization' (Venkataraman, 1972). Many cyanobacteria such as *Nostoc, Anabaena, Plectonema, Tolypothrix, Gloeotrichia* etc. are known to be good free nitrogen fixers. Algalization increases the rice yields by 10 - 40%; 25-30 Kg N/ha per cropping season is derived from rice fields following algalization (Subba Rao, 1982). It further helps in improving soil salinity, reduces soil compaction and C:N ratio. Various growth promoting substances like IAA, NAA, amino acids, vitamins are secreted by these cyanobacteria, making them potent biofertilizer. These biofertilizers can also be utilized for reclamation of saline soil. This practice is much in use in India (U.P., T.N., A.P., Maharashtra, Haryana), Russia, China and Egypt.

(c) Phosphate - solubilizing microorganisms

There are phosphate - solubilizing microorganisms like *Pseudomonas* and *Bacillus* (both bacteria), and *Penicillium* and *Aspergillus* (both fungi) that help convert insoluble phosphates in soil into soluble and usable forms. They do so by secreting various organic acids like formic acid, glycolic acid, acetic acid etc. The inocula of such microorganisms are available in packets. These can be mixed with farmyard manure and applied to soil. Some of the important products available based on phosphate - solubilizing microorganisms are 'Microphos', 'Superphos' and 'Phosphobacteria'.

(d) Mycorrhizal biofertilizers

Mycorrhiza (fungus root) is symbiotic,

non-pathogenic or weakly pathogenic association of fungi and the roots of plants. The basis of symbiosis is that fungus derives food from photosynthesizing plants as carbohydrates and in return fungus provides nutrients to the plants from soil by absorbing through hyphal mat, thereby enhancing the crop productivity. A detailed account of mycorrhizas with regard to their role in plant growth has been reviewed recently (Dwivedi and Dwivedi, 2001). In recent years, the application of artificially produced inoculum of mycorrhizal fungi has increased its significance in the fields of horticulture, agriculture and forestry, owing to its multiple role in plant growth and ability to resist against edaphic, climatic stresses and pathogens. Ectotrophic mycorrhizas are known as characteristic mycorrhizas of the forest trees associated with several basidiomycetous fungi viz., species of *Boletus, Amanita, Russula* etc. 'Mycorhiz' biofertilizer is prepared using mycelial inoculum of *Pisolithus tinctorius* in vermiculite-peat moss substrate. More than a million seedlings of pine were produced using mycorhiz biofertilizer in the U.S.A. (Marx and Schenck, 1983). The vesicular-arbuscular mycorrhiza (VAM) is a common type of symbiosis in at least 80% of higher plant genera and occur in all groups of plants. The fungi involved belong to the genera *Glomus, Gigaspora, Acaulospora, Sclerocystis* etc. The most commonly used method for the mass production of VAM fungi is the pot culture technique which requires the host plant (like strawberry, citrus, sorghum, onion etc.), mycorrhizal fungi and natural soil. Various fruit crops such as citrus, mango, banana, pomegranate and all vegetables (like tomato, chilli, brinjal) are benefited by the application of VAM inoculations. VAM fungi increases water uptake in plants either by providing additional water via hyphal proliferation in soil or enhanced osmotic adjustment. Also, they increase selective absorption of nutrient elements (P, Zn and Cu). Some organic acids, enzymes and phytosiderophores abound in the root exudates of mycorrhizal biofertilizers affecting nutrient dynamics of the rhizosphere region (Smith and Read, 1997). These biofertilizers enhance the tolerance level against heavy metals, high salt and drought. Ectomycorrhiza causes suppression of root pathogens like *Pythium, Phytophthora* and *Fusarium* spp. (Marx, 1972).

Mycorrhizal legumes

Both nodulation and N_2 - fixation depend on the balanced supply of mineral nutrition, specially phosphorus. Mycorrhizal legumes have the benefit of harnessing the resources of phosphorus, which otherwise is unavailable to the plants. Phosphate uptake by plants is enhanced by VAM through two different modes: (a) physical mode, based on increased number of absorption sites, (b) as evinced from kinetic analysis, there is a presence of a pathway which has greater affinity for phosphorus in mycorrhizal roots (Cress *et al.*, 1979). Polyphosphate granules are translocated from outer to inner mycelia, and contained inside vacuoles (Cox and Tinker, 1976). These polyphosphate granules are then carried to arbuscles by cytoplasmic streaming and bulk flow. Recently, mycorrhizal colonization in soybean roots is shown to be stimulated by rhizobial Nod factors (Xie *et al.*, 1995).

4. BIOPESTICIDAL PLANT PROTECTION

Due to the deleterious effects of synthetic pesticides on the plants and other life supporting systems, there is an urgent need to use alternative agents for pests control using sources that are harmless compared to the synthetic pesticides. Plant-origin and microbe-origin antipathogenic substances need encouragement for safety of crop plants from diseases and for unpolluted environment. Many non-pathogenic microbial forms can be used for biocontrol of the diseases. Following is the list of some of the important examples of microbial interaction elucidating the biopesticidal management for improved crop production.

(i) *Agrobacterium radiobacter* strain-84 controls the crown gall of stone fruit plants and this strain is widely used commercially for biocontrol.

(ii) *Pseudomonas fluorescens* interacts and cause lysis of hyphae of *Gaeumannomyces graminis* var. *tritici* causing 'take-all' of wheat.

(iii) Seed dressing with antibiotic-producing microorganisms can be helpful to control soil-borne plant diseases due to antibiosis between antagonists and pathogens. Control of *Rhizoctonia solani* by dressing seeds of wheat with *Bacillus subtilis* was reported, and crop yield was increased by 40%. The decline of 'take-all' disease of wheat by dressing seeds of wheat, before sowing in soil, with cysts of soil amoebae has been reported (Dwivedi, 1986).

(iv) The siderophores- the microbial iron (III)- transport agent produced by fluorescent pseudomonads are also responsible for biocontrol. These are growth promoting factors and some are strong antibiotics. Siderophore works as biocontrol agent of *G. graminis* var. *tritici*, and root rot of wheat caused by *Pythium* spp.

(v) Most potent biocontrol agents amongst fungi are *Peniophora gigantea, Cephalosporium roseum, Trichoderma viride, Aspergillus flavus, Chaetomium* spp., *Talaromyces flavus.* Mycoparasitic interactions have been reported between *Rhizopus oryzae* vs. *Bipolaris spicifera, Cephalosporium roseo-griseum* vs. *Aspergillus flavus* (Chaturvedi and Dwivedi, 1982), *Fusarium oxysporum* vs. *Aspergillus luchuensis* (Chaturvedi and Dwivedi, 1985), *Fusarium solani* vs. *Mucor spinosus* (Dubey and Dwivedi, 1986) etc.

(vi) About 9000 species of nematodes are known and more than one hundred fungi are reported to attack them in crop field soil. *Arthrobotrys robusta* can control *Ditylenchus myceliophagus* and *Meloidogyne.* Some soil inhabiting species of *Fusarium; Cephalosporium, Paecilomyces lilacinus* infect eggs of nemotodes.

(vii) Soil solarization technique treats several soil-borne plant pathogens. Soil solarization is a technique of mulching crop field soil with transparent plastic sheets during summer season for 4 - 6 weeks. Several soil-borne pathogens viz., *Sclerotium rolfsii* causing foot-rot of barley (Singh *et al.*, 1990), *Fusarium oxysporum* f. sp. *psidii, F. solani, Rhizoctonia solani* and *Macrophomina phaseolina* causing root diseases of guava (Dwivedi, 1989) etc. have been worked out for the effect of soil solarization on them. The temperature of soil under mulched condition during these studies increases up to 55°C in the experimental plots in summer inhibiting the propagules of pathogens.

(viii) Destructive effect of mycophagous soil amoebae on fungal propagules / hyphae is a fascinating subject in the field of soil microbiology and plant pathology. Members belonging to *Thecamoeba, Mayorella, Leptomyxida, Saccamoeba* etc. are found to be mycophagous. Based on pot-bioassay study, it is suggested that the biological control of take-all disease of wheat is possible by using soil amoebae (Dwivedi, 1986). A detailed account on biopesticidal effect on crop production can be found elsewhere (Dwivedi, 1993).

REFERENCES

Chaturvedi, A.P. and Dwivedi, R.S. 1982. Formation of resting body by *Cephalosporium roseo-griseum* Saksena inside the conidiophore of *Aspergillus flavus* Link. Ex. Fries during mycoparasitism. *Plant and Soil*, 69, 121-122.

Chaturvedi, A.P. and Dwivedi, R.S. 1985. Mycoparasitic behaviour of *Fusarium oxysporum* Schlechtendahl towards *Aspergillus luchuensis* Inui. *Plant and Soil*, 84, 419-422.

Cox, G. and Tinker, P.B. 1976. Translocation and transfer of nutrients in vesicular-arbuscular mycorrhiza. I. The arbuscule and Phosphorus transfer. A quantitative ultrastructural study. *New Phytol.*, 77, 371-378.

Cress, W.A., Throneberry, G.O. and Lindsey, D.L. 1979. Kinetics of phosphorus absorption by mycorrhizal and non - mycorrhizal tomato roots. *Plant Physiol.*, 484.

Dubey, R.C. and Dwivedi, R.S. 1986. Destructive mycoparasitic behaviour of *Fusarium solani* (Mart.) App. And Woll. against *Mucor spinosus* van Tieghem. *Microbios. Lett.*, 32, 123-127.

Dwivedi, R.S. 1986. Role of soil amoebae in "Take-all" decline of wheat. *Indian Phytopathol.*, 39, 550-560.

Dwivedi, R.S. 1993. Biopesticidal plant protection for sustainable crop productivity. Proc. 80th Ind. Sci. Congr. Pt. III, Presidential Address Botany, 1-35.

Dwivedi, S.K. 1989. *Soil fungal communities of guava orchards with emphasis on wilt disease syndrome.* Ph.D. thesis, Banaras Hindu University, Varanasi.

Dwivedi, S.K. and Dwivedi, P. 2002. Mycorrhizae in ecosystems: an ecofriendly approach for improved plant growth. In: *Proc. Intl. Symp. on microbial biotechnology for sustainable development and productivity* (ed. Rajak, R. C.).

Dwivedi, R.S., Dubey, R.C. and Dwivedi, S.K. 1989. Biology of the rhizosphere with reference to microbial interactions. In: *Plant-Microbe Interactions* (ed. Bilgrami, K.S.), ISCA Symp. Botany, Narendra Publ. House, Delhi, pp 217-238.

Ghai, S.K. and Thomas, G.V. 1989. Legume - Rhizobium interaction for biological Dinitrogen Fixation: its use as green manure. In: *Plant-Microbe Interactions* (ed. Bilgrami, S.K.), ISCA Symp. Botany, Narendra Publ. House, Delhi, pp 47-60.

Marx, D.H. 1972. Ectomycorrhizae as biological deterrents to pathogenic root infections. *Annu. Rev. Phytopathol.*, 10, 429-454.

Marx, D.H. and Schenck, N.C. 1983. *Challenging Problems in Plant Health* (eds Kommendahl, T. and Williams, P.H.), 75th Ann. Publ. Am. Phytopathol. Soc., pp. 334-347.

Singh, P. K. 1979. Use of *Azolla* in rice production in India. In: *Nitrogen and Rice*, International Rice Research Institute, Manila, Philippines.

Singh, R.K., Shukla, R.P. and Dwivedi, R.S. 1990. Effects of solar treatment on germination of sclerotia (*Sclerotium rolfsii* Sacc.) and on other soil mycoflora. *Biology and Fertility of Soils*, 10, 152-154.

Smith, S.E. and Read, D.J. 1997. *Mycorrhizal Symbiosis* (2nd Ed.), Academic Press, London.

Subba Rao, N.S. 1982. *Advances in Agricultural Microbiology* (ed. Subba Rao, N.S.), Oxford & IBH Publ. Co., New Delhi, pp. 219-242.

Tilak, K.V.B.R. 1991. *Bacterial Biofertilizers*, ICAR, New Delhi.

Venkataraman, G.S. 1972. *Algal biofertilizers and rice cultivation*, Today and Tomorrow Printers and Publishers, New Delhi.

Xie, Z-P., Staehelin, C., Vierheilig, H., Wiemken, A., Jabbouri, S., Broughton, W.J., Vogeli-Lange, R. and Boller, T. 1995. Rhizobial nodulation factors stimulate mycorrhizal colonization of nodulating and non-nodulating soybeans. *Plant Physiol.*, 108, 1519-1525.

CHAPTER – 20

MYCORRHIZAL FUNGI AS DETERRENTS OF ROOT PATHOGENS

D.A. Sumana, A.N, Balakrishna and D.J. Bagyaraj

Dept. of Agricultural Microbiology, University of Agricultural Sciences, G.K.V.K. Campus Bangalore –560 065, India

1. INTRODUCTION

Plant roots provide an ecological niche for many of the microorganisms that abound soil. German botanist Albert Bernard Frank in 1885 introduced the Greek word mycorrhiza which literally means"fungus root" . Most plant roots form mycorrhizal associations of one kind or the other with certain fungi in soil. These mycorrhizal fungi perform the function of root hairs. The symbiosis is so well balanced that, although many of the host cells are invaded by the fungal endophytes there is neither visible tissue damage nor low degree of pathogenicity towards its host. Mycorrhizal association generally enhances the growth and vigour of the host plants. In addition the mycorrhizal plants have greater tolerance to toxic heavy metals, root pathogens, drought, high soil temperatures, soil salinity, adverse soil pH and to tranplantation shock. Because of their wide–spread occurrence in nature and their numerous benefits to plants. These fungi are currently attracting much attention in agricultural, horticultural, forestry research. Though there are different mycorrhizal associations, the most common type occurring in all ecological situations, especially in tropics, is the vesicular arbuscular mycorrhiza (VAM) (Bagyaraj 1989, Barea and Jeffries 1995).

Increased plant growth because of VAM colonization is well documented (Bagyaraj and Varma 1995). The increased plant growth is attributed to enhanced uptake of diffusion limited nutrients, hormone production, biological nitrogen fixation, drought resistance and suppression of root pathogens.

Biological control can be defined as the directed, accurate management common components of ecosystems to protect plants against pathogens. Biological control of plant pathogens is currently accepted as a key practice in sustainable agriculture because it is based on the management of a natural resource, i.e., certain rhizosphere organisms. Thus biological control preserves environmental quality by reduction in chemical inputs and is characteristic of sustainable management practices. (Barea and Jeffries 1995). Several workers have reported that VAM fungi can act as biocontrol agents for alleviating the severity of disease caused by root pathogenic fungi,bacteria and nematodes.

It is evident that an increased capacity for nutrient acquisition resulting from mycorrhizal association could help the resulting stronger plants to resist stress. However, VAM symbioses may also improve plant health through a more specific increase in protection (improved resistance and/or tolerance against biotic and abiotic stresses) (Bethlenfalvay and Linderman 1992, Barea and Jeffries 1995). The study of a possible role for VAM symbiosis in protection against plant pathogens began in the 1970s, and a great deal of information has been published on the subject, however, we still know very little about the underlying mechanisms (Linderman 1994, Azcon-Aguilar and Barea 1996).

In this paper a brief account of VAM fungi is given, followed by role played by them in controlling the root pathogens highlighting the mechanisms involved and suggesting the possible future line of work, wherever necessary.

2. VESICULAR ARBUSCULAR MYCORRHIZA

Vesicular arbuscular mycorrhiza are geographically ubiquitous and occur over a broad ecological range. They are commonly found in association with agricultural crops, most shrubs, most tropical tree species and some temperate tree species. The VAM has been observed in 1000 genera of plants representing some 200 families. There are at least 300,000 receptive hosts in the world flora and 150 species of VAM fungi (Bagyaraj 1996). The first extensive survey conducted in a tropical forest in Java brought out that 69 of the 75 species examined were colonized by VAM (Janse 1896). A survey conducted by us in a forest nursery in India indicated that nearly 70% of the tree species surveyed were colonized by VAM (Byra Reddy et al. 1994). According to Gerdemann (1975) it is easier to list the plant families that do not form VAM than to list those that do. Families not forming VAM include Pinaceae, Betulaceae, Orchidaceae, Fumariaceae, Commelianaceae, Urticaceae and Ericaceae. Families that rarely form VAM include Brassicaceae, Chenopodiaceae, Polygonaceae and Cyperaceae. Most of the VAM fungi have a broad host range. The fungi associated are non–septate zygomycetous fungi belonging to the genera *Glomus, Gigaspora, Acaulospora, Entrophospora, Sclerocystis and Scutellospora.* Many are cosmopolitan in distribution, but some may be strictly tropical like *Acaulospora foveata* and *A. tuberculata.* The VAM endophytes are not host specific, although evidence is growing that certain endophytes may form preferential association with certain host plants (Bagyaraj 1992).

These fungi are obligate biotrophs. This comes in the way of inoculum production in a large scale. However methods to produce substrate based inoculum using a suitable host is now available (Bagyaraj 1992). Hence VAM fungi are recommended for nursery raised/containerised plants which are later transplanted to the field site. Most of the forest tree species fall under this category. The role of VAM fungi in improving plant growth is now well documented. Improved plant growth of economically important tree species like

Leucaena leucocephala, Tamarindus indica, Acacia nilotica, A. mangium, A. senegal, Calliandra calothyrsus, Tectona grandis, Casuarina equisetifolia, Dalbergia sissoo, Azadiracta indica, Mangifera indica etc. has been reported by many workers (De la Cruz *et al.* 1990, Colonna *et al.* 1991, Bagyaraj and Varma 1995, Rajan *et al.* 1999). There are several success stories where VAM inoculated plants withstood transplant shock and established better in highly degraded eroded lands and mine spoils (Lambert and Cole 1980, Khan 1981).

The improved plant growth is attributed to increased nutrient uptake, especially phosphate, due to the exploration by the external hyphae of the soil beyond the root-hair zone. The phosphate uptake is more significant in soils deficient in phosphorus. The increased growth of plants inoculated with VAM fungi is not only attributed to improved phosphate uptake but also to better availability of other elements like Zn, Cu, K, S, Al, Mn, Mg, Fe, etc. Allen et al. (1982) illustrated that VAM affects directly the levels of plant hormones like cytokinins and gibberellin-like substances. Plants colonized by VAM fungi can tolerate a wide range of soil water regimes and also improve water relations of many plants.

Anatomical and other physiological studies have brought out that mycorrhizal plants have increased rates of respiration, photosynthesis and increased amounts of sugar, amino acids, RNA, etc. and larger and/or more number of chloroplasts, mitochondria, xylem vessels, motor cells, etc. (Dyala Doss *et al.* 1988). Changes in the root exudation and altered rhizosphere organisms (which also affect plant growth) may result because of colonization of roots by VAM fungi (Mada and Bagyaraj 1993)

Mycorrhizal colonization also allows introduced populations of beneficial soil organisms like rhizobia, *Azotobacter, Azospirillum* and phosphate solubilizing bacteria to maintain higher numbers than in non-mycorrhizal plants and to exert synergistic effects on plant growth (Bagyaraj 1989). It is apparent from the investigations of VAM-plant pathogen interactions that VAM can usually (though not always) deter or reduce the severity of diseases caused by soil-borne pathogens (Linderman 1994, Azcon-Aguilar and Barea 1996).

3. EFFECT OF VAM ON ROOT PATHOGENS

Most of the studies on VAM-root pathogens suggest that VAM decreased or mitigated the disease severity. Several review articles have appeared on this topic. (Schonbeck 1979, Dehne 1982, Bagyaraj 1984, Schenck 1987, Hooker *et al.* 1994, Linderman 1994, Azcon-Aguilar and Barea 1996). The main conclusions that can be drawn are: 1. VAM associations can reduce damage caused by soil – borne plant pathogens, 2. the abilities of VAM symbiosis to enhance resistance or tolerance in roots are not equal for the different VAM fungi so far tested, 3. protection is not effective for all pathogens, and 4. protection is modulated by soil and other environmental conditions. Thus it can be expected that interactions between different VAM fungi will vary with the host plant and the culture system. Consistent reduction of disease symptoms has been described for fungal pathogens such as *Phytophthora parasitica, P. cactorum, P. vignae, Gaeumannomyces graminis* var. *tritici, Fusarium oxysporum, Chalara (Thielaviopsis) basicola, Rhizoctonia solani, R. bataticola, Sclerotium rolfsii, Pythium ultimum, P. splendens, Dothiorella gregania, Botrytis fabae, Ganoderma pseudoferreum and Aphanomyces* spp., bacteria such as *Pseudomonas syringae and P. solanacearum* and nematodes such as *Meloidogyne arenaria, M. incognita, M. hapla, M. javanica, Tylenchulus semipenetrans, T.*

vulgaris, Pratylenchus brachyurus, P. zeae, Helicotylenchus dihystera and *Radophilus similis* (Bagyaraj 1996, Azcon Aguilar and Barea 1996, Rabie 1998, Srivastava *et al.*, 2001). Further integrating VAM fungi and castor cake in the management of root-knot nematode on egg plant (aubergine) have shown that seedlings colonized with mycorrhiza were least infected by *M. incognita* when transplanted in soil which was amended with castor cake. Significant increases in colonization of *G.fasciculatum* on roots of aubergine and chlamydospore densities of mycorrhiza indicated favourable effects of castor cake amendment on the growth of *G. faciculatum* (Rao *et al.*, 1998). Similarly integration of the VAM fungus *G. mosseae* and oil cakes of castor, pongamia and margosa showed that integration of VAM fungi with margosa cake reduced *M. incognita* populations in soil and reduced root galling, egg mass production and fecundity and produced vigorous tomato seedlings with increased root colonization by *G. mosseae* (Reddy *et al.*, 1998). Another study with staggered inoculation of VAM fungi and root pathogen brought out that application of VAM fungi two weeks earlier than pathogen was better than simultaneous application of VAM fungi and the pathogen in reducing the severity of disease. Application of the pathogen two weeks earlier than VAM fungi was not effective in controlling the disease (Suresh and Bagyaraj 1984). Vast (1997) studied the interaction of VAM fungi of *Arabica* coffee and the chemical control nematodes. It was found that when VAM fungi was inoculated well in advance resulted in an enhanced tolerance of *Arabica* coffee cultivars susceptible to endoparasitic nematodes of *Pratylenchus* and *Meloidogyne* that are wide spread in *Arabica* plantations.

Most of the VAM-root pathogen interaction studies have been conducted in crop plants. But the information available on forest tree species is scanty (Chakravarty and Mishra 1986, Mohan and Verma 1996). There are large number of root pathogens causing diseases of forest tree species, especially in the nursery (Mohamed Ali and Sharma 1996). Availability of healthy stock of seedling for planting in the field is an important aspect of forest management. Hence there is a need to carry out more studies on the role of VAM fungi in the control of root pathogens of forest tree species.

4. MECHANISMS OF SUPPRESSION OF ROOT PATHOGENS BY VAM FUNGI

Studies conducted so far suggest that the mechanisms of suppression may be due to morphological, physiological and biological alterations in the host.

(a) Morphological alterations

It has been demonstrated that VAM colonization induces remarkable changes in root system morphology, as well as in the meristematic and nuclear activities of root cells (Atkinson *et al.* 1994). This might affect rhizosphere interactions and particularly pathogen–infection development. The most frequent consequence of VAM colonization is an increase in branching, resulting in a relatively larger proportion of higher order roots in the root system (Hooker *et al.* 1994). However, the significance of this finding for plant protection has not yet been sufficiently considered. More attention needs to be given to root system morphology in the future because it could modify the infection dynamics of the pathogen as well as the pattern of resistance of VAM roots to pathogen attack.

Thickening of the cell walls through lignification and production of other polysaccharides in mycorrhizal plants preventing penetration and growth of pathogens like *Fusarium oxysporum* and

Phoma terrestris have been demonstrated. A stronger vascular system observed in mycorrhizal plants will increase the flow of nutrients, impart greater mechanical strength and diminish the effect of vascular pathogens. Histopathological studies of nematode galls caused by *Meloidogyne* spp. showed that galls in mycorrhizal plants had fewer giant cells or syncytia, which are needed for the development of nematode larvae, compared to nonmycorrhizal plants. The nematodes in mycorrhizal plants were smaller and took a longer time to develop into adults (Bagyaraj 1996).

Pioneering observations by Dehne (1982) illustrated how fungal root pathogens and VAM fungi, although colonizing the same host tissues,usually develop in different root cortical cells,indicating some sort of competition for space. Cordier et al (1996) working with VAM–*Phytophthora* interaction observed that the pathogen does not penetrate cortical cells containing arbuscule; suggesting that localized competition for infection/colonization site does occur between the pathogen and the VAM fungus. Contrary to this in a recent study epicotyl of mycorrhizal pea plants showed a reduction in disease severity caused by *Aphanomyces eutriches* although this part of the plants was not mycorrhizal. This suggested an induced systemic factor may be responsible for increased resistance in mycorrhizal plants (Bodker et al. 1998).

(b) Physiological and Biochemical Alterations

(i) Improved nutrient status of the host plant

Increased nutrient uptake made possible by VAM symbiosis results in more vigorous plants. The plant itself may thus be more resistant to or tolerant of pathogen attack.Though this is true, later studies using non – mycorrhizal plants with tissue ‘P’ concentration similar to that of mycorrhizal plants, indicates that non–P mediated mechanisms are also involved (Bodkar *et al.* 1998).

(ii) Competition for host photosynthates

It has been proposed that the growth of both the VAM fungi and root pathogens depends on host photosynthates and that they compete for the carbon compounds reaching the root (Linderman 1994). When VAM fungi have primary access to photosynthates, the higher carbon demand may inhibit pathogen growth. However, there is little or no evidence that competition for carbon compounds is a generalized mechanism for pathogen biocontrol activity of VAM symbiosis.

Colonization of a plant by VAM fungi alters the host physiology and in turn the root exudation pattern (Smith *et al.* 1994, Mada and Bagyaraj,1993). Decreased root exudation in mycorrhizal plants, because of increased membrane phospholipid content, possibly helps in reducing the infection by root pathogens.

(iii) Higher levels of phenols and amino acids

Higher concentration of ortho–dihydroxy phenols present in mycorrhizal plants compared to non–mycorrhizal plants was found to be inhibitory to the root rot pathogen *Sclerotium rolfsii.* Higher levels of the amino acid arginine in the root extracts of mycorrhizal plants were found to reduce *Thielaviopsis basicola* chlamydospore production (Sharma *et al.* 1992). Increased phenylalanine and serine in tomato roots due to inoculation with *Glomus fasciculatum* was found to be inhibitory to root knot nematodes (Bagyaraj 1996).

(iv) Activation of plant defence mechanisms

During their life cycle plants evolve a number of defence responses elicited by various signals, including those associated with pathogen attack (Huynh *et al.* 1992).

The activation of specific plant defence mechanisms as a response to VAM colonization is an obvious basis for the protective capacity of VAM fungi. Among the compounds involved in plant defence (Bowles 1990) studied in relationship to VAM formation are phytoalexins, enzymes of the phenylpropanoid pathway, chitinases, -1,3- glucanases, peroxidases, pathogenesis related (PR) proteins, callose, hydroxyproline-rich glycoproteins (HRGP) and phenolics (Gianinazzi Pearson et al.1994).

Electrophoretic analysis of soluble extracts from VAM roots has demonstrated that the host plant produces a number of new proteins (endomycorrhizins) in response to VAM colonization (Gianinazzi Pearson and Gianinazzi 1995). New polypeptides are synthesized during VAM infection and others disappear (Dumas Gaudot *et al.* 1994). However, this altered pattern of protein synthesis in the plant is not necessarily related to defence reactions. This is a research area deserving further attention.

Phytoalexins, low–molecular–weight, toxic compounds usually accumulating with pathogen attack and released at the sites of infection, are not detected during the first stages of VAM formation but can be found at later stages of the symbiosis (Morandi et al. 1984). This was true for phytoalexin glyceollin of soyabean and medicarpin of *Medicago trancatula* (Harrison and Dixon, 1993).

There seems to be a similar low activation of the phenylpropanoid–related enzymes. In particular, both phenylalanine ammonium–lyase (PAL), the first enzyme of the phenyl propanoid pathway, and chalcon isomerase, the second enzyme specific for flavanoid/isoflavanoid biosynthesis, increased in amount and activity during early colonization of plant roots by *Glomus intraradices*, but then decreased sharply to levels at or below those in uninoculated controls (Lambais and Mehdy 1993; Volpin et al. 1995). These results suggest that VAM fungi initiate a host defence response, which is subsequently suppressed.

Increased chitinase and beta-1-3- endo glucanase activity because of VAM colonization has also been reported (Lambais and Mehdy 1993). Higher peroxidase activity was found to be associated with the cell wall of epidermal and hypodermal cells of mycorrhizal roots, which can contribute to higher resistance of cells to invasion by pathogens (Gianinazzi and Gianinazzi Pearson 1992). Pathogenesis related proteins and HRGP are synthesized only locally and in very low amount in response to VAM colonization. In summary, it can be said that plant defence mechanisms are activated to a very little extent by VAM colonization. However these compounds could sensitize the roots to pathogens and enhance mechanisms of defence to subsequent pathogen infection (Azcon Aguilar and Barea 1996). Current research using molecular biology techniques and immunological and histochemical analyses will probably provide more information about these mechanisms.

(c) Biological alterations

Mycorrhizal plants harbour higher population of microorganisms in the rhizosphere thus making it difficult for the pathogen to compete and gain access to the root (Secilia and Bagyaraj 1987). Caron (1989) reported a reduction in *Fusarium* population in the soil surrounding mycorrhizal tomato roots as compared with the soil of non-mycorrhizal controls. Meyer and Linderman (1986) found that the number of sporangia and zoospores formed by cultures of *Phytophthora cinnomomi* was reduced by the application of extracts of rhizosphere soil from VAM plants. Secilia and Bagyaraj (1987) isolated more pathogen-antagonistic actinomycetes from the rhizosphere of VAM plants than from non- mycorrhizal controls, an effect that also depended on the VAM fungus involved.

Recently, microorganisms producing siderophores, which are low molecular weight chelating agents that have high affinity for ferric iron and thus fungistatic to many pathogens,were observed in higher numbers in the rhizosphere of mycorrhizal plants (Bagyaraj 1996). These studies merit further attention in order to elucidate how microbiota–mediated changes may be involved in biological control by VAM associations.

The prophylactic ability of some VAM fungi could be exploited in association with other rhizosphere microorganisms known to be antagonistic to root pathogens that are being used as biological control agents (Linderman 1994, Azcon Aguilar and Barea 1996). Preliminary studies suggest that microbial antagonists of fungal pathogens, either fungi or Plant growth promoting rhizomicroorganisms (PGPR) do not antagonize VAM fungi. Moreover they can improve the development of the mycosymbiont and facilitate VAM formation. (Linderman 1994). This has been shown particularly for *Trichoderma spp.* and *Pseudomonas spp.* Certain bacteria have recently been shown to stimulate colonization by VAM fungi. These have been called mycorrhiza helper bacteria (MHB). Inoculation of neem seedlings with a VAM fungus plus MHB was found to significantly improve growth and nutrition compared to those inoculated with VAM fungus alone (Sumana and Bagyaraj, Unpublished). A similar observation was also made by us in tissue cultured *Ficus benjamina* plantlets (Jayanthi and Bagyaraj, Unpublished). Therefore, the management of these interactions improving plant growth and health, in an integrated approach, should be one of the targets for future research.

5. CONCLUSION

There are large numbers of root pathogens causing diseases of forest tree species, especially in the nursery. Availability of healthy stock of seedlings for planting in the field is an important aspect of forest management. Soil drenching with pesticides has to be done to prevent root diseases. Use of chemicals is not only hazardous to human health but also to soil health. Biocontrol which preserves environmental quality is an alternative approach to control root diseases. Like most instances of biocontrol, mycorrhizae cannot offer complete immunity against any root disease. They would only impart a degree of resistance against soil–borne pathogenic fungi, bacteria and nematodes. The diversity of interactions between VAM and plant pathogenic microorganisms shows that each pathogen–VAM fungus–plant combination is unique which is further influenced by the environment and hence generalizations regarding such interactions are difficult to make. The mechanism of suppression of root pathogens by VAM fungi is only partly understood. More studies are needed in this area. Current research using molecular biology techniques and immunological and histochemical analyses will probably provide more information about these mechanisms. However, the possibility of biologically controlling the root pathogens looks promising. Most of these studies have been conducted with crop plants and hence there is a need for exploiting this potential with forest tree species. Many root pathogens can currently be controlled only with expensive physical or chemical soil treatments. Mycorrhiza offers an alternative approach and we should pursue their potential as biological control agents despite the obstacles. The main obstacle is the inability to produce large quantities of the inoculum, as these fungi are obligate symbionts. But techniques are now available to produce VAM inoculum to the extent needed for applying to nursery raised and containerized seedlings. In tree crops addition of small quantities of suitable mycorrhizal inoculum to the nursery bed would result in healthier seedlings which would not only be able to withstand transplant shock better but also challenge infection by root pathogens and grow better when planted in the field site. Other beneficial microorganisms like nitrogen fixers, MHB, PGPR etc can be tested for their compatibility with VAM. Mycorrhizal technology can thus play an important role in production of low cost quality seedlings for afforestation and reforestation programmes in forestry.

REFERENCES

Allen, M.F., Moore, T.S., Jr. and Christensen, M., 1982. Phytohormone change in *Bouteloua gracilis* infected by vesicular-arbuscular mycorrhiza II: Altered levels of gibberllin- like substances and abscisic acid in the host plant, Canadian Journal of Botany, 60: 468-471.

Atkinson, D., Berta, G. and Hooker, J.E., 1994. Impact of mycorrhizal colonization on root longevity and the formation of growth regulators. In: Impact of arbuscular mycorrhizas on sustainable agriculture and natural ecosystems, Birkhauser, Basel, pp. 89-99.

Azcon Aguilar, C. and Barea, J.M., 1996. Arbuscular mycorrhizas and biological control of soil borne plant pathogens—an overview of the mechanisms involved, Mycorrhiza, 6: 220-223.

Bagyaraj, D.J., 1984, Biological interactions with VA mycorrhizal fungi. In: VA mycorrhiza, CRC press, Boca Raton, pp. 131-153.

Bagyaraj, D.J., 1989, Competitions among VAM fungi and their interactions with soil organisms. In: Recent Advances in Microbial Ecology, Japan Scientific Society Press, Japan, pp. 237-241.

Bagyaraj, D.J., 1989, Mycorrhizae. In: Tropical Rain Forest Ecosystems, Elsevier Science Publishers, Amsterdam, pp. 537-546.

Bagyaraj, D.J., 1992. Vesicular arbuscular mycorrhiza – Application in agriculture. Methods in Microbiology, 24: 359-374.

Bagyaraj, D.J., and Varma, V., 1995, Interaction between arbuscular mycorrizal fungi and plants : Their importance in sustainable agriculture and in arid and semiarid tropics. In: Advances in Microbial Ecology, Academic Press, London, pp. 119- 142.

Bagyaraj, D.J., 1996. Mycorrhizal symbiosis in tropical trees in relation to control of root pathogens. In: Impact of diseases and insect pests in tropical forests, Kerala Forest Research Institute, Peechi, India, pp. 239-245.

Barea, J.M. and Jeffries, P., 1995, Arbuscular mycorrhizas in sustainable soil plant systems. In: Mycorrhiza—structure, function, molecular biology and biotechnology, Springer, Heidelberg, pp. 521-529.

Bethlenfalvay, G.J. and Linderman, R.G., 1992, Mycorrhizae in sustainable agriculture, ASA special publication No. 54, Madison, Wisconsin. p. 124.

Bodker, L., Kjoller, R. and Rosendahl, S., 1998. Effect of phosphate and the arbuscular myccorrhizal fungus *Glomus intraradices* on disease severity of root rot of peas (*Pisum sativum*) caused by *Aphanomyces euteiches*, Mycorrhiza, 8: 169-174.

Bowles, D.J., 1990. Defence related proteins in higher plants, Annual Review of Biochemistry, 59: 873-907.

Byra Reddy, M.S., Krishna Naik, L., Bhasker V. and Bagyaraj, D.J., 1994. Status of mycorrhizal association in some tropical tree species, Journal of Soil Biology and Ecology, 14: 51- 57.

Caron, M., 1989. Potential use of mycorrhizae in control of soil- borne diseases, Canadian Journal of plant pathology, 11: 177-179.

Chakravarty, P. and Mishra, R.R., 1986. The influence of VA mycorrhizae on the wilting of *Albizzia procera and Dalbergia sissoo.* European Journal of Forest Pathology, 16: 91- 97.

Colonna, J.P., Theon, D., Ducousso, M. and Badji, S., 1991. Comparative effects of *Glomus mosseae* and P fertilizer on foliar mineral composition of *Acacia senegal* seedlings inoculated with *Rhizobium,* Mycorrhiza, 1: 35-38.

Cordier, C., Gianiazzi, S. and Gianinazzi Pearson, V., 1996. Colonization patterns of root tissues by *Phytophthora nicotianae* var. *parasitica* related to reduced disease in mycorrhizal tomato, Plant and Soil, 185: 223-232.

Dyala Doss, D., Bagyaraj, D.J. and Syamsunder, J., 1988. Morphological and histological changes in the roots of *Eleusine coracana colonized* by VA mycorrhiza, Proceedings of Indian National Science Academy, 54: 291-293.

Duhne, H.W. 1982. Interaction between vesicular arbuscular mycorrhizal fungi and plant pathogens. Phytopathology, 72: 1115- 1119.

De la Cruz, R.E., Lorilla, E.B. and Aggangan,N.S.,1990, Growth of *Acacia auriculiformis* and *A.mangium* in a marginal grassland in response to inoculations with VA mycorrhiza and/or *Rhizobium.* In: Fast growing trees and Nitrogen Fixing Trees, Gusta Fischer, Stuttgart, p. 321.

Dumas Gaudot, E., Guillaume, P., Tahiri-Alaoui, A., Gianinazzi Pearson, V. and Gianinazzi, S., 1994. Changes in polypeptide patterns in tobacco roots colonized by two *Glomus* species, Mycorrhiza, 4: 251-221.

Frank, A.B., 1885. Uber die auf Wurzelsymbiose beruhende Ernahrung gewisser Baume durch unterirdische Pilze. Berichte der Deutschen Botanischen Gesellschaft, 3: 128-145.

Gerdemann, J.W., 1975. Vesicular-arbuscular mycorrhizae. In: The Development and function of roots, Academic Press, New York, pp. 576-591.

Gianinazzi, S. and Gianinazzi Pearson, V., 1992. Cytology, histochemistry and immuno-cytochemistry as tools for studying structure and function in endomycorrhiza, Methods in Microbiology, 24: 109-139.

Gianinazzi Pearson, V., Gollotte, A., Dumas Gaudot, E., Franken, P., and Gianiazzi, S., 1994. Gene expression and molecular modifications associated with plant responses to infection by arbuscular mycorrhizal fungi. In: Advances in molecular genetics of plant-microbe interactions, Kluwer, Dordsecht, pp.179-186.

Gianinazzi Pearson, V. and Gianinazzi, S., 1995. Proteins and protein activities in endomycorrhizal symbiosis. In: Mycorrhiza- structure, function, molecular biology and biotechnology, Springer, Heidelberg, pp.251-266.

Harrison, M.J. and Dixon, R.A., 1993. Isoflavonoid accumulation and expression of defense gene transcripts during the establishment of vesicular-arbuscular mycorrhizal associations in roots of *Medicago truncatula.* Molecular plant Microbe Interactions, 6: 643-654.

Hooker, J.E., Jaizme-Vega, M. and Atkinson,D., 1994, Biocontrol of plant pathogens using arbuscular mycorrhizal fungi. In: Impact of arbuscular mycorrhizas on sustainable agricuture and natural ecosystems, Birkhauser, Basel, pp. 191-200.

Huynh, Q.K., Hironara, C.M., Levine, E.B., Smith, C.E., Borgmeyer, J.R., and Shah, D.M., 1992. Antifungal proteins from plants. Purification, molecular cloning and antifungal properties of chitinases in maize seeds. Journal of Biological Chemistry, 26: 6635-6640.

Janse, J.M., 1896. Les endophytes radicaux de quelques plantes javanaises. Annales du Jardin Botanique de Buitenzorg, 14: 53- 212.

Khan, A.G., 1981. Growth of endomycorrhizal onions in unsterilized coal waste. New phytologist, 87: 363-370.

Lambais, M.R. and Mehdy, M.C., 1993. Suppression of endochitinase, -1,3- glucanases and chalcone isomerase expression in bean vesicular–arbuscular mycorrhizal roots under different soil phosphate conditions. Molecular Plant Microbe Interactions, 6: 75-83.

Lambert, D.H. and Cole, Jr. H., 1980. Effects of mycorrhizae on establishment and performance of forage species in mine spoil. Agronomy Journal, 72: 257-260.

Linderman, R.G., 1994. Role of VAM fungi in biocontrol. In: Mycorrhizae and Plant Health, ASP, St. Paul, pp. 1-26.

Mada, R.J. and Bagyaraj, D.J., 1993. Root exudation from *Leucaena leucocephala* in relation to mycorrhizal colonization. World Journals of Microbiology and Biotechnology, 9: 342-344.

Mohammed Ali, M.I. and Sharma, J.K., 1996, Impact of seed microflora on seed germination and seedling vigour of some important indigenous tree species of Kerala. In: Impact of diseases and insect pest in tropical forests, Kerala Forest Research Institute, Peechi, India, pp. 33-51.

Mohan, V. and Verma, N., 1996, Interactions of VA-mycorrhizal fungi with rhizosphere and rhizoplane mycoflora of forest tree species in arid and semiarid regions. In: Impact of diseases and insect pests in tropical forests, Kerala Forest Research Institute, Peechi, India, pp. 222-231.

Morandi, D.B.J.A. and Gianinazzi-Pearson, V., 1984. Isoflavonoid accumulation in soybean roots infected with vesicular arbuscular mycorrhizal fungi. *Physiology and plant pathology*, 24: 357-364.

Meyer, J.R. and Linderman, R.G., 1986. Selective influence on populations of rhizosphere or rhizoplane bacteria and actinomycetes by mycorrhizas formed by *Glomus fasciculatum.* Soil Biology and Biochemistry, 18: 191.

Rabie, G.H., 1998. Induction of fungal disease resistance in *Vicia faba* by dual inoculation with *Rhizobium leguminosarum* and vesicular-arbuscular mycorrhizal fungi. Mycopathologia, 141: 159-166.

Rajan,S.K., Reddy,B.J.D. and Bagyaraj, D.J. 2000. Screening of arbuscular mycorrhizal fungi for their symbiotic efficiency with *Tectona grandis.* For.Ecol.Manage. 126: 91-95.

Rao, M.S., Reddy, P.P. Sukhada, M., Nagesh, M. and Pankaj, 1998. Management of root-knot nematode on egg plant by integrating endomycorrhiza (*Glomus fasiculatum*) and castor (*Ricinus communis*) cake. Nematologia-Mediterranea. 26: 217-219.

Reddy, P.P., Nagesh, M., Divappa, V. and Kumar, M.V.V., 1998. Management of *Meloidegyne incognita* on tomato by integrating endomycorrhiza, *Glomus mosseae* with oil cakes under nursery and field conditions. Zeitschrift-fur-Pflanzenkrankheiten-und-Pflanzenschutz, 105: 53-57.

Schonbeck F., 1979. Endomycorrhiza in relation to plant diseases. In: Soil-borne plant pathogens, Academic Press, New York., pp. 271-280.

Schenck,, N.C., 1987. Vesicular-arbuscular mycorrhizal fungi and the control of fungal root disease. In: Innovative approaches to plant disease control. Wiley, New York, pp. 179-191.

Secilia. J. and Bagyaraj, D.J., 1987. Bacteria and actinomycetes associated with pot cultures of vesicular arbuscular mycorrhizas, Canadian Journal of Microbiology. 33: 1069-1073.

Sharma, A.K., Johri, B.N. and Gianinazzi. S., 1992. Vesicular- arbuscular mycorrhizae in relation to plant disease. World Journal of Microbiology and Biotechnology, 8: 550-563.

Smith, S,E, Gianinazzi Pearson, V., Koide, R. and Cainey, J.W.G., 1994, Nutrient transport in mycorrhizas: structure, physiology and consequences for efficiency of the symbiosis. In: Management of mycorrhiza in agriculture, horticulture and forestry, Kluwer, Dordecht, pp. 103-113.

Srivastava, A.K., Ahmed, R., Kumar, S. and Sukhada Mohandas, 2001. Role of VA Mycorrhiza in the management of guava in the alfisols of Chotanagpur. Indian Phytopath. 54 : 32-34.

Suresh, C.K. and Bagyaraj, D.J., 1984. Interaction between a vesicular arbuscular mycorrhiza and root knot nematodes and its effect on growth and chemical composition of tomatoes. Nematologia Mediterranea, 12: 31-39.

Vast, P., 1997. Interaction of endomycorrhiza of Arabica coffee and the chemical control of nematodes (*Pratylenchus coffeae* and *Meloidogyne konaensis*). Dix-septieme colloque Scientifique International sur le Café, Nairibi, Kenya, 20-25 juillet. pp. 564-571.

Volpin, H., Phillips, D.A., Okon, Y. and Kapulnik, Y., 1995. Suppression of an isoflavonoid phytoalexin defence response in mycorrhizal alfalfa roots. Plant Physiology, 108: 1449- 1454.

CHAPTER – 21

MICROBIAL DEGRADATION OF XENOBIOTICS IN MARINE ENVIRONMENT

Upal Roy and Saroj Bhosle

Department of Microbiology, Goa University, Taleigao, Plateau, Goa- 403206

1. INTRODUCTION

Ever since the foundations of microbiology were laid by the fundamental work of Pasteur (Kieslich, 1976), an indepth study has been undertaken in elucidating and understanding the significance of microbial transformations related to simple and complex structures involving carbohydrates, simple aliphatics, aromatic rings, steroids etc. The biotransformation may result in complete mineralisation in case the organism utilizes the compound as a sole source of carbon or the compound may be cometabolically converted to more oxidized states, in presence of a simple carbon source. Such transformations are brought about by constitutive or substrate induced enzymes leading to simple one step reaction or reactions involving a few sequential steps, resulting in a fundamental alteration of the compound to a less toxic product or an industrially important compound.

2. XENOBIOTICS IN MARINE ENVIRONMENT

Organic compounds designated as xenobiotics are chemicals (not naturally occurring compounds) which endanger the flora and fauna of marine ecosystem. These include fungicide, herbicides, pesticides, insecticides, chloro-organic chemicals, biphenyl compounds etc. Petroleum or crude oil, which contains many volatile and non-volatile chemical compounds, causes major problems when spilled over the oceans. Many of the problems associated with marine oil pollution are due to great variability of petroleum (Miller, 1973), which is found in four different states: as films, in solution, as emulsion or in the form of tar balls.

Other xenobiotic compounds which are most prevalent in marine environment are organotin, chlorophenol, chlorobiphenyl, etc. which mainly enter due to human activities. As a result of man's agricultural and industrial activities, various xenobiotics are found in estuaries near urban areas. The fate of such compounds in this marine ecosystem is of concern due to their impact on marine life and threat to human health (Kaiser, Dixon & Hodson, 1984). The biodegradation of xenobiotics and their basic units in marine ecosystem are discussed here with, emphasis being mainly on the biochemical aspect of transformation by microorganisms.

3. MICROBIAL DEGRADATION OF AEROMATIC RING

In the natural environments the low biodegradability of xenobiotic may be due to the inablity of microorganisms to effectively metabolize compounds with uncommon chemical structure or properties which are alien to existing enzyme system. Nevertheless microbial communities exposed to xenobiotic compound can adapt to these chemicals and metabolize them completely.

Immediately on exposure of the microbial community to the compound no biodegradation occurs. However, after a contact period of varied time intervals mineralisation starts. In laboratory experiment such adaptation to organic compounds has been demonstrated for bacteria (Aelion *et al.*, 1987; Barkay & Pritchard, 1988; Madsen *et al.*, 1991).

In members of the community of an ecosystem such adaptive responses either enriches specific organisms, resulting in an increase of the degradative capacity of the community (Beachem, 1987; Spain & Veld, 1983) or increase the ability to solubilize a substrate as seen in aquatic bacteria. Adaptation mechanism may also involve the selection of mutant with enzymatic specificity of novel reactions which were not present, at the onset of community, exposed to the introduced compound (Barkay & Pritchard, 1988).

Although the search for biodegradation of unusual xenobiotic has been the concern of microbiologist and biotechnologist, the biodegradation of the basic aromatic ring structures has been a subject widely studied and hence the xenobiotic compounds whose structures are based on aromatic nucleus are known to follows biodegradative pathways initiated with specific reaction followed by enzymes of aromatic ring cleavage pathways.

Most of the reaction used by aerobic bacteria for degrading aromatic hydrocarbons belong to the same general categories, only the reaction that prepares the benzene ring for fission tends to be specialized. The required ring fission substrate, catechol, is formed from the hydrocarbon (benzene) by simultaneous introduction of both the atoms of dioxygen to give a dihydrodiol as the first step, with rare exceptions (Kiyohara & Nagao, 1977). The benzene nucleus is one of the most abundant units of chemicals structure in the biosphere. The ring is strongly stabilized by resonance and its degradation, crucial for the operation of the earth's carbon cycle, is almost exclusively brought about by microorganisms (Dagley, 1987b).

Microbial substrate is degraded until one of the dehydric phenols is formed (Stanley Dagley, 1986). After this one of the nine pathways is selected to direct the flow of carbon into the Kreb's cycle. Ring fission of catechol is catalysed by dioxygenase and is termed ortho fission when it occurs as shown in reaction 1,2 and 3, where as reaction 4,5,6 and 7 are example of meta fission. The former mode of ring opening is also referred to as intradiol cleavage and latter as extradiol cleavage with reactions 8 and 9 termed as distal extradiol cleavages. The destination of each route is the Kreb's cycle.

Many of the aromatic hydrocarbons are degraded via catechol as intermediates. In fact both benzene and toluene are oxidized to catechol and benzoic acid respectively by some bacteria. The benzene nucleus of catechol is now cleaved by either reaction 1 and 4, namely ortho and meta - fission respectively, depending on the organism. In tolune the methyl substituent is oxidised to a carboxyl group which is then released as carbon dioxide during catechol formation followed by ring cleavage, but some bacteria oxidize the benzene ring first,resulting in the formation of cis of toluene which now undergoes dehydrogenation to 3-methylcatechol (Gibson *et al.*, 1970a). Most bacteria fail to mineralize methylcatechols by the ortho fission pathway, for example is 4-methylcatechol, a oxidised product of *p*-xylene (1,4, -dimethylbenzene). The only prokaryote that has been clearly shown to use ortho fission for degrading 4-methyl catechol completely is *Gordona rubra.* However degradation of methyl substituted benzene hydrocarbon by ortho fission depend on the type of lactone formed during the pathway (Dagley, 1985).

Extradiol cleavage (meta cleavage) of aromatic ring in catechol is brought about by the enzyme 2,3 dioxygenase, which occurs even in presence of methyl substituent in the substrates. Catechol 2,3-dioxygenase isolated from phenol grown *P. putida* also readily cleaved the benzene nucleus of methyl catechol (Stanley Dagley, 1986; Gibson *et al.*, 1970a) involving the hydrolysis of the ring fission product, while in pathway(b) the methyl catechol is cleaved and product formed is further oxidised. Report have shown that the catechol 2,3-dioxygenases of some organism can function with two methyl groups in the nucleus, one at C-3 and the other C-4.

The most significant reaction is the hydrolytic step-in sequence (a) where in acetic acid, containing the methyl substituent, is released and the enol form of vinyl pyruvate (2-ketopent-4-enoate, III) remains. When bacteria catalyzes a similar hydrolytic reaction for a substrate containing some other substituent, growth could be supported at the expense of vinyl pyruvate, a normal meta fission metabolite of catechol e.g biphenyl or p-chlorobiphenylmethane is oxidized to a catechol with a substituent in the 3-position. The substituents, in these case, phenyl and p-chlorobenzyl groups, I, respectively, will be removed as benzoic acid and p-chlorophenylacetic acid. Aromatic hydrocarbons especially their halogenated derivatives are metabolized with difficulty and hence products may accumulate which are not directly predicted from a specfic pathway.

4. DEGRADATION OF XENOBIOTICS IN MARINE ENVIRONMENT

Sea is considered as a medium for microbial growth with concentration of nutrients varying from place to place and from time to time at any given locality. Oceans contain every kind of life forms beginning from virus, a bacterium to mammals, but microbes are the major degraders of organic matters. Degradation pathway of few of these compounds are elaborated here with .

(a) Simple aromatics and aliphatic hydrocarbons

The lower aromatics that is Benzene, Xylene, Toluene present in petroleum have an immediate toxic effect on flora of marine ecosystem. Besides higher aromatic compounds are also mutagenic because they are normally nucleic acid intercalating agents, and are known to be potential carcinogen in higher organism as well as human being at even low concentrations.

Petroleum is a fossil fuel and organic in nature. Branched alkanes isoprenoid derivatives, cyclic aliphatic are commonly present, however, alkenes may be present in significant quantities in some petroleum products. The asphatics fraction comprise of a complex fraction, which include compounds containing nitrogen, sulfur and oxygen and termed as NSO and heavy metals such as Pb, Cu, Hg, Cd, Vn, Fe, Ni has been found that low molecular weight aromatic hydrocarbon (benzene, toluene, naphthalene) are degraded to CO_2 by microorganism but higher molecular weight aromatic components are relatively resistant to microbial attack (Floodgate, 1984; Pellizari *et al.*, 1996).

(b) Polyaromatic Hydrocarbons (PAHs)

Polyaromatic Hydrocarbons (PAHs) which contain carbon, hydrogen, with the carbon atoms arranged in a series of adjoining six membered benzene rings are ubiquitous environmental pollutants. Besides their presence in petroleum, PAHs are formed during the incomplete combustion of almost any organic material and have been isolated from air, water and soil sample (Cerniglia, 1984). The recalcitrant nature of PAHs is dependent on the number of ring and the complexity of the structure. Linear structures are the least stable followed by the clustered structures while the angular ring structures are the most stable (Kodama *et al.*, 1970).

Degradation of PAH by microorganism has been found to occur via varied pathways such as then followed for Alkane (Singer & Finnerty, 1984), Xylene/Toluene, (Williams, 1978), Napthalene/salicylate, (Singer & Finery, 1984), Phenanthrene (Cerniglia, 1984), Quinolin, Dibenzothiophene (Kodama *et al.*, 1970; Kilbane, 1989).

(c) Chlorophenol and Polychlorinated Biphenyls (PCBs)

The pollution of coastal sea water by chlorophenols or PCBs affect the biota of this environment because the compounds have high toxicity, are recalcitrant and are accumulated in sediments or at different levels in the tropic chain (Kilbane, 1989). Studies of the fate of chlorophenol in microorganisms or mesosomes, have indicated that primary degradation of chlorophenol in aquatic environment occurs via microbial activity and photo-lysis (Lee *et al.*, 1982; Lee & Ryan, 1979). Repeated additions of chlorophenols to marine or fresh water resulted in an increase in their degradation rates, suggesting an increase in the population of chlorophenol degrading microbes (Lee & Ryan, 1979; Piganetllo *et al.*, 1983). Microbial metabolism of phenol and chlorophenols have been studied in both fresh water and marine environments and is shown to be occuring via oxidative dechlorination and hydroxylation. The metabolism of pentachlorophenol produce trichlorophenol while, chlorocatechols are intermediates in the metabolism of mono chlorophenols (Kaufmann *et al.*, 1978) followed by the ring cleavage pathways.

Photolysis studies have been carried out on a number of chlorophenols. Photolysis of pentachlorophenol in distilled water resulted in a number of degradation products including tetrachlorophenols, chlorinated dihydroxybenzenes and nonaromatic fragments such as dichloro-maleic acid (Hwang *et al.*, 1986).

PCBs are the major pollutants which are generally considered to be resistant to microbial degradation. Such chlorinated biphenyls may contain up to 10 chlorine atoms, the aerobic heterotrophs are found to be better oxidizers of biphenyls containing less chlorine atoms (Abramowicz, 1990). Benzopyrene is one of the most intensely studied polycyclic aromatic hydrocarbon since it is an ubiquitous environmental pollutant which metabolically activated, turns to be potentially cytotoxic, mutagenic and carcinogenic (Cerniglia, 1984). Studies with *Achromobacter* showed that chlorinated benzoic acid was formed as a result of oxidative pathway (Furukawa *et al.,* 1979). Hence, the aerobic organisms are important targets for a PCB bioremediation schemes, although the microbial degradation of PCB based on sequential anaerobic, acidic reaction scheme is also prevalent in the marine ecosystem (Brown *et al.*, 1984; Berkaw *et al.* 1996). In view of remediation based on biological systems, several hybrid PCB degrading strains have been engineered (Havel & Reineke, 1991), among these, the hybrid strains *Pseudomonas putida* JHR and *Pseudomonas* species strain UCR2 have been reported to grow on environmentally important 2-, 2,4-, and 2,5- substituted PCB (Havel & Reineke, 1991). Further introduction of a dehalogenase gene into PCB cometabolizing *Comamonas testosteroni*, has resulted in a recombinant pathway which enables the recombinant variants to grow on di chlorinates and completely mineralize *ortho*- and *para* -substituted monochlorobiphenyls (Pellizari *et al.*, 1996).

(d) Quinolene

Quinilenes have a tetra cyclic structure which gives a different dimension to the degradative pathways. The microbial metabolism of quinolene reviewed by Kaiser *et al.* (1984), have shown that *Pseudomonas sp.* are the predominant potent degraders of quinolene which is also transformed by many other microorganism under aerobic and anaerobic condition. Bacteria are known to initiate attack on these ring systems using water as a source of oxygen for hydroxylation (Gibson & Subramanian, 1984). Initial step of transformation of quinoline is by hydroxylation at position 2 of the heterocyclic aromatic ring, leading to 2-hydroquinoline. Further transformation includes additional hydroxylation steps which leads to the formation of 2,6-dihydroxyquinoline and a trihydroxyquinoline.

(e) Dibenzothiophene

One of the major sulfur containing compound found in marine ecosystem is known to be Dibenzothiophene. High level of this compound has been found in a marine isopod after exposure to an oil spill. An extensive analysis of the fate of oil from Amoco Cadiz wreck (off the coast of Northern France in 1978) showed that alkylated Dibenzothiophene and Phenanthrene compounds persisted years longer than other hydrocarbon fraction and serve as a major aromatic marker in the sediments (Rodrigues, 2000).

Several metabolic pathways and intermediates of Dibenzothiophene have been identified in *Pseudomonas beijirinkia.* The organisms appear to use common pathway of Dibenzothiophene oxidation. The first two oxidation product are cis-Dibenzothiophene diol and Dibenzothio-phene-5-oxide.Two pathways for degradation have been proposed.

The carbon destructive pathway which leads to the overall degradation of Dibenzothiophene and the second pathway is known as 4S pathway. This pathway involves the oxidation of Dibenzothiophene to Dibenzothiophene-5-oxide and later it converted into Dibenzothiophene sulfone, will be followed by ring cleavage, which converts Dibenzothiophene sulfone into 2-hydroxy biphenyl sulfonate. Sulfur is removed as sulfate. This pathway play a major role in mineralisation of thiophenes in marine ecosystems (Kodama *et al.*, 1970).

(f) Organotin

Incorporation of tributyltin, antifouling agents, paints from ships, pose the greatest immediate impact upon the harbor and coastal aquatic biota (Blunden *et al.*, 1984). It is also known that tributyltin is highly toxic to various non target aquatic organism at low concentrations. For example, tributyltin in ppb levels is acutely toxic to amphipod larvae, lobster larvae and zoeal shore crabs, sheep shead minnows (Ward *et al.*, 1981) and missed shrimp (Seligman, 1984). At sub-pp. levels tributyltin causes sublethal effects in zoeal mud crab.

The degradation of butyltin in the marine environment is widely assumed to follow debutylation reaction (Blunden *et al.*, 1984; Kimmel *et al.*, 1977).

Bu_3Sn — Bu_2Sn^{2+} — $BuSn^{3+}$ — Sn(IV)

The butyl groups may be oxidized to CO_2 by microbial activity. Environmental methylation of butyltin may also occure due to the presence of mixed butyl-methylation species in sediments. The toxicity of many of the products of this methylation is still unknown. The source of methylation of methylcarbonion (eq-iii) presently is uncertain but may be due to biogenic or anthropogenic methyltin species or to intermediate oxidative methylation from algal metabolites (Mathlas *et al.*, 1986).

A few possible reactions include the following

(i) $2Bu_3Sn^+ - Bu_2Sn^+ + Bu_4Sn$

(ii) $Bu_2Sn^{2+} + Bu_3Sn^+ - BuSn^{3+} + Bu_4Sn$

(iii) $Bu_3Sn^+ + Me- —- -Bu_3MeSn$

5. MICROORGANISMS INVOLVED IN DEGRADATION PROCESSES

Protozoans, Actinomycetes , yeasts and other fungi, microalgae are main interactive organism with petroleum, they are known to come in contact with petroleum either due to surrounding or by the food which is coated with petroleum.

Most of the bacterial isolates which have been obtained from sea comprises of just a fraction of the existing population. The most frequently encountered genera are *Pseudomonas*, *Achromobacter*, *Havobacterium*, *Acinetobacter*, *Vibrio*, *Bacillus*, *Arthobacter*, *Nocardia*, *Corynebacterium* and *Micrococcus* (Aelion *et al.*, 1987).

There are no detailed reports of extensive degradation of aromatic hydrocarbons by fungi using pathways such as those described in bacteria. However, D.T. Gibson and colleagues have studied intensively the reaction used by a filamentous fungus, *Cunninghamella elegant*, for initial oxidation of naphthalene, biphenyl benzo (a) pyrene, benzo (a)anthracene and methylchloroanthrene (Gibson & Subramanian, 1984).

The hazards and damage caused by pollutants chemicals can be controlled using microbial bioremediation resources. For development appropriate organisms which exist actually in that environment cannot be isolated or enriched. Although microorganisms can biodegrade wide variety of organic substance including those that are recalcitrant and persistent in the ecosystem for a considerable long time, yet the activity of the organisms in the ecosystem present are at a very slow rate. It is therefore essential that, these are amplified either by increasing activities of existing pathways or evolving new degradative pathway to achieve faster degradation of xenobiotics (Goldstine *et al.*, 1985). Development of new strains having enhanced and multiple dissimilarity traits (Balajee & Mahadevan, 1989), including hybrid pathways will be an essential research and development activity for controlling pollution.

In order to evolve such versatile bacterial strain to control pollution, it is important to understand the biochemistry of degradative pathways, their regulation, chemical inducers, operons etc. Such studies on marine micro organisms (Chandramohan, 1997)) with relation to their biodegradable capabilities for xenobiotics have to be undertaken and concentrated on in order to achieve a pollution free marine / aquatic ecosystems.

REFERENCES

Abramowicz, D.A. 1990. Aerobic and anaerobic biodegradation of PCBs: a review. Critic. Rev. Biotechnol.10:241-48.

Aelion, C.M., C.M.Swindall and F.K.Pfainder.1987. Adaptation to and biodegradation of xenobiotics compound by microbial communities from a prestine aquifer. Appl. Environ. Microbiol.53.2212-2217.

Balajee. S and Mahadevan A. 1989. Potential role of engineered microbes to detoxify chemical pollutants. MIRCEN j.5. 119-134.

Barkay, T. and h. Pritchard. 1988. Adaptation of aquatic microbial communities to pollutant stress, Microbial. Sci.5.165-69.

Beachem, I.R.1987. Silent genes in prokaryotes .FEMS Microbiol. Rev, 46.409-417.

Berkaw, M., K.R. Sowers and H.D. May 1996. Anaerobic ortho dechlorination of polychlorinated biphenyls by estuarine sediments from Baltimore Harbor. Appl. Environ. Microbiol. 62, 2534-39.

Blunden, S, J, Hobbs, and L.A: Smith, P.J. 1984. In "Environmental Chemistry ", Bown, H.J.M Ed.: Royl Society of Chemistry: London, 51-77.

Blunden, S.J., Hobbs, L.A; Smith, P.J. 1984. In"Environmental Chemistry"Bowen, H.J.M, ed; Royal Society of Chemistry: London.51-77.

Brown, J.F, R .E . Wagner, D. L. Bedard, M.J. Brennan, J.C. Carnahan, R.J. May and T.J. Tofflemire. 1984, PCB transformation in upper Hudson sediments Northwest. Environ. Sci.3, 167-79.

Carl. E, Cerniglia, 1984. Microbial transformation of aromatic hydrocarbons, chapter-3, In" Petroleum Microbiology" ed by Roland. M. Atlas, Macmillan Publishing Company, N.Y.112.

Cerniglia, C.E., 1984. Microbial transformation of aromatic hydrocarbon, chapter 3. In "Petroleum Microbiology" Ed by Roland. M. Atlas. Macmillan Publishing Company, N.Y.113.

Chandramohan D. 1997. Recent advances in marine microbiology: The Indian scenario.J. Mar. Biotechnol. 5.73-81.

Dagley, S. 1978b. Determination of biodegrability. Q. Rev. Biophys. 11.577-602.

Dagley, S. 1985. Microbial metabolism of aromatic compound, In "Comprehensive Biotechnology" M. Moo-Young, Ed.Vol-1.Pergamon Press, Oxford. 483-505.

Fleming, B.I. 1995. Organochlorine in perspective. Tappi. J. 78.93-98.

Furukawa, K., Tomizuka,.N., and Kamibayashi. A., 1979. Effect of chlorine substitution on the bacterial matabolism of various Polychlorinated Biphenyls, App. Environ. Microbiol., Vol-38,2.301-310.

Gibson, D.T, Hensley, M., Yoshioka, H., and Mabry, T.J.1970a. Formation of (+)-cis -2,3 dihydroxy-1-methylcyclohexa-4,6-diene from toluene by *Pseudomonas putida*. Biochemistry 9.1626-30.

Gibson, D.T., and Subramanian, V, 1984, Microbial degradation of aromatic hydrocarbons. In "Microbial Degradation of organic Compounds" D. T. Gibson. ed; Dekker, New York. 181-252.

Goldstine R.M, Mallary L.M and Alexzander M. 1985. Reason of possible failure of inoculation to enhance biodegradation. Appl. Environ. Microbiol. 50. 977-983.

G.D. Floodgate, 1984. The fate of petroleum in marine ecosystem, Chapter-8, In "Petroleum Microbiology" ed by Roland. M. Atlas. Macmillan Publishing Company, N.Y.355-397.

Havel, J., and W. Reineke 1991. Total degradation of various chlorobiphenyls by co-cultures and vivo constructed hybrid *Pseudomonas.* FEMS microbiol. letters 78,163-170.

Hwang, H.M., Hodson, R.E and Lee. R .F. 1986. Degradation of phenol and chlorophenol by sunlight and microbes in estuarine water. Environ. Sci. Technol.20.1002-1007.

J.T. Singer and W.R. Finery. 1984. Genetics of hydrocarbon utilizing microorganism, In "Petroleum Microbiology". Chapter-7. ed by Roland. M. Atlas. Macmillan Publishing Company, N.Y.310.

Kaiser, K.L.E.; Dixon, D.g.; Hodson, P .V. 1984. In QSAR in Environmental Toxicology, Kaiser. K.L.E. Ed; Reidel Dordrecht, Holland.189-206

Kaufmann, D.D. 1978. In Pentachloro phenol; Rao. K.P, ed, Environmental Science Research Series, Plenum:NewYork, vol 12.27-39.

Kilbane, J.J. 1989. Desulfurization of coal: the microbial solution. Trends. Biotechnol. 7.97-100.

Kimmel, E.C., Fish, R.H and Casida J.E, 1977. Bioorganotin chemistry. Metabolism of organotin compounds in microsomal monooxygenase system in mammals. J. Agri. Food. Chem.25.1-9.

Kiyohara, H, and Nagao, K. 1977. Enzymatic conversation of 1-hydroxy-2-napthonate in phenanthrene grown *Aeromonas* S45PL. Agri. Biol. Chem. 41.705-707.

Klaas Kieslich, 1976. In "Microbial Transformations of non-steroid compounds. John Wiley Sons, George Thieme Publishers.1-2.

Kodama, K., Nakatani, S., Umehera, K., Shemizu, K., Minoda, Y. and Yamada K., 1970.Microbial conversion of petro-sulfur compounds III , isolation and identification of products from dibenzothiophene. Agri. Biol. Chem., 34:1320-1324.

Lee, R. F., Hinga, K., Almquist, and G. 1982. In marine mesocosoms: Biological and chemical research in experimental, Grice, G.D., Reeve, M.R., and Eds. Springer-Verlag, New York 123-135.

Lee, R.F. and Ryan, C. 1979. In Proceedings of the Workshop. Microbial degradation of pollutants in marine environment, Bourquin, A. W., Pritchard, P. H., eds. U.S. Environmental Protection Agency. Gulf Breeze FL, EPA-600/9-79-1012, and 443-450.

Madsen, E.L., J.L. Sinclair and W.C. Ghiorse.1991. In situ biodegradation : microbiological pattern in a contaminated aquifer. Science 252.830-833.

Matthlas, C.L. Bellama., J.M. Olson G.J. and Brinckman F.E.1986.Comprehensive method for determination of aquatic butyltin and butylmethylation species at ultratrace levels using simultaneous hybridization/extraction with gas chromatography- flame photometric detection. Environ. Sci. Technol.20. 609-615.

Miller, J.W., 1973. Proceedings of the joint conference on prevention and control of oil spills, American Petroleum Institute, Washington D.C, 195-204.

M.E. Singer and W.R. Finnerty. 1984.Microbial metabolism of straight chain and branched alkane, chapter-1, In "Petroleum Microbiology" Ed by Roland. M. Atlas. M. Atlas. Macmillan Publishing Company, N.Y.22.

Pellizari, V.H., S. Bezborodnikov, J.F. QuensenIII and J.M. Tiedje, 1996. Evaluation of strains isolated by growth on naphthlene and biphenyl for hybridization of genes to deoxygenase probes and polychlorinated biphenyl degrading ability. Appl. Environ. Microbiol. 62, 2053-58.

Pigantello, J.J., Martinson, M. M., Steiert, J. G., Carlson, R. E. and Craford, R.L. 1983, Biodegradation and photolysis of pentachloro phenol in artificial freshwater streams. Appl. Environ. Microbiol. 46.1024-1031.

P.A Williams, 1978. Microbial genetics relating to hydrocarbon degradation. Chapter- 6. In " Development in degradation of hydrocarbons-1, R.J. Watkinson Ed. Applied science publishers Ltd., London.148-154.

Rodrigues, J. 2000. Studies on the degradation of Dibenzothiophene by a marine organism and its role in mineralization of crude oil. Ph.D. thesis submitted in Department of Marine Science and Biotechnology, Goa university, Goa.

Seligman, P.F 1984. " Fate and effects of Organotin antifouling Leachates in the Marine Environment "Bethesda, MD, progress report prepares for energy research and development office. David W. Taylor Ship Research and Development Center, 1-47.

Spain, J.C., and P.A. VanVeld.1983. Adaptation of natural microbial communities to degrade of xenobiotics comp: effects concentration ,exposure time inoculam and chemical structure. Appl. Environ. Microbiol.45. 428-435.

Stanley Dagley. 1986. "Biochemistry of aromatic hydrocarbon degradation in *Pseudomonas,* capter-15, In "The Bacteria", Vol-X.Ed. Sokatch. Academic Press INC. 527-555.

Ward, G.S.; Cramm, G.C; Parrish, P.R; Trachman, H; Slesinger. 1981.In "Aquatic toxicology and hazard assessment; Fourth Conference", Branson, D.R; Dickson, K.L ed.American Society for testing and Materials: Philadelphia. ASTM STP 737,183-200.

CHAPTER – 22

BIODIVERSITY OF CYANOBACTERIA

B.V. Singh, O.N. Tiwari, Dolly Wattal Dhar and P.K. Singh

National Centre for conservation and Utilisation of Blue Green Algae, IARI, New Delhi - 110012

1. INTRODUCTION

Blue Green Algae constitute a heterogenous assemblage of organisms ranging from unicellular organisms to branched filamentous forms with prokaryotic cellular organisation. Although, they are predominantly hydrophytes, but forms inhabiting other habitats are not rare. Soil inhabiting forms are known as edaphic algae and those colonizing the moist soil surface are referred to as saprophytes. The latter play an important role in soil conditioning and vitality of the soil. Blue green algae or cyanobacteria are photosynthetic, oxygen evolving and many of which can fix atmospheric nitrogen and hence, help to maintain the nitrogenous fertility of the soil (De, 1939).

The blue-green can survive prolonged periods of drought. Successful revival has been possible even after many years of preservation in dried form. Many forms are able to survive very high temperatures and remain dormant in the soil. Forms which do not form resting propagules are able to resist the dry and warm conditions because of the presence of mucilage, viscous cytoplasm, aggregation of several trichomes and pigment adaptation.

The ability of blue green algae to withstand adverse ecological conditions, their capacity to thrive well in hostile environments and to respond to the onset of dry conditions by entering into a dormant resistant state, have distinguished this group as pioneers of plant succession. Both in xeroseres and hydorseres, these appear as the first plant community leading to the establishment of higher plant associations.

Blue green algae play an important role in primary and secondary plant community succession by acting as an integral part of the system. The present strategy is to increase the number of these microbes in the soil to make the soil productive through increasing the availability of plant nutrients.

2. GENERAL ASPECTS OF CYANO-BACTERIAL TAXONOMY

The classification of cyanobacteria has undergone several transitionary changes from its inception in the 1800s.

(a) Traditional approach (Geitler concept)

This is the oldest system of classification based upon general morphological features. This has been modified time to time by Geitler (1932), Fritsch (1945), Desikachary, (1959), Starmach (1966) and Anagnostidis and Komarek (1988). Among these, the classification system given by Fritsch, (1945) has provided the basis for later systems and was one of the first voluminous treatise on this aspect.

Fritsch (1945) divided cyanobacteria into five orders

1. Chroococcales

Unicellular or colonial (palmelloid) forms, pseudoparenchymatous colonies, pseudofilaments, without trichomes, reproduction by binary fission, formation of exocyte and nannocyte, multiple fisson (resulting in non motile monocytes or motile planocytes).

2. Chaemaesiphonales

Unicellular or colonial lithophytes, epiphytes or epilithic, exhibiting marked polarity, reproduction by endospores and exospores, generally unicellular as well as multicellular.

3. Pleurocapsales

Heterotrichous filamentous forms, lacking heterocyst, reproduction by endospore, hormogonia absent, prostrate or erect system.

4. Nostocales

Non-heterotrichous filamentous forms, often showing false branching- hetero-cystous and non-heterocystous, reproduc-tion by hormogonia, hormocyst, planococci and akinetes.

5. Stigonematales

Heterotrichous filamentous forms mostly with heterocyst showing true branching and usually connection between adjacent cells, multiplication mostly by hormogonia and rarely by akinetes.

(b) The botanical approach

Chronologically, the botanical approach was the first to put stamp on cyanobacterial taxonomy. The classical taxonomy of the cyanobacteria is based on the morphological features and their nomenclature is ruled by the Botanical Code, i.e. basically two groups - simple filamentous (Oscillatoriaceae) and heterocystous species (Nostocaceae and Stigonemataceae). The starting point for valid publication of names is the monographs written by Gomont, (1892), and Bornet and Flahaut, (1888). In Geitler determination keys published in 1932, about 1300 species and 145 genera were recognized. Sometimes, on single character basis such as presence or absence of sheath or slight deviation in cell dimension, species have been designated (Komarek and Anagnostidis, 1989).

The problem of the morphological variability has promoted Drouet (1986) to revise the taxonomy profoundly, due to existence of ecophenes - organisms-sharing the same genotype but expressing distinct morphologies under the influence of environmental factors. He drastically reduced the number of species of 62 by selecting certain morphological features. Classical taxonomists were quite critical of this approach (Desikachary, 1973, Anagonistidis and Komarek 1985). Later, DNA-DNA hybridisation showed that taxa placed by Drouet in the same species were genotypically different (Stam, 1980). Anagnostidis and Komarek (1990) have published a new and insightful taxonomic treatise. This revision is based on definitions of smaller, more coherent genera and the authors have tried to integrate all the biochemical, ultrastructural and molecular characters, which have been investigated.

(c) The bacteriological approach

Classical botanical taxonomy was confronted in the seventies with a rather different approach. Since cyanobacteria are bacteria, it was felt that their taxonomy should be treated accordingly and their nomenclature be governed by Bacteriological Code. Stanier and collaborators pioneered the use of physiological and genotypic characters and parameters such as pigment composition, fatty acid analysis, heterotrophic growth, nitrogenase activity were used as the important tool. Therefore, the basis of bacteriological taxonomy of the cyanobacteria was published by Rippka *et al*, (1979). This taxonomic system still relied on the strains at the generic level. The five major sectors recognized by Rippka; coincide broadly with orders of other classification systems. So far only a limited number of taxa have been worked out in detail and incorporated in Bergey's Manual of Systematic Bacteriology.

This radical new scheme of classifying cyanobacteria developed by Rippka *et al.*, (1979) recognises 22 genera placed in five sections, each distinguished by some particular structure and developmental characters.

Section I : Unicellular forms reproducing either by binary fission or by budding. *Gloeobacter, Gloeothece, Synechococcus, Chaemaesiphon, Gloeocapsa* and *Synechocystis* are the six genera included in this section.

Section II : Rapid binary fission of a vegetative cell within a fibrous wall layer, unaccompanied by growth, leading to formation of small, spherical bodies called 'baeocytes is characteristic of the organisms placed in this section, examples are *Dermocarpa, Xenococcus, Dermocarpella, Myxosarcina, Chroococcidiopsis* and *Pleurocapsa.*

Section III-IV-V : Filamentous or trichomentous forms; the trichome in Section III is made up of solely vegetative cells and is uniseriate. The members of section IV and V are collectively distinguished from section III by their ability of cellular differentiation leading to the formation of heterocyst and akinetes. *Lyngbya, Phormidium, Plectonema, Oscillatoria* and *Spirulina* are included in Section III but some workers including Stam (1980), proposed the merger of some of these genera into smaller numbers. Section IV includes *Anabaena, Nodularia, Cylindrospermum, Nostoc, Scytonema* and *Calothrix.*

Concerning the classification of cyanobacteria in major subgroups, it now seems clear that out of the five sections (orders) distinguished by Rippka and Herdman (1992), only the sections IV (Nostocales) and V (Stigonematales) form a consensus group, whereas members of sections I (Chroococcales), II (Pleurocapsales), and III (Oscillatoriales) do not represent coherent evolutionary linkages. In particular the genera *Synechococcus* (Chroococcales) and *Leptolyngbya* (Oscillatoriales) appear to be clearly polyphylectic, and numerous other genera may also be polyphylectic or paraphylectic

(Turner, 1997), e.g., *Synechocystis, Cyanothece, Gloeocapsa* and *Gloeothece* (Chroococcales), or *Oscillatoria* and related genera such as *Lyngbya, Phormidium* or *Plectonema* (i.e., simple filaments without heterocysts or akinetes). Thus, it appears that the unicellular or simple filamentous habit of a cyanobacterium is often poorly correlated with its phylogenetic association.

(d) Fossil record of the cyanobacteria

The great antiquity of the cyanobacteria is well documented though the fossil record is fragmentary and based towards the formations where preservation of the morphology was possible. The earlier unicellular and filamentous forms attributed to the cyanobacteria were found in sedimentary rocks formed 3500 million years ago. Endolithic forms that reproduce with baeocytes were observed in rocks formed 500 million years ago. Heterocystous forms and forms with true branching seem to have appeared more recently, after the Precambrian (Schopf and Walter, 1982; Schopf and Packer, 1987).

Although the taxonomy of cyanobacteria seems to be in a constant state of flux, it should be emphasized that as more and more refined and molecular methods for differentiating cyanobacteria develop, there will be changes but these only reflect the increase in our knowledge about these organisms. Besides these major approaches to classification, modern tools such as biochemical characteristics and DNA/RNA based methods can also prove to be extremely useful.

(e) Chemotaxonomic markers

Molecular information on phylogenetic relationship of organisms can be obtained by chemotaxonomic studies (Fatty acids, quinones, carotenoids etc.) and analysis of macromolecules (nucleic acids and proteins).

One of the very useful easy and rapid methods for analysis and distinction of taxa is through the use of biochemical parameters, for which determination methods are easily available. There is consistency and variation due to factors such as growth conditions which have not always been systematically investigated (Holton, 1981). Moreover, since the molecules employed are typically synthesized through complex pathways, their presence or absence can have different causes. As in other fields of biology, molecular studies, especially sequence comparisons of the small ribosomal subunit RNA (16S RNA), have provided valuable new data to decipher genetic information in the past years. Such data are most useful for attempts to reconstruct the evolution of cyanobacteria and to improve their taxonomy (Wilmotte 1994, Turner, 1997). Together with the information obtained from other approaches e.g., investigations on ultrastructure, chemotaxonomy, numerical analyses of phenotypic features, etc. this has resulted in many unexpected new insights into phylogenetic relationships of taxa at all levels. A relatively slow but steady increase has been taking place in the numbers of genera that have been redefined on the basis of new genetic, physiological, biochemical, morphological and • ultrastructural information (Castenholz, 1992, Rippka and Herdman 1992).

The chemotaxonomic markers presented below have been shown to be useful, or seem worthy of further study.

1. Lipid composition

Kenyon and collaborators (Kenyon 1972 and Kenyon *et al.* 1972) studied the fatty acid composition of 66 cyanobacterial strains. A uniform fatty acid composition was observed for the groups *Anabaena* and *Calothrix*, but not for other taxonomic groupings. The lipid analysis of Prochloron sp. indicated a closer relationship with the cyanobacteria than with eucaryotic algae (Perry *et al.* 1978).

2. Polyamines

The use of polyamines as chemotaxonomic marker in bacteria systematics has recently been reviewed (Hamana and Matsuzaki 1992). The dominant polyamine in the cyanobacterial strains hitherto tested was either spermidine or sym-homospermidine.

3. Carotenoids

The results of the analysis of the carotenoid composition compiled by Hertzberg *et al.*, (1971) indicated that the variations in patterns have probably been useful for species identification but not for determinations at a higher taxonomic utility.

4. Amino acid/Enzyme patterns

The enzymology and regulatory patterns of aromatic amino acid pathway in strains from the five sections defined by Rippka *et al.*, (1979) showed a biochemical diversity which allowed the distinction of sub-groups within several genera including *Synechococcus, Synechocystis, Anabaena, Nostoc* and *Calothrix.* This confirms the heterogeneity of these genera and suggested that biochemical diversity could provide useful taxonomic markers.

(f) Macromolecular methods

Macromolecules, nucleic acid and proteins, are copies or translations of the genetic information and thus, may be the best tools to infer phylogenetic relationships (Murray *et al.* (1990). The 16s RNA has given the most detailed hypothesis on the evolutionary relationships within cyanobacteria.

1. Protein electrophoresis and isozyme pattern

Isozyme patterns can be useful for taxonomic studies at the genus and species level. Studies have indicated that considerable intra and inter-specific heterogeneity of enzyme is observed in cyanobacteria. Shared bands in zymograms can help in clustering of related strains.

2. Phycobiliproteins

Schenk and Kuhfittig (1983) investigated the phycobiliprotein pattern in twenty-one cyanobacterial species by means of polyacrylamide disc gel electrophoresis and observed distinctive heterogeneity even at the strain level. This method may therefore, be useful for distinguishing between identical strains but its utility in establishment of classification schemes needs to be investigated.

3. Immunological studies

Immunological comparisons of proteins have been used for bacterial systematics (Schleifer and Stackbrandt 1983) but have rarely been used for studies on cyanobacteria.

4. Restriction Fragment Length Polymorphism **(RFLP)**

The RFLP technique is generally useful to identify and classify organisms at the population or the species level. It has been productively used to study marine *Synechococcus* sp. strains and cyanobacterial symbionts.

1. The marine *Synechococcus* sp. strains — The taxonomy of small, planktonic, phycoerythrin containing *Synechococcus* sp. strains is problematic. Wood and Townsend (1990) tested the genetic homogeneity of a group of eight strain that showed cross- reactivity to the antiserum directed against *Synechococcus* sp. 7803. The probes were derived from several heterologous genes of *'Anabaena'* sp. strains WH 7803. In the tree topology obtained, the eight strains were distributed into four different branchings. Moreover, the genetic distances between terminal taxa on different branches within the serogroup were as large as the distances found between members of the serogroups and two freshwater *Synechococus* sp. strains that show no cross- reactivity to the antiserum.

2. The symbiotic cyanobacteria in Azolla - A filamentous, nitrogen fixing symbiont lives in the leaf cavities of the water form *Azolla* sp. and is generally identified as *Anabaena azollae*. In total, the symbiosis involving seven *Azolla* sp. have been studied. The probe were derived from heterologous genes from *Anabaena* sp. strain PCC 7120 (Franche and CohenBazire, 1987; Meeks *et al.*, 1988) or genomic and plasmid DNA clones of freshly isolated symbionts (Planzinski *et al.* 1990, 91). The results have shown that (I) the leaf cavities of *Azolla* sp. probably harboured one major symbiont, accompanied by minor species; (ii) There is a good correlation between the genotypic relationships among the freshly isolated symbionts and the classification of their hosts; (iii) Only the minor constituents of the symbiosis could be established as free living culture.

3. The symbiotic cyanobacteria in Cycads and *Gunnera* sp.- The symbionts associated with cycads and angiosperms are classifically identified as members of the genus *Nostoc*. Lindblad *et al.* (1989) used heterologous probes from '*Anabaena*' sp. strain PCC 7120 to characterize the symbionts freshly isolated from fie cycad species from Central America. The results suggest the most cycad species contained one dominant symbiotic strain associated with one or several minor symbionts.

In the case of angiosperms symbiosis, twelve free living cultures derived from symbionts of eight *Gunnera* species growing in Sweden, New Zealand and USA were characterized by Zimmerman and Bergman (1990). Two methods were used : The RFLP analysis using heterologous probes from '*Anabaena*' sp. strains PCC 7120 and the immuno-staining of the protein profiles. Three *Nostoc* sp. isolates from *Gunnera* sp. plants cultivated in a green house in Sweden were identical whereas a fourth Swedish plant grown outdoors harboured a different symbiont. Six different genotypes were observed among the other isolates.

4. DNA base composition

DNA base composition is one of the few molecular characters that have been determined for almost 200 cyanobacterial strains (Herdman *et al.* 1979). It is a 'one-way' taxonomic marker. Although large difference in DNA base composition indicates that the strains cannot be closely related, similar G+C percentages do not give conclusive evidence concerning genotypic relationships, as cyanobacteria as a group have a wide G+C content. This method therefore, lacks Sensitivity.

5. DNA Fingerprinting

Three short tandemly repeated repetitive (STRR) sequences from *Calothrix* sp. strain PCC 7601 hybridized only with the digested genomic DNA of the heterocystous strains tested. Manzel *et al.* (1990) suggested that the banding patterns could be used to distinguished strains at the species and genus level, and can be used to generate species-specific or strain-specific fingerprints.

6. DNA-DNA Hybridization

Wayne *et al.* (1987) proposed the use of

DNA-DNA hybridization as a criterion for the definition of eubacterial species. Stam (1980) pioneered the use of this method for cyanobacteria. Relationships between different strains of *Pseudanabaena*, LPP group, *Anabaena* and Unicellular forms have been elucidated using filter hybridization technique.

7. Protein and Protein-coding Gene Sequence analysis

A number of cyanobacterial protein and protein-coding genes have been sequenced (Tandeau de Marsac and Houmard 1987) but generally this has been performed in the course of genetic studies and not with an evolutionary framework in mind. Therefore, molecular information is scattered over a few organisms and gives a poorly resolved tree topology.

8. 16s rRNA

The 16s RNA has been shown to be an adequate macromolecule to study the evolution of eubacteria. In cyanobacteria the first global evolutionary scheme of cyanobacteria was published in 1988 by Giovannoni and co-workers and helped in establishing coherence with the phylogenetic clustering of strains according to established schemes of classification.

An unexpected outcome of the molecular phylogeny obtained is an association of the Pleurocapsalean cyanobacterium *Chroococcidiopsis thermalis* as a sister taxon to the heterocyst forming Nostocalean and Stigonematalean cyanobacteria, distinctly apart from other Pleurocapsalean cyanobacteria (Turner 1997).

It is hoped that more laboratories will become interested in ribosomal RNA sequences and other molecular techniques to determine cyanobacterial genotypic relationship. However, this molecular information should be integrated with other characteristics of the strains leading to a polyphasic taxonomy that not only be of practical use but will reflect as much as possible the evolutionary relationships of the organisms (Wilmotte and Golubic 1991).

At present, the older morphological system (botanical approach) continues to the be used by most ecologists working with natural collections. Anagnostidis and Komarek (1990) continued their discussion of classification principles in the cyanobacteria with the last part of the series on Stigonemetales.

So far, several hundred different cyanobacteria have been obtained in pure culture and partially characterized, but it must be considered that many (perhaps more than 50%) of the identifications of strains in culture collections may be incorrect (Rudi *et al.* 1997). Clearly, an enormous amount of work remains to be done before a satisfactory system of classification of this important group of photosynthetic prokaryotes will be achieved.

3. MORPHOLOGICAL AND REPRODUCTIVE DIVERSITY

Cyanobacteria exhibit an array of thallus types that is unparalleled among the prokaryotes and which approximates the morphological variation occurring in eukaryotic algal groups. A number of cyanobacterial genera are so similar in appearance to green algae, in particular, that misidentification can result. Green algae can be distinguished from similar appearing cyanobacteria by the presence of definite green-pigmented regions (Chloroplasts) and by the presence of starch, which reacts with an iodine-iodide solution to produce a blue-black product visible with the light microscope. In contrast, pigmentation of cyanobacterial cells is typically distributed homogenously throughout the cell, and blue-black starch iodine complexes do not form. In view of the results of molecular studies showing that unicells, colonies, and unbranched filaments do not form distinct, monophyletic groups, the following diversity has been recognized.

1. Unicellular and colonial forms lacking specialized cells or reproduction

Synechococcus is a tiny (1 μm in diameter), cylindrical, unicell that is an important primary producer in the plankton of fresh and marine waters, and has also been collected from surfaces of algae and plants, films of algae on sandy beaches, outflow channels of hot springs and thermal pools, and endolithic habitats. Cell lengths are typically 2-3 times their width. There is no mucilaginous sheath. Cells may exhibit swimming motility. Both marine and freshwater forms of *Synechococcus* have been associated with "whiting" events, the production of suspended fine grained carbonates that eventually contribute to sedimentary carbonate deposition (Hodell *et al.* 1998). Such later-summer calcite precipitation and cell sedimentation also increase the flux of organic carbon to the sediments of lakes (Hodell and Schelske 1998). A specialized cell surface layer functions as a template for carbonate deposition and is shed after it has become coated with calcium carbonate. It is then replaced by a new surface layer and the process is repeated (Thompson *et al.*, 1997).

Prochloron is a unicellular form that is extensively described in a book edited by Lewin and Cheng (1989). It lives primarily in association with marine didemnid ascidians, colonial tunicates also known as sea squirts. Ascidians are small flattened animals that attach themselves to coral rubble, mangrove roots, sea-grass blades, and the shallow marine substrates in tropical regions throughout the world. *Prochloron* occurs only in coastal waters within the temperature range of 21°-31°C, and ceases photosynthesis below 20°C. The alga inhabits the animals cloacal cavities and upper surfaces and sometimes is also found intracellularly. Although, the algae release glycolate, a photosynthetic product that contributes to the animal nutrition, the didmnids are probably not dependent upon the algae, and they require additional foods. In contrast, *Prochloron* probably does require its host in ways that are not yet understood; by itself the alga has been difficult to maintain in the laboratory. Some forms of *Prochloron* can fix nitrogen in the light, but only in association with their didemnid hosts (Paerl, 1984). *Prochloron* is unusual, but not unique, among cyanobacteria in that possesses chlorophyll b in addition to chlorphyll a, and lacks phycobiliproteins. It thus, appears grass green, a color not generally found among cyanobacteria. Green pigmentation is believed to be adaptive in low N tropical coastal waters. Production of phycobiliproteins requires about three times as much combined nitrogen as does production of the chlorophyll a/b protein complex. The green pigmentation may also allow enhanced absorption of the orange-red wavelengths of light present in shallow waters (Alberte, 1989). Pigments occur in stacked thylakoids, unusal for cyanobacteria. The central portion of *Prochloron* cells is occupied by large central space, which is also unusual for cyanobacteria. In other ways, including cell wall and storage biochemistry, and absence of organelles, *Prochloron* resembles other cyanobacteria. At one time *Prochloron* was thought to represent the ancestor of green algal and higher plant chloroplasts, but molecular sequencing studies have since shown this to be unlikely. Recently it has been demonstrated that the chlorophyll a/b or binding proteins of *Prochloron*, *Prochlorococcus* and *Prochlorothrix* are closely related to cyanobacterial proteins, and not to eukaryotic chlorophyll a/b or a/c light-harvesting proteins, suggesting independent evolutionary origin of these organisms (La Roche *et al.*, 1996).

Prochlorococcus is one of the most numerous components of open-ocean plankton and was discovered using flow cytometric methods (Chisholm *et al.* 1992). Superficially it is similar to *Synechococcus*, but lacks phycobiliproteins and contains chlorophyll b in addition to chlorophyll a, much like *Prochloron*. However, molecular sequencing studies indicate that there is not a particularly close relationship between these two chlorophyll b-containing taxa. In the case of *Prochlorococcus*, green pigmentation may be adaptive in harvesting the blue light that penetrates relatively deeply into ocean waters. Flow cytometry and SSU rDNA sequencing were used to establish that high and low-irradiance ecotypes of *Prochlorococcus* occur in the North Atlantic Ocean (Moore *et al.*, 1998).

Chroococcus occupies freshwaters, marine waters, and some times moist terrestrial locations. *Chroococcus* occurs as single cells or colonies of 2, 4, 16 or, less frequently 32 hemispherical cells resulting from the adherence of daughter cells after division. The algae may be floating, mixed with tangles of filamentous algae, or adherent to substrates. Many species are deep blue-green in colour but a central, paler region is often distinguishable in cells. A gelatinous sheath may or may not be noticeable. The cells of one species, *C. giganteus*, are unusually large for cyanobacteria, about 5-60 μm in diameter.

Gloeocapsa lives in freshwater lakes or on moist soil and the terrestrial surfaces. It can form black bands on high intertidal seacoast rocks, and live beneath rock surfaces in arid regions. It is a phycobiont in certain lichens (Rambold *et al.* 1998). *Gloeocapsa* forms colonies having more cells than *Chroococcus*. The cells are oval or ellipsoidal, with rounded ends. Each cell has a distinct mucilaginous sheath that is surrounded by older sheath material. These sheath layers can be used to deduce the order of cell divisions that have given rise to the colonies. Sometimes the sheath is coloured yellow or brown but is most often colourless.

Aphanocapsa is a free-floating member of the freshwater plankton, partcularly in bog waters. Very large, usually irregularly shaped colonies consists of dozens of spherical cells, evenly distributed within a colourless or sometimes yellowish, gelatinous sheath. The individual cells are quite small, one to several micrometers in diameter, and can be pale to bright blue-green in colour.

Aphanothece occurs in the plankton of hard or soft freshwaters, or on muddy bottoms of fresh and marine waters, and sometimes in moist terrestrial sites. Very large, even macroscopic colonies contain many cylindrical cells embedded in transparent mucilage. The mucilage of individual cells is usually indistinct. One species grows within the mucilage of other colonial cyanobacteria.

Merismopedia is found floating or sedentary in fresh and marine waters, including sand flats. It is a typical constituent of bog-water communities. The ovoid or speherical cells are compactly arranged in orderly rows inflate, rectangular colonies one cell thick. The colony is held together by a mucilaginous matrix, but this sheath does not extend far beyond the cell limits, and the sheath of individual cells are indistinct.

Microcystis is a member of freshwater phytoplankton communities, but may sometimes occur as granular masses on lake bottoms. It consists of colonies of small cells (a few micrometers in diameter) that are evenly distributed throughout a gelatinous matrix. When Young, the colonies are spherical, but older colonies are typically irregular in shape and frequently perforated. In contrast to *Aphanocapsa*, whose colonies are similar, there are numerous gas vesicles in the cells of *Microcystis*, giving them a black appearance. Low numbers of *Microcystis* may occur in soft or hard oligotrophic waters. *Microcystic aeruginosa* and *M. flos-aquae* can be very common in nuisance blooms in eutrophic lakes. The highest photosynthetic rates for *Microcystis* occur at surface irradiance levels, and it is resistance to photo- inhibition (Paerl, *et al.*, 1985). Thus, *Microcystis* may be particularly well adapted to grow at the air-water interface where it may be able to escape carbon limitation. Phosphorus limitation does not reduce the level of gas vacuolation as much in *Microcystis* as it does in *Aphanizomenon*. The gas vacuoles of *Microcystis* are often too strong to be collapsed by turger pressure. These factors help to explain this organisms's frequent tendency to form thick surface growths. Potassium chloride, applied at a concentration of 4-8 mM, inhibits growth of *M. flos-aquae* and may represent a potential control strategy (Parker *et al.*, 1997).

Coelosphaerium occurs in the plankton of freshwaters, often co-occurring with *Microcystis*, *Coelosphaerium* colonies are spherical or irregular in shape, and like *Microcystis*, the cells may be so full of gas vesicles as to give them a black appearance. Unlike *Microcystis*, the cells of *Coelosphaerium* are distributed in a single layer, forming a hollow balloon like structure. Some species occur in soft waters or bogs; these do not typically produce gas vesicles. *Coelosphaerium naegelianum* is a common constituent of nuisance surface blooms in eutrophic waters.

Gomphosphaeria occurs in hard-or softwater lakes as a colony of spherical or heart-shaped blue-green cells mounted at the ends of gelatinous branching stalks, the whole enveloped by colourless mucilage. The heart shaped appearance of dividing cells results from inward growth of the outer wall, while ingrowth of the wall closest to the stalk is delayed. *Gomphosphaeria* can be confused with very similar colonies of the green algae *Dityosphaerium* and *Dimorphococcus*, which however, lack a mucilaginous envelope.

2. Filamentous cyanobacteria lacking spores, heterocysts, or akinetes

Prochlorothrix filaments are composed of elongated cells. *Prochlorothrix* is notable for having chlorophyll b in addition to chlorophyll a, and for the absence of phycobiliproteins, as is also the case for *Prochloron* and *Prochlorococcus* (Turner *et al.* 1989). However, *Prochlorothrix* does not appear to be closely related to either. It can readily be cultured in the laboratory (Matthijs *et al.* 1989).

Oscillatoria has been found in a wide variety of environments, including hot springs, marine habitats and lakes of temperate, tropical and polar regions, as well as moist terrestrial substrates. The name reflects this organism's ability to rotate, or oscillate. In addition, gliding motility also occurs. Disk shaped cells that are typically wider than long are attached end to end to form filaments. The end cells may be rounded or distinctive in ways that are used to define some of the species. Breakage of filaments between separation disks releases shorter hormogonia during vegetative reproduction. Filaments may sometimes adjoin in parallel to form thin films, but a readily identifiable sheath is not present, in contrast to two very similar genera, *Lyngbya* and *Phormidium*, which possess identifiable sheaths. *Phormidium* forms mats that do not readily dissociate and do those off *Oscillatoria*, and sheath of *Phormidium* tend to be looser than the relatively rigid, distinct sheaths of

Lyngbya, which can form thick nuisance growths in warm, high pH lakes. However, the extent to which these three genera should be separated is controversial. The planktonic species, *O. rubescens* and *O. agardhii* var. *isothrix* were changed to *Planktothrix rubescens* and *P. mougeotii*, respectively. Some other species of *Oscillatoria* have been transferred to *Limnothrix* and *Trichodesmium*.

Trichodesmium is common in tropical open-ocean waters sometimes forming visible blooms. Colonies consisting of aggregated filaments may be visible with the naked eye, from 0.5- 3 mm in diameter. Because it fixes significant amounts of nitrogen (30 mg m^2 day^{-1}) it is major source of combined nitrogen in these oligotrophic water. The presence of abundant gas vesicles reduce algal density. The biology of *Trichodesmium* has been reviewed by Capone *et al.*, (1997).

Spirulina is found in marine and fresh waters, including lakes high in sodium carbonate (natron lakes) and bog lakes, and on moist mud. *Spirulina* looks like a highly coiled version of Oscillatoria and exhibits similar oscillatory and gliding motility. The genus *Arthrospira* should probably be included in *Spirulina*.

Microcoleus occurs on the surface of salty sand in marine salt marshes and lagoons, in freshwater lakes and on crust communities. Several trichomes occupy a single, relatively stiff communal sheath. *Microcoleus chthonoplastes* occurs in marine microbial mats around the world (Stal, 1995).

3. Exospore-producing cyanobacteria

Chamaesiphon is a common inhabitant of the surfaces of aquatic plants, algae, and nonliving substrates such as stones or shells in freshwater lake and streams. The cells are cylindrical, with an open, widened sheath at the top, from which are budded off a sequence of rounded spores. The pigmentation is often greyish green.

4. Baeocyte (endospore) producing cyanobacteria

Chroococcidiopsis occurs in extremely arid, cold, or hot terrestrial environments as unicells. Reproduction is by subdivision of the cytoplasm into numerous small baeocytes. These escape by breakage of the parental cell wall and each can develop into a mature vegetative cell. A number of other unicellular and filamentous baeocyte producing genera have been identified.

5. Heterocyst and akinete producing cyanobacteria

Nostoc primarily occurs in terrestrial habitats, frequently in association with fungi in lichens and with bryophytes and vascular plants of various kinds. It occurs on moist rocks and cliffs, on alkaline soils, and in wet meadows and at the edges of shallow lakes. Nostoc is frequently abundant in flooded rice paddies, where it contributes to the fertilization of soils. Characteristically bent or linked filaments of round cells are held in a firm mucilaginous matrix to form colonies that can reach 50 cm in diameter. The sheaths can be coloured yellow, brown, or black. Transiently motile hormogonia are produced; these typically have heterocysts at the ends. Akinetes are characteristically produced midway between heterocytes, with production of additional akinetes proceeding centrifugally. These features help to distinguish *Nostoc* from *Anabaena*, whose akinetes are produced from cells adjacent to heterocysts.

Anabaena is primarily planktonic in freshwaters and marine waters such as the Baltic Sea, where some forms can be nuisance bloom-formers, and toxin-producers. Like *Nostoc, Anabaena* consists of filaments of spherical cells resembling closely string beads embedded in mucilaginous matrix. A number of features are used to distinguish the two genera; looser, more indistinct sheaths in *Anabaena*, with less constricted filaments; habitat differences; and differences in the location of akinetes. In addition, hormogonia of *Anabaena* are said to be more consistently motile than those of *Nostoc*, but this characteristic may best observed in cultures. *Anabaena flos- aquae* is one of the more common species, often found in bloom. Several species, such as the common. *A. spiroides* var. *crassa*, are distinctively twisted or helically coiled.

Aphanizomenon consists of aggregations of filaments that can be so large as to be visible with the naked eye. Large blooms may form in the Baltic Sea and in eutrophic fresh waters.

Cylindrospermum occurs in freshwaters including soft, acid lakes, often forming dark green patches on submerged vegetation and on moist soil. Filament of vegetative cells, whose profiles are square or rectangular, are terminated by a basal heterocyst, and enclosed in loose mucilage. A large, highly ornamented akinete may develop from the lowermost vegetative cell.

Cylindrospermopsis is an important nitrogen fixing planktonic bloom former in eutrophic temperate and tropical freshwaters around the world. It is often unrecognized because it does not appear in major taxonomic keys. Rectangular cells occur in linear or coiled filaments having a basal heterocyst. Akinetes form basely, but not immediately adjacent to the heterocyst, as in *Cylindrospermum*. (Chapman and Schelske, 1997).

Tolypothrix is planktonic or found entangled among submergent vegetation in freshwater lakes, including soft or acidic waters. Filaments are enclosed in a sheath of variable consistency and are highly false branched, usually at a heterocyst. Single false branches are formed by the continued growth of the filament on one side of the heterocyst but not the other. Sometimes double false branches occur when both ends of an interrupted filament continue to grow. The frequency of false branched gives these algae a wooly appearance.

Scytonema forms dark tufted mats in masses of other algae or vegetation in lakes of various types or on terrestrial surfaces, including stones, wood, and soil. It is a phycobiont in several orders of lichens (Rambold *et al.*, 1998). Often, but not always, double false branches occurs. This results from the outgrowth from both ends of a filament that has been interrupted by death of a cell or, less commonly, by heterocyst differentiation. Heterocyst walls may be darkly pigmented.

Gloeotrichia occurs in freshwater habitats, attached to submerged substrates. Akinetes occur adjacent to basal heterocyst (sometimes appearing in chains). *Gloeotrichia* colonies often detach from substrates, becoming planktonic.

6. True branched cyanobacteria

Stigonema most commonly inhabits moist rocks and soil, but some species are aquatic, attached to submerged wood or entangled among other algae. It may form brown tufted mats or cushions on submerged portions of lake macrophytes. *Stigonema* is the primary phycobiont in certain lichens (Rambold *et al.* 1998). The filaments are more than one cell in width (multiseriate) and are true-branched. The branches arise by division of cell in a direction perpendicular to that of the main filament axis. The branches may also be pluriseriate. *Fischerella* occurs in the same habitats as *Stigonema*, and these two taxa are very difficult to distinguish. One difference is that the branches of *Fischerella* are not pluriseriate.

4. CYANOBACTERIAL-PLANT ASSOCIATIONS

A number of liverwort, hornwort, moss, fern, cycad, and angiosperm species are closely associated with nitrogen-fixing cyanobacteria - typically species of *Nostoc* (Dodds *et al.* 1995). The prevalence of cyanobacterial associates within ventral cavities of liverworts and hornworts suggests that early land plants may also have relied upon such associations for a supply of fixed nitrogen. Photosynthesis is primarily the job of the plant partner- the cyanobacteria mainly function in providing fixed nitrogen in exchange for a supply of fixed carbon from the plant.

The aquatic fern *Azolla*, commonly used as a biofertilizer in rice paddies, harbours nitrogen-fixing cyanobacteria (*Anabaena*) in sealed cavities on the undersides of leaves. *Azolla* does not grow on nitrogen-free media without its algal partner, and nitrogen-fixation activity of the alga is as much as 20 times higher within *Azolla* than in free living cyanobacteria. Cycads typically produce aboveground roots (Known as coralloid roots because their branching pattern resembles that of corals) that contain *Nostoc* colonies. Specialised root cells mediate the exchange of nutrients between algae and plant. The tropical flowering plant *Gunnera* harbours *Nostoc* colonies in species mucilage-filled glands at the bases of petioles but how the algae get into the glands is not known (Grilli Caiola, 1992).

5. RELATIONSHIP BETWEEN EVOLUTION AND TAXONOMY

Different evolutionary schemes based on the morphology of the cyanophyceae have been proposed by botanists and are reviewed by Desikachary (1973). Simple morphology was either assumed to be primitive (Progressive evolutions) or derived by simplification from more complex forms. Schwabe (1960) proposed a "reticular system" where different morphological types evolved from one thermophilic ancestor with a complex morphology and were connected through intermediate taxa. On the bacteriological side, the shortcomings of morphological features were so evident that phytogenetic schemes were dismissed as pure speculations (Woese, 1987).

6. CONCLUSIONS

The practice of identification of cyanobacteria is not easy. In countries with high reputation in phycology (Europe, North America, India) many taxa have been identified that belong to the variation range of other species or represent only ecomorphoses, status or developmental stages. On the other hand, in biotypes with greater diversity of species, the convergent, ecologically different but apparently morphologically similar species are sometimes wrongly identified. The assertion of many authors that the majority of cyanophyte species have cosmopolitan distribution arises mainly from misinterpretation.

REFERENCES

Alberte R.S. 1989. Physiological and cellular features of *Prochloron* p. 31-52 In : *Prochloron*. A. Microbial Enigma, (Eds.) R.A. Lewin and L. Cheng. P. 31-52, Chapman & Hall. London, U.K.

Anagnostidis K. and Komarek J. 1985. Modern approach to the classification system of cyanophytes. 1-Introduction. *Algological Studies* 38/39 : 291-302.

Anagnostidis K. and Komarek J. 1988. Modern approach to the classification system of cyanophytes. III-Oscillatoriales. Algological Studies, 50/53 : 327-472.

Anagnostidis, K. and Komarek J. 1990. Modern approach to the classification system of cyanophytes. V-Stigonematales. *Algological Studies* 59 : 1-73.

Borner, E. and Flahaut C. 1886-1888. Revision des Nostocacees heterocystees. *Ann. Sci. Nat. Bot.*, 7 Ser 3 : 323-381; 4 : 343-373; 5 ; 52-129; 7 : 178-262.

Capone D.G., Zehr, J.P., Paerl H.W., Bergman B. and Carpenter E.J. 1997. *Trichodesmium*, a globally significant marine cyanobacterium. *Science* 276 : 1221-1229.

Castenholz R.W. 1992. Species usage, concept and evolution in the cyanobacteria by exchange of genetic material among phyletically related strains. *J. Bacteriol.*, 180 : 3453-3461.

Chapman, A.D. and Schelske C.L. 1997. Recent appearance of *Cylindrospermopsis* (cyanobacteria) in five hypereutrophic Florida lakes, J. Phycol. 33 : 191-195.

Chisholm S.W., Frankel S.I., Goericke R. Olson, R.J., Waterbury J.B., West-Johnson L. and Settler E.R. 1992. *Prochlorococcus marinus* Nov. gen. et. Nov. sp. An oxyphototrophic marine prokaryote containing divinyl chlorophyll b. *Archiv. fur Microbiologie*, 157 : 297-300.

De P.K. 1939. The role of blue green algae in nitrogen fixation in rice fields. *Proc. Roy. Soc.*, London, Series B 127 : 121-139.

Desikachary T.V. 1959. Cyanophyta. Indian Council of Agricultural Research, New Delhi.

Desikachary T.V. 1973. Status of classical taxonomy. In : Carr N.G. and Whitton B.A. (Eds.) The biology of blue green algae, pp. 473-486. Backwell Scientific, Oxford.

Dodds, W.K. Gudder D.A. and Mollenhauer D. 1995. The ecology of *Nostoc*. J. Phycol. 31 : 2-18.

Drouet F. 1986. Revision of the classification of the Oscillatoriaceae. Monograph Acad. Natur Sci. Philadephia. Fulton Press. Lancaster Pennsylvania.

Franche, C. and Cohen-Bazire G. 1987. Evolutionary divergence in the *nif* HDK gene region among nine symbiotic *Anabaena azollae* and between *Anabaena azollae* and some free living heterocytous cyanobacteria. Symbiosis. 3 : 159-178.

Fritsch, F.E. 1945. Structure and reproduction of algae. pp. 2- 939. Univ. Press. Cambridge.

Geitler, L. 1932. Cyanophyceae. In : Rabenhorst's Kryptogamen- flora, Volume 14. Akademische Verlagsgesellschaft, Leipzig.

Giovannoni S.J. Turner S., Olsen G.J. Barns S., Lane D.J. and Pace N.R. 1988. Evolutionary relationships among cyanobacteria and green chloroplasts. *J. Bacteriol.* 170 : 3584-2592.

Gomont M. 1892. Monographie des Oscillartoriales. Ann. Sci. Nat. Bot. 7, Ser 15 : 263-368 : 16 : 91-264 Reprinted 1962 J. Cramer, Weinheim, Germany.

Grilli Caiola M. 1992. Cyanobacterium symbioses with bryophytes and tracheophytes. p. 231-254. In : Algae and Symbioses. (Ed.) Reisser W., Biopress, Bristol, UK.

Hamana, K. and Matsuzaki S. 1992. Polyamines as a chemotaxonomic marker in bacterial systematics. CRC Crit Rev. Microbiol. 18 : 261-283.

Herdman M., Janvier M., Waterbury, J.B. Rippka, R., Stanier T.Y. and Mandel M. 1979. Deoxyribonucleic acid base composition of cyanobacteria. J. Gen. Microbiol. 111 : 63-71.

Hertzberg S., Liaaen-Jensen S. and Siegelman H.W. 1971. The carotenoids of blue green algae. Phytochem. 10 : 3121-3127.

Hodell D.A. and Schelske C.L. 1998. Production, sedimentation and isotopic composition of organic matter in lake Ontario. *Limnol and Oceangr.*, 43 : 200-214.

Holten R.W. 1981. Some possible chemotaxonomic approaches to understanding cyanobacterial taxonomy and phylogeny. *Ann. New York Acad. Sci.* 361 : 397-408.

Kenyon, C.N. 1972. Fatty acid composition of unicellular strains of blue green algae. *J. Bacteriol.*, 109 : 827-834.

Kenyon, C.N. Rippka R., and Stanier R.Y. 1972. Fatty acid composition and physiological properties of some filamentous blue green algae. Arch. Microbiol., 83 : 216-236.

Komarek, J. and Anagnostidis K. 1989. Modern approach to the classification system of cyanophytes 4-Nostocales. Arch. Hydrobiol., 82 : 247-345.

LaRoche J., vander Staay G.W.M., Partensky P., Ducret A., Aebersold R., Li R., Golden S.S., Hiller R.G., Wrench P.M., Larkum A.W.D. and Green B.R. 1996. Independent evolution of the prochlorophyte and green plant chlorophyll a/b light harvesting proteins. *Proc. National Academy Science* 92 : 15244-15248.

Lewin, R.A. and Cheng L. 1989. Collection and handling of *Prochloron* and its hosts. In : Prochloron, a Microbial Enigma (eds.) R.A. Lewin and L. Cheng, Chapman & Hall, London, U.K.

Lindblad P., Haselkorn R., Bergman B. and Nierzwicki-Bauer SA 1989. Comparison of DNA restriction fragment length polymorphism of *Nostoc* strains in and form cycads. *Arch. Microbiol*, 152 : 20-24.

Matthijs, H.C.P., Burger - Wiersma T. and Mur L.R. (1989). A status report on *Prochlorothrix hollandica* - a free living prochlorophyte pp. 83-87. In : Prochloron, (Eds) R.A. Lewin and L. Cheng, Chapman & Hall, London, U.K.

Manzel D, Houmard J. CasteltAM and Tandeau de Marsac N 1990. Highly repeatitive DNA sequences in cyanobacterial genomes. J. Bacteriol. 172 : 2755-2761.

Meeks J.C., Joseph C.M. and Haselkorn R. 1988. Organisation of the nif gene in cyanobacteria in symbiotic association with *Azolla* and *Anthoceros*. Arch. Microbiol. 150 : 61-71.

Moore L.R. Rocap G. and Chisholm S.W. 1988. Physiology and molecular phylogeny of co-existing *Prochlorococcus*, ecotypes. Nature 393 : 464-467.

Murray, R.G.E., Bremner D.J., Colwell R.R., De Vos P., Goodfellow M., Grimont P.A.D., Pfennig N., Stackebrandt E. and Zavarzin G.A. 1990. Report of the adhoc committee on approaches to taxonomy within the proteobacteria. *Int. J. Syst. Bacteriol.*, 40 : 213-215.

Pearl, H.W. 1984. Nitrogen fixation (nitrogenase activity) attributable to a specific *Prochloron* (Prochlorophyta) ascidian association in Palam, Micronesia. Marine biology, 81 : 251-254.

Pearl, H.W., Bland P.T., Bowles N.D. and Haibach M.E. 1985. Adaptation to high intensity, low wavelength light among surface blooms of the cyanobacterium *Microcystis aeruginosa*. Appl. Environ. Microbiol., 49 : 1046-1052.

Parker, D.L., Kumar H.D., Rai L.C. and Singh J.B. 1997. Potassium salt inhibit growth of the cyanobacteria *Microcystis* spp. in pond water and defined media. Implication for control of microcystin producing aquatic blooms. Appl. Environ. Microbiol., 63 : 2324- 2329.

Perry, G.J. Gillan F.T. and Johns R.B. 1978. Lipid composition of prochlorophyte. *J.*

phycol., 14 : 369-371.

Plazinski J., Zheng Q., Taylor R., Croft L., Rolfe B.G. and Gunning B.E.S. 1990. DNA probes show genetic variation in cyanobacterial symbionts of the *Azolla* fern and a closer relationship to free living *Nostoc* strains than to free living *Anabaena* strains. *Appl. Environ Microbiol.*, 56 : 1263-1270.

Plazinski, J., Croft L., Taylor R., Zheng Q., Rolfe B.G. and Gunning B.E.S. 1991. Indigenous plasmids in *Anabaena azollae* : their taxonomic distribution and existence of regions of homology with symbiotic genes of *Rhizobium. Can. J. Microbiol*, 37 : 171-181.

Rambold, G., Friedl T. and Beck A. 1998. Photobionts in Lichens : Possible indicators of phytogenetic relationships. *The Bryologist*, 101 : 392-397.

Rippka, R. and Herdman M. 1992. Pasteur Culture Collection of Cyanobacteria. Catalogue and Taxonomic Handbook. Vol. I. Catalogue of strains 103 pp. Institute Pasteur Paris.

Rippka R., Deruelles J.B., Waterbury J.B., Herdman M. and Stanier R.Y. 1979. Generic assignments, strain histories and properties of pure cultures of cyanobacteria. *J. Gen. Microbiol.*, 111 : 1-61.

Rudi, K., Skulberg O.M. larsen F. and Jakobson K.S. 1997. Strain characterization and classfication of oxyphotobacteria in clone cultures on the basis of 16S rRNA sequences from the variable regions V6, V7, V8. Appl. Environ. Microbiol. 63 : 2593-2599.

Schenk, H.E.A. and Kuhfittig G. 1983. Elektromorphe. Phycochromoprotein - Muster : Eine Taxonomische identifikations - methode fur Cyanobakterien-Species. Biochem. System Ecol. 11 : 163-174.

Scheifer, K.H. and Stackebrandt E. 1983. Molecular systematics of prokaryotes. *Ann. Rev. Microbiol.*, 37 : 143-187.

Schopf J.W. and Packer B.M. 1987. Early archean (3.3 billion to 3.5 billion years old) microfossils from Warsanoona groups, Australia, Science, 237 : 70-73.

Schopf J.W. and Walter M.R. 1982. Origin and early evolution of cyanobacteria. The geological evidence. In : N.G. Carr and B.A. Whitton (eds) The Biology of cyanobacteria. pp. 543-564. Blackwell, Oxford and University of California Press. Berkley.

Schewabe, G.H. 1960. Uber den thermobionten kosmopoliten Mastigocladus laminosus cohn. Schweiz. Z. Hydrol. 22 : 757-792.

Stal, L.J. 1995. Physiological ecology of cyanobacteria in microbial mats and other communities. New Phytologist, 131 : 1- 32.

Stam W.T. 1980. Relationships between a number of filamentous blue green algal strains (Cyanophyceae) revealed by DNA-DNA hybridization. Arch. Hydorbiol., Suppl. 56, Algological Studies 25 : 351-374.

Tandeau de Marsac N. and Houmard J. 1987. Advances in cyanobacterial molecular genetics. In : P. Fay and C. Van Baalen (eds.) The Cyanobacteria, pp. 251-302. Elsevier, Amsterdam.

Thompson, J.B., Schultze-lam S., Beveridge T.J. and DesMarais D.J. 1997. Whiting events : Biogenic origin due to photo- synthetic activity of cyanobacterial picoplankton. *Limnol Oceanography, 42 : 133-141.*

Turner S., Burger-Wiersma T., Giovannoni S.J., Mur L.R. and Pace N.R. 1989. The relationship of a prochlorophyte, *Prochlorothrix hollandica,* to green chloroplasts.

Nature, 337 : 380-382.

Wayne, L.G., Brenner D.J., Colwell R.R., Grimont P.A.D., Kandler O., Krichevsky M.I., Moore L.H. Moore W.E.C, Murray R.G.E., Stackebrandt E., Starr M.P. and Triiper H.G. 1987. Report of the adhoc committee on reconciliation of approaches to bacterial systematics. *Int. J. Syst. Bacteriol.*, 37 : 463-464.

Wilmotte A. 1994. Molecular evolution and taxonomy of the cyanobacteria. In : Bryant DA (ed.) The Molecular Biology of Cyanobacteria, pp 1-25. Kluwer Academic Publishers, Amsterdam, Netherlands.

Wilmotte, A. and Golubic S. 1991. Morphological and genetic criteria in the taxonomy of cyanophyta / Cyanobacteria. *Arch. Hydrobiol. Suppl.* 92 : Algological Studies, 64 : 1-24.

Woese C.R. 1987. Bacterial evolution. *Microbiol. Rev.* 51 : 221-271.

Wood, A.M., and Townsend D. 1990. DNA polymorphism within the WH 7803 serogroup of marine *Synechococcus* spp. (cyanobacteria). J. Phycol. 26 : 576-585.

Zimmerman, W.J. and Bergman B. 1990. The *Gunnera* symbiosis DNA restriction fragment length polymorphism and protein comparisons of Nostoc symbionts. Microb. Ecol. 19 : 291-302.

CHAPTER – 23

CITRIC ACID FERMENTATION — A CURRENT PERSPECTIVE

Sanjay Gupta & Chandra B. Sharma*

Department of Microbiology, SBS (PG) Institute of Biomedical Sciences & Research, Balawala, Dehradun - 248 161, INDIA

1. INTRODUCTION

Citric acid is one of the bulk chemicals mainly used by the food and pharmaceutical industries. In nature it is formed as an intermediate in tricarboxy acid cycle. Citrus fruit juice is a rich natural source for citric acid, and in early years of the twentieth century citric acid was commercially isolated from lemon juice. By the middle of the 20th century, fermentation process for citric acid production on a commercial scale was developed using sugar beet molasses and hydrocarbon as substrates and fungal cells as microbes. *Aspergillus niger* cells were extensively used for citric acid fermentation from sugar beet molasses. However, the fermentation technology for citric acid has to date been a well guarded secret. Currie (1917) conducted fundamental investigations on the acid producing properties of *Aspergilli* and opened the way for the industrial production of citric acid. The recent developments on citric acid fermentation technology have been reviewed by Mattey (1992), Kubicek *et al.*, (1994) and Gupta & Sharma (2000).

2. BIOLOGICAL FUNDAMENTALS

Since early studies, it has been obvious that citric acid fermentation is an extremely complex process. A successful process depends on both an appropriate strain and on an optimal fermentation parameter. Citric acid, being a product of the primary metabolism is unlikely to be excreted under natural conditions in noticeable amounts. Thus any appreciable excretion must be regarded as a result of the severe irregularity of metabolism caused by genetic deficiencies or drastic metabolic imbalance. It is for this reason that only few classes or even genera of microorganisms -

bacteria and/or fungi - have been reported to excrete substantial amounts of citric acid into the fermentation medium under certain conditions. Table I enumerates the spectrum of microbial cultures employed for citric acid production. Selection of a particular microbial strain, however, remains a tedious task, especially when commercially competent citric acid yields are to be achieved. *Aspergillus niger* have dominated others in both laboratory and industrial scale production of citric acid. The major advantages in using this organisms for producing citric acid are (1) the ease with which it can be handled (2) the cheap raw materials that it can utilize for citric acid production and (3) high and consistent yields thereby making the process economical.

3. NUTRITIONAL CONSIDERATION FOR THE PRODUCTION OF CITRIC ACID

Citric acid production by *A. niger* depends on a number of specific nutritional conditions i.e. source of carbon supplies, of nitrogen, phosphate, trace metal ions, pH, temperature and DOT (dissolved oxygen

Table I. Spectrum of microbial cultures employed for production of citric acid

Microorganisms		
Fungus	Bacteria	Yeast
Aspergillus niger	*Bacillus licheniformis*	*Candida sp.*
A. awamori	*Bacillus subtilis*	*Debaryomycus sp.*
A. fonsecaeus	*Brevibacterium flavum*	*Endomyces sp.*
A. luchensis	*Arthrobacter paraffines*	*Hansenula sp.*
A. wentii	*Corynebacterium sp.*	*Kloeckera sp.*
A. saitoi	*Aerobacter sp.*	*Nocardia sp.*
A. usami	*Klebsiella sp.*	*Nematospora sp.*
A. fumaricus	*Pseudomonas sp.*	*Pichia sp.*
A. phoenicus	*Micrococcus sp.*	*Rhodotorula sp.*
A. lanosus		*Saccharomyces sp.*
A. flavus		*Sporobolomyces sp.*
Penicillium janthinellum		*Torula sp.*
P. resitricum		*Torulopsis sp.*
Trichoderma viride		*Trichosporon sp.*
Mucor piriformis		*Zygosaccharomyces*
Ustulina vulgaris		
Botrytus sp.		
Ascochyta sp.		
Absidia sp.		
Talaromyces sp.		
Acremonium sp.		
Eupenicillium sp.		

Table II. Conditions favouring citric acid production

High sugar concentration
Low concentrations of phosphate
Low pH, below 2.0
High oxygen tension
Absence of trace metals Mn^{2+}, Cu^{2+}, Fe^{3+}, Zn^{2+}

tension) (Kubicek and Rohr, 1986, Gupta and Sharma, 1994). The medium constituents which have been found to have an effect on citric acid fermentation are listed in Table II.

(a) Carbon, Nitrogen and Phosphate

Nutritional conditions for citric acid production by fungal strain vary from strain to strain also depend on the type of the process. This has been so since a strain that produce citric acid efficiently under surface culture conditions fails to do so under submerged culture conditions. Also, strains that can use one carbon source efficiently fail to show good acid production when cultured in a medium containing another. Because of these conditions, the best substrate for each organism has to be determined.

In general only sugars which are rapidly taken-up by fungus are useful carbon sources for the citric acid production. In most cases sucrose or molasses are used but glucose or fructose have also been found to be applicable. Polysaccharides are not generally applicable since their hydrolysis by the organism seems to be rate limiting for the sugar transport and therefore, the glucose would be taken-up too slowly for an overflow metabolism. The concentration of the sugar in the medium has been shown to be important, and maximum citric acid production is usually achieved at sugar concentrations as high as 14 to 22% (w/v).

Several sources of crude carbohydrates have also been used for industrial production of citric acid. These are beet and sugar cane molasses, unrefined sucrose, cane juice, citrus molasses, starch hydrolysate, hemicellulose and cellulose hydrolysate (Kubicek & Rohr, 1986, Hang & Woodams, 1998; Kirimura *et al.* 1999). From the above substrates only molasses is used widely on commercial scale but it was found that molasses is very complex system for citric acid production by *Aspergillus niger*. There are several inhibitory substances e.g. acetic acid and certain peptides other than metal ions which have to be removed prior to citric acid fermentation (Gupta 1995). The average composition of molasses is given in Table III.

Table III. Composition of sugarcane molasses (Gupta, 1995)

Component	% (w/w)
Total carbohydrates	51.60 ± 4.0
Sucrose	38.24 ± 0.2
Protein	2.1 ± 0.2
Non sugar organic matter	11.0 ± 0.5
Sulfated ash	13.86 ± 3.2
Sodium	0.2 ± 0.1
Potassium	2.5202 ± 0.50
Iron	0.0234 ± 0.0005
Copper	0.0015 ± 0.0003
Zinc	0.0018 ± 0.0003
Calcium	0.7740 ± 0.20
Phosphorus	1.4860 ± 0.35
Manganese	0.007 ± 0.0001
Water	27.47 ± 4.0

*Values are average of three determinations of the pooled samples from different batch preparations.

Conventionally, nitrogen is supplied in the form of ammonium sulfate or nitrate. Physiologically the ammonium compounds are generally preferred since during their consumption the pH decreases, which is a prerequisite for citric acid fermentation. In general, a high concentration of nitrogen leads to greater vegetative growth and delays the onset of the production phase. It is therefore necessary to correctly determine the nitrogen source and concentration essential for the maximal citric acid production by different fungal strains under different fermentation conditions. In general the concentration of ammonium ions during citric acid fermentation should not be limiting and thus varies over a wide (0.3 - 1.5g $NH_{4+}^{/L}$).

Xu *et al.*, (1989) found that urea was the best nitrogen source followed by yeast extract, peptones, ammonium sulfate, malt extract and sodium or ammonium nitrate.

Phosphate is an essential requirement for the fungal growth. The concentration of phosphate in the fermentation medium has a profound effect on the amount of citric acid produced. In order to achieve citric acid accumulation the fungal growth must be restricted. Firstly Perquin in 1938 showed the citric acid overflow due to the phosphate deficiency in the medium. Later, Szucs (1944) reported that in the presence of assimilable phosphorous, submerged citric acid fermentation is not possible. Omar *et al.*, (1992) and Pappagianni *et al.*, (1999b) reported the cultivation of *Aspergillus niger* for citric acid production under phosphate limitations. Although there are not too many reports on this aspect, in general a phosphate concentration of about 0.1% - 0.2% in the fermentation medium appears to be adequate. The combined effect of NH_4NO_3 and K_2HPO_4 on citric acid production is shown in Table IV.

Table IV. Effect of NH_4NO_3 and K_2HPO_4 on citric acid production

NH_4NO_3 % (w/v)	K_2HPO_4 (w/v) 0.05	0.10	0.20
	Citric acid (KgM^{-3})		
0.20	85.65	96.45	89.41
0.25	88.40	98.46	88.45
0.30	93.64	104.65	100.36
0.35	91.44	94.64	94.28

(b) pH

The maintenance of low pH value is extremely important for the successful progress of fermentation. In general a pH values less than 3, citric acid is major fermentation product whereas at the higher pH values substantial amount of gluconic acid and oxalic acid may also be produced (Roukas & Harvey 1988). Most of the total citric acid produced by *Aspergillus niger* during a fermentation period is excreted not during the exponential growth phase but during a secondary production phase which is characterized by the low extra cellular pH and reduced growth rate (Wallrath *et al.* 1992; Pappgianni *et al.*, 1999b). A low initial pH has the main advantage of preventing contamination, suppressing oxalic acid formation and making the sterilization operation more efficient. Inorganic acids such as HCI or H_2SO_4 and NaOH are used to adjust the pH of the fermentation media.

(c) Aeration

The citric acid fermentation is essentially an aerobic process and the organism needs an abundant supply of oxygen beyond that required for growth. The organism depends on the terminal oxidation of the reduced coenzymes of the respiratory chain by oxygen. Therefore it is absolutely necessary to ensure an adequate supply of oxygen during the fermentation. Three ways in which aeration can be ensured are : (1) surface cultivation where oxygen diffuses through the surface which is maintained in a static conditions (2) by submerged aeration in which sterile air or oxygen pumped from beneath the surface of the culture (3) shake culture, which is a process of both surface and submerged aeration. This process is however not feasible on an industrial scale. Industrial production of citric acid by *Aspergillus niger* is susceptible to regulation by dissolved oxygen tension (DOT) of the medium. Its production is stimulated by increase in aeration. It can further be increased by supplementing pure oxygen instead of air (Krishansen & Sinclair, 1978). It was further reported by Olsvik & Krishtiansen (1992) that if the air supply is stopped for a few minutes the citric acid production is irreversibly retarded and does not completely recover even after the resumption of aeration. Further Gupta & Sharma, (1994) and Sakurai *et al.* (1997) reported an increase in productivity of citric acid in the presence of increase in dissolved oxygen tension (DOT) in a fermenter.

(d) Trace metals

Trace metals are very important for citric acid accumulation due to their opposite effect at low and high concentrations. It was established that growth requirements of *A. niger* include carbon, hydrogen, oxygen, nitrogen, potassium, phosphorous, magnesium and sulphur. However, citric acid production is very sensitive to the concentration of these metals in the fermentation media. Infact, successful citric acid production depends to a great extent on the control of the concentration of trace elements. The importance to the citric acid fermentation of the correct levels in the medium of the trace elements zinc, iron, copper and manganese has long been appreciated. Tomlinson *et al.*, (1950) reported that in surface fermentation iron, zinc, manganese and copper in distinctly low concentration are all essential for citric acid accumulation. This optima varied from strain to strain. Noyes (1969) reported the possibility of eliminating the negative effect of Fe^{3+} with high concentration of Cu^{2+}. The critical nature of Mn^{2+} was also reported by Clark *et al.* (1966) and Sanchez-Marroquin *et al.* (1970). They further reported that manganese have the most inhibitory influence on citric acid accumulation (Fig. 1). The physiological and metabolic changes in *Aspergillus niger* grown in citric acid producing medium in presence of different heavy metal ions i.e. Zn^{2+}, Cu^{2+}, Mn^{2+}, Fe^{3+}, were also worked out (Jernjec *et al.* 1990; Benuzzi & Segonia, 1997; Lerra & Callieri, 1999; Pappaginanni *et al.* 1999b). In general, Mn^{2+} deficiency reduces the activity of pentose phosphate pathway, the activity of tricarboxylic acid cycle enzymes, some impairment of anabolism, definite changes in cell wall composition and cell morphology were reported indicating possible alteration in membrane integrity leading to the inhibition of citric acid production.

Other trace elements such Al^{3+}, Ba^{2+}, Co^{2+}, Ni^{2+} etc, have been reported to have an effect on fungal morphology and citric acid production (Kumamoto & Kamura, 1972; Banik & Ghosh, 1998). The exact role of some of these metals in citric acid production is not known. Table V show the effect of metal ions on citric acid production from sugarcane molasses by fermentation with *Aspergillus niger* KCU520 (Gupta 1995).

Table V. Effect of metal ions on citric acid production from sugarcane molasses by fermentation with *Aspergillus niger* (KCU 520)

Metal Ions (mg/L)	Citric acid production (KgM^{-3})	Percent of Control
Control	98.90	100.00
Mn^{2+} (1.0)	21.75	21.99
Fe^{3+} (10.0)	92.15	93.17
Cu^{2+} (15.0)	95.75	96.81
Zn^{2+} (1.0)	76.23	77.08
Mn^{2+} (1.0) + Cu^{2+} (5.0)	72.44	73.24
Mn^{2+} (1.0) + Cu^{2+} (10.0)	91.46	92.47
Mn^{2+} (1.0) + Cu^{2+} (15.0)	88.40	89.38
Mn^{2+} (1.0) + Fe^{3+} (8.0)	86.14	87.10
Mn^{2+} (1.0) + Fe^{3+} (10.0)	86.50	87.50
Mn^{2+} (1.0) + Fe^{3+} (12.0)	76.35	77.20
Mn^{2+} (1.0) + Zn^{2+} (0.5)	22.50	23.53
Mn^{2+} (1.0) + Zn^{2+} (1.0)	20.10	20.32
Al^{3+} (5.0)	82.56	83.47
Co^{2+} (5.0)	80.65	81.54
Ni^{2+} (5.0)	76.42	77.26

(e) Additives and stimulants

Effect of various regulators on citric acid production has been reported in literature. Millis *et al.* (1963) established

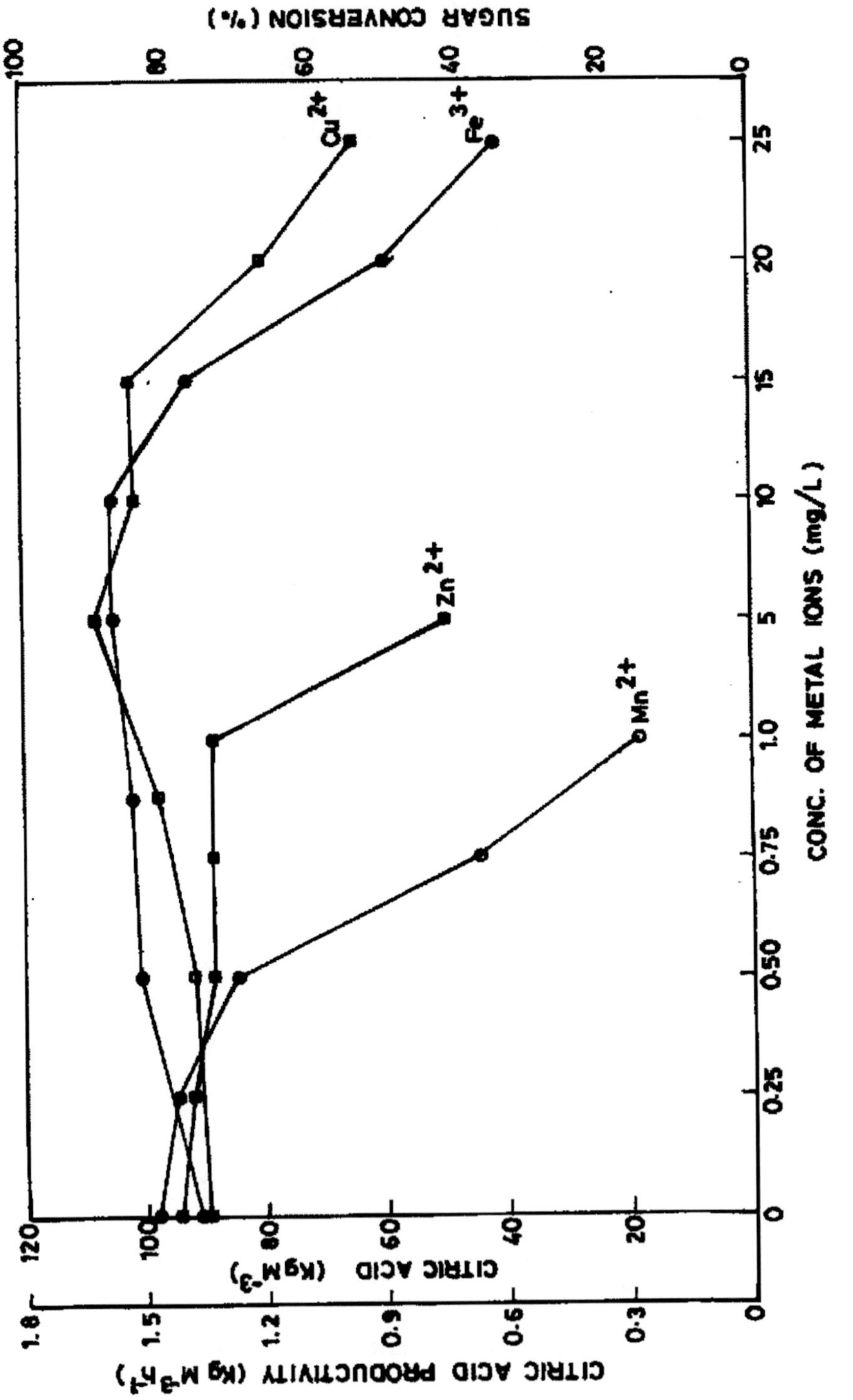

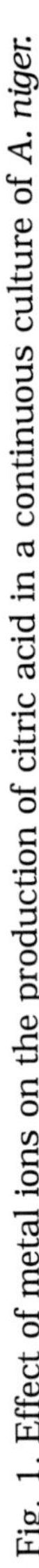

Fig. 1. Effect of metal ions on the production of citric acid in a continuous culture of *A. niger.*

that fatty materials, especially those containing a high proportion of unsaturated fatty acids, greatly enhanced the citric acid yield. Gold & Kieber (1967) reported that use of small quantities of starch is beneficial for citric acid fermentation, Bruchmann (1966) reported that by addition of small quantities of H_2O_2 in fermentation medium a high citric acid yield was achieved. However, in disagreement to the enhancement of citric acid fermentation by the addition of vegetable oils, starch or H_2O_2 (Rohr *et al.* 1983), the citric acid production by *A. niger* KCU 520 was only marginally (10-15%) increased (Garg, 1990). Lipids may provide additional acetyl-Coa through beta-oxidation, for citric acid production. Further, lipids alter the surface tension of the medium and thus improve the aeration efficiency. The other widely applicable theory for lipid action on citric acid fermentation is that unsaturated fatty acids may serve as alternate hydrogen receptor during citric acid fermentation, thus allowing moulds to metabolize actively for longer periods of time thereby improving the citric acid yield (Millis *et al.* 1963). Garg (1990) reported in agreement to this theory, by addition of lipids the cells were found to be fairly active, to produce citric acid for a longer period (Table VI).

Small amounts of (0.1%) H_2O_2 in the fermentation medium also enhanced the citric acid yield appreciably, while higher concentrations were found to be inhibitory. Earlier explanation to this effect was given as inhibition of aconitate hydratase enzyme by H_2O_2 (Bruchmann, 1966). But later this explanation was found to be incorrect as the aconitate hydratase was active during fermentation. An alternative explanation may be that H_2O_2 may provide more dissolved oxygen to the medium, a favourable condition for citric acid accumulation (Rohr *et al.* 1983; Garg 1990). The retardation of fungal growth by H_2O_2 at higher concentration (1% or above) might be due to the conversion of sugar to sugar acids. These sugar acids not only lower the pH of the medium but also are inhibitory to citric acid fermentation (Rohr *et al.* 1983).

Table VI. Effect of various regulators on citric acid fermentation by *A. niger* KCU 520 (Garg, 1990)

Regulator	Optimum concen-tration (%)	Yield of citric acid (%) surface culture[1]	Submerged culture[2]
Control	—	82.3	69.0
Groundnut oil	2.0	94.2	86.8
Hydrogen peroxide	0.1	93.8	84.6
Starch	0.3	89.7	79.1
Control[3] without extra air supply	—	39.0	31.2

1. Sterlized air was circulated over the surface of the cell mass cake at a rate of 2 lit/min.
2. Aeration was provided by shaking the culture flask in an orbital shaker at 150 rpm
3. Extra air supply was stopped by eliminating any aeration facility in surface culture and plugging the culture flasks with stoppers in case of submerged fermentation.

The effect of other chemicals with no nutritive value in improving citric acid production has also been examined. For example, inhibitors of metabolism such a CaF_2, NaF, KF have been found to accelerate the citric acid production. Similarly, malic hydrazide, under submerged culture conditions, has been found to increase the citric acid production by *A. niger.* Other chemical additives such as aromatic amides, esters of dichloroacetic acid, sodium sulphite and crysllic acids, were found to be protecting against the contamination by other microorganisms. In addition, various complex components such as mycelial digest, Bakers yeast, rice bran etc. were also found to be useful in increasing citric acid yields (Cf. Garg, 1990; Kubicek *et al.* 1994).

4. BIOCHEMICAL BASIS OF CITRIC ACID ACCUMULATION

The filamentous fungus *Aspergillus niger* is able to accumulate high concentrations of citric acid from hexoses or disaccharides when cultured under a particular set of conditions. During citric acid production, conversion of carbon substrate to citric acid is the main metabolic activity. This semi steady-state situation is particularly attractive to investigate the flux control in the metabolic pathway. The effect of a certain change in enzyme activity on the flux can be determined by measuring substrate consumption and product formation.

The metabolic pathway for the biosynthesis of citric acid from hexoses in *A. niger* involved both glycolytic and tricarboxylic acid cycles. However, the percental involvement during the growth of fungus is about 2 :1 in favour of glycolytic pathway and about 4:1 in case of citric acid fermentation, indicating a major role of glycolytic pathway for citric acid accumulation (Fig. 2) (Mattey, 1992; Kubicek *et al.* 1994). Following uptake, hexoses are degraded mainly via glycolysis-yielding pyruvate. Part of this pyruvate is then converted to actyl-CoA, part to oxaloacetate. Finally these two compounds are condensed to citric acid, which is secreted and accumulated in the medium.

The regulation of glycolysis and the events which lead to its stimulation in this fungus have therefore received considerable attention. The presence of an excess of carbon sources enabling fast metabolism was shown to trigger arise in the intracellular concentration of fructose-2, 6- phosphate, a strong activator of glycolysis, suggesting a pivotal role of this compound in *Aspergillus niger* citric acid accumulation. The phosphofructokinase (PFK) is regarded as the key regulatory enzyme for citric acid accumulation by regulating the glycolytic flow in *Aspergillus niger* which is activated by NH_{4+}, AMP and fructose 2, 6-biphosphate and inhibited by citrate and phospho-enol-pyruvate (Ruijter *et al.* 1997). This over expression of enzyme by *Aspergillus niger* - a high citric acid accumulation rate is attributed to an "energy overflow" metabolism, which is related to the development of a high glycolytic flow and citrate accumulation under conditions of supplying high sugar concentration (Kubicek *et al.* 1994, Leangon *et al.* 1999) and limited growth because of reduced respiration.

The biochemical rationale have been proposed regarding the role of TCA cycle with respect to citric acid accumulation. Two key enzymes that have been examined is detail are aconitase and isocitrate dehydrogenase. The activity of these enzymes have been shown to decrease to a very low levels during the period of citric acid accumulation while the activity of the condensing enzyme has been found to increase (Ramakrishnan *et al.*, 1955; LaNauze 1966; Rohr Kubicek 1981; Gupta and Sharma, 1995).

Aconitase, one of the two key enzymes in citric acid production, is sensitive to a high concentration of the metallic ions such as Fe^3. Restricting the activity of this enzyme during production stage appears to be the key to success in citric acid production.

Isocitrate dehydrogenase (ICDH) and Isocitrate lyase (ICL) break down isocitrate acid. ICL has been shown to be absent from *Aspergillus niger* when citric acid was accumulating. The case of ICDH was found to be more complex, there being two enzymes, with NADH and NADPH specifities. *A. niger* has only a low activity of NAD^+ - specific isocitrate dehydrogenase whereas three NADP+ dependent enzymes have been found, two of which being

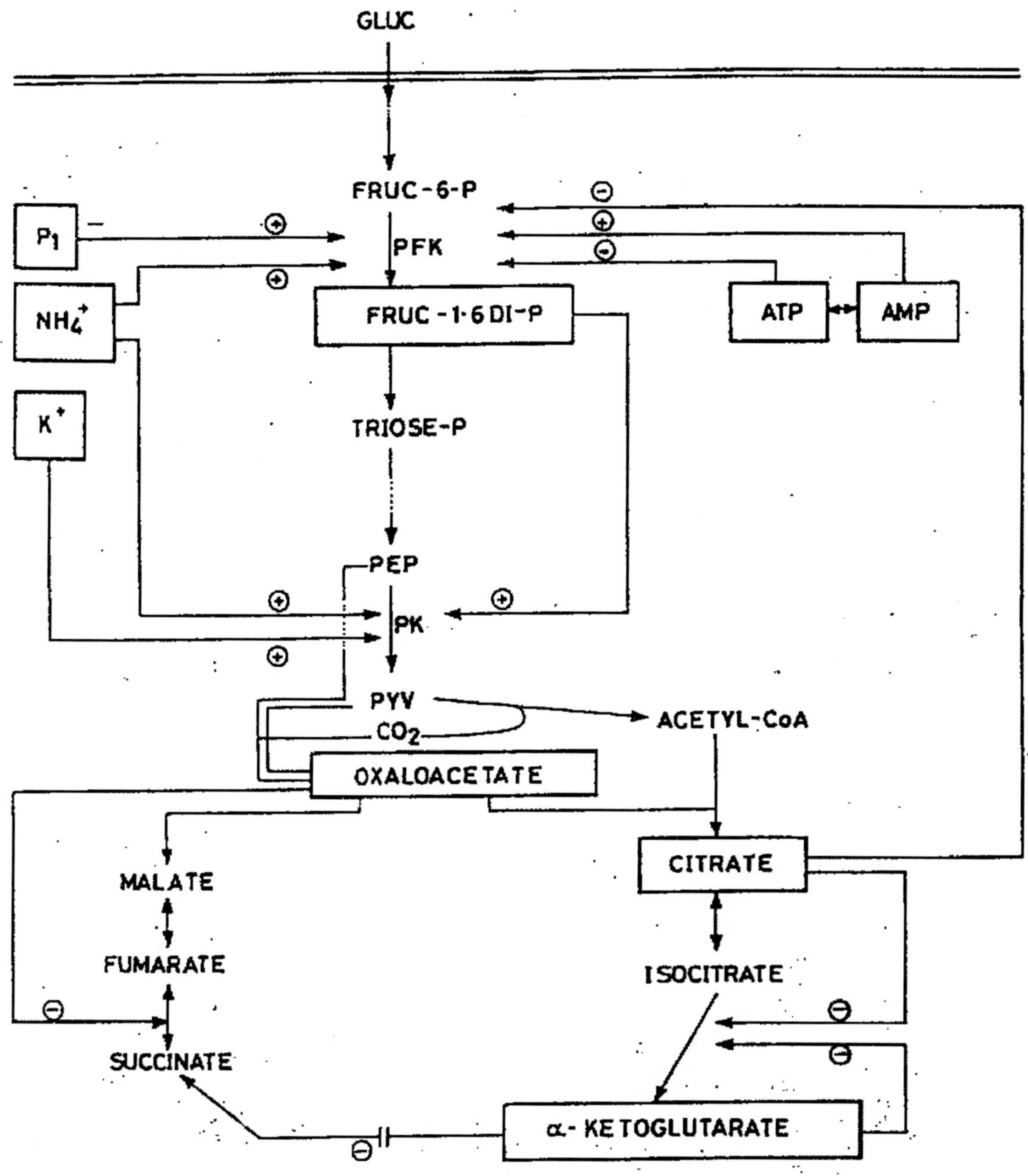

Fig. 2. Metabolic regulation of citric acid accumulation in *Aspergillus niger*. (Rohr *et al.*, 1983). (+) Activation and (–) Inhibition

involved in the TCA cycle. Both mitochondrial NADP+ specific isocitric dehydrogenase are inhibited by physiological concentration of citrate (Mattey, 1992). This could provide a mechanism by which citrate once accumulated to a certain extent could inhibit its own further breakdown and so enhances the citric acid production.

There are certain environmental factors, which are a prerequisite for citric acid accumulation, should exert their influence by stimulating glycolysis or by contributing to most of the influencing factors. This has never been studied on the biochemical basis. Only recently have the action of the manganese and some aspects of regulation by oxygen been studied in more details.

The physiological and metabolic changes in *A. niger* grown in a citric acid producing medium under manganese deficiency are listed in Table VII. Fig. 3 shows the Mn^{2+} ions requiring steps in metabolic cycle of *A. niger*. Various metabolic and physiological alterations occurs due to Mn^{2+} deficiencies. In contrast in the presence of manganese cell growth increases, sugar consumption is diminished and acidogenesis decreases drastically. An investigation of key enzymes of glycolysis, pentose phosphate pathway, TCA cycle, nitrogen metabolism and glycogenesis indicates that manganese deficiency was accompanied by repression of anabolic and TCA cycle enzymes with an exception to citrate synthase (Rohr *et al.* 1983). More importantly Mn^{2+} deficiency disturbs the protein breakdown which results in the elevated intracellular pool of NH_4^+. Since NH_4^+ are potent positive regulator of PFK enzyme, they some what helps in neutralising the effect of citrate inhibition.

Table VII. Physiological alterations in manganese deficient medium

Lower levels of pentose phosphate and tricarboxylic acid cycle enzyme.
Higher level of pyruvate and oxaloacetate.
Elevated levels of amino acids during iodiophase.
Higher pool of ammonia
Low triglyceride and phospholipids level
Elevated cell wall chitin but lower b-glucan and galactose polymers
Lower levels of protein and nucleic acids during idiophase

The critical influence of oxygen has been studied with special reference to the operation of oxidative mechanisms from the equation.

Glucose+3/2 O_2 .. . . . Citric acid + $2H_2O$

It is evident that oxygen is theoretically a substrate which is metabolically needed for the reoxidation of the glycolytic NADH during citric acid fermentation. Also a short interruption in the supply of the oxygen causes a rapid drop in citric acid production without a negative effect on growth-irreversibly inactivates the alternative respiratory branch. The metabolic activity of the fungus found to play a significant role during fermentation. The “catabolic reduction charge” [NADH/(NAD^++ NADH)] was higher during the citric acid accumulation. The anabolic reduction charge [NADPH/{$NADP^+$ + NADPH)], on the other hand remain same. The energy overflow metabolism occur under the conditions of high glycolytic activity and limited growth because of reduced respiration. A decrease in respiration is caused in this system by deficiency of trace elements especially manganese.

From the present state of knowledge the mechanism of citric acid accumulation can be interpreted as follows : A large intracellular pool of NH_4^+ caused by the

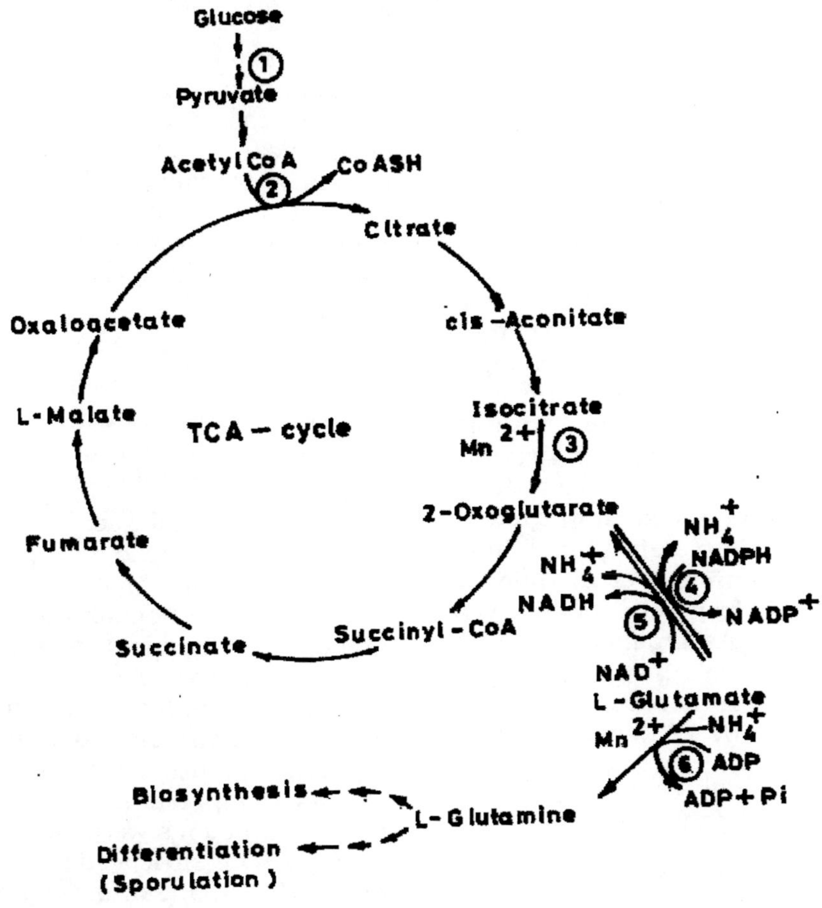

Fig. 3. Metabolic pathway showing Mn^{2+} IONS requiring steps of enzymes of *A. niger* (Cf. Punekar *et al.*, 1983)

negative effect of the severe limitations of manganese on protein turnover and an increased respiratory activity which in part not coupled to ATP synthesis-stimulate metabolic flux through glycolysis without significant metabolic control. This situation together with the constitutive nature of pyruvate carboxylase, the peculiarities in the operation of the TCA cycle and the kinetic properties of the enzyme metabolizing oxaloacetate results in a rise in the cellular concentration of citrate. This in turn stimulates citric acid accumulation by inhabiting isocitric dehydrogenase.

5. COMMERCIAL PRODUCTION PROCEDURES

(a) Production strains

Efficient industrial fermentation require a potent strain. Since citric acid manufactures keep their strains and methods of strain selection, information on the origin and selection programs for industrial strain is not generally available. Most of the wild type *Aspergillus niger* strains are inefficient and as such as unsuitable for commercial citric acid fermentation. Some of the workers obtain certain mutants by exposure of spores to ultraviotet light (Das & Roy, 1981; Garg & Sharma 1991; Gupta & Sharma 1995; Devassy 2000), X-rays (Gardner *et al.* 1956) hertzian waves, radium salts gamma rays (Islam & Begum 1996). Few mutants were found to have improved acid producing capacity by two or more stage mutagenesis and the combined treatment by physical and chemical mutagenic agents (cf. Berry *et al.*, 1977). Polyploidy and heterokinesis parasexual hybridization in *Aspergillus niger* was also used for obtaining active strains for citric acid production. Kirimura *et al.* (1986) developed diploid strains of *Aspergillus niger* by protoplast fusion technique. They have also improved the citric acid yield by haploidization of diploid *Aspergillus niger* (Kirimura *et al.* 1988) and also reported the formation of intergenic fusants between *Aspergillus niger* and *Trichoderma viridae* (Kirimura *et al.*, 1990). More recently they have reported the citric acid production by 2-deoxy glucose resistant mutant strain of *Aspergillus niger* (Kirimura *et al.*, 1992; Sarangabin *et al.*, 1994; Parvez *et al.*, 1998).

(b) Inoculum development

Spore propogation for inoculation is also a very important part of fermentation production strategy. Large numbers of spores can be propogated, either by growing the mould in a liquid sporulation medium or dipersing the spores in the medium or harvesting in dry form.

In submerged fermentation, morphology of the mycelium and also the shape of hyphae of inoculum are important indicators for citric acid fermentation. It was found that hyphae when inoculated to fermenter should be abnormally short, stubby, forked and bulbous (Honicker *et al.*, 1989; Omar *et al.*, 1992; Pappagianni *et al.*, 1999c).

(c) Raw Material

Raw material is a basic need in any fermentation industry. Various cheap carbohydrate sources can be utilized for citric acid production. However it is necessary to evaluate the suitability or raw material as well as the necessary operational parameters in fermentation trials. Raw materials may be classified into two groups :

(1) Raw materials with low ash content i.e. cane or beet sugar, starch hydrolysates.

(2) Raw material with high ash content and high amounts of other non-sugar substance i.e. cane or beet molasses.

However, molasses has found wide spread use for citric acid production. The principal reasons for this are, its low price compared to other sources of sugar and presence of various essential nutrients besides sucrose. These include minerals, organic compounds and vitamins. Mainly, beet and sugarcane molasses are used for citric acid production by *A. niger*. However, molasses represents a complex system. Various problems were faced in its efficient utilization for citric acid fermentation (Rohr *et al.*, 1983). One of the major problems is the variation in its composition and mineral contents. Hence, molasses of different origins give different yield of citric acid. There seems to be involved more than one factor in the regulation of citric acid yield during fermentation. In any case, it would be necessary to evaluate the stability of the stock material and also to standardize the method of pretreatment of molasses, especially the clarification procedure used to reduce or eliminate trace metal concentration in molasses (Garg, 1990; Garg & Sharma, 1991; Gupta 1995).

(d) Clarification of substrate

The problem of clarification of carbohydrate source is very old.

Hexacyanoferrate (HCF) was first used as a means of precipitating unwanted metals in molasses solution (cf. Rohr *et al.*, 1983). Clark *et al.*, (1966) had done a complete study on the effect of HCF on molasses medium. They showed that by HCF treatment most of metals, known to interfere with citric acid production, were efficiently precipitated without affecting nitrogen, carbon and phosphorous contents of the medium. Table VIII shows the precipitation of Cu^{2+} Fe^{3+} Zn^{2+} and Mn^{2+} from sugarcane molasses by HCF treatment (Garg 1990). Several other methods were used to remove contaminating metallic ions from the fermentation media including treatment with aluminum hydroxide and passage of medium over ion exchange resins. Addition of media to chelating agents or specific precipitants including morphiline, quanternary ammonium or cyclic ammonium compounds, small amount of milk powder, camphor or tannic acid, gelatin, EDTA, activated charcoal, polyethylene amine were also found to give higher yield of citric acid.

The effect of lower alcohols, especially methanol, on citric acid production is of great interest. It is most widely used clarification method with the exception of Cu addition and HCF precipitation. Addition of various lower alcohols in small concentration reversed the negative effect of trace metals on citric acid fermentation (Rohr *et al.*, 1983).

(e) Fermentation process

The development of processes for citric acid fermentation can be divided in to three phases. In the first phase, citric acid production was confined to stationary or surface culture conditions. The second phase, consisted of the development of submerged fermentation process for citric acid production. The third phase which is of recent origin, involves the development of solid-state culture, continuous culture and multistage fermentation techniques for citric production. More recently, immobilized biocatalyst technology has rapidly emerged as a novel means to produce citric acid.

(f) Surface fermentation process

In the classical process for the manufacture of citric acid, the culture medium is kept in shallow pans and the fungus develop as a mycelium mat on the surface of the medium. This process is termed as surface or shallow pan process. In this process, the cost of production of citric acid is low. It is still used by many

Table VIII. Precipitation of Cu^{2+}, Fe^{3+}, Zn^{2+} and Mn^{2+} from diluted sugarcane molasses (total carbohydrate 14% w/v) by HCF treatment

HCF Conc.	Metal ions precipitated by HCF							
%	Cu^{2+}	ppm	Fe^{3+}	ppm	Zn2+	ppm	Mn^{2+}	ppm
0.000	0	0.00	0	0.00	0	0.00	0	0.00
0.025	59	1.18	53	37.10	52	1.30	60	0.90
0.050	74	1.48	66	46.20	67	1.67	70	1.14
0.075	81	1.62	82	57.40	82	2.05	90	1.35
0.100	88	1.76	96	67.20	95	2.37	98	1.47
0.125	90	1.80	97	67.90	96	2.40	98	1.47
0.150	90	1.80	96	67.20	95	2.37	89	1.45
0.175	87	1.74	95	66.50	96	2.40	98	1.47
0.200	88	1.76	94	65.80	95	2.37	96	1.44

manufacturers. The system consist of a fermenter in which shallow trays are stacked one over the other and an effective air circulation system by which humidified air is circulated in a laminar flow manner. The medium can be inoculated either by liquid spore suspension or by spraying spores through air circulation (Rohr *et al.*, 1983; Kubicek *et al.* 1994).

(g) Submerged Fermentation process

The advantages of the submerged fermentation process over the liquid surface fermentation in terms of investment and operating cost induced investigators to attempt to develop a submerged fermentation process for citric acid production. Both conventional stirred reactors and tower fermenters are being used for this process. Behaviours of inoculum pellet formation is an important factor which governs acid yield. According to Kristiansen & Charley (1981) batch submerged fermentation after inoculation with *Aspergillus niger* leads to continuous fermentation system. Further support for the improved production in submerged fermentation were reported by Gupta & Sharma (1995); and Pappagianni *et al.* (1999).

(h) Solid state Fermentation process

The simplest process for citric acid fermentation has been developed in Japan. It largely resembles the solid-state fermentation based on the know-how of the koji process. The raw materials are starch containing residual pulps from starch manufactures, eg. sweet potato, fibrous residues, rice or wheat bran and others. Further, it was showed by Chaudhary *et al.* (1978) that can or beet pulp was impregnated with sugar solution like sucrose or molasses. Garg & Sharma (1991); Lu *et al.*, (1997); Pintadu *et al.*, (1998); Roukas (1998), reported higher yields of citric acid by solid state fermentation techniques than by surface and submerged fermentation using sugar cane baggase, carod pod pulp, cellulosic material as solid support.

(i) Continuous fermentation process

The fermentation industry appears to have been fairly reluctant to adopt continuous culture as a production technique despite the claims that it has many advantages over batch culture. Continuous culture was superior to batch culture in production. Attempts to produce citirc acid in a continuous culture have however been explored. Kristiansen and Charley (1981) reported the production of citric acid by *Aspergillus niger* in a single stage continuous culture under nitrogen limiting conditions. In recent years continuous citric acid production has also been reported on a laboratory scale with *Aspergillus niger* cells immobilized calcium alginate gel (Honecker *et al.* 1989; Gupta & Sharma, 1994), polyacrylamide gel (Horitsu *et al.* 1985; Garg & Sharma, 1992) and on polyurethane foam (Lee *et al.* 1989) . In addition Kautola *et al.* (1991) and Rymowiczs *et al.*, (1993) have used alginate immobilized Yarrowia lipolytica yeast in a continuous air-life bioreactor. Unfortunately the citric acid productivity in this processes has been low varying between 0.12 and 0.95 $KgM^{-5}h^{-1}$. Gupta & Sharma (1994, 1995, 2000) reported continuous production of citric acid using a combination of submerged immobilized and surface stabilized cultures of *Aspergillus niger*, optimizing performance parameters (Table IX) using unclarified molasses as carbon source. The continuous process shows the culture stability over a period of 30 days with an increased productivity of 1.70 $KgM^{-3}h^{-1}$ corresponding nearly to 90% conversion of molasses sugars to citric acid, which in comparison to batch submerged and batch surface was very high (Table X).

Table IX. Optimum performance parameters for the continuous citric acid production by fermentation with Aspergillus niger GS-III using unclarified sugarcane molasses as carbon source

Parameter	Unclarified molasses
Dilution rate	0.015 h^{-1}
Flow rate	6 ml/h
Sugar concentration	12% (w/v)
Temperature	32°C
pH	4.5
Inoclum size	10% (w/v)
Spores per bead	2 x 10^5
Bead size	7 mm
Acclimatization period	6 days
Medium depth	20 mm
Aeration rate	3.01/mm
NH_4NO_3 concentration	0.25% (w/v)
K_2HPO_4 concentration	0.10 (w/v)
$MgSO_4$ $7H_2O$ concentration	0.03% (w/v)
Fe^{3+} tolerance	25 mg/L
Cu^{2+} tolerance	18 mg/L
Zn^{2+} tolerance	10 mg/L
Mn^{2+} tolerance	1.5 mg/L

Table X. Comparison of various Fermentation systems used for citric acid production by *A. niger* GS-III

Fermentation system	Period (days)	Total citric acid (KgM^3)	Productivity (KgM^{-3} h^{-1})	Sugar Conv. (%)
Submerged batch culture	14	78 ± 9.0	0.229	60 ± 5.0
Surface batch culture	12	90 ± 9.0	0.313	70 ± 5.0
Calcium-alginate immobilized multiple cycle	7X3*	180 ± 12.5	0.354	60 ± 5.0
Solid state surface multiple cycle	5X6*	447 ± 22.0	0.620	62.6 ± 6.0
Continuous culture	30	1232 ± 32.0	1.717	90.2 ± 3.5

*Shows number of cycles operated with same culture

(j) Immobilized cell fermentation process

Recently immobilized biocatalyst technology has emerged as a novel means to utilize enzymes as efficient and heterogeneous catalysts for the multitude of industrial analysis and medical applications. Physical entrapment in porous polymeric carriers is by far the most widely used technique for whole cell immobilization. The first successful industrial application for the immobilized cell process was done by Chibata *et al.*, (1974), while this technology has been replaced by other convenient methods. It is still preferred method for initial laboratory scale studies. There are only few reports regarding citric acid production from the immobilized cells in polycrylamide gel (Horitsu *et al.*, 1985; Garg and Sharma, 1992). The application of the alginate for the purpose of whole cell immobilization was also reported. The immobilization of living microbes in Ca-alginate is now widely used, since the method is very simple and also the gel material is non-toxic and cheap (Kautola *et al.* 1991; Garg & Sharma, 1992; Rymowics *et al.* 1993; Gupta & Sharma 1994, 2000). Other methods of immobilization are also reported for citric acid production by different workers i.e. agarose (Khare *et al.*, 1994). Polyurethane foam (Vassilev & Vassileva, 1990), porous cellulose beads (Sakurai *et al.*, 1997).

6. SUMMARY AND CONCLUSION

The need for citric acid has been steadily increasing because of its use in industries other than food, there has been significant improvement in past two decades in producing this organic acid using *Aspergilli* and other fungi. Although the methodology of culturing this organism has not undergone any drastic change, the exploitation of solid-substrate fermentation immobilized cell fermentation and continuous fermentation processes had considerably improved the productivity, product accumulation and yield upto 1.7 - 1.8 $KgM^{-3}h^{-1}$, 110-120 KgM^{-3} and 90% respectively (Gupta & Sharma, 2000). Further the cultures were used continuously for 4-5 weeks without any noticeable loss in the productivity of citric acid improving the culture stability.

One improvement that is wanting is the development of cultures that will produce citric acid in unclarified molasses medium in a shorter period of time. The first report in this line by Gupta & Sharma (1995), developing a new mutant strain of *Aspergillus niger* GS-III, showing resistance to manganese ions inhibition of citric acid production, opens a new line of development of mutants resistant to metal ions inhibition of citric acid production. Further development, stabilization culture stability of these mutants (Gupta & Sharma, 2000; Devassy, 2000) can be exploited at industrial level, to produce citric acid from crude raw materials without any pretreatment in a shorter time without loss in efficiency.

REFERENCES

Benuzzi, D.A. & Segonia, R.F. 1997. Effect of copper concentration on citric acid productivity by *Aspergillus niger* strain. Appl. Biochem. Biotechnol, 61 : 393-417.

Berry, D.R., Chmiel, A. & Al-Obaidi, Z. 1977. Citric acid production by *Aspergillus niger*. In : Genetics & Physiology of Aspergillus (J.E. Smith & J.A. Pateman eds). Academic Press, London pp 405-426.

Bruchmann, E.E. 1966. Action of hydrogen peroxide on the accumulation of citric acid by *Aspergillus niger*. Naturwissenschaften, 53 : 226-227.

Chaudhary K., Lakshminarayan K., Ethiraj, S. & Tauro, P. 1978. Citric acid production by *Aspergillus niger* from Indian cane molasses by solid state Fermentation. J. Ferment. Technol. Jpn. 56 : 554-557.

Chibata, I. Tosa T. & Sato, T. 1974. Immobilized aspartase containing microbial cells. Preparation and enzymes properties. Appl. Biochem. Biotechnol., 27 : 878-885.

Clark, D.S., Ito, K. & Horitsu, H. 1966. Effect of manganese and other heavy metals on submerged citric acid fermentation of molasses. Biotechnol, Bioeng. 8 : 465-471.

Das, A. & Roy, P. 1981. Rapid strain selection for citric acid production. In : Advances in Biotechnology (M. Mooyoung, ed) Vol. 1, pergamon Press. Toronto, pp 51-55.

Devassy, B. 2000. Isolation of a high citric acid yielding mutant strain of *Aspergillus niger* resistant to metal ions inhibition. M.Sc. Thesis. C.C.S. Univ. Meerut, India.

Gardner, J.F., James L.V. & Rubbo, S.D. 1956. Production of citric acid by the mutants of *Aspergillus niger*. J. Gen. Microbiol., 14 : 228-237.

Garg, K. 1990. Bioconversion of molasses to citric acid. Ph.D. Thesis. University of Roorkee, Roorkee, India.

Garg, K & Sharma C.B. 1992. Continuous Production of citric by immobilized whole cells of *Aspergillus niger* J. Gen. Appl. Microbiol. 38 : 605-615.

Garg K. & Sharma, C.B. 1991. Repeated batch production of citric acid from sugarcane molasses using recylced solid-state surface culture of *Aspergillus niger*. Biotechno Letts. 13 : 913-916.

Ghosh, P. & Banik. A.K. 1998. Effect of chemical nutrients on aconitase activity during citric acid fermentation by a mutant strain of *Aspergillus niger*. Acta Microbiologica Polonica, 47 (3) 253-260.

Gold, W. & Kieber R.J. 1967. US Patent 3326774.

Gupta, S. & Sharma, C.B. 1994. Continuous production of citric acid from sugarcane molasses using a combination of submerged immobilized and surface stabilized cultures of *Aspergillus niger* KCU520. *Biotechnol. Letts.* 16, 599-604.

Gupta S. 1995. Bioprocess development for citric acid production. Ph.D. Thesis, University of Roorkee, Roorkee, India.

Gupta, S. & Sharma C.B. 1995. Citric Acid fermentation by the mutant strain of the *Aspergillus niger* resistant to manganese ions inhibition. *Biotechnol. Letts.* 17 : 269-274.

Gupta, S. & Sharma C.B. 2000. A high performance fermentation process for citric acid production from sugarcane molasses by *Aspergillus niger*. Biochem. - Environ. Agric. (In press)

Hang, Y.D. & Woodams, E.E. 1998. Production of citric acid from corn cobs by *Aspergillus niger*. *Bores. Technol.*, 65 : 241- 245.

Honecker, S., Bisping, B., Vang, Z. and Rehm. H.J. 1989. Influence of sucrose concentration and phosphate limitation citric acid production by immobilized cells of *Aspergillus niger*. Appl. Microbial Biotechnol. 31 : 17-24.

Horitsu, H., Adachi, S., Takahashi, Y., Kawai, K. and Kawano, Y. 1985. Production of citric acid by *Aspergillus niger* immobilized in polyacrylamide gel. Appl. Microbial Biotechnol. 22, 8-12.

Islam, M.R. & Begum, R. 1996. Citric acid fermentation by *Aspergillus niger*. Further induction of high yielding mutants with the help of gamma radiation. Bangladehs, J. Microbial. 13, 71-74.

Jernejc, K., Cimmerman, A., Vendramin, M. & Perdih, A. 1990. Lipids of a citric acid producing *Aspergillus niger* strain grown in copper and manganese supplemented media. Appl. Microbial. Biotechnol. 32 : 699-703.

Kautola, H., Rymowicz, W., Linko, Y.Y. & Linko, P. 1991 : Production of citric acid with immobilized yarrowia lipolytica. Appl. Microbiol. Biotechnol. 35 : 447-449.

Khare, S.K. Jha K. & Gandhi A.P. 1994. Use of Agarose entrapped *Aspergillus niger* cells for the production of citric acid from soy whey. Appl. Microbiol. Biotechnol. 38 : 441-448.

Kirimura K., Watanabe, T., Sumagaua, T. & Usami, S. 1999. Citric acid production from xylan & xylan hydrolysate by semi solid culture of *Aspergillus niger*. Biosec. Biotechnol. Biochem. 63 : 226-231.

Kirimura K., Sarangbin, S., Rugoaseel, S. & Usami S. 1992. Citric acid production by 2- deoxyglucose-resistant mutant strains of *Aspergillus niger*. Appl. Microbiol. Biotechnol. 36 : 573-577.

Kirimura, K. Nakajima I., Lee, S.P., Kawabe S. & Usami S. 1988. Citric acid production by the diploid strains of *Aspergillus niger* obtained by protoplast fusion. Appl. Microbiol. Biotechnol. 27 : 504-506.

Kirimura, K. Yaguchi, T. & Usami, S. 1986 : Intraspecific protoplast fusion of citric acid producing strains of *Aspergillus niger*. J. Ferment Technol. 64 : 473-479.

Kritiansen, B. & Charlcy, R. 1981. Continuous process for production of citric acid. In : Advances in Biotechnology (M. Moo-Young ed.) Vol. I, Pergamon Press. Toronto, pp. 221-227.

Kritiansen, B. & Sinclair C.G. 1979. Production of citric acid in continuous culture. Biotechnol. Bioengg. 20 : 1711-1722.

Kubicek, C.P. & Rohr. M. 1986. Citric Acid. CRC Crit. Rev. Biotechnol. 3 : 331-367.

Kubicek, C.P., Witteveen, C.F.B. & Visser, J. 1994. Regulation of organic acid production by *aspergilli*. In : The genus *Aspergillus* (K.E. Powell et al., Eds) pp. 135-144, Plenum Press, New York.

La Nauze, J.M. 1966. Aconitase and isocitric dehydrogenase of *Aspergillus niger* in relation to citric acid production. J. Gen. Microbiol., 44 : 73-81.

Leangon, S., Maddox, I.S. & Brooks, J.D. 1999. Influence of the glycolytic rate on production of citric acid and oxalic acid by *Aspergillus niger* in solid state fermentation. World J. Microbiol. Biotechnol. 15 : 493-495.

Lee, Y.H., Lee C.W. & Chang, H.N. 1989. Citric acid production by *Aspergillus niger* immobilized on polyurethane foam. Appl. Microbiol Biotechnol. 30 : 141-143.

Lu, M.Y., Brooks, J.D. & Maddox, I.S. 1997. Citric acid production by solid state fermentation in a packed bed reactor using *Aspergillus niger*. Enzyme Microbiol. Technol. 21 : 392-405.

Mattey, M. 1992. The production of organic acids. Crit. Rev. Bitotechnol. 12 : 87-122.

Mills, NF, Trumpy B.H. & Palmer B.M. 1963. The effect of lipids on citric acid production by an *Aspergillus niger* mutant. J. Gen. Microbiol. 30 : 365-379.

Noyes, R. 1969. Citric acid production processes. Noyes Develop. Corp. Park Ridge. New Jersey. USA.

Olsvik, E.S. & Kristiansen B. 1992. Influence of oxygen tension, biomass concentration and specific growth rate on the rheological properties of filamentous fermentation broth. Biotechnol. Bioeng. 40 : 1293-1299.

Omar, S.H., Honecker, S. & Rehm. H.J. 1992. A comparative study on the formation of citric acid and polyols and on morphological changes of three strains of free and immobilized *Aspergillus niger*. Appl. Microbiol, Biotechnol, 36 : 518-524.

Pappagianni, M., Mattey, M. & Kritiansen B. 1999a. Influence of glucose concentration on citric acid production and morphology of *Aspergillus niger* in batch culture. Enzyme. Microbiol. Technol. 25 : 710-713.

Pappagianni, M., Mattey, M. & Kritiansen B. 1999b. *Aspergillus niger* morphology and citric acid production in submerged batch fermentation : Effect of culture pH, phosphate and manganese levels. Food Tech. Biotechnol. 37 : 165-180.

Pappagianni, M., Mattey, M. & Kristiansen B. 1999c. Hyphal vacuolation and fragmentation in batch and hot batch culture of *Aspergillus niger* and its relation to citric acid production. Process Biochem. 35 : 359-368.

Parvez, S. Rajoka, M.I. & Ahmed, M.N. 1998. Citric acid production from sugarcane molasses by 2-deoxyglucose resistant mutant strain of *Aspergillus niger*. Folia Microbiologica 43 : 59-62.

Perra, L.M. & Callieri, D.A. 1999. Influence of calcium on fungal growth and citric acid production during fermentation of a sugarcane molasses based medium by a strain of *Aspergillus niger*. World J. Microbiol. Biotechnol. 15 : 647-649.

Pintadu, J., Lonsane, B.K., Roussas, S. & Graim-Perrand, I. 1998. On line monitoring of citric acid production in solid-state culture by respirometry. Process Biochem. 33 : 513-520.

Ramakrishnan, C.V., Steel, R. and Lentz, C.P. 1955. Mechanism of citric acid formation and accumulation in *Aspergillus niger*. Arch. Microbiol. 55 : 270-273.

Rohr, M. and Kubicek, C.P. 1981. Regulatory aspects of citric acid fermentation by *Aspergillus niger*. Process Biochem. 16 : 34- 37.

Rohr, M., Kubicek, C.P. and Kominnek, J. 1983. Citric acid. In : Biotechnology (H.J. Rehm and G. Reed, eds.) Vol. 3, Verlag Chemie, Weinheim pp. 420-454.

Roukas, T. and Harvey, L., 1988. The effect of pH on production of citric and gluconic acid from beet molasses using continuous culture. Biotechnol. Lett. 10 : 289-294.

Rymowicz, W., Kautola, H., Wojtatowicz, M., Linko, Y.Y. and Linko, R. 1993. Studies on citric acid production with immobilized Yarrowia lipolytica in repeated batch and continuous air-lift bioreactors Appl. Microbio. Biotechnol. 39 : 1-4.

Roukas, J. 1998. Citric acid production from carob pod extract by cell recycle of Aspergillus niger ATCC 9142, Enzyme Microbiol Technol. 22 : 91-113.

Ruijter, G.J.G., Panneman, H. & Visser, J. 1997. Overexpression of phosphofructokinase and pyruvate kinase in citrict acid producing *Aspergillus niger*. Biochimica, Biophysica, Acta, 1334, 317-326.

Sakurai, A, Itoh, M. Sakabibura, M., Saito, H. & Fugita M. 1997. Citric acid production by *Aspergillus niger* immobilized on porous cellulose beads. J. Chem. Tech. Biotech. 70 : 157-160.

Sanchez-Marroquin, A. Carreno, R. and Ledezma, M. 1990. Effect of trace elements on citric acid fermentation by *Aspergillus niger*. Appl. Microbiol. 20 : 888-892.

Sarangbin, S., Morikawa, S., Kirimura K. and Usami, S. 1994. Formation of autodiploid strains in *Aspergillus niger* and their application to citric acid production from starch. J. Ferment. Bioengg. 77 (5) : 474-478.

Szucs, J. 1944. Method of producing citric acid by fermentation. US patent 2353771.

Tomlinson, N., Campbell, J.J.R. & Trussel 1950. The influence of zinc, copper, iron and manganese on the production of citric acid by *Aspergillus niger*. J. Bacteriol. 59, 217-227.

Vassilev, N. & Vassileva, M. 1990. Influence of reactor mode on citric acid production by *Aspergillus niger* immobilized on polyurethane foam. In : Bont. Jam de, Visser, J., Mattianson, B. (Eds.) physiology of immobilized cells. Elsevier, Amsterdam pp. 331-334.

Wallrath, J., Schmidt, M & Weiss H. 1992. Correlation between manganese deficency, loss of repiratory chain complex I activity and citric acid production in *Aspergillus niger*. Arch. Microbiol., 158, 435-438.

Xu, D.B., Kubicek C.P. & Rohr, M. 1989. A comparison of factors influencing citric acid production by *Aspergillus niger* grown in submerged culture and on filter paper. Appl. Microbiol. Biotechnol, 30, 444-449.

CHAPTER – 24

MICROBES IN THE MANAGEMENT OF PLANT PARASITIC NEMATODES

P.C. Trivedi

Department of Botany, University of Rajasthan, Jaipur - 302 004, India

1. INTRODUCTION

Indiscriminate use of pesticides for the past few decades has done great harm to human beings, animals, vegetation and to the environment as a whole. Pests, still account for about 35 per cent loss of agricultural yield notwithstanding losses due to heavy application of chemicals. Biological control of pests is one important means for solving prevalent pest problems in both agricultural as well as other environmental situations. Bio-intensive integrated pest management in which, greater emphasis is given to biocontrol, is perhaps the need of the hour for tackling many such situations.

Plant parasitic nematodes form an important niche in agro- ecosystems. They feed on plant organs as ectoparasites, semi-endoparasites and endoparasites and cause reduced plant productivity. About 2,500 species are known to parasitize the lower as well as higher plants at global level. In India, so far, about 600 species belonging to about 85 genera have been either described or reported to be associated with more than 700 plant species growing in diverse agroclimatic zones. Many of them have been proved to be highly pathogenic and cause considerable economic yield losses in cereals, pulses, vegetables, fruits, ornamentals and forest crops. Almost every crop is being parasitized by one or more phytonematode species. Despite this, their role as a limiting factor in agricultural production has not been duly recognised perhaps, due to their microscopic size, subterranean habitat, slow rate of multiplication and above all, the symptoms produced by them are generally not clear and not easy to pinpoint.

2. NEMATODE AND CROP LOSSES

There are numerous estimates of the economic importance of nematodes in crop production on a world-wide and on individual country basis, but precise values cannot be determined. It was because of the nematode's "small size and hidden way of life" and lack of definite information on it's occurrence and pathogenicity. The assessment of the society of Nematologists Committee on crop losses, indicates annual losses in the United States due to plant parasitic nematodes to the tune of $ 1,038,374,000 in field crops, $ 225,145, 900 in fruit and nut crops, $ 266,989,100 in vegetable crops and $ 59,817,634 in ornamental crops (Anon, 1971).

In a world-wide survey conducted by Sasser (1989), the 10 most important genera of plant parasitic nematodes revealed were *Meloidogyne, Pratylenchus, Rotylenchulus* and *Helicotylenchus* This order of importance of the various genera was found to be fairly representative for most regions of the world. The estimated overall annual yield loss of the world's major crops due to damage by plant parasitic nematodes is 12.3 per cent. The 20 crops that represent a miscellaneous groups important for food or export value were reported to have an estimated yield loss of 14 per cent. Losses for the 40 crops in developed countries average 8.8 per cent compared with 14.6 per cent for developing countries. Global crops losses due to nematodes on 21 crops, 15 of which are life-sustaining, were estimated at 77 billion dollars annually based on 1984 production figures and prices. The United States portion, alone is 5.8 billion dollars. These figures are staggering, and the real figure, when all crops throughout the world are considered, probably exceeded 100 billion dollars annually.

Like many other countries, plant parasitic nematodes are significant impediments in agricultural productivity in India. Root-knot nematodes, reniform nematode, ear-cockle nematode and several ectoparasitic nematodes are distributed throughout the country. Though, adequate information from different regions of the country is yet not available or collected, some localized but intense nematode problems like cereal-cyst nematode in Rajasthan on wheat and barley, potato-cyst nematode on potato in Nilgiri hills of Tamil Nadu, burrowing nematode on banana in Karnataka and Kerala and on coconut, arecanut and black pepper in Kerala are of major concern. Yield loss estimates, available in India for a number of crops are not very precise and adequate but available information indicates that nematodes are important factor in low productivity of a number of crops in different regions of the country.

3. BIOLOGICAL SUPPRESSION OF NEMATODES

At the current rate of population growth on this earth, the present population of six billion is expected to become seven billion in just 10 years. The demand for food, shelter and clothing would remain ever-increasing. Increase in agricultural production is expected to be a top priority area in overall development process management of plant pathogens and pests is, therefore, an utmost necessity in order to increase agricultural production. The management approach and strategies should however, be effective, economically free from any kind of pollution hazards and acceptable to the users. Biological control of plant parasitic nematodes has become most significant and intensive area of research of late, because of environmental concerns and non-availability of effective nematicides.

Biological control is defined as the action of parasites, predators or pathogens in maintaining another organism's population density at a lower average than would occur in their absence (Jatala, 1986). Many natural enemies attack nematodes in soil and reduce their population. Exploitation of these enemies for practical use in the management of plant parasitic nematodes at present seems to be practically demanding and relevant approach in view of the greater awareness of pollution free environment. Exploitation of biocontrol agents and their artificial application and marketing for the large scale use of the growers places them under the category of the newly coined word "Biopesticides'. Biopesticides are different from biological control. In biological control, biocontrol agent is released into population to establish itself and control the target pest, but biopesticides refer to microbial pathogens that uniformly kill or suppress the target pest in a manner similar to chemical pesticides. The biopesticides in general show advantages of high degree of specificity on the target pests. They cause no adverse effect on non-target or beneficial organisms. The approach is also free from pest resistance and residue problems. Several predators like : predacious fungi, predacious nematodes and some other predacious organisms such as : tardigrades, turbellerians, enchytraeids, insects, mites, parasites like : viruses, bacteria, fungi and some other organisms have shown some promise as biocontrol agents. They constitute natural biresources for management of plant parasitic nematodes (Khan, 1998).

The current status of development in biological nematode control utilizing a number of soil antagonists has been reviewed extensively (Jatala, 1986; Kerry, 1984; Swarup and Gokte, 1986; Trivedi, 1992, 1997).

Attributes of successful biocontrol agents

An ideal biocontrol agent, which would be successful in practical application is supposed to possess the following attributes —

(i) *Biologically competitive* — Bio-agent should be biologically competitive with other existing or introduced parasites and predators in the host ecosystem.

(ii) *Not to be pathogenic* — Application of an organism to field soils should not be pathogenic to the agricultural crops, human beings and to the other animals.

(iii) *High nematode reducing potential* — A successful biocontrol must suppress nematode population in large quantities so that after a given period, the nematode population is reduced to a level below the damage threshold.

(iv) *Ecological compatibility* — Efficacy of biological suppressor of nematodes may be increased by successful manipulation of the environment. The agent must be capable of adapting to or tolerating varying soil environment and structural conditions, spread enough to locate and kill its prey or to establish itself to retard the activity of the intended prey.

(v) *High reproductive potential* — Without high reproductive potential sufficient infective propagation units are not produced to maintain a high population. There must be a balance in between its reproductive and infective potential. Rapid and numerical increase is most important attributes of a successful biological agents for effective commercial production. Biological agent is one that can be readily and inexpensively mass-produced.

(vi) *Efficient dispersal* — Biological agents should have a high dispersal capability.

(vii) *Synchronized life cycle* — Temporal compatibility between pest and pest suppressors must exist to bring them together contemporaneously.

(viii) *Searching capacity* — Any

organism that can actively search for its prey will be more effective than passive one.

(ix) *Genetic stability* — For long term effectiveness, genetic stability of biocontrol agent is necessary.

(x) *Long term storage* — The infective units or propagules of a suitable biocontrol agent should have the characteristic of withstanding long-term storage, until their use.

(xi) *Resistance and survival capacity* — Resistance and survival capacity must be higher in the biocontrol agents so that they can survive in the extreme soil environments, even in the absence of its host.

(xii) *Capable of producing deleterious compounds* — Biocontrol agent should have the capability of producing antibiotics or other deleterious compounds which will either inhibit nematodes or antagonist produced should be such that increase the chance of the agent in its survival and infection later.

(xiii) *Host specificity* — Monophagous spp. are efficient regulators compared to polyphagous ones.

(xiv) *Chemical compatibility* — Compatible in its application with fertilizers and pesticides etc.

4. BIOCONTROL AGENTS

Biocontrol agents fall in two broad categories – Predators and Parasites. Endozoic and opportunistic fungi, bacteria, protozoans, viruses and rickettsial have been recognised as parasites of nematodes, while trapping fungi, nematodes, turbellarians, enchytreids, mites, collembolans, tardigrades and protozoans have been recognised as predaceous forms.

Parasites

(A) Viruses

A limited information is available on viruses, affecting viability of nematodes. Loewenberg *et al.*, (1959) reported a virus disease in *Meloidogyne incognita* causing sluggishness of second stage juveniles of the nematodes. These diseased juveniles failed to incite root galling. However, they did not observe virus particles in diseased nematodes. The phenomenon of swarming of virus particles was first observed by Hollis (1958) and is known to occur in at least 13 species of plant parasitic nematodes. (McBride and Hollis, 1966). Many workers (Zuckerman *et al.*, 1973; Ibrahim and Hollis, 1973; Ibrahim *et al.* 1973) reported virus infection in plant parasitic nematodes, *Dolichodorus meterocephalous* and *Tylenchorhynchus martini*. However, conclusive evidence that the virus like particles associated with the nematodes can cause a disease that results in a slow death of nematodes is still lacking.

(B) Rickettsias

Rickettsias were found in cyst nematode species e.g. *Heterodera goettingiana* and *Globodera rostochiensis* (Shepherd *et al.*, 1973) and in *H. glycines* (Endo, 1979). Rickettsias are minute, unicellular, rod-shaped organisms which divide by binary fission. Such organisms have been recorded in *Xiphinema index* associated with yellows disease of grapevines (Rumbos *et al.*, 1977). In general, only minor pathological effects on host cells were observed and it is questionable if these organisms are parasites or symbiotes?

(C) Bacteria

Interest of nematologists on bacterial antagonists of plant parasitic nematodes was evoked by the report of Mankau and Prasad (1972) that *Pasteuria penetrans* was a potential bio- control agent. Bacterial antagonists of plant parasitic nematodes are grouped under the categories of obligate parasites, antagonistic rhizosphere and other soil bacteria.

(a) *Obligate parasites* — The obligate endospore forming bacterial parasites belong to *Pasteuria* spp. and currently three species viz. *P. penetrans*, *P. thornei* and *P. nishizawe* are known to parasitize nematode (Sturhan *et al.*, 1994). *P. penetrans* is the most widely prevalent species and has been reported from more than 200 host nematodes belonging to diverse genera (Sayre & Starr, 1988, Sturhan, 1985) from all over the world.

The *P. penetrans* group and other members of the genus are highly host specific and in some instances even population specific. Certain populations of these obligate parasites have a host-range restricted to a small group of nematodes (Stirling, 1985, Bhattacharya & Swarup, 1988, Oostendorp *et al.*, 1990; Singh & Dhawan, 1994; Sayer *et al.*, 1991; Bouijate & Ciancio, 1995; Sharma & Davies, 1996). The two tropical root-knot nematodes, *Meloidogyne incognita* and *M. Javanica* are the major hosts, frequently met with. The life cycle of the bacterium is perfectly synchronized with that of its hosts (Mankau and Imbriani 1975). The bacterium causes sterility in the parasitized nematode as a result of failure of the reproductive system to develop (Mankau & Imbriani, 1975).

Pasteuria penetrans is multiplied for experimental purposes using the host *Meloidogyne* sp. on a suitable host plant. The nematode - infected roots containing the endospores of the bacterium are powdered, sieved through fine mesh and used as spore powder (Stirling & Wachtel 1980; Sharma & Stirling, 1991). Attempts to multiply *P. penetrans in vitro* on synthetic media have so far not been successful, eventhough the endospores germinate in some favourable media, they fail to complete the life cycle (Bishop & Ellar, 1991). The biocontrol potential of *P. penetrans* has been established against *Meloidogyne* spp. by a number of workers in different countries, mostly through pot culture experiments (Mankau, 1973; Sayre *et al.* 1984; Bhattacharya & Swarup, 1988, Stirling *et al.*, 1990; Walia & Dalal, 1994; Melki *et al.*, 1995; Tzortzakakis *et al.*, 1997, Sosamma & Koshy, 1997); a limited number if field observations also go to prove its bioefficacy as a biocontrol agent (Brown *et al.*, 1985).

(b) *Rhizosphere Bacteria* — Rhizosphere bacteria mainly florescent *Pseudomonas* (Oostendorp and Sikora, 1989; Spiegel *et al.*, 1991) and certain others like *Bacillus subtilis* and *B. cereus* (Oka *et al.*, 1993), *B. sphaericus* (Recke & Sikora, 1992), *Anthrobacter* (Kloepper *et al.*, 1988), *Serratia* (Kloepper *et al.*, 1988) and *Agrobacterium* (Racke & Sikora, 1992) are antagonistic to nematode. Among the rhizobacteria, *P. fluorescence* have been investigated for its antagonism to nematodes more extensively than others and found effective against *Meloidogyne* spp. (Gokte & Swarup, 1988), *Heterodera avenae, H. cajani* and *H. zeae* (Gokte & Swarup, 1988) and *H. schachtii* (Oostendorp and Sikora, 1989). Bare root dip treatment of tomato seedlings in a suspension of *P. fluorescence* was reported effective against *M. incognita* (Shanti & Sivakumar, 1995).

(c) *Antagonistic soil Bacteria* — Soil bacteria like *Clostridium butyricum* (Hollis & Rodrigues-Kabana, 1966, 1967), *Disulfovibrio desulficans* (Rodrigues-Kabana *et al.*, 1965), *Chromobacterium* sp. (Wilt and smith, 1970) and *Bacillus thuringiensis* var. *thuringiensis* (Prasad & Tilak, 1972) producing butyric acid, hydrogen sulphide, cyamide and exotoxins respectively, have been demonstrated to be antagonistic to nematodes. Among these *B. thuringiensis* is considered a potential biocontrol agent. Under natural conditions *B. thuringiensis* suppressed root-knot nematode in soil (Prasad & Tilak, 1972). *In vitro* studies established that the exotoxin produced by the bacterium during growth and sporulation had high nematicidal activity against juveniles and eggs of *Meloidogyne* spp. (Prasad *et al.*, 1972).

(D) FUNGI

Nematode - Destroying fungi — Nematode destroying fungi of diverse biology and affinities are ubiquitous in most soils and undoubtedly, in many instances, play an important role in regulating nematode population dynamics. More than 200 nematode destroying fungal species have been reported (Jatala, 1986; Rodriguez-Kabana, *et al.*, 1987). They are divided into three groups - *Predaceous, Parasitic* and *Opportunistic fungi.*

(a) *Predaceous fungi* — A lot of informations have been available regarding predaceous fungi antagonists to nematodes but reports are conflicting and inconclusive regarding their practicability in most cases. Most predacious fungi belong to Hypomycelidae; Denteromycetes; Zygomycetes and few other groups. The structure involved in trapping the nematodes vary greatly and have been grouped as follows —

(i) *Sticky branches* — Short lateral branches of the mycelium usually a few cells long (Sayre, 1971) anastomose to form simple loops e.g. *Monacrosporium cionopagum.*

(ii) *Sticky networks* — The mycelium curl around and anastomoses along with similar branches from the parent hypha. The loops thus produce complex three dimensional structure. Nematodes become emeshed in these adhesive networks and struggle for life for a while, but the hyphae readily penetrate the nematode and consume the body contents e.g. *Dactylella vermicola.*

(iii) *Sticky knobs* — These consists of spherical or subspherical lobes on one or two celled lateral hyphae with the terminal sticky cells (Sayre, 1971) e.g. *Dactylella dreshsteri.*

(iv) *Constricting rings* — These traps are formed by short hyphal branches curling back on themselves and anastomosing forming three or four celled rings, borne on one or two celled lateral hyphae e.g. *Arthrobotrys dacryoides, Dactylella bembicode.* The induction of trap formation is directly correlated with nematode mobility and composition of the medium (Jansson & Nordbring-Hertz, 1980).

(v) *Non-constricting rings* — The development of the rings is similar to the constricting rings. Nematodes simply become wedged in these traps without the aid of adhesive structure e.g. *Dactylaria candida.*

Recently two commercial formulations of predaceous fungi were produced : Royal 300, a preparation of an isolate of *Arthrobotrys robustus* for the control of *Ditylenchus myceliophagus* on mushrooms and Royal 350, a different isolate of *Arthrobotrys* for control of *Meloidogyne* on tomatoes (Cayrol *et al.*, 1978, Mankau 1981; Jatala, 1986).

(b) *Endoparasitic Fungi* — These fungi most commonly invade the nematode host by adhesive spores that first attach to the nematode cuticle or by spores that are ingested to the nematode and lodged in the alimentary canal. The most frequently encountered hyphomycetes genera are *Acrostalagums, Harposporium* and Meria, and Zoosporium forms are *Catenaria, Myzocytium* and *Haploglossa.* These organisms are unable to parasitize the stylet bearing phytonematodes because the spores cannot be ingested (Cook, 1968). Jansson *et al.*, (1985) showed that *Meria coniospra* is capable of inducing a reduction of root-knot nematode galling on tomatoes in greenhouse pot trials. *Hirsutella rhossiliensis* has been shown by Sayre *et al.* (1987) to have some ability to suppress

Criconemella xenoplax on Peach seedlings in greenhouse experiments. Fungi *Catenaria auxilliaris* was observed as parasite of *Heterodera schachtii* and *Nematophthora gynophila* on *H. avenae* and it plays a role of regulating population dynamics in some soils (Crump *et al.*, 1983; Kerry, 1980; 1977; Kerry *et al.*, 1980). Very few of these fungi have been considered seriously as potentially useful agents for biological control because of their obligate parasitic nature and little known biology.

(c) *Opportunistic fungi* — In recent years, a group of fungi, which are saprophytic deriving their nutrition through decomposition of organic matter, has emerged as promising biocontrol agents for management of endoparasitic nematodes. They are termed as "opportunistic fungi" because they parasitize some stages of nematodes whenever they get an opportunity to come in contact. By adaptation they are not nematode feeders. They infect, colonise and consume reproductive structures of *Globodera, Heterodera, Meloidogyne and Rotylenchulus* at sedentary stages of their life cycle. When the females of *Globodera* and *Heterodera* break in through the root cortex and mature into the cysts and become exposed on the root surface, they become infected and colonized by these fungi. Eggs of *Meloidogyne and Rotylenchulus* released into the soil are rapidly colonized and destroyed.

The opportunistic fungi can survive host free periods and can be grown axenically belong mostly to genera *Cylindrocarpon, Exophila, Fusarium, Gliocladium, Paecilomyces, Phoma* and *Verticillium.*

Paecilomyces lilacinus has been the subject of many recent researches for evaluating its biocontrol potential against nematodes. Fungus was found efficient in root-knot nematode control (Jatala *et al.*, 1980; 1983; Pandey & Trivedia, 1992); Trivedi, 1990; Sharma & Trivedi, 1989; Sharma *et al.* 1988). This fungus can be easily grown on a number of agricultural wastes and organic materials for field dispensing. In Philippines, it is marketed as 'Biocon'.

Verticillium chlamydosporium is another excellent example parasitizing *M. arenaria* (Morgan *et al.*, 1981), *Heterodera glycines* (Gintis *et al.*, 1983), *H. avenae* (Sharma & Trivedi, 1995; Bhardwaj & Trivedi, 2000), *H. cajani* (Bhardwaj & Trivedi, 1998; 2000).

Many of the known fungal enemies of nematodes, particularly the parasitic and predaceous forms are lacking in some of the Key desirable characteristics of biocontrol agents and this may account for the lack of success where they have been tested. This does not appear to be the case with cyst and egg colonizing opportunistic soil fungi, which also have the added advantage of being easily growing in axenic cultures and introduced into soil. The opportunistic fungi are also not free from weaknesses. The most serious weakness is related to their biology as the control is opportunity dependent. All isolates of these fungi, particularly of *P. lilacinus* and *V. chlamydosporium* are not equally effective as biocontrol agents or are not parasitic on the nematodes. Some are known to cause allergy in humans.

VAM fungi

A number of reviews and numerious research papers have appeared on nematode - VAM interaction using a variety of crop plants as test-host in the recent past decades from different parts of the world (Caron, 1989; Carling *et al.*, 1989; Hasan & Jain, 1991; Hussey & Roncadori 1982; Jain, 1994; Schonbeck, 1987; Trivedi 1995).

Many workers have demonstrated a reduction in nematode population or disease incidence (Bagyaraj *et al.*, 1979; Cooper & Grandison, 1986; Jain & Hasan, 1988; Kassab & Taha 1990; Sikora & Schonbeck, 1975; Sharma 1992; Sharma & Trivedi, 1994) but in cases where nematode population remain unaffected (O'Bannon, *et al.*, 1974; Cason *et al.*, 1983) or even increase under the influence of mycorrhiza (Atilano *et al.*, 1981; Kasaab & Taha, 1990) are not uncommon. The survey of literature reveals that the *Glomus* sp. are most predominant genera of VAM fungi in India and one or other species, especially *Glomus fasciculatum* has shown great promise in the management of nematodes (Jalali & Chand, 1990, Trivedi, 1995). Potential of VAM fungi as biocontrol - cum - biofertilizer is yet to be fully acknowledged.

Predacious Organisms

A. Protozoans

Under laboratory conditions many protozoans have been known to attack nematodes. An amoeboid organism *Theratomyxa weberi*, attacking *Heterodera trifolii*, *M. incognita*, *Aphelenchus avenae*, *Aphelenchoides rutgerri* and *Rotylenchulus reniformis* was observed (Sayre, 1973). Doncaster and Hooper (1961) noted that *Urostyla* sp. can catch nematodes but sometimes the prey escaped after destroying the protozoan itself. The extent to which these organisms can be efficient and economically useful is not known.

B. Tardigrades

Tardigrades commonly known as the water bear, always made distinct tear in cuticle and remained firmly attached to prey during feedings. Predation of nematodes by tardigrades was observed by several investigators (Doncaster & Hooper, 1961; Hutchinson & Streu, 1960, Sayre & Powers, 1966). Sayre (1973) has recorded considerable reduction in population of *M. incognita* and *Ditylenchus dipsaci* due to *Hypsibius myrops*.

C. Turbellarians

These are carnivorous flatworms that feed on nematodes and other soil organisms. The feeding of *M. incognita* by *Adenoplea* sp. was observed by Sayre and Powers (1966). They concluded that although addition of *Adenoplea* sp. decreased root galling index, it was insufficient to be of any practical value.

D. Collembolans

Collembolans are active predators, found in large numbers on organic matters around roots (Sayre, 1971). Brown (1954) and Murphy and Doncaster (1957) found *Onychiurus armatus* to be the most active collembolans on *Heterodera cruciferae*.

E. Mites

The feeding of mites on nematodes was first observed by Linford and Oliveira 1938. Rodriguez *et al.*, (1962) Cultured *Macrocheles* sp. on *Rhabditis* sp. and found more than 30 species of mites devouring nematodes whole out of the 41 species which were identified as Predators (Muraoka & Ishibashi, 1976). *Lasioseius salpatus* (Imbriani & Mankau, 1983) and *Tyrophagus putrescentiae* fed voraciously on plant and soil nematodes.

F. Enchytraeids

Schaerbtemberg and Tendle (1951) have reported control of *Heterodera schachtii* by enchytraeids worm but it is doubtful if these can be of much use as of biocontrol agents, mainly because of their saprophagous nature.

G. Nematodes

Nematodes in the Mononchida, Dorylaimida and Diplogasteroidea by largely Predaceous although, the actual feeding habits of most species is unknown.

(a) *Mononchiala* — The interest in using mononchs for controlling plant parasitic nematodes was initiated long back. Small (1979) found a significant reduction in *Globodera rostochiensis* and *M. incognita* in presence of *Prionchalus punctatus*. in pots. Considerable work on mononches has been done on factors influencing predation, food choice, strike rate of Predation, effect of chemicals on their survival (Bilgrami & Jairajpuri, 1985a, b; Bilbrami *et al.*, 1983, 1986) to establish predator - prey relationship of mononchs.

(b) *Dorylaimida* — The *Dorylaims* and *Nygolaims* are one of the most ubiquitous group of nematodes occurring in all types of soil and habitats. The detection of *Eudorylaimus obtusicaudatus* feeding on the eggs inside the *H. schachtii* cysts or the increase in the population of predators (*Thormia*) in the presence of citrus nematodes and decrease in absence of them in pot trials (Boosalis and Mankau, 1985) indicate their potential in nematode biocontrol.

(c) *Diplogasteroidea* — They possess useful characteristics of an efficient biocontrol agents and thus be exploited in future for managing nematodes.

5. CONCLUSIONS AND FUTURE PROSPECTS

Natural enemies of nematodes undoubtly play a significant role in reducing their population densities. Among the bacterial antagonists of plant parasitic nematodes, the obligate parasite *Pasteuria penetrans*, the plant growth promoting bacteria *Pseudomonas fluorescence* and the nematicidal strains of *Bacillus thuringiensis* are considered potential biocontrol agents, against plant parasitic nematodes. *Pasteuria penetrans* has all the desirable qualities of a biocontrol agent except limitation by its host specificity and its non- availability for *in-vitro* mass culturing. The opportunistic fungi *Paecilomyces*, *Verticilium* and *Trichoderma* seem to be most promising. These fungi spread rapidly, are ubiquitous and are present abundantly in the rhizosphere. They also have the added advantage of being easily grown in axenic culture and can be introduced into soil. In France, two commercial agents 'Royal 300' and 'Royal 350', are available for control of *Ditylenchus myceliophagous* on mushrooms and *Meloidogyne* spp. on vegetables. The parasitic fungus, *P. lilacinus* has been produced in Philippines as 'Biocon' which can be used for the control of several nematode species.

The importance of VAM fungi such as *Glomus mosseae* and *G. fasciculatum* in the management of phytonematodes such as root-knot, burrowing and reniform nematodes is well documented on major crops. VAM fungi are obligate parasites, which hinders their effective utilization in integrated nematode management programmes.

Fortunately, the expansion of integrated pest management strategy has encouraged management approach with integrated multiple control practice, one of which may be the biological control. Research workers should pay due attention on the following aspects for the nematode management using potential biocontrol agents in a concerned ecosystem —

(1) Identification and selection of effective strains of natural enemies/antagonists;

(2) To develop standardized and efficient rearing, culturing, storage, handling, release and evaluation procedure;

(3) To understand the biology, ecology, physiology, genetic behaviour of biocontrol agents;

(4) To identity most efficient host

genotypes - symbiont strains;

(5) To develop mass culture technique for field application;

(6) To demonstrate and assess the full benefits of biocontrol agents under field conditions.

(7) To educate public for effective utilisation of biocontrol agents.

In future, use of these microorganisms for the management of nematode pests, required extensive studies by dedicated scientists with optimistic approach to work out their practical feasibility.

REFERENCES

Anonymous, 1971. Estimated crop losses due to plant parasitic nematodes in the United States. suppl. *J. Nematol*, special Publication No. 1 (USA), P.7.

Atilano, R.A., Menge, J.A. and Van Gundy, S.D. 1981. Interaction between *Meloidogyne arenaria* and *Glomus fasciculatum* in grape. *J. Nematol.* 213 : 52-57.

Bagyaraj, D.J., Manjunath, A. and Reddy, D.D.R. 1979. Interaction of vesicular-arbuscular mycorrhizae with root-knot nematode in tomato. *Plant Soil* 51 : 397-403.

Bhattacharya, D. and Swarup, G. 1988. *Pasteuria penetrans*, a pathogen of the genus *Heterodera*, its effect on nematode biology and control. *Indian J. Nematol.* 18 (1) : 61-70.

Boosalis, M.G. and Mankau, R. 1965. Parasitism and Predation of soil microorganism. *In* Ecology of soil-borne plant pathogens, ed. K.P. Baker, W.C. Snyder, pp. 374-379, Berkeley, Univ. Calib. Press. 571 PP.

Brown, S.M., Kepner, J.L. and Smart, G.C. 1985. Increased crop yields following application of *Bacillus Penetrans* to field plots infested with *Meloidogyne incognita. Soil Biology and Bio Chemistry.* 17 : 483-486.

Bilgrami, A.L., Ahmad, I. and Jairajpuri, M.S. 1983. Some factors influencing predation by *Mononchs aquaticas. Revue Nematol.* 6 : 325-326.

Bilgrami, A.L., Ahmad, I. and Jairajpuri, M.S. 1986. A study of the intestinal contents of some mononchs. *Revue Nematol.* 9 : 191-194.

Bilgrami, A.L. and Jairajpuri, M.S. 1985a. Cannibalism in *Mononchs aquaticus. Indian J. Nematol.* 14 : 202-203.

Bilgrami, A.L. and Jairajpuri, M.S. 1985b. The responses of *Mononchs aquaticus* to chemicals and pH. *Indian. J. Nematol.* 14 : 171-174.

Bishop, A.H. and Ellar, D.J. 1991. Attempt to culture *Pasteuria penetrans in vivo. Biocontrol Science and Technology.* 1 : 101-114.

Bharadwaj, P. and Trivedi, P.C. 1998a. Biocontrol of *Heterodera avenae* infecting wheat using different inoculum sequence of *verticillium chlamydosporium. J. Indian Bot. Soc.* 77 : 5-8.

Bharadwaj, P. and Trivedi, P.C. 1998b. Efficacy of *verticillium Chlamydosporium* as a biocontrol agent for the management of cyst nematode *Heterodera cajani. J. Indian Bol. Soc.* 77 : 139-145.

Bharadwaj, P. and Trivedi, P.C. 2000a. Assessment of nematophagous fungi and neem cake compatibility against *Heterodera cajani* on cowpea. *Indian phytopath.* 53 (4) : 478-479.

Bharadwaj, P. and Trivedi, P.C. 2000b. Dynamics of nematophagous fungus *Verticillium chlamydosporium* & its potential in the management of *Heterodera avenae* population. *In* Microbs : Agriculture, Industry and Environment (Eds. D.K. Matheshwari, R.C. Dubey, G. Prasad & Navneet). Bishan Singh Mahendra Pal Singh, Dehradun, PP 125-133.

Carling, D.E., Roncadori, R.W. and Hussey, R.S. 1989. Interaction of VAM fungi, root knot Nematode and phosphorus fertilization on soybean. *Plant Dis.* 73 : 730-733.

Caron, M. 1980. Potential use of mycorrhizae in control of soil borne disease. *Can. J. Pl. Path.* 11 : 177-179.

Caron, T.K.M., Hussey, R.S. and Roncadori, R.W. 1983. Interaction of vesicular-arbuscular mycorrhizal fungi and phosphorus with *meloidogyne incognita* on tomato. *J. Nematol.* 15 : 410-417.

Cooper, K.M. and Grandison, G.S. 1986. Interaction of vesicular arbuscular mycorrhizal fungi and root-knot nematode on cultivars of tomato and white clover susceptible to *Meloidogyne hapla. Annals of Applied Biology* 108 : 555 - 556.

Cayrol, J.C., Frankowski, J.P., Laniece, A.D., Hardemare, C. and Talon, J.P. 1978. Center les nematodes en champignoniere mise au point d'une methode de lutte biologique al aide d'un hyphomycetes predateur; *Arthrobotrys robustus* Souehe "auipoles" (Royal 300). *Revue de Horticulture* 184 : 23-30.

Cook, R.C. 1968. Relationship between nematode destroying fungi and soil borne phytonematode. Phytopathology 58 : 909- 913.

Crump, D.H., Sayre, R.H. and Young, L.D. 1983. Occurrence of nematophagous fungi on cyst nematode populations. *Plant Disease* 67 : 63-64.

Doncaster, C.C. and Hooper, D.J. 1961. Nematodes attacked by Protozoa and tardigrades. *Nematologica* 6 : 333-335.

Endo, B.Y. 1979. The ultrastructure & distribution of an intercellular bacterium like microorganisms in tissues of larvae of the soybean cyst nematode, *Heterodera glycines. J. Ultrastruct. Res.* 67 : 1-4.

Gokte, N. and Swarup, G. 1988. On the potential of some bacterial biocides against root-knot nematode and cyst nematodes. *Indian J. Nematol.* 18 : 152-153.

Gintis, B.O., Moran-Jones, G. and Rodriguez-Kabana, R. 1993. Fungi associated with several developmental stages of *Heterodera glycines* from an Alabama soybean field soil. *Nematropica* 13 : 181-200.

Hasan, N. and Jain, R.K. 1991. Biological control of nematodes in forage-crops-current status and future prospects. *In* : Recent Advances in Nematology (Ed. Dwivedi, B.K. & Hussain, S.I.), Bioved Research Society, Allahabad.

Hollis, J.P. 1958. Induced swarming of a nematode as a means of isolation. *Nature* 182, 956.

Hollia, J.P. and Rodriguez-Kabana, R. 1966. Rapid kill of nematodes in flooded soil. *Phytopathology* 6 : 1015-1019.

Hollis, J.P. and Rodriguez-Kabana, R. 1967. Fatty acids in Louisiana rice fields. Phytopathology 57 : 841-847.

Hussey, R.S. and Roncadori, R.W. 1982. Vesicular-arbuscular mycorrhizae may limit nematode activity and improve plant growth. Plane Disease 66 : 9-14.

Hutchinson, H.T. and Streu, M.T. 1960. Tardigrades attacking nematodes. *Nematologica* 5 : 149.

Ibrahim, I.K.A. and Hollis, J.P. 1973. Electron Microscope studies on the cuticle of swarming and non-swarming *Tylenchorhynchus martini. J. Nematol.* 5 : 275-281.

Ibrahim, I.K.A., Joshi, M.M. and Hollis, J.P. 1973. The swarming virus disease of *Tylenchorhynchus martini.* Int. Congr. Plant. Pathol. Minneapolis Minn., Sept. 1973.

Jain, R.K. and Hasan, N. 1988. Role of VAM Fungi and nematode activities in forage production. *Acta Bot. Indica* 16 : 84-88.

Jain, R.K. and Hasan, N. 1994. Vesicular-arbuscular mycorrhizal (VAM) Fungi, A potential biocontrol agent for nematodes. (*In* : Vistas in seed Biology - Ed. Singh, T. & Trivedi, P.C.), Printwell Publishers, Jaipur, P. 49-60.

Jalali, B.L. and Chand, H. 1990. Current trends in mycorrhizal research. Proc. of the national conference on Mycorrhiza HAU, Hisar, Haryana, India, pp. 210.

Jansson, H.G., and Nardbring - Hertz, B. 1985. Interactions between nematophagous fungi and plant parasitic nematodes, attraction, induction of trap formation and capture. *Nematologica* 26 : 383-389.

Jatala, P. 1986. Biological Control of plant parasitic nematodes. *Annual Review of Phytopathology*, 24 : 453-489.

Jatala, P., Devaux, A.J. and campos, R. 1980. Field application of *Paecilomyces lilacinus* for controlling *Meloidogyne incognita* on potatoes. *J. Nematol.* 12 : 226-227.

Jatala, P. Salas, R. Kaltenbach, R. and Bocangel, M. 1983. Multiple application and long term effect of *Paecilomyces lilacinus* in controlling *Meloidogyne incognita* under field applications. *J. Nematol.* 13 : 445.

Kassab, A.S. and Taha, A.H.Y. 1990. Interaction between plant parasitic nematodes, vesicular-arbuscular mycorrhiza, rhizobia and nematicide on Egyptian Clover. *Ann. Agric. Sci.* 35 : 509-520.

Kerry, B.R. 1977. Observations on fungal parasites of female and eggs of the cereal cyst nematode, *Heterodera avenae* & other cyst nematodes. *Nematologica* 23 : 193-201.

Kerry, B.R. 1980. Biocontrol. Fungal parasites of female cyst nematode by parasitic fungi. *J. Nematol.* 12 : 253-259.

Khan, M.W. 1998. Biopesticides in the management of plant parasitic nematodes. *In* "Trends in microbial Exploidation" Eds. Raj. B., Upadhyay, R.S. and Dubey, N.K.; International Society for Conservation of Natural Resources, Varanasi, PP 138-148.

Linford, M.B. and Oliveire, J.M. 1938. Potential agents of biological control of Plant parasitic nematodes. *Phytopathology* 28 : 14.

Loewenberg, J.R., Sullivan, T. and Schuster, M.L. 1959. A virus disease of *Meloidogyne incognita*, The Southern root-knot nematode. *Nature* 184, 186.

Mankau, R. 1975. *Bacillus penetrans* causing a virulkent disease of plant parasitic nematodes. *J. Invertor Pathol.* 26 : 333-339.

Mankau, R. 1981. Microbial control of nematodes. *In.* Plant Parasitic Nematodes. Ed. B.M. Zuckerman, R.A. Rohde. 3 : 475-495, N-York, Academic Press.

Mankau, R. and Imbriani, J.L. 1975. The life cycle of an endoparasite in some Tylenchid nematode. *Nematologica* 21 : 89-94.

Mankau, R. and Prasad, N. 1972. Possibilities and Problems in the use of a sporozoan endoparasite for biological control of plant parasitic nematodes. *Nematropica*, 2 : 7.

Melki, T., Tzorzakakis, E.A. and Gower, S.R. 1995. The use of *Pasteuria penetrans* in the management of root-knot nematodes in protected crop. *Nematologica* 41 : 321.

McBridge, J.M. and Hollis, J.P. 1966. Phenomenon of swarming in nematodes. *Nature* 211 : 545-546.

Morgan-Jones, G., Godoy, G. and Rodriguez-Kabana, R. 1981. Verticillium Chlamydosporium fungal parasite of Meloidogyne arenaria females. Nematropica, 11 : 115-120.

Murphy, P.W. and Doncaster, C.D. 1957. A cultural method for soil neiofauna and its application to the study of nematodes Predators. *Nematologica* 2 : 202-214.

O,Bannon, J.H., Inserra, R.N., Nemic, S. & Vovlas, N. 1979. The influence of *Glomus mosseae* on *Tylenchulus semipenetrans* infected and uninfected citrus limon seedlings. *J. Nematol.* 11 : 247-250.

Oka, K., Chal, I. and Spiegel, Y. 1993. Control of root-knot nematode *Meloidogyne javanica* by *Bacillus cereus*. Biological Science & Technology. 3 : 115.

Oostendorp, M. and Sikora, R.A. 1989. Seed treatment with antagonistic rhizobacteria for suppression of *Heterodera schachtii* early root infection of sugarbeet. *Revue de Nematologie* 12 : 77-83.

Oostendorp, M., Dickson, D.W. and Mitchell, D.J. 1990. Host-range and ecology of isolates of *Pasteuria* spp. from the southern United States. *J. of Nematol.* 22 : 525-531.

Pandey, Rohini & Trivedi, P.C. 1992. Fungus *Paccilomyces lilacinus* in controlling root-knot nematode and their histological interactions in *Capsicum annuum. Proc. National Acad. Sci.* 62(B) : 275-283.

Prasad, S.S.V. and Tilak, K.V.B.R. 1972. Aerobic spore-forming bacteria from root-knot nematode infested soil. *Indian J. Microbiology* 11 : 59-60.

Prasad, S.S.V., Tilak, K.V.B.R. and Gollakota, K.G. 1972. Role of *Bacillus thuringiensis* var. *thuringiensis* on the larval survivability and egg hatching of *Meloidogyne* spp. the causative agent of root knot disease. *J. of Invertebrate Patho.* 20 : 377.

Racke, J. and Sikora, R.A. 1992. Isolation formulation and antagonistic activity of rhizosphere bacteria toward the potato cyst nematode *Globodera pallida. Soil Biology & Biochemistry* 24 : 531-526.

Rodriguiz-Kabana, R., Jordan, J.W. and Hollis, J.P. 1965. Nematodes : Biologicla Control in rice fields. Role of hydrogen subfide. *Science* 148 : 524-526.

Rodriguiz-Kabana, R., Margan-Jones, G., Goday, G. and Gintis, B.O. 1984. Effectiveness of species of *Gliodadium, Paccilomyces* and *Verticillium* for control of *Meloidogyne arenaria* in field soil. *Nematropica* 14 : 155-170.

Rumbos, I., Sikora, R.A. and Nienhaus, F. 1977. Rickettsia-like organisms in *Xiphinema index* found associated with yellow disease of grapevines. *Z. Pflanzenkr. Pflanzenschutz.* 84 : 240-243.

Sayre, R.M. 1971. Biotic influences in soil environment. *In* : Plant Parasitic Nematodes, ed. B.M. Zuckerman, W.F. Mai, R.A. Rohde, I. 235 - 266, N. York, Academic Press.

Sayre, R.M. 1973. *Theratromyxa weber* An *Amoeba* Predatory on plant parasitic nematodes. *J. Nematol.* 5 : 258-264.

Sayre, R.M. 1980. Bicontrol. *Bacillus penetrans* and related parasites of nematodes. *J. Nematol.* 12 : 260-270.

Sayre, R.M. and Powers, E.M. 1966. A predacious soil turbellarian that feeds on free-living and plant parasitic nematodes. *Nematologica* 12 : 619-629.

Sayre, R.M. and Starr, M.P. 1988. Bacterial diseases and antagonisms of nematodes. *In* : Diseases of Nematodes. vol. I. Poinar G.O. Jr. & Jansson (eds.) Hans-Borge. CRC Press, Inc. Boca Raton, Florida, U.S.A. PP. 69-101.

Schaerbfenberg, B. and Tendle, H. 1951. Angew. *Entomol.* 32 : 476-488.

Schonbeck, F. 1979. Endomycorrhiza in relation to plant disease. In : Soil borne Plant Pathogens. Int. Symp. Soil borne Pl. Pathogens. 4th Muncik 1978, N. York, Academic Press, pp. 271-280.

Sharma, R.D. and Davis, K.G. 1986. Comparison of two sympatric *Parteuria* populations isolated from a tropical vertisol soil. *World Journal of Microbiology and Biotechnology* 12 : 361-366.

Sharma, A. and Trivedi, P.C. 1989. Influence of inoculum levels of *Paecilomyces lilacinus* on the biocontrol of root-knot nematode, *Meloidogyne incognita.* Inter. Nematol. Network Newlett. USA 6(2) : 27-29.

Sharma, J.L., Trivedi, P.C., Sharma, M.K. and Tiagi, B. 1988. Fungus *Paccilomyces lilacinus* as a biocontrol agent of root-knot nematode, *Meloidogyne incognita. The Indian Zoologist* 12 : 259-262.

Sharma, Widhi and Trivedi, P.C. 1994. Interaction of root-knot nematode with VAM on *Abelmoschus esculentus* and their effect on net yield. *In* : Vistas in seed Biology vol. II (Eds. Singh, T. & Trivedi, P.C.) Printwell Publishers, Jaipur, PP. 61-81.

Sosamma, V.K. and Koshy, P.K. 1997. Biological control of *Meloidogyne incognita* on black pepper by *Pasteura penetrans* and *Paccilomyces lilacinus. J. of Plantation crops* 25 : 75-76.

Shepherd, A.M., Clark, S.A. and Kempton, A. 1972. An intracellular microorganism associated with tissues of *Heterodera* spp. *Nematologica* 19 : 31-34.

Small, R.M. 1979. The effect of predatory nematodes on populations of plant parasitic nematodes in pots. *Nematologica* 25 : 94-103.

Stirling, G.R. 1984. Biological control of *Meloidogyne javanica* with *Bacillus penetrans. Phytopathology* 74 : 55-60.

Stirling, G.R. 1985. Host specificity of *Pasteuria penetrans* within the genus *Meloidogyne Nematologica* 31 : 203.

Stirling, G.R. and Wachtel, M.F. 1980. Mass producion of *Bacillus penetrans* for the biological control of root-knot nematode. *Nematologica* 26 : 308-312.

Stirling, G.R., Sharma, R.D. and Perry, J. 1990. Attachment of *Pasteuria penetrans* spores to the root-knot nematode *Meloidogyne javanica* in soil and its effect on infectivity. *Nematologica* 36 : 246-252.

Sturban, D., Winkelheide, R., Sayre, R.M. and Wergin, W.P. 1994. Light and electron mincrospic studies of the life cycle and development stages of *Pasteuria* isolate parasitizing the pea cyst nematode, *Heterodera goettingiana. Fundamental and applied Nematology* 17 : 29-42.

Singh, B. and Dhawan, S.C. 1994. Effect of soil texture on attachment of bacterial spores of pasteuria penetrans to the second stage juveniles of Heterodera cajani Indian J. Nematol. 22 : 72-74.

Swarup, G. and Gokte, N. 1986. Biological control. In Plant Parasitic Nematodes of India Problems & Programs. Ed. G. Swarup and D.R. Das Gupta, IARI, N. Delhi, PP. 476-489.

Spiegel, Y., Cohn, E., Galper, S. Sharon, E. and Chet. I. 1991. Evaluation of a newly isolated bacterium, *Pseudomonas chitinolytica* sp. for controlling the root-knot nematode, *Meloidogyne Javanica. Biocontrol Science and Techno.* 1 : 115-125.

Shanti, A. and Sivakumar, C.V. 1995. Biological potential of Pseudomonas fluorescens against root-knot nematode *Meloidogyne incognita* on tomato. J. of Biological control

9 : 113-115.

Trivedi, P.C. 1992. Biocontrol agents in nematode management - A review. *In* Recent Advances in Nematology, (Ed. B.K. Dwivedi) Bioved Research Society, Allahabad PP 101-120.

Trivedi, P.C. 1995. Biocontrol potential of VAM fungi in the management of plant parasitic nematods. Platinum Jubilee volume. *I. Indian Bot. Soc.* 74 A : 165-171.

Trivedi, P.C. 1998. Potential of mycorrhizal fungi in the management of nematode - Anoverview. *In* Ecofriendly pest suppression (Eds. R.D. Gautam and D. Prasad), Westvill Publi. House, N. Delhi PP 153-169.

Tzorzakakis, E.A., Channer, A.G.R. De and Gowen, S.R. 1997. Studies on the potential use of *Pasteuria penetrans* as biocontrol agent of root-knot nematodes (*Meloidogyne* spp.). *Plant Pathology* 46 : 44-45.

Walia, R.K. and Dalal, M.R. 1994. Efficacy of bacterial parasite Pasteuria Penetrans application as nursery soil treatment in controlling root-knot nematode, *Meloidogyne Javanica* on tomato. *Pest Management and Economic Zoology* 2 : 19-21.

Wilt, G.R. and Smith, R.E. 1970. Studies on interactions of aquatic bacteria and aquatic aumatodes. Water Resource Research Insti. Bulletin 701 : 1-6.

Zuckerman, B.M., Dicklow, M.D. and Acosta, N. 1993. A strain of *Bacillus thuringiensis* for the control of plant parasitic nematodes. Biocontrol Science and Technology 3 : 41- 46.

INDEX